S0-BDC-075

HISTORY OF MATHEMATICS

VOLUME I

GENERAL SURVEY OF THE HISTORY OF ELEMENTARY MATHEMATICS

BY

DAVID EUGENE SMITH

DOVER PUBLICATIONS, INC.
NEW YORK · NEW YORK

This new Dover edition, first published in 1958, is an unaltered and unabridged republication of the last edition. It is published through special arrangement with GINN AND COMPANY.

Manufactured in the United States of America

Dover Publications, Inc.
180 Varick Street
New York 14, N. Y.

PREFACE

This work has been written for the purpose of supplying teachers and students with a usable textbook on the history of elementary mathematics, that is, of mathematics through the first steps in the calculus. The subject has come to be recognized as an important one in the preparation of teachers of mathematics and in the liberal education of students in colleges and high schools. Although several works upon the history of mathematics are already available, the author feels that a book written from somewhat different standpoints will be found helpful to those who are beginning the study of the subject in our universities, colleges, and normal schools, and because of this belief the present work has been prepared.

A history of mathematics may be constructed on several general plans, each of which may be justified by the purpose in mind. For example, it may be arranged solely with a view to the chronological sequence of events, or as a series of biographies, or according to the leading branches of mathematics, or as a source book of material for study, or with respect to national or racial achievements, or in various other ways, each of which may have certain advantages.

The general plan adopted in the preparation of this work is that of presenting the subject from two distinct standpoints, the first, as in Volume I, leading to a survey of the growth of mathematics by chronological periods, with due consideration to racial achievements; and the second, as in Volume II, leading to a discussion of the evolution of certain important topics. To attempt to fuse these two features and thus to carry them along together has often been attempted. It characterizes, for example, the monumental treatise of Montucla and, to a large extent, that of Cantor. For the teacher, however, this plan is not satisfactory, and the excellent work of Tropfke is an example of the tendency to break away from the mere chronological recital of facts. Long experience in teaching the subject in colleges has convinced the author that a general historical presentation is desirable for the purpose of relating the development of mathematics to the development of the race, of revealing the science as a great stream rather than a static mass, and of emphasizing the human element, but that this ought to lead to a topical presentation by which the student may understand something of the life history of the special subject which he may be studying, whether it be the elementary theory of numbers, the methods of calculation, the solution of equations, the functions of trigonometry, the common symbolism in use, the various types of elementary geometry, the early steps in the calculus, or one of the various other important topics of elementary mathematics. The general plan can best be understood by a glance at the table of contents in each volume.

Perhaps the chief objection to the general arrangement set forth in Volume I is that the reader may occasionally feel that a mere statement of the subjects in which some particular mathematician was interested is not very illuminating, and that a more extended statement of his achievements would have greater significance. In most cases, however, a further elaboration of the record would destroy the possibility of successfully carrying out the plan of showing the growth of the several leading branches of elementary mathematics by themselves, as in Volume II, at least without a large amount of wearisome repetition. Of the two evils the lesser has been chosen.

In Volume I, which forms the general survey by periods, attention has been given to geographical and racial considerations as well as to chronological sequence. While it is evident that no race or country has any monopoly of genius, and while the limits of successive centuries are only artificial boundaries with no significance in the creation of the masterpieces of any science, nevertheless linguistic and racial influences tend to develop tastes in mathematics as they do in art and in letters, and certain centuries stand out with interesting prominence.

The student will therefore find it to his advantage to give some attention to the geographical distribution of scholars as well as to the general periods in which they lived. While it is impossible to grade countries according to any definite scale of excellence, and while the world has always seen more or less of the migration of scholars from one country to another, it is possible in a general way to give prominent positions to those national groups which have contributed most to the advancement of the science in each period under discussion.

In this treatment of the subject an attempt has been made to seek out the causes of the advance or the retardation of mathematics in different centuries and with different races, but always with the consciousness that the world has no certain prescription for the creation of the genius and that the causes of any series of historical events are usually very intangible. The effort has also been made to introduce enough of the anecdote to relieve the monotony of mere historical statement and to reveal the mathematician as a human being like others of his race.

While the footnote is often condemned as merely an apology for obscurity or as an exhibition of pedantry, it would be difficult, in a work of this kind, to dispense with its aid. There are two principal justifications for such a device: first, it enables an author to place the responsibility for a statement that may be open to question; and second, it encourages many students to undertake further study, either from secondary sources or, what is more important, from the original writings of the men who rank among the creators of mathematics. With these two points in mind, footnotes have been introduced in such a way as to be used by readers who wish for further aid, and to be neglected by those who wish merely a summary of historical facts. For the student who seeks an opportunity to study original sources, a slight introduction has been made to this field. The text of the book contains almost no quotations in foreign languages, the result being that the reader will not meet with linguistic difficulties in the general narrative. In the notes, however, it is frequently desirable to quote the precise words of an author, and this has been done with reference to such European languages as are more or less familiar. It is not necessary to translate literally all these extracts, since the text itself sets forth the general meaning. Students who have some general knowledge of Latin, French, or German will have little difficulty, and in many cases will have much interest, in seeing various statements in their original form. For special reasons a few notes have been given in Greek, but in every case the meaning is evident from the text.

The footnotes have also permitted of the insertion of various biographical items which would merely burden the text, but which have considerable value to the student. In a general way it may be said that it is a matter of no moment where a man was born or on what day; but a work of this nature must be more than a book to be read,—it must be a work for future reference, and for this reason there may properly be made available, to be used if thought necessary, certain material which will aid the student in his later research. It would be possible to place all such supplementary material at the end of the book, but this would be merely an invitation to ignore it entirely.

No selection of names is ever satisfactory, even to the writer who makes it. In this work there are often included in one period names which would not be considered a century later; while others are omitted, particularly in the last three centuries, that would have been given prominence had their possessors lived at an earlier date. The criterion of selection has been the contribution of the individual to the development of elementary mathematics, his reputation as a scholar, and in particular his work in the creating of tendencies to further the study of particular branches of the subject. For the latter reason certain names have been included which would not otherwise have been considered. In Volume II a few minor names of arithmeticians have been mentioned in connection with the peculiar use of certain terms and the like, with no biographical notes, the latter being of little or no consequence.

In connection with dates before the Christian era the letters B. C. are used; in connection with dates after the beginning of this era no distinguishing letters are added except in a few cases near the beginning of the period, in which the conventional letters A. D. have occasionally been inserted to avoid ambiguity. With some hesitation, but for a purpose which seems valid, dates are frequently given in parentheses after proper names. It is well recognized that a precise date, like 1202 after the name Fibonacci, is of no particular value in itself. It makes no difference, in ordinary cases, whether Fibonacci wrote his *Liber Abaci* in 1202, in 1180, or in 1220, or whether *abacus* is spelled *abbacus*, as in some manuscripts, or in the more correct Latin form. On the other hand, two things are accomplished by a free use of such dates. In the first place, a reader is furnished with a convenient measuring instrument; he does not have to look in the index or a chronological table in order to see approximately where the particular writer belongs in the world's progress. The casual reader may well be pardoned if he does not recall where Bede, Alcuin, Gerbert, Jordanus, Fibonacci, and Roger Bacon stood chronologically with respect to one another, and in reading a technical history of this kind there is no reason why he should not be relieved of the trouble of consulting an index when he meets with one of these names in the text. In the second place, it needs no psychologist to confirm the familiar principle that the mind comes, without conscious effort, to associate in memory those things which the eye has frequently associated in reading. At the risk, therefore, of disturbing the minds of those who are chiefly interested in the literary aspect of a general statement of the progress of mathematics, many important dates have been repeated, especially where they have not appeared in the pages immediately preceding.

In quoting from other writers the rule has been followed of making the quotation exact in spelling, punctuation, and phraseology. In carrying out this rule it is inevitable that errors should occasionally enter into the transcription, particularly in the case of old dialects; but the effort has been made to give the language precisely as it appears in the original. This accounts for the fact that certain French words in a quotation will sometimes appear without the modern accent, and that a word like Lilāvati may appear with any one of three spellings, depending upon the translator to whose work reference is made, or upon the author using the word.

Use has been made of such international symbols as *s.a.* (*sine anno*, without date of publication), *s.l.* (*sine loco*, without place of publication), *s.l.a.* (without place or date), *c.* (*circa*, about), and *seq.* (*sequens*, following), and of the abbreviations *ed.* (edition, edited by, or edition of), *vol.* (volume), and *p.* (page).

At the close of each chapter there has been given a page of topics for discussion, so arranged as to command more attention than they would have received had they been given in scattered form. These topics are not limited to questions to be answered from the text, but have purposely been made general, suggesting somewhat more extended fields for study. The student will find it to his advantage if he is thus led to consult encyclopedias, general histories, and such works as are suggested in the bibliographical notes and as are available in libraries to which he may have access. It is by no means expected that an elementary work like this should contain the material for an extended study of any of these topics.

In the selection of illustrations the general plan has been to include only such as will be helpful to the reader or likely to stimulate his interest. It would be undesirable to attempt to give, even if this were possible, illustrations from all the important sources, for this would tend to weary the reader. On the other hand, where the student has no access to a classic that is being described or even to a work which is mentioned as having contributed to the world's progress in some humbler manner, a page in *facsimile* is often of value.

In general the illustrations have been made from the original books or manuscripts in the well-known and extensive library of George A. Plimpton, Esq., who has generously allowed this material to be used for the purpose; from the author's collection of books, manuscripts, mathematical portraits and medals, and early mathematical instruments; from manuscripts in various other libraries; and from such works as those by Professor Breasted.

Long experience in the use of books of reference has led the author to believe that a single index is more convenient than a series of indexes by names, subjects, and titles. Furthermore, readers who have used works like those of Cantor and Tropfke, for example, know the annoyance of a long list of page references after a given name, many of them of no particular significance. In this work, therefore, only a single index is given in each volume, and in each entry the page references are only such as the reader will find of particular value. In each case the first reference after a proper name relates to the biography of the individual, if one is given; the others relate to his leading contributions and are arranged approximately in order of importance.

DAVID EUGENE SMITH

CONTENTS

BIBLIOGRAPHY

The extent of a bibliography in a work of this kind is a matter of judgment. It can easily run to great length if the writer is a bibliophile, or it may have but little attention. The purpose of giving lists of books for further study is that the student may have access to information which the author has himself used and which he believes will be of service to the reader. For this reason the secondary sources mentioned in this work are such as may be available, and in many cases are sure to be so, in the libraries connected with our universities, while the original sources are those which are of importance in the development of elementary mathematics or which may be of assistance in showing certain tendencies.

The first time a book is mentioned the title, date, and place of publication are given, together, whenever it seems necessary, with the abbreviated title which will thereafter be used. To find the complete title at any time, the reader has only to turn to the index, where he will find given the first reference to the book. The abbreviation *loc. cit.* (for *loco citato*, in the place cited) is used only where the work has been cited a little distance back, since any more general use of the term would be confusing. The symbolism "I, 7" has been used for "Vol. I, p. 7" in order to conserve space, although exceptions have been made in certain ambiguous cases, as in the references to Heath's *Euclid*, references to Euclid being commonly by book and proposition, as in the case of Euclid, I, 47.

Although the number of works and articles on the history of mathematics is very great, the student will be able, in the initial stages of his investigation, to consult relatively few. For his convenience the books that he may most frequently use are here listed, special reference being made to those in English, French, and German which are likely to be found in college, university, and city libraries. The student will also find it advantageous to consult the leading encyclopedias.

Allman, G. J., *Greek Geometry from Thales to Euclid*, Dublin, 1889. Referred to as Allman, *Greek Geom.*

Ball, W. W. Rouse, *A Short Account of the History of Mathematics*, 6th ed., London, 1915. A readable survey of the general field. Referred to as Ball, *History.*

Bretschneider, C. A., *Die Geometrie und die Geometer vor Eukleides*, Leipzig, 1870. Referred to as Bretschneider, *Die Geometrie.*

Cajori, F., *A History of Elementary Mathematics*, rev. ed., New York, 1917. Referred to as Cajori, *Elem. Math.*

—— *A History of Mathematics*, 2d ed., New York, 1919. Referred to as Cajori, *History.*

Cantor, M., *Mathematische Beiträge zum Kulturleben der Völker*, Halle, 1863. Referred to as Cantor, *Beiträge.*

—— *Vorlesungen über Geschichte der Mathematik*, 4 vols., Leipzig, 1880–1908, with various revisions. The standard general history of mathematics. Referred to as Cantor, *Geschichte.*

Encyklopädie der Mathematischen Wissenschaften, Leipzig, 1898–, with a French translation. Referred to as *Encyklopädie.*

Gow, James, *A Short History of Greek Mathematics*, Cambridge, 1884. Referred to as Gow, *Greek Math.*

Günther, S., and Wieleitner, H., *Geschichte der Mathematik*, 2 vols., Leipzig, 1908–1921. Referred to as Günther-Wieleitner, *Geschichte.* The second volume is the work of Dr. Wieleitner.

Hankel, H., *Zur Geschichte der Mathematik in Alterthum und Mittelalter*, Leipzig, 1874. Referred to as Hankel, *Geschichte.*

Heath, Sir Thomas Little, *A History of Greek Mathematics*, 2 vols., Cambridge, 1921. Referred to as Heath, *History.* Although the following special works by the same author are referred to in the footnotes, they are so important that it seems advisable to include them in this general bibliography.

—— *Apollonius of Perga*, Cambridge, 1896. Referred to as Heath, *Apollonius.*

—— *Archimedes*, Cambridge, 1897. Referred to as Heath, *Archimedes.*

—— *Aristarchus of Samos*, Oxford, 1913. Referred to as Heath, *Aristarchus.*

—— *Aristarchus of Samos. The Copernicus of Antiquity*, London, 1920. Referred to as Heath, *Aristarchus* (abridged).

—— *Diophantus of Alexandria*, 2d ed., Cambridge, 1910. Referred to as Heath, *Diophantus.*

—— *Euclid in Greek, Book I*, Cambridge, 1920. Referred to as Heath, *Euclid in Greek.*

—— *The Thirteen Books of Euclid's Elements*, 3 vols., Cambridge, 1908. Referred to as Heath, *Euclid.*

—— *Greek Mathematics and Science*, pamphlet, Cambridge, 1921. Referred to as Heath, *Address.*

Hilprecht, H. V., *Mathematical, Metrological, and Chronological Tablets from the Temple Library of Nippur*. Philadelphia, 1906. Referred to as Hilprecht, *Tablets*.

Libri, G., *Histoire des Sciences Mathématiques en Italie*, 4 vols., Paris, 1838–1841. Valuable on account of its style and its extensive notes. Referred to as Libri, *Histoire*.

Loria, G., *Guida allo Studio della Storia delle Matematiche*, Milan, 1916. Very valuable for its bibliography of the history of mathematics.

Marie, M., *Histoire des Sciences Mathématiques et Physiques*, 12 vols., Paris, 1883–1888. Biographical, convenient for reference, but inaccurate. Referred to as Marie, *Histoire*.

Mikami, Y., *The Development of Mathematics in China and Japan*, Leipzig, 1913. Referred to as Mikami, *China*. See also Smith-Mikami.

Miller, G. A., *Historical Introduction to Mathematical Literature*, New York, 1916. Serves a purpose in English similar to that of Loria's *Guida* in Italian. Referred to as Miller, *Introduction*.

Montucla, J. E., *Histoire des Mathématiques*, 2d ed., 4 vols., Paris, 1799–1802. Although written in the 18th century, it is a classic that is well worth consulting, particularly for its style. Referred to as Montucla, *Histoire*.

Pauly (A.)-Wissowa (G.), *Real-Encyclopädie der Classischen Altertumswissenschaft*, Stuttgart, 1894–. Best reference work for classical biography and antiquities. Referred to as Pauly-Wissowa.

Poggendorff, J. C., *Handwörterbuch zur Geschichte der exacten Wissenschaften*, 4 vols., Leipzig, 1863–1904. Referred to as Poggendorff.

Smith, David Eugene, *Our Debt to Greece and Rome. Mathematics*, Boston, 1922. Referred to as Smith, *Greece and Rome*.

—— *Rara Arithmetica*, Boston, 1908, being a bibliography of early arithmetics. Referred to as *Rara Arithmetica*.

Smith, D. E., and Karpinski, L. C., *The Hindu-Arabic Numerals*, Boston, 1911. Referred to as Smith-Karpinski.

Smith, D. E., and Mikami, Y., *History of Japanese Mathematics*, Chicago, 1914. Referred to as Smith-Mikami.

Tannery, P., *La Géométrie Grecque*, Paris, 1887. Referred to as Tannery, *Géom. Grecque*.

—— *Mémoires Scientifiques*, edited by J. L. Heiberg and H. G. Zeuthen, 2 vols., Paris, 1912.

—— *Pour l'Histoire de la Science Hellène de Thalès à Empédocle*, Paris, 1887. Referred to as Tannery, *Histoire*.

Tropfke, J., *Geschichte der Elementar-Mathematik in systematischer Darstellung*, 2 vols., Leipzig, 1902, 1903; 2d ed., 1922–. The best history of elementary mathematics. Referred to as Tropfke, *Geschichte*.

Zeuthen, H. G., *Histoire des Mathématiques dans l'Antiquité et le Moyen Age*, translated by J. Mascart, Paris, 1902. Referred to as Zeuthen, *Histoire*.

In matters of biography the student will find in W. Smith, *Dictionary of Greek and Roman Biography*, London, 3 vols., 1862–1864 (referred to as Smith's *Dict. of Greek and Roman Biog.*), a work of exceptional value, particularly with reference to the Greek mathematicians; and in the *Dictionary of National Biography* he will find the British mathematicians treated in a scholarly manner. For French biographies, Michaud's *Biographie Universelle* (1854–1865), Hoefer's *Nouvelle Biographie Générale* (1857–1866), *La Grande Encyclopédie*, and Larousse's *Grand Dictionnaire Universel du XIX siècle français* are very satisfactory. For German biographies the *Allgemeine Deutsche Biographie*, the Brockhaus *Conversations-Lexikon*, and the Meyer *Grosses Konversations-Lexikon* are helpful. Various earlier and less frequently used works are referred to in the footnotes.

Such special works as those of Matthiessen, Braunmühl, and Dickson are mentioned in the notes from time to time, as well as various other sources of information that may be found in the larger libraries.

Of the journals devoted to the history of mathematics those to which the student will most frequently refer are Boncompagni's *Bullettino di Bibliografia e Storia delle Scienze Matematiche e Fisiche*, Rome, 1868–1887 (referred to as Boncompagni's *Bullettino*), and Eneström's *Bibliotheca Mathematica*, Leipzig, three series, 1885–1915 (referred to as *Bibl. Math.*).

PRONUNCIATION, TRANSLITERATION, AND SPELLING OF PROPER NAMES

General Question. The question of the spelling and transliteration of proper names is always an annoying one for a writer of history. There is no precise rule that can be followed to the satisfaction of all readers. In general it may be said that in this work a man's name has been given as he ordinarily spelled it, if this spelling can be ascertained. To this rule there is the exception that where a name has been definitely anglicized, the English form has been adopted. For example, it would be mere pedantry to use, in a work in English, such forms as Platon and Strabon, although it is proper to speak of Antiphon and Bryson instead of Antipho and Bryso. When in doubt, as in the case of Heron, the preference has been given to the transliteration which most clearly represents the spelling that the man himself used.

In many cases this rule becomes a matter of compromise, and then the custom of a writer's modern compatriots is followed. An example is seen in the case of Leibniz. This spelling seems to be gaining ground in our language, and it has therefore been adopted instead of Leibnitz, even though the latter shows the English pronunciation better than the former. Leibniz himself wrote in Latin, and the family spelled the name variously in the vernacular. There seems, therefore, to be no better plan than to conform to the spelling of those recent German writers who appear to be setting the standard that is likely to be followed.

There is also the difficulty of finding a satisfactory solution in the case of men who were themselves polyglots, who lived in polyglot towns, or who made their homes in more than a single country. This is seen, for example, in the case of the Bernoullis. Jacques Bernoulli lived in Basel, a Swiss city where German was chiefly spoken and where the common spelling of the name of the place is the one here given. He was of Belgian descent, but he usually wrote either in Latin, in which his first name was spelled Jacobus, or in French, in which he would naturally use the name of Jacques. To call him James, as various English writers have done, would merely confuse an American reader, while to adopt the German Jakob would be to use a form which Bernoulli himself did not adopt in writing. The fact that he preferred to use French as his means of correspondence, when not writing in Latin, makes it desirable to speak of him as Jacques and to follow a similar usage with respect to his brother, Jean Bernoulli.

Another difficulty arises when we consider the Graeco-Latin forms of names in the Renaissance period. In general, if a man commonly used such a form, as was the case with Grammateus, Regiomontanus, and Dasypodius, this form has been used in the text, with the family name given in a footnote. In a case on the border line, like that of Schoner, however, the vernacular form, spelled as the man seems himself to have preferred, has been adopted. It must also be understood that early writers were often not uniform in spelling their own names. Thus, we have Recorde and Record, Widman and Widmann, and Scheubel and Scheybel, and in these cases all that the historian can do is to endeavor to choose that spelling which the writer seems himself to have most commonly used.

A further difficulty is encountered with certain names in regard to which the possessor was himself undecided as to his preference. A typical case is that of Leonardo Pisano Fibonacci,—Leonardo the Pisan, son of Fibonacci, or of Bonacci, or of Bonacius. It would be proper to write his name "l. pisano," since this form appears in one of the early manuscripts; or Bigollo, since he used this nickname; or Leonardo of Pisa, although this combines Italian and English; but the form Fibonacci has been chosen for general use, chiefly because Fibonacci's Series is so frequently mentioned in mathematics. It would not be difficult to show a lack of consistency in many cases, as when the common form of Gemma Frisius (Gemma the Frieslander) is preferred to the family name of Renier, with various spellings. In the case of a name like that of Pacioli, where different forms are used in the various works of the individual, the one seemingly preferred by the majority of historians has been chosen. In the case of a name like Joannes or Johannes the effort has been made to use the form which the possessor used, or at least the one which was the more commonly employed by his contemporaries when referring to him.

The greatest difficulty in transliteration arises with respect to oriental names. In the first place, we have no international system of transliteration that is generally accepted; and in the second place, it is difficult to know the name which the writer himself preferred. An Arab scholar may have as many as a dozen parts to his name; a Japanese or Hindu writer may have an intimate name and also an official name; a Chinese mathematician may be known only by an ideogram, the pronunciation of the name being lost or varying in different parts of his own country; and certain of these names may have found their way into medieval Latin and have been distorted almost beyond recognition.

If a name is fairly familiar in English, like Omar Khayyam, it has been retained, even if the form is open to criticism. If it has taken an English form but is not so familiar, as in the case of Savasorda, the attempt has been made to use the distorted name and also to adopt the best modern transliteration of the real name from which this is derived. In the cases of such less familiar names as seem to deserve mention, these will neither be read aloud nor be kept in mind by most readers, and hence an abridged form has been given in the text, in as good transliteration as seems possible, the full form being placed in a footnote. The Arabic *al-* has been used instead of *el-* or *ul-*, simply because it is the most common form in English. As a matter of fact, the Arabic pronunciation, like that of the Chinese, is by no means standardized.

The pronunciation of proper names has been given in cases where it is likely to be helpful to the student, and in many cases the accent has been indicated when the name first appears in the text. In such cases the English pronunciation has been taken whenever the name has become thoroughly anglicized, but otherwise the pronunciation has been given as nearly as possible as it stands in the vernacular. In the case of Greek names the original form has usually been given in the notes, partly because of the differences in accent and partly because the Greek alphabet is well enough known to allow the original and frequently interesting form to be understood.

Arabic Names. The standard authority on the transliteration and pronunciation of Arabic names is Suter, a Swiss writer, whose "Die Mathematiker und Astronomen der Araber und ihre Werke" appeared in Volume X of the *Abhandlungen zur Geschichte der Mathematischen Wissenschaften,* and in "Das Mathematiker-Verzeichniss im Fihrist," in Volume VI of the same work. The rules given by this writer have, in general, been followed, except that *j* has been used for *ǧ*, *y* for *j*, *v* for *w*, *kh* for *ch,* and *al-* for *el-*, to conform to English pronunciations and custom. Although the reader will seldom need to pronounce the names, it will be helpful to be able to do so if necessary. The following is a summary of the scheme of transliteration and pronunciation employed:

b, **d**, **f**, **g**, **h**, **j**, **l**, **m**, **n**, **p**, **s**, **sh**, **t**, **th**, **w**, **x**, **z** as in English.
a as in *ask*; **â** as in *father*, the form *â* being used instead of *ā* in Arabic words, partly to conform to the Suter list.
e as in *bed.*
i as in *pin*; **î** as in *pique.*
o as in *obey.*
u as in *put*; **û** as in *rule.*
d, **s**, **ṭ**, **ẓ** as in English but made with the tongue spread so that the sounds are produced largely against the side teeth.
n is generally pronounced by Europeans as simple *n*.
ḏ like *th* in *that*; **ṯ** like *th* in *thin.*
ġ is a voiced consonant formed below the vocal chords; it is sometimes compared to a guttural *g* and sometimes to a guttural *r*.
h retains its consonant sound at the end of a word.
ḥ may be compared to the German hard *ch*, as in *nach.*
k as in English; **kh** is the hard German *ch*, as in *nach.*
q like *c* or *k* in *cook.*
r stronger than in English.
v like the English *w*; **y** as in *you.*
' represents the *spiritus lenis* and may be taken simply as separating distinctly two vowels, like the break between the *e*'s in *reentrant.*
A final vowel is shortened before *al* (which then becomes '*l*) or *ibn* (whose *i* is then silent).
In *al* the final *l* often takes the sound of a following consonant, as in *al-Rashid* (*ar-Rashid*).
The accent is on the last syllable containing a long vowel or a vowel followed by two consonants, except that a final long vowel is not usually accented. Otherwise the accent falls on the first syllable.

Hindu Names. The transliteration of Hindu names has changed greatly within a century, and even yet is not internationally standardized. In general, in quoting from earlier English writers, the forms which they used have been followed. Thus, there will be found in the notes various references to Taylor's *Lilawati*, this being the name of the book as the translator used it; or, when the actual title is mentioned, to Colebrooke's translation of the *Lílávatí*, this being the form which this author used; but the modern form *Lilāvati* appears in the text. The effort has been made to follow the best current practice of English orientalists, and in determining the form and pronunciation of Sanskrit words the following equivalents have been used:

b, d, f, g, h, j, l, m, n, p, v, w, x, z as in English.
a like *u* in *but*; thus, *pandit*, pronounced *pundit*; **ā** as in *father*, the form *ā* being used instead of *á* in Hindu words.
e as in *they*.
i as in *pin*; **ī** as in *pique*.
o as in *so*.
u as in *put*; **ū** as in *rule*.
c like *ch* in *church* (Italian *c* in *cento*).
ḍ, ṇ, ṣ, ṭ like *d*, *n*, *sh*, *t* made with the tip of the tongue turned up and back into the dome of the palate.
h preceded by *b*, *c*, *t*, *ṭ* does not form a single sound with these letters but is a more or less distinct sound following them, somewhat as in *abhor*; **ḥ** is final consonant *h*.
k as in *kick*.
ṁ, ṅ like the French final *m* or *n*, nasalizing the preceding vowel.
ś, English *sh*.
y as in *you*.
' in some transliterations is used to indicate the *spiritus lenis*, a break between two letters.
The accent is as in Latin: if the penult is long, it is accented; if it is short, the antepenult is accented.

Japanese Names. Modern Japanese scholars have carefully transliterated into the Roman alphabet the names of all their leading mathematicians. The letters are pronounced as in English except that *i* is pronounced like *e* in *feel*; *e* as in *grey*; *ai* as in *aisle*; and *ei* like long *a*; but *i* and *e* also take a short sound as in English. Japanese names have only a slight accent.

Chinese Names. There is no uniform system of transliterating and pronouncing Chinese names and terms. The author's colleague, Professor Hirth, in his *Ancient History of China*, followed in general the plan adopted by the Royal Geographical Society of London and the United States Board on Geographic Names, and the present text follows in the main the rules which he has laid down. Briefly stated, the scheme of pronunciation is as follows:

a as in *father*.
e, é as in *men*. The accent simply shows that it does not form part of a diphthong.
i as in *pique*. When followed by *n* or by a vowel it is short as in *pin*.
ï, used when *i* is intonated with the adjoining consonant, as in *jï*, or is but faintly heard, when it follows *e*, as in *leï*.
o as in *mote*.
ö like the French *eu* in *jeu* or like the German *ö*.
u like *oo* in *boot*. When preceding *n*, *a*, or *o* it is short.
ü like the French *u*. When preceding *n*, *a*, or *é* it is short.
ai like *i* in *ice*.
au like *ow* in *how*.
eï somewhat like *ey* in *they*.
óu a diphthong with the two vowels distinctly intonated.
ui like *ooi* contracted into a diphthong.
The initials **k, p, t, ch, ts,** and **tz** are not so hard as in English. When pronounced as hard as possible they are followed by (') as in *k'an*.
ch like *ch* in *church*. When followed by *ï*, the vowel blends with it.
f, h, l, m, n, sh, w, ng as in English.
j like *j* in French.
ss like *ss* in *mess*. When followed by *ï*, the vowel disappears.
y like *y* in *you*.

Names from Other Languages. In the case of Russian names there has been chosen the transliteration which represents most effectively the English equivalent sounds. For example, the spelling Lobachevsky has been preferred to the German form, Lobatschewski, or to other forms which are not appropriate to our language. The same may be said with respect to other foreign names where the Roman alphabet is not in use or is supplemented by other letters.

HISTORY OF MATHEMATICS

GENERAL SURVEY OF ELEMENTARY MATHEMATICS

CHAPTER I

PREHISTORIC MATHEMATICS

1. In the Beginning

In the Beginning. When we attempt to trace the beginnings of the history of mathematics, we are confronted by the necessity of defining our terms. What do we mean by "history," and what does the term "mathematics" signify? If we consider history as a narration of recorded events, we shall have one course laid out for us; but if we look upon it as a relation of incidents which probably happened even before the advent of the human race, then our course is a different one. In the latter definition of the term we may properly include the so-called prehistoric, and this course will be followed in the present chapter even though our discussion of the prehistoric period will necessarily be brief. The history of mathematics may, therefore, have its inception in that nebulous period which we vaguely call "the Beginning."

Meaning of Mathematics. When we attempt to define "mathematics" we find ourselves encircled by unexpected limitations, and these limitations are still more in evidence when we change the term to "elementary mathematics." If mathematics means that "abstract science which investigates deductively the conclusions implicit in the elementary conceptions of spatial and numerical relations," as the Oxford

Dictionary defines it, then the history of mathematics cannot, strictly speaking, go back much earlier than the time of Tha'les (*c.* 600 B.C.), a relatively modern writer if we consider the antiquity of the race. Such a limitation, however, would not be a satisfactory one, for it would withdraw from our consideration those early steps in the development of the science which have great interest to the student and which are of value in considering the education of the individual.

It is well, therefore, to discard such niceties of definition and to take a broader view of the case, seeking to tell the story of the genesis of mathematics even before the period in which the science, as defined above, began to exist. Such a procedure will lead us back not only to the days when the human race was young, but to the ages immediately antedating its appearance upon the earth, and even farther. If one should wish to reach an absolute zero from which to begin his narrative, he would soon find himself lost in a maze of perplexities. For our purposes a brief statement of the mathematics in the Beginning will suffice.

Cosmic Figures. When we consider the birth of a solar system like ours, and point a telescope at a nebula in the actual throes of such a stupendous effort, we are impressed by the fact that one of the great cosmic forms is the spiral, a curve not scientifically studied until late in the evolution of human intelligence. We are also struck by the fact that the heavenly bodies obey various physical laws which we commonly express in mathematical language, and that they finally tend to move in elliptic paths about some larger mass or to trace a parabola seemingly from the Infinite into our solar system and back again. As they cool we find that the minerals which compose these heavenly bodies acquire certain habits, the molecules of carbon combining to produce a diamond crystal in the shape of a regular octahedron, those of silica forming hexagonal prisms with pyramidal ends, and those of water arranging themselves in snow crystals of certain fundamental types. In the Beginning, therefore, the spiral curve and the mathematical laws of physics are found everywhere present. Then come the conics to determine the paths of the cooling bodies, followed by the regular or semiregular polyhedrons as representing the mathematical habits inherited by various molecules.

SPIRAL NEBULA

The birth of a solar system

In considering these early evidences of mathematics in the great cosmic plan it is natural to make claims which some may feel to be extravagant. For example, since the square of $a+b$ is always and everywhere equal to $a^2+2ab+b^2$, we may assert that here is a type of those mathematical truths which have no beginning in time and which shall have no end. Such an assertion is perfectly true, and, this being the case, the history of mathematics may be considered as a record of the discovery of existing laws in this science and of the invention of better symbols as needed from time to time for their expression.

SNOW CRYSTAL

Photograph under a microscope below freezing temperature

Herbert Spencer once spoke of the properties of space as "eternal, uncreated"—as "anteceding either creation or evolution," adding these impressive words:

It is impossible to imagine how the marvellous space-relations discovered by the Geometry of Position came into existence. The consciousness that without origin or cause Infinite space has ever existed and must ever exist, produces in me a feeling from which I shrink.

Advent of Life. Passing from the preorganic era of the earth's existence, when mathematics was manifest in the spirals of the nebulas, in the courses of the planets and the comets, and in the crystallizing habits of the minerals, all of which give meaning to the statement, often attributed to Plato, that "God eternally geometrizes," we find new interest in the history of mathematics with the advent of life upon our planet. We now enter upon what seems like a period of experiment in plant and animal forms, a period roughly estimated as some ten million years in extent, the last half million of which may possibly mark what is only the beginning of the experiments with human beings.

It needs only the most casual observation to show the presence of mathematical forms in plant life. They appear, for example, in the phyllotaxy, or leaf arrangement, of plants; in the regular polygons in the structure of the pineapple, the breadfruit, and the watermelon; and in the Golden Section of the fern, the ivy, and other leaves. Such considerations naturally lead us to speculate on the causes that led certain leaves, millions of years ago, to arrange themselves about a stalk in obedience to the law of series first expressed by Leonardo Fibonacci in the 13th century. We know the economic law involved, but how did the plants come to know it? We know the relation of this law to the Golden Section and to the expression $\frac{1}{2}(\sqrt{5}-1)$, and we have discovered that a related esthetic law was recognized by the Greeks in their architecture and their plastic arts; but how came the vegetable world to adopt this principle in its scheme of life? Was it the result of countless experiments in the search for a maximum of efficiency, culminating in a plant habit, or was some other law involved which the future may be expected to reveal?

Recognition of Mathematical Concepts. With the advent of animal life there comes a possibility of the recognition of such mathematical concepts as form, number, and measure, a recognition which did not have to await the coming of the human race. In that remote period in which the lower animal forms were the sole conscious beings of earth we infer that there was a species of counting, or at least of pseudo-counting,—an inference drawn from a study of the animal life of today. Magpies, for example, recognize the size of a group of five or even six objects; the chimpanzee knows that five objects are more than four, and even certain insects show a similar power to recognize the relative sizes of small groups. It is here that we find the first uncertain steps in the development of an arithmetic.

Our mathematical concepts which have certain analogues in the lower animal life are not limited to the domain of number. The spider seems to recognize both regular polygons and similarity of figures in making a web, and the laws of maxima and minima are followed by the bee in building up the hexagonal wax cells of the honeycomb. No beast is so stupid as not to know that a straight line is the shortest path between two points, and few birds fail to observe the principle of symmetry in the structure of a nest. Through long ages the animal world, like the world of plants, has acquired experience and established habits of maximum efficiency, and the results show themselves in a large variety of geometric forms, some of them embodying a high degree of refinement, as in the capping of a cell in a honeycomb. Who can say when or where or under what influences the Epeira, ages ago, first learned to trace the logarithmic spiral in the weaving of her web? As Henri Fabre asks, is such a question "a mere dream in the night of the intricate,—an abstract riddle flung out for our understanding to browse upon?"

Advent of the Human Race. With the advent of the human race there developed an opportunity for mathematics to show itself more consciously. This appeared in art, the world's universal language; in religious mysticism, the expression of our efforts to fathom the unfathomable; and, in a less degree, in commerce, war, and the needs of the pastoral life. Each of these human interests, but especially decorative art, contributed to an appreciation of geometry; each, but especially religious mysticism and commerce, contributed to the development of number; and each lent its aid to the creation of an interest in architecture and in the science of astronomy.

2. Primitive Counting

Early Efforts. When the primitive savage began to develop number names, the process was a tedious one. His needs were simple; the designation of the size of his family, the counting of his enemies, or some similar use of small numbers represented all that his meager life demanded of mathematics. Even after the hunter had given considerable place to the shepherd, number played a minor rôle in the life of the world. The invention of money was still thousands of years in the future, and without some such medium of exchange a large part of our arithmetic would disappear. The herdsman could recognize that one of his sheep was missing without being able to count his flock, and even his dog could do the same, the needs of the two in this respect being quite on a par. For a long time after the advent of man such simple numbers as two and three were sufficient for all purposes not met by nouns of multitude, like lot, heap, crowd, school (of fish), pack (of hounds), and flock. A well-known instance of a deaf-and-dumb boy who acquired a knowledge of numbers from observing his fingers, even before he was taught to count, shows us that the idea of number did not have to await the development of spoken language, and so a savage may appreciate three without having a name for numbers beyond two.

Number and Language. In a general way, however, the development of the number sense keeps pace with the growth of a number language, and this language can be inferred from the speech of certain lower types of savages which have been studied within the last century. For example, the numerals of thirty selected Australian languages extend in no case beyond four, the tribes which use them not having even reached the point of recognizing, as a basis for counting, the fingers of one hand, and in most cases having number names for only one and two. In general, everything beyond two is called "much" or "many." So poor are these particular languages that only cardinal numbers exist, the ordinals being unknown. This paucity of vocabulary is not universal among the Australian aborigines, however, for in one linguistic region the numeral names reach as far as fifteen or twenty. Furthermore, there is an element of doubt in the reports of anthropologists, first because they are not always correctly informed by the natives themselves, and also because the absence of number names does not generally mean that the primitive tribes had no names for groups of two or three. An illustration of the latter consideration is seen in the case of the Andamans, a tribe of Oceanic negritos. Their number names are limited to one and two, but they are able to reach ten by this process: the nose is tapped with the finger tips of either hand, beginning with one of the little fingers, the person saying "one" (*úbatúl*), "two" (*íkpór*), and then repeating with each successive tap the word "*anká*," which means "and this." When the second hand is finished, the two hands are brought together to signify $5 + 5$, and the word "all" (*ardùru*) is spoken. To mention another example, the Pitta-Pitta, a tribe in Queensland, are able to count the fingers and toes without a system of numerals, but only by the aid of marks in the sand, and in various other parts of Australia the natives show habitual uncertainty as to the number of fingers they have on a single hand.

Objective Counting. Judging by other races of low intelligence, the primitive man could count only by pointing to the objects counted, one by one. Here the object is all-important, as was the case with the early measures of all peoples. The habit is seen in the use of such units as the foot, ell,[1] thumb,[2] hand, span, barleycorn, and furlong (furrow long). In due time such terms lose their primitive meaning and we think of them as abstract measures. In the same way the primitive words used in counting were at first tied to concrete groups, but after thousands of years they entered the abstract stage in which the group almost ceases to be a factor. Indeed, so completely has this transformation been effected that it is difficult for us to appreciate how greatly the human mind had developed to render it possible. We say "seven," but we no

[1] Anglo-Saxon *eln-boga* (elbow); Latin *ulna* (forearm), whence the French *aune*.
[2] Latin *pollex*, whence the French *pouce*, the basis for our inch.

longer think of a certain group of objects, nor do we demand such a group in order to count; we think of a word in an endless series, the word coming just after "six" in the series and just before "eight." In the Malay and Aztec tongues, however, the number names mean literally one stone, two stones, three stones, and so on; while the Niuès of the Southern Pacific use "one fruit, two fruits, three fruits," and the Javans use "one grain, two grains, three grains," all these being relics of the concrete stage of counting. When a Zulu wishes to express the number six, he says "taking the thumb" (*tatisitupa*), meaning that he has counted all the fingers of the left hand and has begun with the thumb of the right hand. For seven he says "he pointed" (*u kombile*), meaning that he has reached the finger used in pointing. After the world abandoned the use of objects in counting, or of objects used as an aid to counting, there developed the possibility of an infinity of number names, and hence in the course of time mankind was confronted by the necessity for classifying numbers and for naming them according to some simple plan.

The Radix in Number Systems. Fundamentally the system of number names, although not the names themselves, has been and is the same with all peoples. Some number is selected as the radix, or basis for counting, ten being the general favorite for the reason that we have ten fingers. We therefore say that we count on a scale of ten, or a decimal scale. Ten was not, however, the primitive radix. Very likely two was first used, as in the case of those natives of Queensland who count "one, two, two and one, two twos, much," the last term or some similar expression covering all numbers larger than four. A similar system is found among the African pygmies, who count *a*, *oa*, *ua*, *oa-oa* (two-two), *oa-oa-a* (two-two-one), *oa-oa-oa* (two-two-two), and so on. There may be some relic of this custom in our counting by braces, couples, pairs, or casts, and in the early Syriac numeral forms shown above.

1 2 3 4

EARLY SYRIAC NUMERALS

Showing influence of the scale of two

Early Scales. We have more tangible evidence of an early scale of three than we have of two. There are various instances of the use of three to mean a large number, a number beyond (*trans*) which all groups are simply "many." It is from this use of three that we have the Latin *ter felix*, thrice (very) happy; the Greek *trismeg'istos* (*τρισμέγιστος*), thrice (very) greatest; the English "Thrice is he armed that hath his quarrel just," and the related French *très bien*. A similar use of three is occasionally found among the lower races, as when the native Tasmanians count "one, two, plenty." The scale of three appears in several instances, as in the case of the Yahgan Fuegians, a tribe of Tierra del Fuego, where the number names begin "*kaueli*, *kombaï*, *maten*, *akokombaï* (the other two), *akomaten* (the other three)." The Demaras of Africa are reported to have no number name beyond three, so that this number would form a probable radix if they needed to count further. Some of the early notations, like that of the old Phœnician system, grouped their marks by threes, a possible reminder of their ancient ternary system.

Four has also been used to mean a large number and hence to form a kind of radix. Thus, Horace speaks of those who are *ter quaterque beati* (thrice and four times blessed). A more definite trace of the use of four as a radix is, however, found among certain South American tribes, where the counting begins with one, two, three, four, and then proceeds to four and one, four and two, and so on as far as the simple needs of the people may demand.

Scale of Five. The first scale to be extensively used was the quinary scale. Two, three, and four were but feeble attempts of the race, the five fingers of the hand being brought into use as soon as mankind needed a basis for counting to numbers of considerable size. The left hand was generally used, the right index finger pointing to the first object to be counted and then to a finger, repeating the process until the five fingers had been used and then repeating the operation, sometimes marking the fives by pebbles or sticks. Certain South American tribes have been observed to count by hands,

—one, two, three, four, hand, hand and one, hand and two, and so on. Mungo Park (1771–1806) found a similar system in one of the tribes of Africa, and in one part of Paraguay five is called "the fingers of one hand," ten being "the fingers of both hands," and twenty "the fingers of both hands and feet." With not a little poetic fancy one of the Carib tribes speaks of ten as "the children of the hands." Even in our more prosaic tongue we speak of the smaller numbers as digits (*digiti*, fingers). Sometimes a more primitive scale becomes blended with the scale of five, as in the case of the Yukaghirs of Siberia. These people count "one, two, three, three and one, five, two threes, one more, two fours, ten [with] one missing, ten." A similar blending of scales is frequently found, and indeed in our own language we appear to have special names for the numbers up to twelve,[1] then going back to the decimal scale in our word *thirteen* (three ten).[2]

Scale of Ten. Of the scales from six to nine inclusive, few traces exist. As soon as the race found that it could count by the aid of the fingers of one hand, it naturally passed to the use of the fingers of both hands, thus forming a scale of ten, or to the use of the fingers and toes, which gave the scale of twenty. There is, however, an occasional trace of the intermediate scales, as when the Bolans, or Buramans, of the west coast of Africa counted by sixes. A relic of this radix also appears in South Bretagne, where the word *triouech* (three sixes) is still used for eighteen.

One reason for the wide adoption of the scale of ten is that few characters are needed in writing numbers with such a radix, as would also be the case with such convenient scales as eight or twelve. The chief reason, however, is the fact that man has ten fingers. There is no way of telling when the world adopted this scale, but its wide geographic range leads to the belief that it must have been before the general migrations of

[1] This may be only apparent, however, for possibly eleven is *ein* + *lif* (one over ten), and twelve is *twa* + *lif* or *zwo* + *lif* (two over ten).

[2] An interesting example of the persistence of the scale of five is seen in the German peasant calendars as late as 1800. See K. Brunner, "Ein Holzkalender aus Pfranten," *Zeitschrift des Vereins für Volkskunde*, XIX, 249.

the race. An examination of more than seventy African languages showed that ten was used as a radix in every case, and although this may be due directly to the fingers, it is quite as likely to be due to a common linguistic source. There is an interesting theory that the decimal scale originated very late, and soon after the scale of five was discarded, the theory being based on the fact that Homer, Æschylus, Plutarch, and Apollonius used the word *pempa'zein* (πεμπάζειν, literally, "to five") to indicate counting. This conjecture puts the adoption of the scale too late to account for other known facts, however, and hence it is hardly worthy of serious consideration. Nevertheless, it is certain that such traces of the scale of five persisted long after the scale itself was discarded.

Scales above Ten. In Italy the early Etruscans seem to have used a combination of the scales of ten and twenty, as was the case when our ancestors said "two score years and ten" or "two score and twelve"; but among the Romans ten was the sole radix in general counting, being held in great honor, as witness the lines of Ovid:

> A year was when the moon had filled out its tenth circle;
> This number [ten] was at that time [held] in great honor
> . . . because [there are] that many fingers by which we
> are accustomed to count.[1]

There is reason to believe that the scale of twelve, the duodecimal scale, was favored in prehistoric times in various parts of the world, but chiefly in relation to measurements. While it may have been suggested by the approximate number of lunations in a year, it was undoubtedly its divisibility by two, three, and four, thus allowing for simple fractional parts, that made it attractive. Its popularity is seen in the number of inches in a foot, the number of ounces in the ancient pound, the number of pence in a shilling, the number of lines in an

[1] Annus erat, decimum cum luna receperat orbem;
Hic numerus magno tunc in honore fuit
. . . quia tot digiti, per quos numerare solemus.
Fasti, III, 121

inch, and the number of units in a dozen.[1] As already stated, we have one relic of the scale of twelve in our own system of counting, for we do not begin the usual form of the decimal naming of numbers when we reach ten, but we say "eleven, twelve" instead of "oneteen, twoteen." Another relic is found in the old Frisian language, which has "twelvety" (*tolftich*) for 120, and still others are found in various parts of Europe.

Scale of Twenty. The scale of twenty—the vigesimal, or vicenary, scale—was not uncommon in prehistoric times. It is a reminder of the barefoot days of the race, when men counted toes as well as fingers. Some trace of it is still found in the Malayan languages, but it was early carried to parts of the world far removed from the tropics. That it was at one time in favor with the ancient Celtic peoples is evident from the present French use of *quatre-vingt* (four-twenty, for four twenties) instead of *huitante*, for eighty; *quatre-vingt-dix* (four twenty-ten) instead of *nonante*, for ninety; and such older forms as *six-vingt* for 120, *sept-vingt* for 140, and *huit-vingt* for 160, and even up to *quinze vingt*.[2] Further traces are found in the Gaelic use of "one, ten, and two twenties" for fifty-one; in the Danish use of "the mean between twice 20 and thrice 20" (that is, $2\frac{1}{2} \times 20$) for 50; in the Welsh "one and fifteen over twenty" for thirty-six; and in the Breton "eleven and three twenties" for seventy-one. The Maya of Yucatan and the Aztecs of Mexico had elaborate systems in which twenty was prominent; and while the Greenlanders have no such elaborate system, they use the expressions "one man" for twenty, "two men" for forty, and so on. There is a similar usage among the Tamanacs of the Orinoco, who speak of twenty as "one man" and of twenty-one as "one to the hands of the next man." Of the various other evidences of the scale of twenty mention may be made of the system used by the Vey tribe in Africa. In this system 16 is read $10 + 5 + 1$, and so on to 19. Thereafter and to 99 the

[1] Dozen, from *duodecim*, that is, *duo* (two) + *decem* (ten).

[2] The Hospice des Quinze Vingt, founded by Louis IX, 13th century, still exists in Paris.

number twenty is in evidence, 30 being $20+10$, 40 being 2×20, 50 being $2\times 20+10$, 70 being $3\times 20+10$, and 99 being $4\times 20+10+5+4$.[1]

The English Score. In our own language the use of the score is so common as to suggest the predilection of our ancestors for counting by twenty. The King James version of the Bible shows how general was the use of the word only three centuries ago, and in earlier times this use was even more pronounced. An instance of this is seen in the ancient poem known as *Morte d'Arthur*: "Att Southamptone on the see es sevene skore chippes," and a further illustration, showing the French influence, is seen in an item of the accounts of the Scottish exchequer for 1331, the sum of £6896 5s. 5d. appearing as follows:

$$\text{vj}^{\text{m}}\quad \text{viij}^{\text{c}}\quad \text{iiij}^{\text{xx}}\quad \text{xvj}\quad \text{ƚj}\quad \text{v}\bar{\text{s}}\quad \text{vđ}$$

That twenty was once a limit of counting may be inferred from our use of "score" to mean an indefinitely large number, as when we say "a score of times" and in the similar use of *vingt* by the French.

In this connection we may also consider the prehistoric custom among certain peoples of using a special number as a kind of radix, although one that did not, like ten, form the basis of a number system. An instance of this custom appears in the Biblical use of forty, as in the case of "forty days and forty nights," and similar instances are found in the languages of the Marquesas Islanders and the Hawaiians.

The extent to which primitive peoples carried their number systems was, of course, determined by their needs. The native Australians, for example, were little given to trade, and so they felt the need of number names only to two, three, or four, and for the Hottentots five was a sufficient limit. Among the Papuan peoples, those of Paraido count to ten, while those of Pauwi, living farther inland, find five sufficiently large for

[1] J. Büttikoffer, in *Internationales Archiv für Ethnographie*, I (Leyden, 1888), 108, commenting on the work of H. Hartert in *Globus*, LIII, 236.

their simpler needs. The Kafirs, however, possessing herds of cattle, count to a hundred or more, while the Nubians and Abyssinians, representing a higher civilization, often use numbers to a thousand or even to a million, without apparent European influence. For a similar reason one of the Polynesian languages has number names to thousands. The Hindus, having need for large numbers in a semireligious way, as appears from their Vedic writings, early developed a numeral system that is practically unlimited, the most extensive to be found among any ancient people. In general it may be said that mathematics met the needs of primitive life, with respect to number, as these needs developed.[1]

[1] Since it is impossible, in a work of this nature, to consider this subject at any length, the student is referred to the following sources as being particularly valuable: L. L. Conant, *Number Concept* (New York, 1896); E. B. Tylor, *Anthropology* (New York, 1898), and *Primitive Culture*, 2 vols. (New York, 1891); M. Cantor, *Mathematische Beiträge zum Kulturleben der Völker* (Halle, 1863); F. Ratzel, *The History of Mankind*, translated from the second German edition by A. J. Butler, 3 vols. (New York, 1904); John Crawford, "On the numerals as evidence of the progress of civilization," *Transactions of the Ethnological Society of London*, II (N.S.), 84; W. C. Eells, "Number Systems of the North American Indians," *Amer. Math. Month.*, XX, 263, 293; Joseph Deniker, *The Races of Man* (London, 1900); R. E. Dennett, "How the Yoruba Count," *Journal of the African Society*, XVI (London and New York, 1916–1917), 242; A. Mann in the *Journal of the Anthropological Institute of Great Britain and Ireland*, XVI (London, 1887), 57; J. S. Hittell, *A History of the Mental Growth of Mankind in Ancient Times* (New York, 1893); H. F. Feilberg, "Die Zahlen im dänischen Brauch und Volksglauben," *Zeitschrift des Vereins für Volkskunde* (Berlin, 1894), pp. 243, 374; Marianne Schmidt, "Zahl und Zählen in Afrika," in the *Mitteilungen der Anthropologischen Gesellschaft in Wien*, XXXXV (1915), 165 (the standard authority on the number systems in Africa, together with a full bibliography); J. Hehn, *Siebenzahl und Sabbat bei den Babyloniern und im alten Testament* (Leipzig, 1907); K. F. Meyer, "Die Sieben vor Theben," *Zeitschrift für Ethnologie*, VII (1875), 105; VIII (1876), 264; F. v. Andrian, "Die Siebenzahl im Geistesleben der Völker," *Mitt. d. Anthrop. Gesellsch. in Wien*, XXXI (1901), 225; J. Fraser, "The Polynesian Numerals One, Five, Ten," *Journal of the Polynesian Soc.*, XI (Wellington, 1902), 1; J. C. Gregory, *The Nature of Number* (London, 1919); *Journal of the Royal Anthropological Institute of Great Britain and Ireland*, various articles as in Vols. XXXVII, XXXIX; A. H. Keane, *Man Past and Present* (Cambridge, 1899); Sir John Lubbock, *Pre-Historic Times as illustrated by Ancient Remains* (London, 1869); W. J. McGee, "The Beginning of Mathematics," *The American Anthropologist*, I (1899), 646; S. G. Morley, *An Introduction to the Study of Maya Hieroglyphics* (Washington, 1915); O. F. Peschel, *The Races of Man* (New York, 1888).

3. Geometric Ornament

Early Art. A further prehistoric stage of mathematical development is seen in the use of such simple geometric forms as were suggested by the plaiting of rushes, the first step in the textile art. From this there developed those forms used in clothing, tent cloths, rugs, and drapery which are usually found among primitive peoples. Since the earliest trace of human art that we have thus far found is seen in representations of animals, these being drawn on bone in the Early Stone Age, one might expect to find such figures in early mural decorations, and this is not only the case but is one means of dating the latter with some degree of approximation. The geometric ornament, however, became in due time a favorite one among nearly all early peoples. This may have been because the plaiting of rushes furnishes an easy medium for the representation of geometric forms, but at any rate such forms as the swastika and the Greek key developed at an early period. Such decorations are not confined to the textiles of the people; they are equally prominent in architecture in all parts of the world. They are found on the early monuments of Mexico, on the architectural remains of Peru, on the huts of the savage, and on the early buildings of the historic period in various parts of the Old World, especially on those devoted to the commemoration of the dead or to the worship of the gods.

EGYPTIAN POTTERY OF THE PREDYNASTIC PERIOD

It shows the earliest stage of geometric ornament on pottery. The Predynastic Period extended from *c.* 4000 to *c.* 3400 B.C. From the Metropolitan Museum, New York

The same instinct that leads to geometric decoration of religious structures shows itself in the decoration of personal ornaments and of articles intended for domestic use. This is seen in the handicraft of the Stone Age, it is found in the rich gold work of early Egypt, and it is equally in evidence in most of the jewelry of modern times. It is not merely the instinct of symmetry that we find in these petrified thoughts of the race; it is quite as much a desire to fathom the mystery and grasp the meaning of the beauty of geometric form.

CYPRUS JUG OF THE PERIOD 3000–2000 B.C.

Pottery of the Early Bronze Age, showing the second stage in geometric ornament. From the Metropolitan Museum, New York

Early Pottery. The early pottery of Egypt and Cyprus shows very clearly the progressive stages of geometric ornament, from rude figures involving parallels to more carefully drawn figures in which geometric design plays a more important part and in which such mystic symbols as the swastika are found. Art was preparing the way for geometry.

4. Mysticism

Religious Mysticism. The beginning of an appreciation of the wonders of mathematics is closely connected with the beginning of religious mysticism. Man wondered at the heavens above him; he wondered at life and he wondered even more at death; all was a mystery. He likewise wondered at the peculiarities of geometric forms and at the strange properties of such numbers as three and seven, the two primes within his limited number realm that were not connected with his common scales of counting. The mystery of form and the mystery of number he connected with the mystery of the universe about him,—the universe in which he felt himself a mere mote in the sunbeam. His sense of wonder at the potency of the sun led him to the orientation of his religious structures; his recognition of a pole star led him to consider a fourfold division of his horizon, and to speak of the four corners of the earth; and it is not impossible that the swastika and the various other cruciform figures of the ancient civilization are a recognition of this tendency. The number four was looked upon as peculiarly significant by certain American aborigines as well as by the early peoples of Asia, Australia, and Africa, and we may have a relic of this attitude of mind when we speak of "a square man" or one who acts "squarely."

PAINTED JUG FROM CYPRUS
1000–750 B.C.

Pottery of the Early Iron Age in Cyprus, showing a third stage in the use of geometric ornament, with the swastika. The Geometric Period of decoration closed, for the Mediterranean countries, just before the time of Thales. From the Metropolitan Museum, New York

Architecture. Just as we find an instinctive appreciation of the beauties of geometric forms as applied to personal ornaments, so we find it as applied to architecture, not merely with respect to decoration as already mentioned, but in the general structure of temples, of altars, and of tombs. In early India, for example, there seems to have been no study of geometry as such except in connection with forms used in the temple, and this was probably the case in other parts of the earth. A desire to adapt symmetry to architecture is seen in the terraced pyramids of Mexico as well as in those of Egypt; and while these buildings are not prehistoric, they doubtless are the outgrowth of prehistoric forms.

Observations of the Stars. As already mentioned, the primitive man seems to have felt that the secret of the stars was closely bound up with the secret of his destiny. It was this that led the Babylonian shepherd and the desert nomad to observe the stars, to speculate upon their meaning, and to take the first steps in what developed into a priest lore in the temples along the Nile and in the land of Mesopotamia. It was this, too, that led the early philosophers and poets to consider the stars as lighted lamps suspended in a vast material vault, or as golden nails fixed in a crystal sphere,—ideas perfectly suited to the childhood of the race. When it was that these observations of the heavens led to angle measure, to the recording of such celestial phenomena as eclipses, and to a naming of the signs of the zodiac and the constellations, we cannot say. One writer of prominence[1] places a recognition of the common constellations as early as 17000 B.C., and while this date seems to be very improbable, even though supported by certain historico-astronomical considerations, it is doubtless true that the period of this recognition and of the observance of certain celestial phenomena is very remote. While there is good reason for thinking that these early steps in astronomy were taken in Mesopotamia, the proof is not sufficiently strong to enable us to say that this was unquestionably the case, nor are we able to fix upon the period within any particular century or even within any particular millennium. Similarly, we are unable to state the time or place in which the early peoples began to recognize the constellations or to give them fanciful names. After attaining a certain degree of success in our research, we are lost in the prehistoric clouds.

Lengthen the story all that we can, it is not possible to extend it back more than an imperceptible distance on the great clock face. For if we represent the period of all life on our planet by one revolution of the minute hand, the period of human life will be covered by only half a minute, and recorded history will be represented by less than two seconds. What we definitely know of the history of mathematics covers a period in world development so short as to seem almost infinitesimal.

[1] G. Schlegel, *Uranographie Chinoise*, 2 vols., II, 796 (Leyden, 1875). For a recent discussion of the whole question, see Léopold de Saussure, "Les origines de l'astronomie chinoise," *T'oung Pao*, Vols. X seq. (Leyden).

TOPICS FOR DISCUSSION

1. Geometric forms that were in existence before the advent of life on the planet.

2. Laws of motion that entered into the formation and perpetuation of our solar system.

3. Geometric forms that appear prominently in the vegetable world and in the bodily structure of certain animals.

4. Geometric forms that appear prominently in the products of the labor of the lower animals, with the question of maximum efficiency in any of these cases.

5. The question of animal counting or pseudo-counting as discussed by psychologists.

6. Evidence of primitive counting without any scale.

7. The world's use of scales below five as shown by a study of our language and of savage tribes.

8. Reasons why the scales of five, ten, and twenty were the leading favorites.

9. Reasons why the scale of twelve would have been a particularly good one.

10. Reasons why three and seven have been particularly notable as mystic numbers, with several illustrations.

11. Circumstances which developed a high degree of skill in counting among certain peoples.

12. Reasons which led primitive peoples to the use of geometric forms in ornament.

13. The effect of religious mysticism upon primitive mathematics.

14. Various stages of geometric ornament in Cyprus, Crete, and the mainland of Greece.

15. Possible influence of geometric decoration upon the study of geometry as a science.

16. Geometric decoration that has persisted in all ages, with a study of the probable causes for this persistence.

17. Causes leading to an interest in astronomy among primitive peoples. Features of the ancient astronomy that are still found either in our present study of the science or in folklore.

18. Evidence of the antiquity of astronomical ideas, particularly in Mesopotamia, Egypt, and China, with probable evidence in the case of India and other parts of the East.

CHAPTER II

THE HISTORIC PERIOD DOWN TO 1000 B.C.

1. General View

Sources of our Knowledge. The period down to the arbitrarily selected date 1000 B.C. overlaps the prehistoric period mentioned in Chapter I, the prehistoric gradually merging into the historic in certain parts of the world but not reaching this stage in certain other parts. Various facts which might properly have found place in Chapter I will therefore be related in this chapter, but in a general way we shall now pass to that period in the evolution of the race in which less use need be made of conjecture in the recital of the story of mathematics, although it cannot be wholly eliminated.

The sources of our knowledge are no longer mere tradition, nor does inference from the study of savage tribes constitute so important a basis for our statements. The sources are now, in general, the relics of human activity, largely in the form of inscriptions or manuscripts which actually date from remote centuries, or of copies of such evidences.

Countries Considered. There are four countries which have left to posterity such an abundance of historical material prior to the beginning of the first millennium of the pre-Christian era as to warrant our special consideration. These countries, considered geographically instead of politically, are Egypt, Mesopotamia, China, and India. Each claims for itself a high degree of antiquity, each claims to have been a pioneer in mathematical development, each is ethnographically somewhat of a unit; and in certain respects the claims of each have reasonable foundations. Each had, at least in considerable areas, a salubrious climate in the warm intervals between the several descents of the ice from the north,—descents which characterize what is commonly called the Glacial Epoch,—and hence each was able to develop an early civilization. Each flourished along one or more important rivers, which not only furnished water for navigation and for domestic purposes but also afforded opportunities for the application of a rude form of mathematics to the irrigation projects which were already in evidence in the early centuries of the historic period. From each, after the ice retreated, a human stream in due time flowed to the more invigorating climate of the north and carried along with it traces of the early mathematical lore which had already begun to develop in the temples, and of the customs established among the primitive traders of the more favored lands.

Antiquity of Mathematics. Although there is evidence showing that the human race has been on the earth for hundreds of thousands of years, the earliest definite remains that have come down to us are from the Early Stone Age, or Paleolithic Age, which seems to have begun in the warm interval after the third descent of the ice. This was not less than 50,000 years before our era and may have been from 50,000 to 75,000 years earlier still. In the remains of this period we find implements which suggest the existence of barter and the need of numbers for counting, although they date from a period thousands of years before there had dawned upon human intelligence any idea of written numerals. We may simply suppose that the presence of such implements and the important discovery of a means for making fire, which seems to have occurred about 50,000 years ago, are evidence of a degree of intelligence high enough to assure some idea of number.

About 15,000 years before our era there is thought to have begun the Middle Stone Age, the period of the fourth descent of the ice. In this period we find the oldest known works of art. These works show such an intellectual advance as to make it quite certain that the world had reached a period when the abstract notion of number must have been in evidence,—a judgment warranted by our knowledge of all primitive peoples today who have reached this stage in art.

The Late Stone Age is relatively a recent period, dating from *c.* 5000 B.C., and by this time there had developed quite elaborate number systems, and the observation of the stars had become a fairly well organized science. That this was the case we know from various historical facts which will be mentioned later.

Advent of Writing. Metal, as distinguished from natural ore, was discovered *c.* 4000 B.C., possibly in the Sinai peninsula, and with this there came a new need for weighing and measuring and a new impetus to a system of barter which was doubtless very old even at that time. About 500 years later, writing is known to have been in use, and the system of ruling over masses of people had become so advanced as to render possible the control of a population of several millions by one government. The bearing of all this upon the development of a number system and upon systematic taxation is apparent. About 3000 B.C. the earliest stone masonry was laid and sea-going ships began to cross the Mediterranean, and a little later the pyramids of Egypt were erected, so that history now enters a period in which mathematics reached out beyond mere counting and into such fields as that of practical geometry, including a primitive kind of leveling and surveying.

2. China

Early Chinese Mathematics. We have no definite knowledge as to where mathematics first developed into anything like a science. Mesopotamia has several strong claims to priority, and so has Egypt. As to China, we have little positive knowledge of its earliest literature, the possibility of corruption of its texts being such as to cast doubts upon its extreme claims. Until native scholars develop a textual criticism commensurate with that which has been developed in the Occident, this uncertainty will continue to exist. The historical period begins with the 8th century B.C., or, at the earliest, with the reign of Wu Wang, the Martial Prince, in 1122 B.C.[1] In beginning with

[1] A. J. Little, *The Far East*, p. 20. Oxford, 1905.

China, therefore, it must not be thought that we should recognize the validity of all the claims that are often advanced for the antiquity of her science.

Basing his opinion upon later historical descriptions of the primitive astronomy of China, Professor Schlegel of The Hague, as already remarked, asserts that the Chinese recognized the constellations as early as 17000 B.C., which was about the close of the Early Stone Age. There is nothing impossible in such a supposition, although it is improbable. The race had developed considerably by that time, and it may well have extended its poetic fancy to the giving of forms to groups of stars which it had looked upon for thousands of years. Professor Schlegel also fixes upon 14700 B.C. as an approximate date of the duodenary zodiac, other scholars asserting that 13000 B.C. is more probable and still others fixing upon 4000 B.C.,—a discrepancy that may well arouse skepticism as to the validity of any of these hypotheses. Schlegel also believes that there is evidence of the extended study of the celestial sphere in China in or about 14600 B.C.

While such claims are generally doubted by competent Sinologists, it is quite likely that the Chinese developed some acquaintance with descriptive astronomy at an early period, and that this development necessitated such knowledge of mathematics as the measure of time and angles and the use of fairly large numbers. Reasonably well-founded tradition gives the probable dates of Fuh-hi,[1] the reputed first emperor of China, as 2852–2738 B.C.,[2] and in his reign there were extensive astronomical observations. In this general period the Chinese are believed to have changed their zodiac into one of twenty-eight animals.

[1] In general, the transliteration of Chinese names is that of Y. Mikami, *The Development of Mathematics in China and Japan* (Leipzig, 1913) (hereafter referred to as Mikami, *China*), and H. A. Giles, *A Chinese Biographical Dictionary* (London, 1898). The transliteration varies greatly with different Sinologists, a title like *I-king* appearing as *Yih-ching*, *Yi-ching*, *Ye King*, *Y-Ching*, and so on. In many cases I have been greatly assisted by my colleague, Professor Friedrich Hirth, one of the greatest living Sinologists.

[2] F. Hirth, *The Ancient History of China*, p. 7 (New York, 1908). Professor Hirth follows Arendt's tables as being the most carefully considered.

Reign of Huang-ti. In the year 2704 B. C.[1] Huang-ti, the Yellow Emperor, began his reign. Under his patronage it is said that Li Shu wrote on astronomy and that Ta-nao established the *Chia-tsŭ*, or sexagesimal system, both of these statements being supported by copies (possibly altered) of ancient records.[2] Even the emperor himself is said to have taken such an interest in mathematics as to write upon astronomy and arithmetic, and in his reign an eclipse of the sun was observed and recorded. Tradition assigns to this period even the decimal system of counting, although it is more likely that some popular work on the subject was written at this time. It was possibly during the reign of the emperor Yau[3] (*c.* 2357–*c.* 2258 B.C.) that two brothers, Ho and Hi, made astronomical observations. They are said to have suffered the displeasure of the emperor through their failure to predict a solar eclipse,[4] an incident showing a state of mathematical advancement quite equal to that in Greece in the time of Thales, some 1500 years later. The story is told in the *Shu-king* (*Canon of History*), an ancient record sometimes attributed to the pen of the emperor himself and sometimes to that of Confucius nearly two thousand years later.[5]

[1] According to Arendt and Hirth. Giles gives 2698 and others give 2697. Huang-ti is said to have died at the age of 111 years.

[2] As stated in Volume II, it is doubtful if the Chinese used anything like a sexagesimal system at this time, although they may have learned from the Sumerians that 60 is a convenient unit for subdivision.

[3] Reputed to have lived nearly a full century. See A. T. de Lacouperie, *The Languages of China before the Chinese*, p. 9 (London, 1887); Hirth, *loc. cit.*, p. 29.

[4] R. Wolf, *Geschichte der Astronomie*, p. 9 (Munich, 1877); *Handbuch der Mathematik, Physik, Geodäsie und Astronomie*, I, 7 (Zürich, 1872).

[5] Hirth, *loc. cit.*, pp. 29, 33, 251, who believes it to be a late work, or at least, if written in the time of Yau or of his immediate successors, to have been greatly modified by later copyists. Some effort has been made to fix the date of the eclipse as May 7, 2165 B.C. Other proposed dates are October 11, 2154 B.C., October 12, 2127 B.C., October 24, 2006 B.C., October 22, 2155 B.C., and October 21, 2135 B.C. The discrepancy between these later dates and those tentatively assigned to Yau, as given above, has little significance in the present state of knowledge as to Chinese chronology. The whole subject is still in the conjectural stage and awaits extended research on the part of capable Sinologists.

It is this emperor Yau and his successor, the emperor Shun, who, it is said, carried farther to the eastward the dominion established by the Bak tribes which had come from western Asia. These tribes had been under the civilizing influences of the people of Susiana, who in turn had received their civilization from Babylon.[1] If this theory proves to be correct, the similarity between certain early forms of astronomy and mathematics in the East and the West is more easily explained.

I-king. Of the "Five Canons" (*Wu-king*) of the Chinese probably the third in point of antiquity is the *I-king*, or *Book of Permutations*.[2] In this appear the *Liang I*, or "two principles" (the male, *yang*, ——; and the female, *ying*, – –) and from these were formed the *Sz' Siang*, or "four figures,"

⚌ ⚍ ⚎ ⚏

3 2 1 0

and the *Pa-kua* (eight-kua) or eight trigrams, the eight permutations of two forms taken three at a time, repetitions being allowed. These *Pa-kua* had various virtues assigned to them and have been used from a very early period until the present for purposes of divination. It was probably Wön-wang (1182–1135 B.C.) who wrote the *I-king*; at any rate it was he who extended the *Pa-kua* into the sixty-four hexagrams now found in this classic.[3]

[1] A. T. de Lacouperie, *loc. cit.*, pp. 9 seq.

[2] It is often called the oldest of the Chinese classics, as in the edition by J. Mohl, *Y-King*; *Antiquissimus Sinarum liber* (Stuttgart, 1834–1839). In the extensive literature on the *I-king* the following works may be consulted: H. Cordier, *Bibliotheca Sinica*, II, cols. 1372 seq. (Paris, 1905–1906); A. T. de Lacouperie, "The Oldest Book of the Chinese," *Journal of the Royal Asiatic Society*, XIV (N.S.) (London, 1882), 781, reprinted in 1892, with an extensive bibliography; T. McClatchie, "The Symbols of the Yih-King," *The China Review*, I (Hongkong, 1872), 151; J. Edkins, "The Yi king of the Chinese," *Journal of the Royal Asiatic Society*, XVI (N.S.) (London, 1884), 360; H. J. Allen, *Early Chinese History*, chap. viii (London, 1906). The first European edition of the *I-king* appeared at Frankfort in 1724.

[3] Hirth, *loc. cit.*, p. 59. The *Pa-kua* are attributed to Fuh-hi by Liu Hui, who wrote *c.* 250 A.D. The Leibniz theory, set forth in his *Philosophia Sinensium*, § 4, that these symbols had some connection with binary numerals, has no historical foundation in the *I-king* as originally written.

It is hardly conceivable by the Western mind that such a set of symbols should last for thousands of years, that it should be the subject of such a large number of books and monographs as have appeared in explanation of its meaning, and that it should be known today to everyone among the hundreds of millions who have come under the influence of the Chinese philosophy, not merely in China but all through the East.

☰	☱	☲	☳	☴	☵	☶	☷
k'ién heaven	*tui* steam	*li* fire	*chön* thunder	*sün* wind	*k'an* water	*kön* mountain	*k'un* earth
7	6	5	4	3	2	1	0
HEAVEN SKY	COLLECTED WATER	FIRE	THUNDER	WIND WOOD	WATER AS IN RAIN MOON	HILLS	EARTH
S.	S.E.	E.	N.E.	S.W.	W.	N.W.	N.

THE PA-KUA, OR EIGHT TRIGRAMS

From the *I-king*, or *Book of Permutations*. On the ordinary diviner's compass these directions are reversed

An examination of the above interpretation of the *Pa-kua*, the one commonly given by Oriental writers, suggests the Pythagorean doctrines with respect to numbers, and as we proceed we shall find still more to strengthen the belief that the West obtained much of its mysticism from the East.

Although there is no historical evidence that the Chinese looked upon the *Pa-kua* as numerals, based upon the scale of two, it is true that if we take —— for one and – – for zero, the successive trigrams, beginning at the right, have values which we may represent by our numerals as 000, 001, 010, 011, 100, 101, 110, and 111. If these are considered as numbers written on the scale of two, their respective values are 0, 1, 2, 3, 4, 5, 6, and 7.

The *Pa-kua* are found today on the compasses used by the diviners in every city and village of China. They are also

THIBETAN "WHEEL OF LIFE"

From a sheet of block printing done at Lhassa. This portion represents the signs of the zodiac, the *Pa-kua*, and, in the center, a magic square

found on fans, vases, and many other objects of the home, and on talismans of various kinds in common use in Thibet and other parts of the Far East.

The lo-shu and ho-t'u. The *I-king* also states that the *Pa-kua* were footsteps of a dragon horse which appeared on a

4	9	2
3	5	7
8	1	6

MAGIC SQUARE

The rows, columns, and diagonals in this particular magic square have 15 as their respective sums

river bank in the reign of the Emperor Fuh-hi, and that the *lo-shu*, in reality the magic square here shown, was written

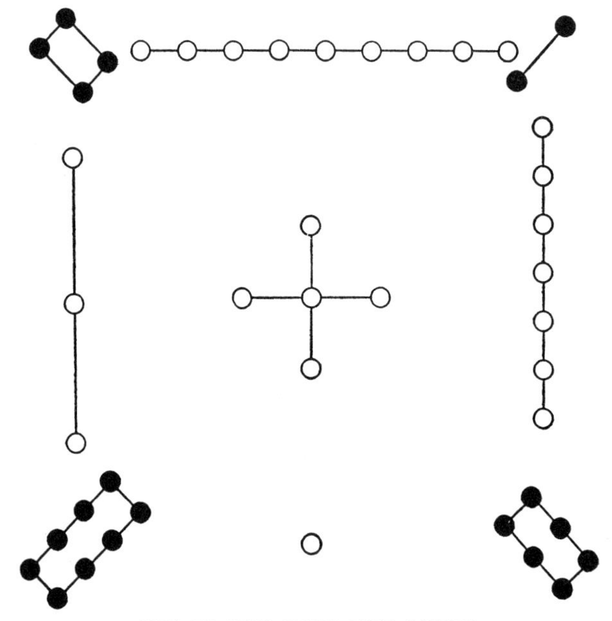

THE LO-SHU FROM THE I-KING

This is the world's oldest specimen of a magic square. The black circles are used in representing feminine (even) numbers, the white ones in representing masculine (odd) numbers upon the back of a tortoise which appeared to Emperor Yu (*c.* 2200 B.C.) when he was embarking on the Yellow River. The *ho-t'u*, also a highly honored mystic symbol, appears in the same work.

It thus appears that the *I-king* is not a work on mathematics, but that it contains the first evidence of an interest in permutations and magic squares that has come down to us. It is

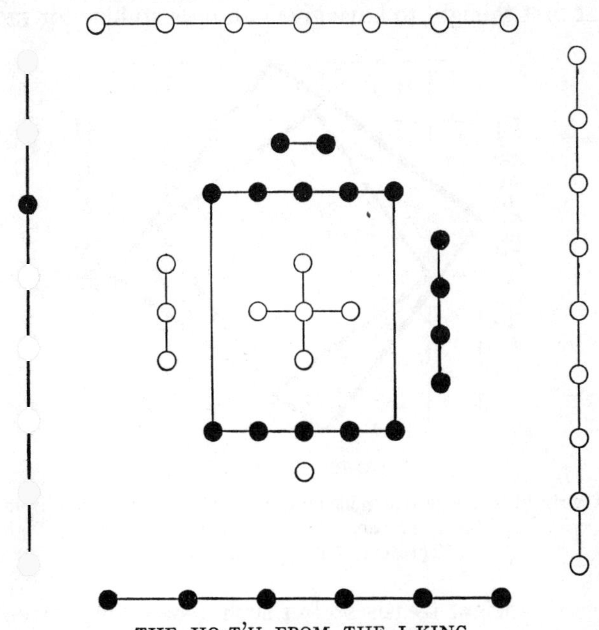

THE HO-T'U FROM THE I-KING

This was never considered so important as the *lo-shu*, lacking as it does the interest of the magic square

reasonable to believe, however, that both these ideas were already ancient when the book was written.

The Chóu-peï. The oldest Chinese work that can be designated as mathematical is the *Chóu-peï*, or the *Chóu-peï Suan-king*,[1]

[1] *Suan-king*, or *Suan-ching*, means "arithmetic classic." Also transliterated in various other ways, such as *Tcheou-pei-swan-king*. See E. Biot, "Traduction et examen d'un ancien ouvrage chinois intitulé Tcheou pei," *Journal Asiatique* (1841), p. 595, with a discussion of dates.

a work relating chiefly to the calendar but containing information referring to ancient mathematics, including some work on shadow reckoning. The author and the date of the work are both unknown, and there is some reason for believing that it has undergone considerable change since it was first written. The fact that Emperor Shï Huang-ti[1] of the Ch'in Dynasty, in 213 B.C., ordered all books burned and all scholars buried, would seem at first thought to have given an opportunity for radically

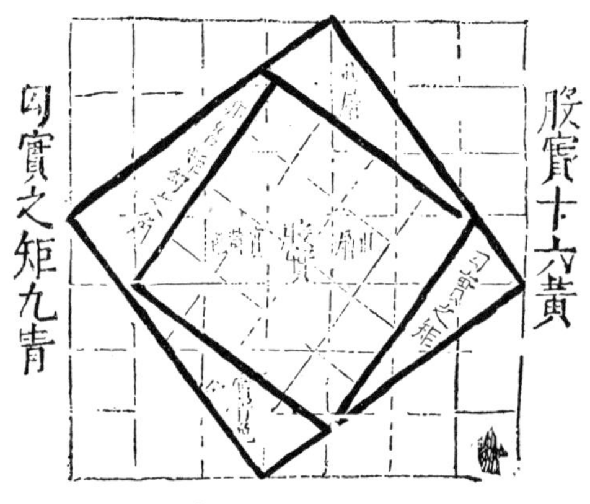

CHÓU-PEÏ SUAN-KING

A work written in the second millenium B.C. This illustration is from a very early specimen of block printing. It shows the figure of the Pythagorean Theorem, but gives no proof

altering all ancient treatises; but such a sweeping decree could not possibly have been executed, and even if every book had been lost there would have been many who could have repeated the ancient classics verbatim from memory. The probability is that we have about as near the primitive form of these classics as we have of the writings attributed to Boethius, Bede, or Alcuin, or of certain Greek authors whose works we assume

[1] Also transliterated Tsin Chi Hoang-ti and Tsin schè huâng ty (born 259 B.C.; died 210 or 211 B.C.). The claims of such writers as Weber and J. B. Biot for a high grade of mathematical learning in China before this time are contradicted by L. Am. Sédillot, "De l'astronomie et des mathématiques chez les Chinois," Boncompagni's *Bullettino*, I, 161.

as known. In any case it is probable that we have in the *Chóu-peï* a very good record of the mathematics of about 1105 B.C., the year of the death of Chóu-Kung, a party to one of several dialogues which the book records.[1] One of these dialogues is between the prince Chóu-Kung and his minister Shang Kao, and relates to number mysticism, mensuration, and astronomy. Among the stories told of the energy of Chóu-Kung is one relating to his habit of rushing several times from his bath, holding his long, wet hair in his hand, to consult with his officials. Tradition also states that he had a wrist like a swivel, on which his hand could turn completely round,—an odd fiction for those who are interested in stories of mathematicians. A few extracts from the *Chóu-peï* will give some idea of the nature of the work:

The art of numbers is derived from the circle and the square.

Break the line and make the breadth 3, the length 4; then the distance between the corners is 5.[2]

Ah, mighty is the science of number.

Forms are round or pointed; numbers are odd or even. The heaven moves in a circle whose subordinate numbers are odd; the earth rests on a square whose subordinate numbers are even.

One who knows the earth is intelligent, but one who knows the heavens is a wise man. The knowledge comes from the shadow, and the shadow comes from the gnomon.[3]

The Nine Sections. Next in order of antiquity among the mathematical works of China is the *K'iu-ch'ang Suan-shu*, or *Arithmetic in Nine Sections*.[4] This is the greatest of the Chinese classics in mathematics, and for many centuries has been held in the highest esteem in the Orient. As to its authorship and the period in which it was written we are ignorant.

[1] Y. Mikami, *China*, p. 4; W. A. P. Martin, *The Lore of Cathay*, p. 30 (New York, 1901); A. Wylie, *Chinese Researches*, Part III, p. 159 (Shanghai, 1897).

[2] This evidently refers to the right-angled triangle whose three sides are in the ratio 3:4:5, a special case of the Pythagorean Theorem.

[3] The gnomon was the index which cast the shadow on the sundial.

[4] In some editions, *K'iu-ch'ang Suan-shu-ts'au-t'u-shuo*.

We know that not long after the burning of the books (213 B.C.) there appeared a mathematician by the name of Ch'ang Ts'ang, that he collected the writings of the ancients, and that he seems to have edited the *K'iu-ch'ang Suan-shu.* There is a tradition, unsupported by positive proof, that the work was originally prepared by direction of the Chóu-Kung, who, as already stated, died in 1105 B.C., and it has even been asserted that it dates back to the reign of Huang-ti in the 27th century B.C.[1] The evidence of tradition, therefore, places it very early, and it seems probable that it existed, at least in great part, in the period of which we are writing, that is, before 1000 B.C.

Topics in the Nine Sections. The work consists, as the title says, of nine sections, books, or chapters. The titles and the sequence of chapters vary somewhat in different editions, but the following list is substantially correct as given in the revision of the work in the 2d or 3d century B.C.:

1. *Fang-t'ien* (*Squaring the farm*), relating to surveying, with correct rules for the area of the triangle, trapezium (trapezoid),[2] and circle ($\frac{1}{2}c \cdot \frac{1}{2}d$ and $\frac{1}{4}cd$), and with the circle approximations $\frac{3}{4}d^2$ and $\frac{1}{12}c^2$, where π is taken as 3.

2. *Su-mi* (*Calculating the cereals*), relating to percentage and proportion.

3. *Shuai-fen* (*Calculating the shares*), relating to partnership and the Rule of Three.[3]

4. *Shao-Kuang* (*Finding length*), relating to the finding of the sides of figures, and including square and cube roots.

5. *Shang-kung* (*Finding volumes*), relating to volumes.

6. *Chün-shu*, or *Kin-shu* (*Alligation*), relating to motion problems (couriers, hare and hound) and alligation.

[1] A. Wylie, "Jottings of the Science of Chinese Arithmetic," *North China Herald*, 1852, and the *Shanghai Almanac* for 1853; K. L. Biernatzki, "Die Arithmetik der Chinesen," Crelle's *Journal*, Vol. LII (1856).

[2] The meanings of the words *trapezoid* and *trapezium* were curiously interchanged in England and America about a century ago, and the error still persists in America. In this work the two words will be given as above, but the meaning will always be the etymological one for trapezium, a quadrilateral with two parallel sides.

[3] A kind of proportion, discussed at length in Volume II.

7. *Ying-pu-tsu*, or *Ying-nu* (*Excess and deficiency*), relating to the Rule of False Position,[1] the terms "excess" and "deficiency" relating to two concepts that are used in this rule.

8. *Fang-ch'êng* (*Equation*), relating to simultaneous linear equations, with some idea of determinants.

9. *Kou-ku* (*Right triangle*), relating to the Pythagorean Triangle.

These four works constitute those Chinese classics involving mathematics which were probably written in whole or in part before the year 1000 B.C. They show a degree of advancement quite as high as that found in the other ancient countries, and they prove that China was among the pioneers in the establishing of the early science of mathematics.

3. India

Early Hindu Mathematics. When we pass from a consideration of Chinese mathematics to the mathematics of India, Babylonia, and Egypt, we meet with the mental product of an entirely different type of people, or rather of two different types. There were two great branches of the human race affecting the Western World on the one hand, and India, Mesopotamia, and certain adjoining regions on the other hand. The first of these branches is supposed to have wandered from the Northern Grasslands, and constitutes what is known as the Indo-Europeans. In the West its members appear as Celts, Romans, and Greeks, and in Asia Minor it has several representative groups. In the East this same stock is seen in the Medes, Persians, and Hindus. The eastern branch is properly designated as Aryan, from which we have the name "Iran" for Persia. The people were generally highly imaginative, and their work in mathematics developed along such lines as the theory of numbers, geometry, and astronomy.

The second great branch is thought to have had its first habitat in the Southern Grasslands of Arabia, and is represented

[1] A primitive method of solving equations, considered at length in Volume II.

by what is known as the Semitic peoples. These include the inhabitants of Assyria, Babylonia, Phœnicia, and the Phœnician colonies. They dwelt in the paths of trade from East to West, and their work in mathematics developed chiefly along the line of computation,—a line which led to extensive numerical work in the field of astronomy as well as in that of commerce.

If the early mathematical achievements of the Chinese are uncertain as to date and importance, much more so is the early progress of the Hindus. Not only are we without any satisfactory records of the remote past of these people, but we are not infrequently confronted by claims that are preposterous and that are so recognized by Hindu scholars themselves. The first edition of the *Sūrya Siddhānta* of the Swami Press at Meerut, for example, says that the work was "Compiled about 2,165,000 years ago," representing a period about four times as long as it is thought the human race has been in existence. With even more absurdity the Laws of Manu are placed as far back as $6 \times 71 \times 4,320,000$ years ago,[1] giving almost an appearance of modesty to the ancient Chaldean claims that their astronomical observations began more than 720,000 years ago. As a matter of fact this well-known work on astronomy, the *Sūrya Siddhānta*, was probably written about the 4th or 5th century of our era. So little sympathy had the early native scholars with those outside their own caste that a general literature is wholly lacking, and it has only been through the labors of those from other lands that an all-round view of scientific progress has been attempted. There is, however, sufficient evidence for the belief that primary schools existed very early in India, and that arithmetic and writing were looked upon as the most important of the seventy-two recognized branches of learning, at least in the elementary stages of education.[2]

[1] On the extravagant ideas in the native Hindu chronology see J. C. Marshman, *Abridgment of the History of India*, p. 2 (London, 1893). See also W. Jones, "On the Chronology of the Hindus," in his *Works*, IV, 1 (London, 1807); M. Elphinstone, *History of India*, p. 136 (London, 1849).

[2] A. Hillebrandt, *Alt-Indien*, p. 111 (Breslau, 1899).

Lack of Authentic Records. As to authentic records, India has none written before the first Mohammedan invasion, *c.* 664 A.D.[1] All that we know of her earlier history is what we can glean from her two great epics, the Mahābhārata and the Rāmayāna, and from coins and a few inscriptions. The Mahābhārata relates the skill in numerals possessed by the ancient heroes, and the inscriptions tell us something of the notation used by the Hindus two thousand years ago, but neither gives us any knowledge of the period closing a thousand years before our era. The Vedas, the sacred writings of India, lead us to understand that in this period some attention was given to astronomy, as was the case in contemporary China, Mesopotamia, and Egypt.

All that we can say, therefore, about this period of Hindu mathematics is that there is some evidence from ancient literature that in very early times India paid attention to astronomy and calculation, just as was the case with other advanced peoples of that period.[2]

4. Babylon

Early Babylonian Mathematics. For our purposes Chaldea and Babylonia are synonymous, each name referring to the land extending from the delta of the Tigris and Euphrates northward to Assyria, the hilly, forest-covered district originally surrounding the ancient capital of Assur (Asshur). Indeed, it is convenient at present to consider as one large group all those Semitic peoples descended from the wanderers from the Southern Grasslands who settled in Assyria, in the region about Nineveh, in Asia Minor, and along the Phœnician coast. We shall also find it convenient to include a non-Semitic tribe, the Sumerians, who dwelt in the land of Sumer at the head of the Persian Gulf, directly in one of the chief paths of world

[1] The so-called Mohammedan period did not begin until 1001 A.D.

[2] G. Oppert, *On the Original Inhabitants of Bharatavarṣa or India*, p. 1 (London, 1893); R. C. Dutt, *A History of Civilization in Ancient India* (London, 1893).

commerce. These people, coming from the mountainous region to the east, early developed a numeral system, and numerals used by them in the 28th century B.C. are known to us through certain inscriptions. Dwelling in a low country formed by alluvial deposits, and thus deprived of stone for monumental purposes, the primitive Sumerians resorted to the use of bricks

NUMERALS OF THE 28TH CENTURY B.C.

Sumerian tablet. The numerals at this time were made with the upper end of the scribe's stylus and appear as curved symbols, and as such can easily be recognized. From Breasted's *Ancient Times*

for the preservation of their records. Upon the surface of clay tablets they pressed with a round and pointed stick, the result being a circular, a semicircular, or a wedge-shaped (cuneiform) character. These inscriptions were a mystery to the modern world until the first half of the 19th century, when Grotefend (1802) suggested and Rawlinson (1847) perfected the key to the rich literature of ancient Mesopotamia. The clay tablets, after being inscribed, were baked by fire or in the sun, and thousands of them are now available for study in various museums. These records of the Sumerians give us the information that nearly 3000 years before Christ their merchants were familiar with bills, receipts, notes, accounts, and systems of measures. In no part of the world have we as clear evidence of commercial mathematics at this early date as is revealed by these Sumerian tablets. Here also we find evidence of an approach to a scientific calendar, although of a later date than similar evidence found in Egypt, and here is probably to be found the first use of a kind of scale of 60 in counting.

Early Calendars. Some knowledge of mathematics must, however, have long preceded the work recorded on these Sumerian tablets. The old Babylonian year began with the vernal equinox, and the first month was named after the Bull. The calendar must, therefore, have been established at a period in which the sun was in Taurus at this equinox, and such a period began about 4700 B.C. A calendar of any kind presupposes a system of numbers and some form of calculation, so that we may safely say that some kind of arithmetic existed in Babylonia in the 4th or 5th millennium B.C. Indeed, so far as the calendar is concerned, it should be said that the Sumerians celebrated the beginning of the year at the vernal equinox as early as 5700 B.C., and possibly even earlier.[1]

Early Babylonia. What is commonly known as Early Babylonia endured from about 3100 to about 2100 B.C. Sargon, the first great ruler, flourished about 2750 B.C., his remarkable career beginning in Akkad, the district just north of Sumer. It was partly due to this proximity of territory that the people of Akkad in particular and of Babylonia in general adopted the business methods, the astronomy,[2] the calendar, the

[1] H. Radau, "Miscellaneous Sumerian Texts from . . . Nippur," in the *Hilprecht Anniversary Volume*, pp. 408, 410. Chicago, 1909.

[2] E. F. Weidner, *Handbuch der Babylonischen Astronomie*, Bd. I. Leipzig, 1915.

measures, and the numerals of the more highly cultivated Sumerians. In Sargon's reign we find a record of eclipses, so that the numeral system must have been well advanced,[1] and for him there was compiled the first great treatise on astrology of which we possess any original fragments.[2]

Among the tablets of about 2400 B.C. that have been deciphered are various specimens dating from the reigns of kings of the third dynasty of Ur[3] and recording the use of a kind of draft or check, the measurement of land in *shars*, the weighing by talents (*gur*), the measurement of liquids by *ka*, the taking of interest, the use of the fractions $\frac{1}{2}$, $\frac{1}{3}$,[4] and $\frac{5}{6}$, and the measurement of both liquids and solids by the *qa* (not identical with the *ka*).

In order to fix clearly in mind the period of which we are speaking there should be mentioned not only the reign of Sargon (*c.* 2750 B.C.) but the remarkable reign of Hammurabi or Hammurapi (*c.* 2100 B.C.), in which the world's first great code of laws, so far as we know, was written, and in which the calendar was reformed. Among the other interesting relics of the time of Hammurabi is the ruin of the oldest known schoolhouse. This was discovered by French archeologists in 1894.[5] In the building were numerous tablets on which the pupils had written their lessons, and it is from such tablets as these that we have part of our knowledge of the arithmetic of the Babylonians.

The general conclusion of archeologists, as will be elaborated on page 40, is that these early Babylonians (in the thousand years of their activity) developed a fair knowledge of computation, of mensuration, and of commercial practice, in spite of an awkward numeral system by which they were handicapped.

[1] See also F. Thureau-Dangin, in the *Hilprecht Anniversary Volume*, p. 156.

[2] G. Bigourdan, *L'Astronomie*, 1920 ed., p. 27. Paris, 1911.

[3] G. A. Barton, *Haverford Library Collection of Cuneiform Tablets*, Part I (Philadelphia, n. d. [1905]) ; *ibid.*, Part II (1909).

[4] These from Barton, *loc. cit.*, Part I. On the taking of interest, the rates running from 20% to $33\frac{1}{3}$%, see E. Huber, "Die altbabylonischen Darlehnstexte," in the *Hilprecht Anniversary Volume*, pp. 189, 217.

[5] For a plan of the building see J. H. Breasted, *Ancient Times*, p. 136 (Boston, 1916) ; hereafter referred to as Breasted, *Anc. Times*.

Early Assyria. As early as 3000 B.C. a Semitic tribe of nomads settled at Assur, and in due time it too adopted the Sumerian calendar and such of the mathematics of trade as had been developed by these people of the south.

ARAMEAN WEIGHT FOUND IN ASSYRIA

The weight is of bronze and the inscription is Aramaic. Fifteen of these lion weights were found in Nineveh and testify to the common presence of Aramean merchants in Assyria. From Breasted's *Ancient Times*

Much later, and after 1200 B.C., the Arameans, or Syrians, established kingdoms in the region to the west of Assyria. They were great merchants, and the Sumerian mathematics of trade, which had worked slowly northward through Babylonia and Assyria, now found place in the new territory. We have bronze weights of this period, showing that whole numbers, fractions, measures, and elementary forms of computation played a considerable part in the daily life of the people.

Early Chaldea. The desert tribe called the Kaldi came into prominence long after the period now under discussion. It gained a foothold in ancient Sumer and finally (606 B.C.) conquered the Assyrians and established the Chaldean empire in the region of Babylonia. Although their empire lasted only to 539 B.C., they made great progress in science. In particular, astrology was extensively cultivated, the equator was probably divided into 360°, the twelve signs of the zodiac definitely appeared, and mathematics flourished as the handmaid of commerce and astronomy. Thus Babylonia became Chaldea, and Chaldea became the patron of science and art.

Early Cuneiform Tablets. Our first important knowledge of Babylonian arithmetic was derived from two tablets found in 1854 at Senkereh, the ancient Larsam or Larsa, on the Euphrates, by a British geologist, W. K. Loftus. These tablets contain the squares of numbers from 1 to 60 and the cubes of numbers from 1 to 32.[1] Their date is uncertain, but the evidence seems to show that they were of about the Hammurabi period (*c.* 2100 B.C.).

Since the discovery of the Senkereh tablets there have been unearthed some 50,000 tablets at Nippur, the modern Nuffar, an ancient city lying to the south of Babylon, and among these are many that relate to mathematics.[2] They are apparently from a large library which seems to have been destroyed by the Elamites about 2150 B.C. or a little earlier, and again about 1990 B.C., and they constitute the most extensive mass of ancient mathematical material ever brought to light. The cylinders include multiplication and division tables, tables of squares and square roots, geometric progressions, a few computations, and some work on mensuration. Neugebauer's studies (1935) of a large number of tablets show that the Sumerians and Babylonians could solve special linear, quadratic, cubic, and biquadratic equations and had some knowledge of negative numbers.

Babylonian Geometry. The tablets found at Nippur and elsewhere also give us some knowledge of the Babylonian geometry. From these it seems that as early as 1500 B.C. the Babylonians could find the area of a rectangle, including that of a square; the area of a right-angled triangle; the area of a trapezium (trapezoid); and possibly the area of a circle, the volume of a parallelepiped, and the volume of a cylinder. There is ground for the belief that they knew the law of expansion of $(a+b)^2$, although we have no knowledge as to whether this was inferred from a geometric figure or from their extensive study of square numbers. There is also some reason to believe that they knew the abacus, since it has been suggested that one of their signs (ŠID) may have been derived from a pictograph of such an instrument.

[1] Apparently from 1 to 60 originally, but part of the tablet is broken off.

[2] H. V. Hilprecht, *Mathematical, Metrological, and Chronological Tablets from the Temple library of Nippur* (hereafter referred to as Hilprecht, *Tablets*). Philadelphia, 1906.

Scale of Sixty. One peculiarity of Babylonian arithmetic is the constant use of the number 60,—a use which finally suggested the development of sexagesimal fractions and which still survives in our division of degrees, hours, and minutes into sixty sub-units. It is generally thought that the Babylonians, interested as they were in watching the stars, early came to believe that the circle of the year consisted of 360 days. It is also thought that they knew that the side of the regular inscribed hexagon is equal to the radius of the circle, this property suggesting the division of 360 into six equal parts, and 60 being thus looked upon as a kind of mystic number. This may, indeed, be the origin of this use of 60, but we find other nations using 40, 20, and even 15 in somewhat the same way, with no apparent reason, so that all such customs may have developed from racial notions which were started by some leader or sect with no particular reason in mind. It is more probable that 60 was chosen because of its integral divisors 2, 3, 4, 5, 6, 10, 12, 15, 20, and 30, thus rendering work with its fractional parts very simple.

Although the subject of fractions with the denominator 60 is discussed in Volume II, a brief mention may be made at this time of an important tablet first described in 1920.[1] It dates from *c.* 200 B.C. and illustrates the Babylonian custom of using either 360 or 60 for the denominator except in the cases of unit fractions and of fractions in which the numerator is 1 less than the denominator. For example, $\frac{60}{360}$ may appear as $\frac{10}{60}$ or as $\frac{1}{6}$ (ŠUŠŠU), $\frac{240}{360}$ as $\frac{40}{60}$ or $\frac{2}{3}$ (ŠINIPU), and $\frac{300}{360}$ as $\frac{50}{60}$ or $\frac{5}{6}$ (PĀRAB).

5. Egypt

Early Egyptian Mathematics. Whatever claims may properly be made for the antiquity of mathematics in various other countries, claims of even greater validity can justly be made for the science in Egypt. Civilization has generally developed along great rivers; the Nile is one of the world's greatest arteries

[1] H. F. Lutz, "A mathematical cuneiform tablet," *American Journal of Semitic Languages*, XXXVI, 249.

of commerce, and its fertile valley is one of the world's greatest gardens. Egypt was a well-protected country, and civilization had a more favorable opportunity for uninterrupted development there than in such lands as Mesopotamia, Phœnicia, India, and China. Furthermore, her art, as shown by wall sculptures, was much farther advanced in the 4th millennium B.C. than it was, say, with the Sumerians, and so there is every reason to feel that her science was also in the lead of that in other lands.

The earliest dated event in human history is the introduction of the Egyptian calendar of twelve months of thirty days each, plus five feast days, in the year 4241 B.C.[1] Such an achievement as the creation of this calendar, a better one than was used in Europe from the time of the Romans until the reform of Gregory XIII (1582), and in some respects better than the one used at present, shows a high development of computation as well as of astronomy. No authentic record of mathematical progress in any other country dates back as far as this; it reaches back even into the Stone Age, more than a thousand years before the earliest stone masonry and long before any people had the slightest idea of an alphabet as we understand the term. An event like this is a silent but powerful witness to the noteworthy arithmetic attainments of its sponsors and to a long series of scientific observations by the temple astronomers. Furthermore, our own calendar may be said to be merely a poor adaptation of this ancient Egyptian one, although containing the great improvement of having as centennial leap years only those of which the numbers representing the hundreds are divisible by four.

Third Millennium B.C. in Egypt. When we approach the year 3000 B.C., toward the close of the Second Dynasty, we find ourselves in a period of rapid development in practical engineering. We have no manuscripts of this period from which to obtain direct information, but in the achievements of the engineers it is possible to recognize a number of interesting

[1] Breasted, *Anc. Times*, p. 45.

facts. Professor Breasted has characterized the development of civilization in the 30th century B.C. in these words:

Hardly more than a generation before this 30th century the first example of hewn stone masonry was laid, and in the generation after this 30th century the Great Pyramid of Gizeh was built. With amazingly accelerated development the Egyptian passed from the earliest example of stone masonry just before 3000 B.C. to the Great Pyramid just after 2900. The great-grandfathers built the first stone masonry wall a generation or so before 3000 B.C., and the great-grandsons erected the Great Pyramid of Gizeh, within a generation after 2900. . . .

One finds it difficult to imagine the feelings of these earliest architects . . . as they paced off the preliminary plan and found an elevation in the surface of the desert which prevented them from sighting diagonally from corner to corner and applying directly a well-known Egyptian method of erecting an accurate perpendicular by means of measuring off a hypotenuse. . . .

The Egyptian engineers early learned to carry a straight line over elevations of the earth's surface, or a plane around the bends of the Nile. In his endeavor to record the varying Nile levels in all latitudes the Egyptian engineer was confronted by nice problems in surveying, even more exacting than those which he met in the Great Pyramid. A study of the surviving nilometers has disclosed the fact[1] that their zero points, always well below lowest water, are all in one plane. This plane inclines as does the flood slope from south to north. The Pharaohs' engineers succeeded in carrying the line in the same sloping plane, around innumerable bends in the river for some seven hundred miles from the sea to the First Cataract.[2]

Accuracy of Early Engineers. Such was the degree of accuracy secured by these early surveyors that Petrie found the maximum error in fixing the length of the sides of the Great Pyramid to be only 0.63 of an inch, or less than $\frac{1}{14000}$ of the total length, and the angle error at the corners to be $12''$, or only $\frac{1}{27000}$ of a right angle.

[1] L. Borchardt, *Nilmesser und Nilstandsmarken.*

[2] J. H. Breasted, "The Origins of Civilization," *The Scientific Monthly*, X, 87.

Speculations on the Great Pyramid. As to the speculations relating to the Great Pyramid it is possible to make only a brief statement in this work. That mathematics, and possibly mathematical mysticism of some kind, played an important part in the design of the structure is admitted by all scholars, but precisely what the dominating principle was we do not know. It has been suggested that four equilateral triangles were put together for the pyramidal surface, but this theory is not borne out by measurements, the base being considerably longer than the sloping edge.

A second theory asserts that the ratio of the side to half the height is the approximate value of π, or that the ratio of the perimeter to the height is 2π. It is true that this would give the value of π as about 3.14, an approximation that may have been known to the pyramid builders; but this was possibly a mere matter of chance. If one searches in any building, or indeed in any given object, for lines having this ratio, they are not difficult to find. Nevertheless, there probably is some mysticism of this kind in the proportions of the structure.

A third theory makes the claim that the angle of elevation of the passage leading to the principal chamber determines the latitude of the pyramid, approximately 30° N., or that the passage itself pointed to what was then the pole star; but even after making all reasonable allowances in favor of this hypothesis, the angular difference is too great to make out a very strong case.

It is also claimed that the pyramids have a constant angle of slope, and it is true that the three at Gizeh vary but little, being approximately 51° 51′, 52° 20′, and 51°; but others have slopes running from about 45° to 74° 10′.

Recent measurements have been so accurately made that it is probable that further study will reveal in the near future whatever mathematical principles actuated the architects. For the present we may simply dismiss the speculations of such men as Charles Piazzi Smyth[1] as interesting rather than scientific.

[1] An English astronomer; born at Naples, 1819; died 1900; astronomer royal of Scotland (1845–1888). *Our Inheritance in the Great Pyramid* (1864), *Life and Work at the Great Pyramid*, 3 vols. (1867).

Testimony of the Wall Reliefs. The wall reliefs of this general period of the Pyramid builders testify to the collection of taxes, probably in the form of grain, and the issuing of receipts by the officials of the king. Nothing so tangible, showing the applications of elementary arithmetic at this early period, has been found in any other region except Sumeria.

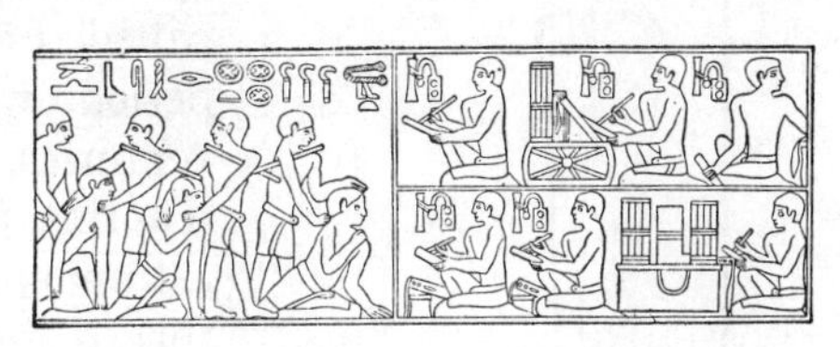

COLLECTION OF TAXES, *c.* 3000 B.C.

Showing the clerks and scribes at the right, with pen and papyrus, and the officials and taxpayers at the left. From Breasted's *Ancient Times*

Reign of Amenemhat III, or Moeris. About 1850 B.C., in the 12th Dynasty, there came to the throne one of the most energetic of all the kings of Egypt, Amenemhat III.[1] In his reign there was carried out an extensive system of irrigation, necessitating a knowledge of leveling, surveying, and mensuration such as had probably never been developed before this time in any other part of the world, except perhaps in Mesopotamia.[2] There is good reason to believe that in this reign, say about 1825 B.C., there was written the original of the oldest elaborate manuscript on mathematics now extant, the Ahmes treatise mentioned a little later. If the conjecture is correct, the unit fraction was already known in Egypt, as also seems to have been the case in Mesopotamia,[3] and the simple equation with a fairly usable symbolism,

[1] The name also appears as Ne-mat-re and as Amenemha. He is the Moeris of Herodotus (II, 148–150), the Marros of Diodorus Siculus (I, 52), and the Mares of Eratosthenes. He is also referred to by Strabo (XVII), Pliny (*Hist. Nat.*, V, 9, 50, and XXXVII, 12, 76), and Pomponius Mela (I, cap. 9). Recent Egyptologists give the date of his reign as 1849–1801 B.C.; others place his reign as *c.* 1986–*c.* 1942.

[2] On the Egyptian work in this line see A. Wiedemann, *Aegyptische Geschichte*, I, 256 (Gotha, 1884); J. Lieblein, "L'Exode des Hébreux," *Proceedings of the Society of Biblical Archæology*, XXI (London, 1899), 55. On the claims of priority in the general development of mathematics, see E. Weyr, *Ueber die Geometrie der alten Aegypter*, p. 4 (Vienna, 1884).

[3] See the author's review of Hilprecht's work in the *Bulletin of the American Mathematical Society*, XIII (2), 392.

arithmetic and geometric series, and the elements of mensuration were already familiar to the élite among the mathematicians of the Nile Valley.

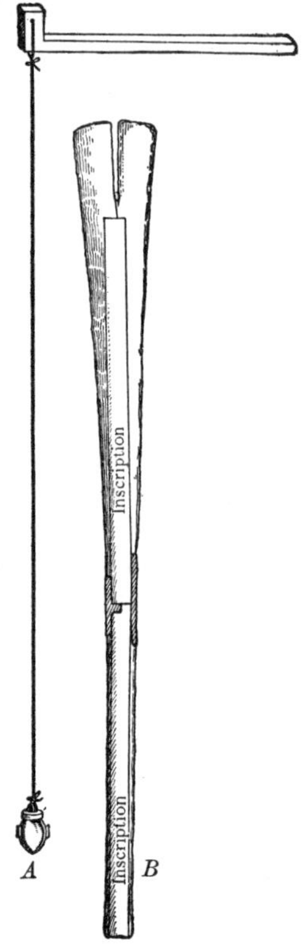

THE OLDEST ASTRONOMICAL INSTRUMENT KNOWN

The original is in the Berlin Museum. Part *A* was a plumb line. By its aid the observer could hold *B* over a given point and sight along the slot to some object like the North Star, thus establishing a meridian line. From Breasted's *Ancient Times*

Mathematics of the Feudal Age. Amenemhat III lived in the so-called Feudal Age of Egypt,—a period which lasted for several centuries, closing about 1800 B.C. To this period belongs the oldest astronomical instrument extant, a forked stick used in sighting for the purpose of obtaining the meridian. Such a work presupposes some ability in calculation and in constructive geometry, and the existence of this ability is still further proved by the Ahmes treatise. There is another treatise, written much earlier than this,[1] in which we find mention of the civil calendar of twelve months of thirty days each, plus five extra days, as already mentioned.

About the close of the Feudal Age a postal service existed in Asia under Egyptian control, requiring some means of payment on the part of those whose convenience it served. At the same time census lists were prepared for use in the taxation of the people, surveys for irrigation projects were made, and the Nilometer served to foretell the beginning and the end of the rise of the river, all of which involved,

[1] That is, in the 30th century B.C. See E. Mahler, "Der Kalender der Babylonier," in the *Hilprecht Anniversary Volume*, pp. 1, 9.

as already stated, the use of a considerable amount of mensuration and computation, and adds to the evidence of an interest in mathematics in this period of Egyptian history.

In the period of the Middle Kingdom (2160–1788 B.C.) business arithmetic was such as to demand bills, accounts, and tax lists. From this period we have various fragments of papyrus rolls which were found in the remains of the libraries of the feudal lords. These are the oldest libraries of papyrus rolls thus far known. Among these remains were found such evidences of the commercial activity above mentioned as are seen in the fragments of papyri found at Kahun and now in London and Berlin.[1]

Ahmes Papyrus. About 1650 B.C. there lived in Egypt a scribe named Aʿh-mosè, commonly called by modern writers Ahmes.[2] He wrote a work on mathematics; or rather he copied an older treatise, for he says: "This book was copied in the year 33, in the fourth month of the inundation season, under the majesty of the king of Upper and Lower Egypt, ʿA-user-Rêʿ, endowed with life, in likeness to writings of old made in the time of the king of Upper and Lower Egypt, Ne-maʿet-Rêʿ. It is the scribe Aʿh-mosè who copies this writing." Another manuscript of the same period, containing a number of lines on fractions, is in the British Museum. It was published in 1927. The actual manuscript[3] of Ahmes has come down to us, having been purchased in Egypt about the

[1] The earliest date in the London fragments is in the reign of Amenemhat III. See W. M. Flinders Petrie, *Kahun, Gurob, and Hawara*, chap. vi by F. L. Griffith (London, 1890). This article places the date *c.* 1986–1942 B.C., which is somewhat later than that given by the earlier writers and a little earlier than that given by some of the latest authorities.

[2] "Aʿh-mosè" was also the name of certain kings. Ahmes I, often known as Amosis or Amasis, came to the throne at the beginning of the 18th Dynasty, when Egypt entered upon a period of empire, and it was he who expelled the Hyksos and pursued them into Palestine.

[3] Since the first edition there have appeared editions of the Rhind Papyrus by T. E. Peet (London, 1923) and A. B. Chace (Oberlin, O., 2 vols., 1927, 1928). The Chace edition is the more elaborate, containing a facsimile of the papyrus, a transcription into hieroglyphic and Latin characters, a complete translation, numerous notes, and an extensive bibliography by R. C. Archibald. The preferred form of the name is given as Aʿh-mosè and the date as between 1750 and 1580.

middle of the 19th century by the English Egyptologist, A. Henry Rhind (whence the name "Rhind Papyrus"), and having later been acquired by the British Museum. It is one of the oldest mathematical manuscripts on papyrus extant.

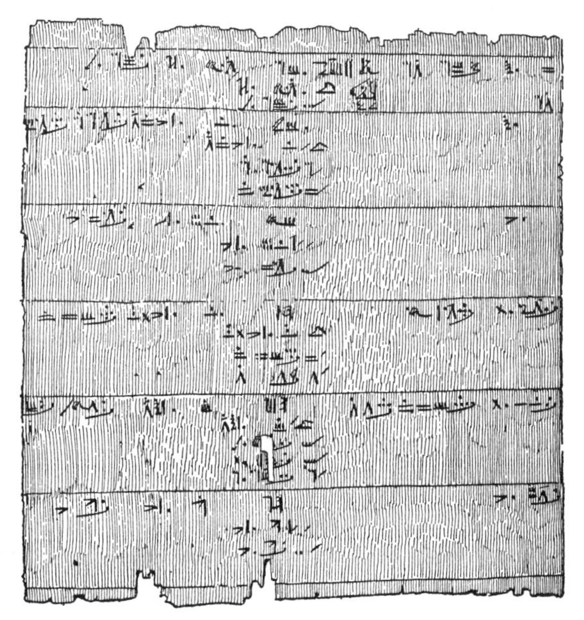

A PAGE FROM THE AHMES PAPYRUS

Written *c.* 1550 B.C. The original is in the British Museum

The Ahmes manuscript is not a textbook, but is rather a practical handbook. It contains material on linear equations of such types as $x + \frac{1}{7}x = 19$; it treats extensively of unit fractions; it has a considerable amount of work on mensuration, and it includes problems in elementary series.[1] The

[1] The British Museum published an inexact facsimile of the papyrus in 1898 under the title *Facsimile of the Rhind Mathematical Papyrus*. The standard works on the subject are those of Chace and Peet, mentioned on page 47, and (less valuable) that of A. Eisenlohr, *Ein mathematisches Handbuch der alten Aegypter*, 2d ed. (Leipzig, 1877). See also F. L. Griffith, *Proceedings of the Society of Biblical Archæology*, 1891, 1894; A. Favaro, *Atti della R. Accad. . . . in Modena*, Vol. XIX.

internal evidence shows the work to be a compendium of the contributions of at least two or three authors.[1]

Evidence of Commercial Mathematics. About 1500 B.C. there was built by Queen Hatshepsut[2] the temple known at present as Der al-Bahri. This is not far from Thebes[3] and in 1904 was uncovered and made known to modern scholars. On the walls of this temple is pictured the receipt of tribute from the land of Punt, probably on the Somali coast of Africa,[4] and mention is made of "reckoning with numbers, summing up in millions, hundreds of thousands, tens of thousands, thousands, and hundreds," showing the extent to which numbers were used in commercial matters even before coins were invented. There are certain inscriptions of the same period in the tomb of Rekhmire, at Thebes, giving the tax list of Upper Egypt, and interesting because of the fact that the highest number is 1000 and that $\frac{1}{2}$ is the only fraction used.[5]

Oldest Sundial. From this period or a little later, but from about 1500 B.C., there dates the oldest sundial extant, an Egyptian piece now in the Berlin Museum, showing that the Egyptians had already developed, as we might have inferred from the other mathematical and astronomical knowledge possessed by them, a good system of timekeeping by means of a primitive sun clock. On this clock the shadow shortened as the forenoon advanced, and lengthened from noon to night.

[1] This is seen particularly in the several rules which are evidently followed in the formation of unit fractions.

[2] Hāt-shepset, Hatasu, or Hatshepsu, also known as Ramaka (Ma-ka-ra). See E. A. W. Budge, *The Mummy*, p. 30 (Cambridge, 1893).

[3] The No Amon (City of Ammon) of the Bible, also known to the Greeks as Diospolis.

[4] J. H. Breasted, *Ancient Records, Egypt* (Chicago, 1906, 1907), II, pp. 104, 114, 210, 211; IV, pp. 362 seq.; hereafter referred to as Breasted, *Anc. Records*. See also official accounts of the same period in facsimile in Golénischeff, *Les Papyrus hiératiques nos. 1115, 1116A et 1116B de l'Ermitage impérial à St. Pétersbourg*, 1913.

[5] Breasted, *Anc. Records*, II, 283. In later inscriptions, as of *c.* 600 B.C., the fractions $\frac{1}{6}$ and $\frac{2}{3}$ appear (*ibid.*, IV, p. 486). Still later, from 19 A.D. to 250 A.D., the papyri tell us of the periodic census introduced apparently by Augustus, with taxes and the records of imports and exports. See A. S. Hunt, "Papyri and Papyrology," *Journal of Egyptian Archæology*, I (1914).

There were six hours in the forenoon and six in the afternoon, from which division of the day came the system of twelve hours later adopted in Europe. Such clocks, in various forms, were afterwards used by the Greeks and gave rise to the sundials of

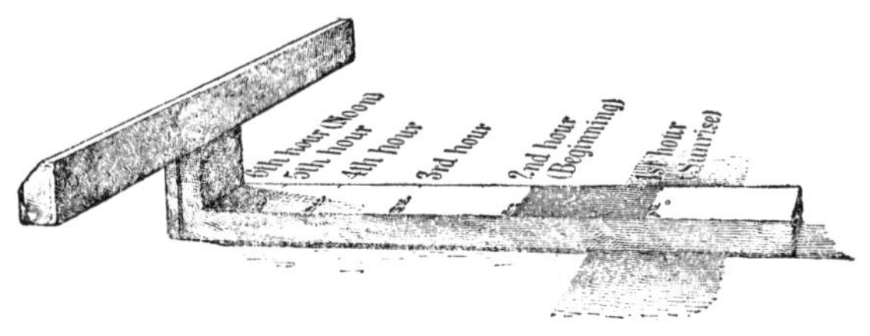

OLDEST SUNDIAL EXTANT

Egyptian specimen, restored after Borchardt, now in the Berlin Museum. Dates from *c.* 1500 B.C. In the morning the crosspiece was turned to the east, and in the afternoon to the west. From Breasted's *Ancient Times*

later times. The clock above shown bears the name of Egypt's greatest general, Thutmose III, who has justly been called her Napoleon.

Practical Problems. By the time of Seti I (*c.* 1350 B.C.) business calculation had come to require larger numbers than those needed in the time of Ahmes. This is seen from the problems in the Rollin papyrus manuscript now in the Louvre,[1] one of which, line for line as in the text, is as follows:

1601 392,325
together bread 107,893 makes in *ten* 364,371
bread 6121 loaves 1800 *thes* makes in *ten* 21,600
together 385,871
rest 6354
quantity of maize sacks 1601 makes in bread 112,090
makes in *ten* 392,306
brought to the magazine bread 114,064 makes in *ten* 385,971

The meaning is that here are two accounts of 1601 sacks of wheat each, the produce varying in the two cases. The weights

[1] M. F. Chabas, *Aegyptische Zeitschrift*, 1869, p. 85. The manuscript was published by W. Pleyte in 1868, and a new translation by Eisenlohr appeared in 1897 in the *Proceedings of the Society of Biblical Archæology*, XIX, 91, 115, 147, 252.

are calculated in *thes* or *ten*, 12 *ten* making 1 *thes*, a *ten* being about 316 grams. In the first case a loaf of unbaked bread weighed 3.63 *ten*, 1.15 kg., or $2\frac{1}{2}$ lb., and after it was baked it weighed 3.37 *ten*, 1.06 kg., or $2\frac{1}{3}$ lb. But the first case also gives the weight of 6121 loaves as 21,600 *ten*, which is at the rate of 3.52 *ten* per loaf when baked, so that the sizes evidently varied. In the second case the bread weighed 3.55 *ten* per loaf, possibly unbaked, and 3.38 *ten* per loaf when delivered. The first account may be represented as follows:

107,893 loaves weigh	364,371 *ten*
6121 loaves weigh	21,600 *ten*
together they weigh	385,971 *ten*
there is left	6,354 *ten*
the total being	392,325 *ten*

The problem in itself is of little moment except as it shows the practical use of large numbers in these early times.

Rameses II divides the Land. At the close of Seti's relatively short reign his son, Rameses II (*c.* 1347 B.C.), known to the Greeks as Sesostris, came to the throne. In his reign a redivision of land took place among the people, and surveying must have attracted much attention.

Herodotus (*c.* 484–*c.* 425 B.C.), referring to information that he had received from the priests, relates the following:

Sesostris also, they declared, made a division of the soil of Egypt among the inhabitants, assigning square plots of ground of equal size to all, and obtaining his chief revenue from the rent which the holders were required to pay him every year. If the river carried away any portion of a man's lot, he appeared before the king, and related what had happened; upon which the king sent persons to examine, and determine by measurement the exact extent of the loss; and thenceforth only such a rent was demanded of him as was proportionate to the reduced size of his land. From this practice, I think, geometry first came to be known in Egypt, whence it passed into

Greece. The sundial, however, and the gnomon, with the division of the day into twelve parts, were received by the Greeks from the Babylonians.[1]

Harris Papyrus. Rameses IV came to the throne *c.* 1167 B.C. and immediately prepared a remarkable document setting forth the great works of his father, Rameses III (1198–1167 B.C.), including a list of his extensive gifts to the gods. The list shows the proportion of the wealth of ancient Egypt held by the temples and is of value in giving the numerals of the period. This document, known as the Harris Papyrus, is still extant[2] and affords the best example of practical accounts that has come down to us from the ancient world.

That surveying played a prominent part in the life of Egypt is seen in an inscription on the tomb of Penno at Ibrim, in Nubia, in the reign of Rameses VI (*c.* 1150 B.C.), in which the boundaries and areas of five districts are given.[3]

Evidence of Egypto-Cretan Relations. Thus we see that before 1000 B.C. Egypt had developed enough knowledge of astronomy to devise an excellent calendar, and that she was in possession of a commercial system requiring extensive work in computation, of an elaborate scheme of leveling and surveying, of a considerable knowledge of what we would now consider as a kind of algebra, and of some ability in mensuration, especially as it related to granaries and to the use of grain products in the making of bread.

Recent excavations have shown the existence of a high degree of civilization in Crete in this period of progress in Egypt, and there is also evidence of amicable relations between these two countries in early times. Our knowledge of the subject is too limited, however, to determine whether it has any bearing upon the history of mathematics. The deciphering of the Cretan inscriptions still awaits the further efforts of scholars.

[1] Herodotus, II, 109.

[2] Breasted, *Anc. Records*, IV, 127 seq. See also S. Birch, *Zeitschrift für Aegyptische Sprache*, pp. 119 seq. (1872).

[3] Breasted, *Anc. Records*, IV, 233. For the Egyptian measures in common use at this period, and thus far identified as to equivalents, see *ibid.*, p. 88.

TOPICS FOR DISCUSSION

1. The countries in which mathematics flourished prior to 1000 B.C., and the reasons for this mathematical activity.

2. Reasons for supposing mathematics to have made some progress in the Late Stone Age, or even earlier.

3. Influences leading to an extension in the use of mathematics in the third millennium B.C.

4. Probable nature of the earliest mathematics of China, and the influences which developed the study of this science.

5. General nature of the early written mathematical works in China, with approximate dates.

6. The first traces of number mysticism in the East.

7. General period in which the *Nine Sections* was written. Nature of the work.

8. Probable nature of the early Hindu mathematics.

9. Influences that developed Babylonian mathematics and the method of recording the science.

10. General nature of Babylonian mathematics.

11. Evidence of early mathematics in Egypt. General nature of the work in the earliest periods.

12. Mathematics of the Feudal Age in Egypt.

13. The Ahmes Papyrus, its origin and general nature.

14. Evidences of development of commercial arithmetic between the time of Ahmes and 1000 B.C.

15. Types of problems in arithmetic, algebra, and mensuration that interested the ancient Egyptians.

16. Comparison of the mathematical progress and interests of China, India, Babylonia, and Egypt in early times.

17. A consideration of the reasons why this period was lacking in power to advance its mathematics.

18. A study of the evidence of mathematics in Crete and Cyprus before 1000 B.C., and the influence of this mathematics upon Greek science.

19. The evidences of interrelation of mathematical ideas in Mesopotamia, Egypt, and the islands of the Mediterranean Sea.

20. Mathematical and astronomical instruments of this period.

21. The degree of accuracy apparently secured by engineers before the year 1000 B.C.

CHAPTER III

THE PERIOD FROM 1000 B.C. TO 300 B.C.

1. The Occident in General

Geographical Limits. For our present purposes we may define the Occident of the period from 1000 B.C. to 300 B.C. as practically identical with Greece and her colonies. Whatever mathematics Rome had in this period was essentially Greek, and most of the Mediterranean world, aside from the hinterland of Phœnicia and Egypt, may therefore be conveniently classified as under the influence of the Hellenic civilization. Phœnicia contributed little that was not commercial, and the golden age of Egypt was already past.

Protected Regions. Philosophy, letters, mathematics, art, and all the finer products of the mind require peaceful surroundings for their development. It is for this reason that mathematics at this time flourished best on the protected islands of the Ægean Sea, on the Greek peninsula, and in the Greek towns of Southern Italy. In all these places invasion was difficult and the rewards of the invader were few. Commercial and intellectual communication with the rest of the world was possible, so that peace without stagnation was, relatively speaking, assured.

It would also be proper to include some mention of the mathematics of Mesopotamia, since this was quite as occidental as oriental; but aside from its use in astronomy the science was not sufficiently in evidence in Babylon at this time to demand our attention.

Chronological Limits. The reason for taking the lower arbitrary limit of 300 B.C. is that a new era in the history of mathematics begins with the founding of the Alexandrian School at about that time. This event led to a reshaping of mathematics either through the efforts of scholars connected with the first great cosmopolitan university or through the works written by those who came under their influence.

2. The Greeks

Birth of Greek Arithmetic. Commercial arithmetic was well advanced in various neighboring states long before it was known in Greece. The merchants of the Phœnician coast (along and across which passed the routes of trade with the Orient), receiving inspiration from Babylon, early developed a fairly good business arithmetic, and in due time became the teachers of this art in Egypt, Asia Minor, and the Ægean isles. The recent excavations in the ancient palace of Knossos show that in early times the commercial arithmetic of Babylon reached even as far west as Crete, and future studies are likely to reveal much valuable information relating to this island. Indeed, in what is called the Early Minoan Period of Crete, Greece was still a forest, thinly peopled by a nomad race, for this was long before the warlike Dorians, about a thousand years before our era, made themselves masters of Peloponnesus and changed the whole tenor of Greek life. Thucydides describes the country at this early period as a theater of frequent migrations, when "each man cultivated his land only according to his immediate needs, with no thought of amassing wealth." Under such conditions, before the coining of money was known, only the most primitive arithmetic was demanded. A little counting and a little rude barter were all for which the ancient Greek civilization had created a need. Even at a much later period than the one we have described, Greece was little inclined to commerce. Her older cities were not generally seaports, and what little navigation she had was concerned with war and piracy rather than with the development of trade.

External Influences. It was only when the Greeks began to come into closer contact with other peoples that they showed any interest in arithmetic. Indeed, contrary to the idea that is commonly expressed, Greece always depended largely upon external influences for her mathematics, and few who advanced this science in her schools were born within her continental area. But when we speak of the early efforts of Greece to put herself in contact with the external world through colonization, it must be understood that we are still ignorant as to how far peninsular Greece was then a colonizer and how far she herself was a colony, since her people lived as much along the Asiatic as along the European coast.

Miletus. We are told by Herodotus (*c.* 484–*c.* 425 B.C.) and Strabo (*c.* 66 B.C.–*c.* 24 A.D.), however, that Miletus, the greatest commercial town of the twelve forming the Ionian confederacy, was an Athenian colony, although there are good reasons for doubting the statement. Situated at a strategic point on the coast of Asia Minor, it in turn became a great colonizing center, and in the 7th century B.C. established no less than ninety towns along the shores of the Black Sea and the Mediterranean, even opening Egypt to her commercial settlements. This fact had a bearing upon the early science of the Greeks, since it was at Miletus that their mathematics had its beginning; and it was here, doubtless, that their commercial arithmetic first developed to any great extent. It was in Lydia, just east of here, that coins were first struck in the West, in the 7th century B.C., and Miletus at once recognized and adopted the new invention, anticipating Athens, indeed, by over half a century. The influence of this movement, particularly in relation to arithmetic, is evident. Without the aid of coins all business calculation must have been very cumbersome, money consisting of bars or ingots of metal that had to be weighed, and small currency being practically nonexistent save in the form of shells or trinkets. We can therefore determine fairly well the time and place of the beginning of any noteworthy business arithmetic among the Greeks, namely, about the 7th century B.C. and along the coast of Asia Minor.

ANCIENT COINS

Coins found in Asia Minor. They are among the earliest known, dating from about 550 B.C. From Breasted's *Ancient Times*

Logistic. At the period of which we have been speaking, the Greek science of numbers, the arithmetic proper, had not yet been invented. Only the art of calculating had made any appeal to these practical people. This branch of the subject went by the name of "logistic," and its beginnings must be sought in prehistoric times. Greek tradition states that it came from the Phœnicians, whose trading instincts are well known, and many comparatively recent writers have felt that this tradition had a foundation in fact. It must not be thought, however, that the interesting properties of numbers were entirely unrecognized before this time. Various curious relations had been the subject of discussion in the Orient for many centuries, and some knowledge of number mysticism had doubtless been acquired by the priestly caste in Greece long before logistic existed as a special subject of study.

Although by the tradesman in Miletus, and later in Corinth and other seaport towns, logistic must have been looked upon as important, it is probable that the ordinary Greek could neither multiply one number by another nor perform any other operation in what we now call arithmetic. There were doubtless schools at that time, for Herodotus (*c.* 450 B.C.) and Diodorus Siculus (1st century B.C.) both speak of them as then known, but logistic was looked upon as a technicality of trade, just as we may today look upon the use of a slide rule or a typewriter. A little later, however, it came more into favor, for Plato refers to it rather than to the theoretical part of the science when he says:

> Very unlike a divine man would be he who is unable to count, "one, two, three," or to distinguish odd and even numbers. . . . All freemen, I conceive, should learn as much of these branches of knowledge as every child in Egypt is taught when he learns his alphabet. In that country arithmetic games have actually been invented for the use of children, which they learn as a pleasure and amusement.

Furthermore, Plato recommends the use of apples and other objects in the presenting of the idea of number, quite as a modern teacher would employ them.

In spite of the extensive use of logistic among the Greek merchants, not a single treatise upon the subject remains. A Greek multiplication table, written on wax at about the beginning of our era, and hence somewhat later than the close of the period under discussion, is still preserved in the British Museum; and this, together with a few examples in addition, subtraction, and multiplication, and an abacus, are all that have come down to us that bears directly upon the practical

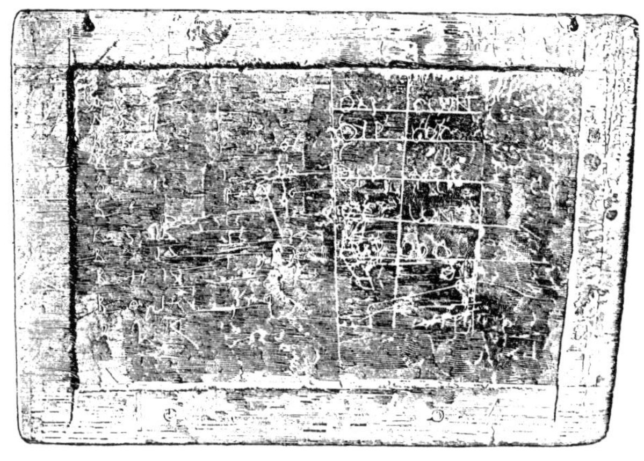

GREEK MULTIPLICATION TABLE ON A WAX TABLET

One of the few examples of the Greek logistic. This specimen is now in the British Museum and dates from about the beginning of the Christian era

computations of the Greeks. To these examples reference will be made when we come to consider the abacus and the various operations.

Arithmetic. Although the precise nature of the Greek logistic, the art of calculating, is very little known, fortunately the same cannot be said of the Greek arithmetic, the theory of numbers. As a subject for philosophers and by them committed to writing, it has come down to us as it was left by the later Greeks, and probably with its details little changed from the original form given to them in the earlier days. This topic, relating to the remote ancestor of our present number theory, will be considered, together with logistic, in Volume II.

Greek Geometry. Although both logistic and arithmetic developed in the Orient as well as in the Occident, geometry as a logical science is purely a product of the western civilization. On the other hand, intuitive geometry is universal, differing as a matter of course in degree of accomplishment in the various parts of the world. Egypt possibly knew the law of the square on the hypotenuse of a right-angled triangle long before Pythagoras, and there is reason to believe that China and India were also familiar with it; but the first proof of the theorem, and apparently the first idea of a geometric proof, are both due entirely to the Greeks. Indeed, we may say that all of our geometry, considered as a logical sequence of propositions, whether relating to two-dimensional or to three-dimensional space and whether limited to circles and straight lines or extended to include conic sections and higher plane curves, had its origin solely in the Greek civilization. So completely was the Greek mathematics given over to geometry that both arithmetic and the science that was much later known as algebra were treated almost entirely from the geometric standpoint. We shall therefore see that, although mathematics among the Greeks included geometry, arithmetic, logistic, music, and a kind of algebra, the central element was geometry.

Centers of Mathematical Activity. Mention has been made of Miletus, and before proceeding farther it is desirable to locate the other centers of mathematical activity in Greece and her colonies. The following cities and countries will be mentioned frequently, the numbers referring to the map on page 60:

Abdera, 16.
Alexandria, 41.
Amisus, 31.
Antinoopolis, 43.
Apameia, 35.
Aquitania, 2.
Athens, 13.
Byzantium, 28.
Cadiz, 1.
Chalcedon, 29.
Chalcis, 37.
Chios, 17.
Clazomenæ, 24.
Cnidus, 22.
Constantinople, 28.
Crete, 20.
Crotona, 9.
Cyprus, 34.
Cyrene, 11.
Cyzicus, 27.
Elea, 8.
Elis, 12.
Gades, 1.
Gerasa, 39.
Jerusalem, 40.
Laodicea, 32.
Larissa, 36.
Medma, 7.
Mendes, 42.
Miletus, 23.
Naples and Pompeii, 6.
Nicæa and Bithynia, 30.
Paros, 19.
Perga, 33.
Pergamum, 26.
Ptolemais, 44.
Rhodes, 21.
Rome, 3.
Samos, 18.
Sicily, 4.
Smyrna, 25.
Stageira, 14.
Syene, 45.
Syracuse and Messina, 5.
Tarentum, 10.
Thasos, 15.
Tyre, 38.

MATHEMATICAL-HISTORICAL MAP OF THE MEDITERRANEAN COUNTRIES IN CLASSICAL TIMES

This map shows the location of places most frequently mentioned in relation to Greek and Roman mathematics, with the names (page 61) of scholars connected with them. The numbers are arranged on the map from left to right. The dates following the names are merely approximate, and those before the Christian era are indicated by the letters B.C., as in the text. France and Spain are not shown, because they are mentioned only with respect to Cadiz (Gades) and Aquitania (in southern France), and to include them would reduce in size the more essential parts of the map. It will be observed that, as stated in the text, mathematics flourished best in territory that was protected by sea, by desert, or by mountainous regions

Cities and Countries by Map Numbers

1. *Cadiz* (Gades): Columella, 25. In southwestern Spain, not shown.
2. *Aquitania:* Victorius (Victorinus), 450. In southern France, not shown.
3. *Rome:* Varro, 60 B.C.; P. Nigidius Figulus, 60 B.C.; Vitruvius, 20 B.C.; Frontinus, 100; Menelaus, 100; Hyginus, 120; Balbus, 100; Domitius Ulpianus, 200; Nipsus, 180; Epaphroditus, 200; Sextus Julius Africanus, 220; Censorinus, 235; Serenus, 200; Porphyrius, 275.
4. *Sicily:* Diodorus Siculus, 1st century B.C.
5. *Syracuse* and *Messina:* Archimedes, 225 B.C.; Julius Firmicus Maternus, 340; Dicæarchus of Messina, 320 B.C.
6. *Naples* and *Pompeii:* Pliny, 75.
7. *Medma:* Philippus Medmæus, 350 B.C.
8. *Elea:* Zeno, 450 B.C.; Parmenides, 460 B.C.
9. *Crotona:* Pythagoras, 540 B.C.; Philolaus, 425 B.C.
10. *Tarentum:* Pythagoras, 540 B.C.; Philolaus (?), 425 B.C.; Archytas, 400 B.C.
11. *Cyrene:* Theodorus, 425 B.C.; Nicoteles, 250 B.C.; Eratosthenes, 230 B.C.; Synesius, 410.
12. *Elis:* Hippias, 425 B.C.
13. *Athens:* Solon, 600 B.C.; Agatharchus, 470 B.C.; Socrates, 425 B.C.; Meton, 432 B.C.; Phaeinus, 432 B.C.; Euctemon, 432 B.C.; Theætetus, 375 B.C.; Plato, 380 B.C.; Speusippus, 340 B.C.; Ptolemy, 150.
14. *Stageira:* Aristotle, 340 B.C.
15. *Thasos:* Leodamas, 380 B.C.
16. *Abdera:* Democritus, 400 B.C.
17. *Chios:* Œnopides, 465 B.C.; Hippocrates, 460 B.C.
18. *Samos:* Pythagoras, 540 B.C.; Conon, 260 B.C.; Aristarchus, 260 B.C.
19. *Paros:* Thymaridas, 380 B.C.
20. *Crete:* Early civilization, particularly at Knossos.
21. *Rhodes:* Eudemus, 335 B.C.; Geminus, 77 B.C.; Poseidonius, 100 B.C.
22. *Cnidus:* Eudoxus, 370 B.C.
23. *Miletus:* Thales, 600 B.C.; Anaximander, 575 B.C.; Anaximenes, 530 B.C.
24. *Clazomenæ:* Anaxagoras, 440 B.C.
25. *Smyrna:* Theon, 125.
26. *Pergamum:* Great library; parchment.
27. *Cyzicus:* Callippus, 325 B.C.
28. *Byzantium* (Constantinople): Proclus, 460; Psellus, 1075.
29. *Chalcedon:* Xenocrates, 350 B.C.; Proclus, 460.
30. *Nicæa* and *Bithynia:* Hipparchus, 140 B.C.; Theodosius, 50 B.C.; Sporus, 275.
31. *Amisus:* Dionysodorus, 50 B.C.
32. *Laodicea:* Anatolius, 280.
33. *Perga:* Apollonius, 225 B.C.
34. *Cyprus:* Early civilization.
35. *Apameia:* Poseidonius, 100 B.C.
36. *Larissa:* Domninus, 450.
37. *Chalcis:* Iamblichus, 325.
38. *Tyre:* Marinus, 150; Porphyrius, 275.
39. *Gerasa:* Nicomachus, 100.
40. *Jerusalem:* Religious.
41. *Alexandria:* Euclid, 300 B.C.; Eratosthenes, 230 B.C.; Apollonius, 225 B.C.; Aristarchus, 260 B.C.; Hypsicles, 180 B.C.; Heron, 50; Menelaus, 100; Ptolemy, 150; Diophantus, 275; Pappus, 300; Theon, 390; Hypatia, 400.
42. *Mendes:* Commercial.
43. *Antinoopolis* (Antinoe): Serenus, 50.
44. *Ptolemais:* Synesius, 410.
45. *Syene:* Eratosthenes, 230 B.C.

Early Greek Appreciation of Geometric Forms. Greece went through the same stages of appreciation of geometric forms that Egypt and Crete went through. This is seen in the use of crude parallels, then of the less crude and more elaborate forms, and finally of the more delicate forms found in the period just preceding the development of the highest type of Greek art. The use of these geometric forms was especially noteworthy just before the time of Thales, what is known as the geometric style in the decoration of vases having reached its climax in the 8th century B.C.

GREEK GEOMETRIC FORMS JUST PRECEDING THE TIME OF THALES

From a specimen of the 8th century B.C. in the Metropolitan Museum, New York

Greek Algebra. Algebra as a science distinct from arithmetic and geometry was invented long after Greece had ceased to be a center of civilization. Certain identities that we now express in algebraic form were well known to the Greeks, however, and were demonstrated with even greater rigor than is the case in our textbooks today, where the work is practically limited to rational numbers. For example, the Greeks proved that

$$(a + b)^2 = a^2 + 2ab + b^2,$$

although they had no algebraic shorthand by which to express the fact, and although they considered only lines and rectangles instead of numbers and products. In like manner they knew such other identities as

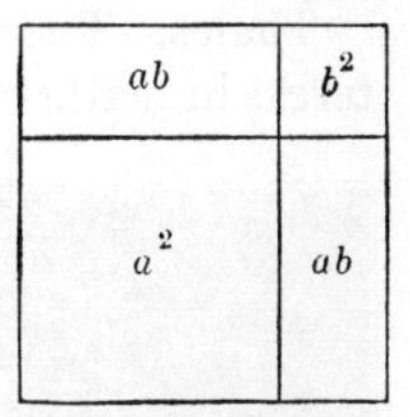

$$(a+b)(a-b)=a^2-b^2,$$

$$a(x+y+z)=ax+ay+az,$$

and $$(a-b)^2=a^2-2ab+b^2,$$

although they considered these also as geometric relations. They could complete the square of the binomial expression

$$a^2 \pm 2ab,$$

but, again, this was looked upon as the filling out of a geometric figure that was made up of a square increased or decreased by twice a rectangle. Greek algebra, as a form of arithmetic distinct from geometry, was developed some time after the period which we are now studying. When we come to consider the life and works of Diophantus (*c.* 275), for example, we shall see that the later Greeks made a remarkable advance in the analytic treatment of this subject. They developed a fairly good symbolism, and they considered algebra, under the name "arithmetic," as entirely distinct from the geometry of which we have been speaking.

3. Origins of Greek Mathematics

The Makers of Greek Mathematics. There are three important periods in the development of Greek mathematics, two of them within the chronological limits now being considered and one immediately following the later of these limits. The periods may be characterized as, first, the one subject to the influence of Pythagoras; second, the one dominated by Plato and his school; third, the one in which the Alexandrian School flourished in Grecian Egypt and extended its influence to Sicily, the Ægean Islands, and Palestine. We shall now consider the names of some of those who made the mathematics of the first two of these important periods, and some of the influences which led them to undertake their epoch-making work.[1]

Thales. The first of the Greeks to take any scientific interest in mathematics in general, and in the union of astronomy, geometry, and the theory of numbers in particular, was Tha'les.[2] Before his time there had been the usual interest of early peoples in the mystery of the heavens, as witness the statement of the poet Archil'ochus[3] that a solar eclipse was observed some time before Thales was born. Not until the time of Thales, however, did the science of mathematics begin in the Greek civilization.

THALES

Ancient bust in the Capitoline Museum at Rome, not contemporary with Thales

Miletus was then, as we have seen, a trading and colonizing center, a city of wealth and influence. Herodotus (*c.* 450 B.C.) tells us that Thales was of Phœnician descent; but his mother, Cleobuline, bore a Greek name, while the name of his father, Examius, is Carian. The name of Thales himself was

[1] The most recent and elaborate works on this period are A. Mieli, *La Scienza Greca*, of which Volume I (Florence, 1916) deals with the history of Greek science before Aristotle; and Sir T. L. Heath, *A History of Greek Mathematics*, 2 vols. (Cambridge, 1921), of which the first volume covers the period from Thales to Euclid.

[2] Θαλῆς. Born at Miletus, *c.* 640 B.C.; died *c.* 546.

[3] Ἀρχίλοχος. Born *c.* 714 B.C.; died *c.* 676.

probably a common one. Indeed, we have in the Metropolitan Museum in New York at present a Cypriote vase of his period which bears the name in the form

—evidently that of the owner.[1] Considering the general recognition of the abilities of Thales, even during his lifetime, it is not impossible that a vase made in Cyprus may have been intended for him, but there is no further evidence that this was the case.

CYPRIOTE VASE WITH THE NAME OF THALES

The name *TA* + *LE* + *SE*, when viewed from above, has the appearance shown in the text. The vase is about contemporary with Thales of Miletus. From the Metropolitan Museum of Art, New York

Stories concerning Thales. Thales was a merchant in his younger days, a statesman in his middle life, and a mathematician, astronomer, and philosopher in his later years. In his mercantile ventures he seems to have been unusually successful, even in dealing with the shrewdest of the Greek trading races. Aristotle (*c.* 340 B.C.) tells us how he secured control of all the oil presses in Miletus and Chios in a year when olives promised to be plentiful, subletting them at his own rental when the season came. Plutarch (1st century) also testifies to his ingenuity in the following anecdote:[2]

Solon went, they say, to Thales at Miletus, and wondered that Thales took no care to get him a wife and children. To this Thales made no answer for the present; but a few days after procured a stranger to pretend that he had left Athens ten days earlier and

[1] Reading from right to left, the characters have the values *ta* + *le* + *se*, corresponding to the Greek Θα + λῆ + ς, Thales. The vase is part of the Cesnola collection found in Cyprus.

[2] In his life of Solon.

Solon inquiring what news there was, the man replied according as he was told: "None but a young man's funeral, which the whole city attended, for he was the son, they said, of an honorable man, the most virtuous of the citizens, who was not then at home, but had been traveling a long time." Solon replied, "What a miserable man is he! But what was his name?" "I have heard it," said the man, "but have now forgotten it, only there was great talk of his wisdom and justice." [After Solon had been drawn on to pronounce his own name and had learned that it was his own son,] Thales took his hand and, with a smile, said, "These things, Solon, keep me from marriage and rearing children, which are too great for even your constancy to support; however, be not concerned at the report, for it is a fiction."

Solon (*c.* 639–559 B.C.), it should be observed, was interested in astronomy and was the one who introduced a "leap month" into the Athenian calendar (594 B.C.).

Trade was then an honorable calling, and Thales seems to have traveled in Egypt on his commercial ventures, and early writers tell of his also visiting both Crete and Asia. He was not the only mathematician to have thus turned trade to profit, for Plutarch has this to say of him: "Some report that Thales and Hippocrates the mathematician traded, and that Plato defrayed the charges of his travels by selling oil in Egypt." In this way Thales may have accumulated the wealth that permitted him to indulge his taste for learning and to found the Ionian School. It was through this indulgence that he acquired such a reputation as to be enrolled as the first among the Seven Wise Men of Greece, and that he was esteemed as the father of Greek astronomy, geometry, and arithmetic.

Arithmetic of Thales. Of the nature of the arithmetic that Thales brought back from Egypt we have little direct knowledge. Iamblichus of Chalcis (*c.* 325 A.D.) tells us that he defined number as a system of units, and adds that this definition and that of unity came from Egypt. This is not much, but it is enough to show that Thales was interested in something besides the merely practical. It is probable that he knew many other number relations, for the Ahmes papyrus contains some work in progressions, and such knowledge would hardly escape so careful an observer as Thales. It is, however, in his work in founding deductive geometry and in his capacity as a teacher of Pythagoras rather than as a discoverer of facts that Thales commands our attention.

Interest in Astronomy. He took much interest in astronomy, and Herodotus (I, 74) tells us that he even succeeded in predicting an eclipse. Some authorities suppose this eclipse to have occurred on May 28, 585 B.C., while others place it about twenty-five years earlier. He could have obtained certain information on this subject from a study of the Chaldean records, but whether this was his source of information we cannot say. At the present time we have numerous cuneiform tablets of the 7th century B.C. which record such prognostications. One of these reads: "To the king, my master, I have written that there was about to be an eclipse. The eclipse has now taken place. This is a sign of peace for the king, my master."

A man like Thales, possessed of an inquisitive mind, coming in contact with scholars from other lands, either on his travels or in the commercial center of Miletus, would lose no opportunity to secure information of this kind and to make use of it in his teaching. Doubtless his scientific training led him to discard the astrological notions of the Chaldeans but to retain whatever of astronomy came to his attention.

Geometry of Thales. In geometry he is credited with a few of the simplest propositions relating to plane figures. The list, according to the most reliable ancient writers, is as follows:

1. Any circle is bisected by its diameter.
2. The angles at the base of an isosceles triangle are equal.
3. When two lines intersect, the vertical angles are equal.[1]
4. An angle in a semicircle is a right angle.[2]

[1] See Proclus, ed. Friedlein, pp. 157, 250, 299 (Leipzig, 1873).

[2] There is a doubt about his knowing this. It is inferred from a statement by Pam'phila (Παμφίλη, a woman historian, 1st century A.D.), but there is no early authority for the statement.

5. The sides of similar triangles are proportional.[1]

6. Two triangles are congruent if they have two angles and a side respectively equal (Euclid, I, 26).[2]

Importance of his Geometry. As propositions in geometry these may seem trivial, since they are intuitive statements; but their very simplicity leads us to believe that it was the fact that Thales was the first to prove them that led Eudemus (*c.* 335 B.C.) and other early writers to mention them. Up to this time geometry had been confined almost exclusively to the measurement of surfaces and solids, and the great contribution of Thales lay in suggesting a geometry of lines and in making the subject abstract. With him we first meet with the idea of a logical proof as applied to geometry, and it is for this reason that he is looked upon, and properly so, as one of the great founders of mathematical science. In the history of mathematics, as in the history of civilization in general, it is the setting forth of a great idea that counts. Without Thales there would not have been a Pythagoras—or such a Pythagoras; and without Pythagoras there would not have been a Plato—or such a Plato.

Philosophy of Thales. In philosophy he is said to have asserted that water is the origin of all things, that everything is filled with gods, that the soul is that which originates motion, and that matter is infinitely divisible; but his basis for belief in these assertions is not very satisfactory. Like most of his contemporaries, he left no written works.

Anaximander. At the death of Thales the leadership of the Ionian School passed over to Anaximan'der,[3] who is generally

[1] He used this proposition in measuring the height of a pyramid by means of the shadow of the pyramid and that of a staff. See Diogenes Laertius, *Vitæ Philosophorum*, ed. Cobet, p. 6 (Paris, 1878); Pliny, *Hist. Nat.*, XXXVI, 17; Plutarch, *Septem Sapientium Convivium*, ed. Didot, III, 174 (Paris, 1841). Pliny's statement that he measured the shadow at the time of day when the shadow "is equal in length to the body projecting it" is not very convincing. This would be too simple and is quite contrary to Plutarch's version.

[2] Eudemus (*c.* 335 B.C.), a disciple of Aristotle, refers this to Thales.

[3] Ἀναξίμανδρος. Born *c.* 611; died *c.* 547 B.C. J. Neuhaeuser, *Anaximander Milesius* (Bonn, 1883).

thought to have been his pupil. Anaximander or some contemporary of his first brought into use in Greece the gnomon, an instrument resembling the sundial[1] and used for determining noon, the solstices, and the equinoxes. Aside from this, Anaximander seems to have had no interest in mathematics. It was about this time that the water clock (clep'sydra),[2] already known to the Assyrians, found its way into Greece, and very likely Anaximander's gnomon came also to be used for determining the time of day.

4. From Pythagoras to Plato

Pythagoras. Of all the interesting figures in the history of ancient mathematics Pythag'oras[3] ranks easily first, partly from the mystery surrounding his life, partly from his own mysticism, partly from the brotherhood which he established, and partly from the unquestioned ability of the man himself.

Early Life of Pythagoras. As with Euclid and Heron, of whom we shall presently speak, so with Pythagoras,—the date and the place of his birth are both unknown. He seems to have been born between the 50th and 52d Olympiads, to use the Greek system of chronology, or between 580 and 568 B.C. of our calendar. Although called a Samian, we are not certain that he was born on the island of Samos, for Suidas, a late medieval writer (*c.* 1000), says that he was born in Italy, and

[1] Really, the pointer which casts the shadow on the dial.

[2] Κλεψύδρα, from κλέπτειν (to hide) + ὕδωρ (water).

[3] Born at Samos (?), *c.* 572 B.C.; died at Tarentum (?), *c.* 501. W. Schultz, *Pythagoras und Heraklit*, Leipzig, 1905; A. Ed. Chaignet, *Pythagore et la philosophie pythagoricienne, contenant les fragments de Philolaüs et d'Archytas*, Paris, 1873; W. Bauer, *Der ältere Pythagoreismus*, Bern, 1897; W. W. Rouse Ball, "Pythagoras," in the *Math. Gazette*, London, January, 1915; G. J. Allman, *Greek Geometry from Thales to Euclid*, Dublin, 1889 (hereafter referred to as Allman, *Greek Geom.*); F. Cramer, *Dissertatio de Pythagora*, Prog., Sund, 1833; W. Lietzmann, *Der pythagoreische Lehrsatz*, Leipzig, 1912; Armand Delatte, "Études sur la littérature pythagoricienne," in the *Bibliothèque de l'École des hautes Études*, Vol. 217 (Paris, 1915). Of the early histories of Pythagoras the one best known is that of Iamblichus, *c.* 325. It first appeared in print at Franeker, 1598. Better editions by Ludolph Kuster (Amsterdam, 1707), and A. Nauck (Petrograd, 1884).

that as a child he migrated to Samos with his father.[1] Nevertheless the weight of authority favors his Samian birth, and a number of coins of the island, struck some centuries after his time, bear his name and figure, and this would hardly have been the case had he merely spent his boyhood there.[2] Various stories are told of his parentage, but we are quite uncertain whether his father was an engraver of seals or a merchant. At any rate he lived after Greece had enjoyed two centuries of commercial activity, and at the dawn of that golden age which began in Athens in the 6th century B.C. and closed for that city at the end of the 5th century B.C.

FIGURE OF PYTHAGORAS

A coin of Samos of the reign of Trajan (98–117), and therefore much later than Pythagoras. It shows the honor in which he was held and the claim of Samos as his native country

Period of Pythagoras. But in whatever land he was born, and in whatever year, and of whatever parentage, Pythagoras lived in stirring times and was himself one of the great makers of the civilization of his period. Samos was just becoming a center of Greek art and culture, Polyc'rates was just ascending the throne, and Anac'reon was beginning to write his famous lyrics in the Samian court. Pythagoras was therefore brought up amid scenes that could hardly fail to stimulate a youth of his native powers and urge him to a high intellectual life. Moreover, the spirit of the times was active in great works. Buddha was just promulgating his doctrine in India, and Confucius and Lao-Tze were laying the foundation for their philosophic cults in China; and, whether or not Pythagoras came into personal touch with the Far East, he lived when the world was ripe for great movements.

[1] See also M. Barbieri, *Notizie istoriche dei Mattematici e Filosofi di Regno di Napoli,* cap. ii (Naples, 1778), who (p. 25) thought that Samos was the modern Crepacore, a town in Southern Italy.

[2] One of these coins is shown in the illustration. There are also a few gems, of doubtful age, which are said to represent Pythagoras. See C. W. King, *Antique Gems and Rings,* I, 212, and II, plate XXXVIII, No. 1 (London, 1872).

The fact that arithmetic and geometry took such a notable step forward at this time was due in no small measure to the introduction of Egyptian papyrus into Greece. This event occurred about 650 B.C.,[1] and the invention of printing in the 15th century did not more surely effect a revolution in thought than did this introduction of writing material on the northern shores of the Mediterranean Sea just before the time of Thales.

Studies and Travels of Pythagoras. Our knowledge of the life of Pythagoras is very limited, the early writers having vied with each other in the invention of fables relating to his travels, his miraculous powers, and his teachings. He seems to have sought out Thales and to have been his pupil. Tradition says that he was initiated by the master into the secrets of Zeus on Mount Ida, and was then told that if he would have further light he must seek it in Egypt. We now lose all definite knowledge of Pythagoras for a considerable period. Appuleius,[2] a Roman writer of about 150 A.D., asserts that he was captured by Cambyses the Persian,[3] that he learned from the Magi, and that he even sat at the feet of Zoroaster himself; but part of the story cannot be true, because Zoroaster probably died about the time that Pythagoras was born, and possibly much earlier than that, for the date is very uncertain. Isoc′rates, a writer of a century after Pythagoras, and Callim′achus, librarian of the Alexandrian library, who lived in the 3d century B.C., both assert that he spent some years in Egypt. Pliny, writing in the 1st century of our era, says that Pythagoras was there in the time of Psammetichus,[4] and Strabo, about the beginning of the Christian era, states that he studied in Babylon. Others claim that he went as far east as India, but we have no definite proof of any of these statements.

[1] In the reign of King Psammet′ichus (Psammitichus) I, soon after 660 B.C., sometimes given as *c.* 640–610 and sometimes as 671–617. See T. Gomperz, *Les Penseurs de la Grèce*, French ed., p. 13 (Lausanne, 1904). This is the Psemtek of the monuments and the first king of the 26th Dynasty.

[2] Usually so spelled in ancient texts, but occasionally Apuleius.

[3] Reigned 529–522 B.C.

[4] That is, Psammetichus III, who reigned 526–525 B.C., when Pythagoras was about 46 years old.

Contact with the East. In spite of the assertion of various writers to the contrary, the evidence derived from the philosophy of Pythagoras points to his contact with the Orient. The mystery of the East appears in all his teachings.[1] His mysticism of numbers is quite like that found earlier in Babylon, and indeed his whole philosophy savors much more of the Indian than of the Greek civilization in which he was born. According to our best evidence the familiar proposition of geometry that bears his name was known, as already stated, in India, China, and Egypt (?) before his time, and all that can be claimed for him in relation to it is that he may have given the earliest general demonstration of its truth.

School of Crotona. When Pythagoras reappeared after his years of wandering, he sought out a favorable place for a school, and finally settled upon Crotona, a town on the southeastern coast of Italy, in a territory called by the Italic Greeks of that time Great Greece.[2] This town was a wealthy seaport, and it was to the young men of well-to-do families that Pythagoras made his appeal. Pretending to have the power of divination, given at all times to mysticism, and possessed in a remarkable degree of personal magnetism, he gathered about him some three hundred of the noble and wealthy young men of Magna Græcia and established a brotherhood that has ever since served as a model for all the secret societies in Europe and America. He divided his disciples into two groups, the hearers and the mathematicians, the latter having passed through a probationary period as members of the former group.

Oral Teaching of Pythagoras. Pythagoras never embodied his doctrines in any treatise. Like Thales, and also like those Oriental teachers from whom he probably learned, he transmitted his theories by word of mouth. This he did through the elect of his brotherhood, thus making known his doctrines freely to all who were deemed worthy to receive them. This

[1] E. W. Hopkins, *The Religions of India*, p. 559 (Boston, 1902); L. von Schroeder, *Indiens Literatur und Cultur*, pp. 718 seq.; *Reden und Aufsätze*, p. 168 (Leipzig, 1913); *Pythagoras und die Inder* (Leipzig, 1884).

[2] Ἡ μεγάλη Ἑλλάς, *Magna Graecia.*

method of imparting knowledge was not due merely to a spirit of mysticism, but was quite as dependent upon a lack of good writing material. Parchment had not as yet been invented, the wax tablet was serviceable only for brief epistles, the clay cylinders of Babylon were subject to similar limitations, and the fragile papyrus of Egypt was probably somewhat rare in Magna Græcia. Pythagoras therefore followed the custom of his time in passing his philosophy along by word of mouth, just as the ancients had transmitted to his generation the songs of Homer. Even in Plato's time there was no bookshop in all Athens where worthy manuscripts could be purchased, nor was there any when Euclid taught in Alexandria. Not until the time of Augustus was the book trade established, making possible the easy and certain transmission of knowledge, and not until fifteen hundred years later was printing known in Europe.

For the doctrines of Pythagoras we are indebted chiefly to Eudemus of Rhodes (*c.* 335 B.C.), whose works, though lost, are known to us through extracts preserved by later writers. We also know of the doctrines of Pythagoras through passages from a work by Philolaus of Crotona (who lived in the 5th century B.C.), from a statement by Archytas of Tarentum (*c.* 400 B.C.), a friend of Plato, and from a number of passages in the works of later writers.

Philosophy of Pythagoras. Pythagoras based his philosophy upon the postulate that number is the cause of the various qualities of matter. This led him to exalt arithmetic, as distinguished from logistic, out of all proportion to its real importance. It also led him to dwell upon the mystic properties of numbers and to consider arithmetic as one of the four degrees of wisdom,—arithmetic, music, geometry, and spherics (astronomy), these forming the quadrivium of the Middle Ages. Aristotle (384–322 B.C.) tells us that Pythagoras related the virtues to numbers, and Plutarch says that he believed that earth was produced from the regular hexahedron, fire from the pyramid, air from the octahedron, water from the icosahedron, and the heavenly sphere from the dodecahedron, in all of which the physical elements are related both to number and to form.

Philolaus probably voiced the teaching of the master when he asserted that five is the cause of color, six of cold, seven of health, and eight of love.

The Chinese say that five represents wind, and two represents earth, and these ideas are also claimed for the Pythagorean system.[1] Here again the resemblance between the mysticism of this school and that commonly found in the Far East leads to the belief that Pythagoras must have come into relations with the wise men of the Orient. Savoring of the East, too, is the description given by Suidas, a late medieval Greek compiler, of a ceremony called Pythagus, in which there is written something in blood on the face of a mirror, at the time of the full moon, the words then being read in the reflection in the circle of the moon; but there is no ancient authority for such a statement.[2]

Shakespeare refers to the acceptance by Pythagoras of the Hindu belief in the transmigration of souls, in these words:

> Thou almost mak'st me waver in my faith,
> To hold opinion with Pythagoras,
> That souls of animals infuse themselves
> Into the trunks of men.
>
> *Merchant of Venice*

Unity and Infinity. From various early writers we judge that Pythagoras asserted that unity is the essence of number, the origin of all things, the divine; that he had the idea of the limited and the unlimited; and that he held that from the latter came the ideas of time, space, and motion. Diogenes Laertius (2d century A.D.) says that he was interested in number, and that the part of mathematics "to which Pythagoras applied himself above all others was arithmetic";[3] and Aristox'enus[4] says that he esteemed this science above all others.

[1] J. Hager, *An Explanation of the Elementary Characters of the Chinese*, p. xv. London, 1801.

[2] J. C. Bulengerus, *De Lvdis privatis ac domesticis veterum*, p. 31. Lyons, 1627.

[3] Diogenes Laertius, VIII, i, 11, p. 207 ed. Cobet. Various references in connection with Greek mathematics may be found in Allman, *Greek Geom.*

[4] Ἀριστόξενος, a philosopher. Born at Tarentum, *c.* 350 B.C.

Geometry of Pythagoras. In the field of geometry Eudemus (*c.* 335 B.C.) informs us that Pythagoras "investigated his theorems from the immaterial and intellectual point of view," and that "he discovered the theory of irrational quantities and the construction of the mundane figures."[1] Favori'nus, a philosopher living in southern France *c.* 125, asserts that he employed definitions in his work in mathematics, this being the first trace that we have of such use.[2] In particular, he defined a point as "unity having position."[3] He or his school knew that the plane space about a point may be filled by six equilateral triangles, four squares, or three regular hexagons,—a fact which had doubtless been inferred as a result of the observation of mosaic pavements long before this time, but which no doubt he was able to prove. It is probable that Pythagoras proved the proposition relating to the sum of the angles of a triangle, that he constructed a polygon equivalent to one given polygon and similar to another, and that he could construct the five regular polyhedrons; and he may possibly have proved the theorem relating to the square on the hypotenuse. It seems likely that he taught that the earth is a sphere in space; at any rate, this theory was accepted by various later philosophers.[4]

Pythagoras on Music. Pythagoras is said to have discovered that the fifth and the octave of a note can be produced on the same string by stopping at $\frac{2}{3}$ and $\frac{1}{2}$ of its length, respectively, and it is thought that this harmony gave rise to the name of "harmonic proportion," since

$$1 : \tfrac{1}{2} = 1 - \tfrac{2}{3} : \tfrac{2}{3} - \tfrac{1}{2}.$$

Although he seems to have derived some knowledge of music from Egypt,[5] he is generally called the inventor of musical science, or the harmonic canon (a mere tradition), but we

[1] *I.e.*, of the five regular polyhedrons. Proclus (c. 412–485), ed. Friedlein, p. 65.

[2] Diogenes Laertius, VIII, i, 25, p. 215 ed. Cobet.

[3] Proclus, ed. Friedlein, p. 95.

[4] On doubts as to the Pythagorean Theorem see G. Junge, *Bibl. Math.*, VIII (3), 62, and H. Vogt, *ibid.*, IX (3), 19. On the astronomical question see P. Duhem, *Le Système du Monde* (Paris, 1913–1917).

[5] J. G. Wilkinson, *Manners and Customs of the Ancient Egyptians*, I, vi. London, 1878.

know nothing of the notes or of the system that he used.[1] With his love for music and number it is natural to believe that he must have taken great pride to himself for connecting the two in the harmonic proportion. He seems to have held that the

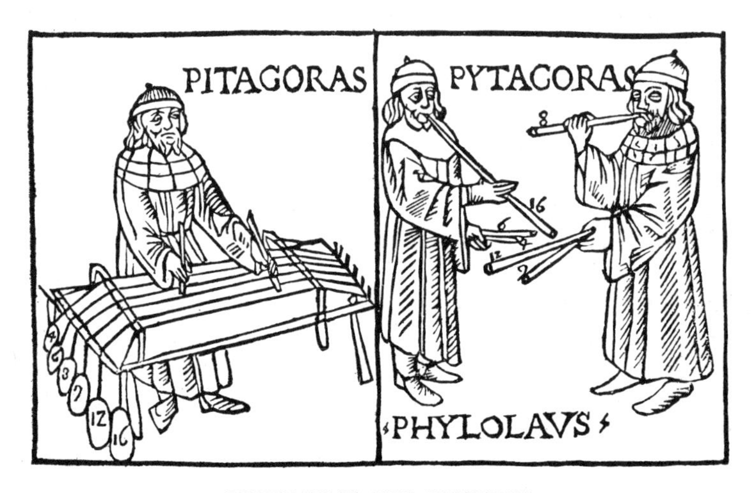

PYTHAGORAS THE MUSICIAN

From F. Gafurius, *Theorica Musice*, Milan, 1492. One of the first crude attempts to portray Pythagoras by means of a woodcut, and the first to portray him as a musician. In the same work he is also shown as a bell ringer

intervals between the heavenly bodies were determined by the laws of musical harmony, and hence arose the doctrine of the harmony of the spheres.

The influence of Pythagoras became so great that the government caused his brotherhood to be dispersed, although the members still spread the doctrines of the sect throughout Greece. Pythagoras died an exile from Crotona, possibly at Tarentum. Two centuries later, however, and during the disasters of the First Samnite War, 343 B.C., the Senate erected his statue in Rome, in response to an order of the Delphic Oracle to pay this honor to "the wisest and bravest of the Greeks," and the people learned to call him the preceptor of

[1] A. Baumgartner, *Geschichte der musikalischen Notation*, p. 11. Munich, 1856.

King Numa, while even the great Æmilian family was, in later years, proud to claim him as one of their honored ancestors.

We shall now consider a few of the other Greek mathematicians who attained prominence before the time of Plato. All of them were influenced by the doctrines of Pythagoras, and it is convenient and proper to consider them in close connection with the Pythagorean School, including in the discussion the members of the Eleatic School mentioned below.

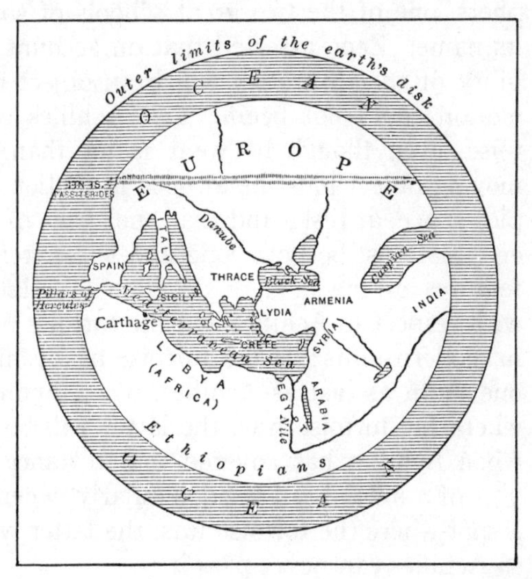

MAP OF THE WORLD BY HECATÆUS, 517 B.C.
Showing the primitive ideas held at the time of Pythagoras. From Breasted's *Ancient Times*

Lesser Writers. Among the contemporaries of Pythagoras was Anaxim′enes of Miletus,[1] probably a pupil of Anaximander. Diogenes Laertius quotes two letters from him to Pythagoras, in one of which he speaks of Thales as his teacher; but his tastes were in the direction of philosophy rather than of mathematics. In this period there also flourished the geographer Hecatæ′us, whose map of the world serves to show how fragmentary was the knowledge then possessed even by the best scientists.

About the time of Pythagoras there also flourished Ameris′tus,[2] a geometer of some prominence and brother of the poet Stesich′orus. He is mentioned by Proclus (*c.* 412–485 A.D.), but nothing is known of his work.

[1] Ἀναξιμένης, born at Miletus, *c.* 585; died *c.* 528 B.C.

[2] The Mamercus of J. Gow, *History of Greek Mathematics*, p. 145 (Cambridge, 1884). Hereafter referred to as Gow, *Greek Math.*

Zeno of Elea.[1] About the time of the death of Pythagoras there was born at Elea[2] the philosopher Zeno, whose work on motion represented a noteworthy advance in the science, even though the mathematical feature is in evidence in only a single instance. It was from Elea that the Eleatic School of philosophers, one of the two great schools of southern Italy, derived its name. Zeno asserted that on account of the infinite divisibility of space through which an object must pass in moving, motion could not begin; that Achilles could not pass a tortoise, even though he went faster than the tortoise; that a moving object must be at once in motion and, because it occupies space, at rest; and that one space of time might, in different relations, be both long and short, reminding us of certain features of the modern doctrine of relativity. His argument with respect to Achilles and the tortoise may be thus expressed in modern units: If the tortoise has a mile the start and goes one tenth as fast as Achilles, when Achilles reaches the point where the tortoise was, the latter will be $\frac{1}{10}$ of a mile ahead; when Achilles has covered that distance, the tortoise will be $\frac{1}{100}$ of a mile ahead; and, similarly, whenever Achilles reaches a spot where the tortoise was, the latter will still be ahead, and so Achilles can never pass it.[3]

Anaxagoras. Among the noteworthy contemporaries of Zeno was Anaxag'oras,[4] the last of the celebrated philosophers of the Ionian School. He was a friend and teacher of Euripides, Pericles, and other great men of his time, but was condemned to death[5] at the age of seventy-two for being favorable to the Persian cause. Although his chief work was in philosophy, where his prime postulate was that "reason rules the world," he was interested in mathematics and wrote on the quadrature of

[1] Ζήνων (*Zenon*). Born at Elea, *c.* 496 B.C. He was living in the time of Pericles (died 429). Heath, *History*, I, 271.

[2] Ἐλέα or Ὑέλη (*Elea* or *Hyele*); Latin, *Velia*; in southern Italy.

[3] F. Cajori, "The History of Zeno's Arguments on Motion," *Amer. Math. Month.*, XXII, 1, 39, 77, 109, 143, 179, 215, 253, 293; "The Purpose of Zeno's Arguments on Motion," *Isis*, III, 7.

[4] Ἀναξαγόρας. Born at Clazomenæ, Ionia, *c.* 499; died *c.* 427 B.C.

[5] Ancient writers are not clear upon this point.

the circle and on perspective.[1] When banished from Athens he remarked, "It is not I who have lost the Athenians, but the Athenians who have lost me."

Agatharchus. About this period (470 B.C.) an Athenian artist, Agathar'chus,[2] applied stereometry to the theory of perspective. He is said to have painted the scenery for a tragedy which Æschylus produced. In his work on drawing he showed how to make use of the notion of projection upon a plane surface.

Socrates.[3] Although we do not commonly think of Soc'rates, the Athenian statesman and philosopher, as a mathematician, yet for his work on induction and for his insistence upon accurate definition he should be mentioned in connection with the early development of a logical geometry. As the teacher of Plato he assisted in the development of that great maker of philosophers and of those who based their mathematics upon sound logic. Socrates has left us no writings of his own, but we have the testimony of Plato, Euclid, and others that they were greatly his debtors. Probably no more noble tribute has been paid to him than that given by Dr. Jowett in his paraphrase of the words of Plato: "And he, Socrates, is a midwife, although this is a secret; he has inherited the art from his mother bold and bluff, and he ushers into light, not children, but the thoughts of men."[4] Xenophon[5] and Diogenes Laertius[6] tell us, however, that he felt that geometry and astronomy were useful merely for measuring fields and telling the time of day,—a view which, if really held by him, has been advanced by men of far less mentality in every generation since that time, and with the same empty results.

Œnopides of Chios.[7] Probably a Pythagorean and certainly one of the leading astronomers of his time, Œnop'ides is

[1] *Anaxagorae Fragmenta* (Leipzig, 1827); better edition by Schorn (Bonn, 1829). See also F. Breier, *Die Philosophie des Anaxagoras von Klazomenä nach Aristoteles* (Berlin, 1840).

[2] Ἀγάθαρχος.

[3] Σωκράτης. Born near Athens, 468 B.C.; died at Athens, 399.

[4] Jowett's *Plato*, IV, 123.

[5] *Memorabilia*, IV, 7.

[6] *Lives of the Philosophers*, II, 32.

[7] Οἰνοπίδης. Born in Chios; fl. *c.* 465 B.C.

thought to have learned the science of the stars and the obliquity of the ecliptic from the priests and temple astronomers of Egypt. He is said to have invented the cycle of 59 years for the return of the coincidence of the solar and lunar years, giving the length of the solar year as 365 days and somewhat less than 9 hours. Proclus (*c.* 460) attributes to him the discovery of two problems of Euclid, one (I, 12) referring to the drawing of a perpendicular to a given line from an external point, and the other (I, 23) referring to the making of an angle equal to a given angle. If this is really the case, it shows how slight had been the advance in demonstrative geometry, even in the century following the death of Pythagoras.

Democritus. Democ'ritus,[1] known to later generations as the Laughing Philosopher, inherited great wealth, spent his fortune in travel, met the learned men of many lands, was a man of remarkable diligence in study, and died in poverty. His works are lost, except for certain fragments.[2] One of his teachers in philosophy is said to have been Leucip'pus,[3] the founder of the atomic theory of the ancient philosophy which asserted that the original characteristics of matter are functions of quantity instead of quality, the primal elements being particles homogeneous in quality but heterogeneous in form. Archimedes tells us, in his work on *Method*, that Democritus was the first to show the relation between the volume of a cone and that of a cylinder of equal base and equal height, and similarly for the pyramid and prism. In spite of the manifest bearing of his work upon an infinitesimal calculus, it seems to have had no influence in this direction among the Greeks. It is said that Plato felt that all the writings of Democritus should be burned. At any rate he had so slight an opinion of the latter that he makes no mention of him in any of his works. Such

[1] Δημόκριτος. Born at Abdera, Thrace, *c.* 460 B.C.; died *c.* 357.

[2] F. W. A. Mullach (F. G. A. Mullachius), *Democriti operum fragmenta.* Berlin, 1843.

[3] Λεύκιππος. The date and place of his birth and of his death are unknown.

treatment at the hands of Plato was perhaps due to the boastful nature of Democritus, who speaks of himself in these words:

I have wandered over a larger part of the earth than any other man of my time, inquiring about things most remote; I have observed very many climates and lands and have listened to very many learned men; but no one has ever yet surpassed me in the construction of lines with demonstration; no, not even the Egyptian harpedonaptæ (Ἀρπεδονάπται), with whom I lived five years in all, in a foreign land.

These harpedonaptæ (literally, "rope stretchers") were the surveyors of ancient Egypt, and the quotation suggests that the logical demonstration of propositions was then practiced in that country as well as in Greece.

Parmenides of Elea. Parmen'ides of Elea[1] taught at Athens in the middle of the 5th century B.C., and among his theories of the universe was the one that the earth is a sphere. His work, however, was that of a philosopher rather than a mathematician. It was in his time that Herodotus (*c.* 450 B.C.) wrote his history, and it is in this work that the idea of a meridian first appears in any literature now extant. From this time on for several centuries the sphericity of the earth was accepted as valid by many philosophers. The theory was revived in the 12th century of our era and was strongly asserted by Roger Bacon (*c.* 1250).

Philola'us,[2] a distinguished Pythagorean, was born at Crotona, or possibly at Tarentum, and according to Plato was a contemporary of Socrates. Although Pythagoras handed down his doctrines by word of mouth, it is stated by Porphyr'ius (fl. *c.* 275) that Ly'sis, who was a prominent philosopher, and an obscure Pythagorean named Archip'pus, put into writing some of the doctrines of the school and transmitted them to their descendants as secret heirlooms. Philolaus, however, was the first to write a treatise on the teachings of Pythagoras and to

[1] Παρμενίδης. Born in Elea; fl. *c.* 460 B.C.
[2] Φιλόλαος. Fl. *c.* 425 B.C.

make it public. Judging from the fragments that have come down to us,[1] his interest was in philosophy rather than in mathematics, although he touches upon the latter field in his description of a gnomon.

Hip'pias of Elis,[2] known both as a statesman and as a philosopher, belonged to the sophists, a class of teachers who traveled from place to place and took money for their services,—a practice quite contrary to the ideas of earlier philosophers. He accumulated wealth by teaching and public speaking, and Plato speaks of him as a vain man, given to arrogance and boasting. He seems to have been possessed of a wide but superficial knowledge. His contribution to mathematics was confined to his invention of a simple device for trisecting any angle, this device being known as the quadratrix. Since it was studied and described at a later period by Deinostratus (Dinostratus, *c.* 350 B.C.), it generally bears the latter's name.[3]

Hippocrates.[4] Various stories are told of Hippoc'rates, among them one that he was an unsuccessful merchant, later becoming a Pythagorean philosopher with a special interest in mathematics. Aristotle speaks of him as skilled in geometry but as otherwise stupid and weak. He is mentioned by ancient writers as the first to arrange the propositions of geometry in a scientific fashion and as having published the secrets of Pythagoras in the

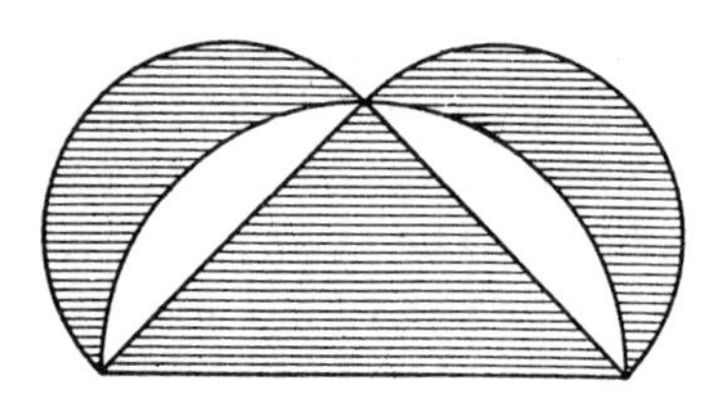

[1] A. Böckh, *Philolaos des Pythagoreers Lehren, nebst den Bruchstücken seines Werkes* (Berlin, 1819); W. R. Newbold, "Philolaus," in *Archiv für Geschichte der Philosophie*, XIX, 176.

[2] Ἱππίας. Born at Elis, on the west coast of Peloponnesus; fl. *c.* 425 B.C.

[3] Allman and Hankel do not believe that this is the Hippias to whom Proclus refers as the inventor of the quadratrix. There is room for the doubt that these eminent writers express, but Cantor, Montucla, and various other historians feel that the evidence is in favor of Hippias of Elis. See also Gow, *Greek Math.*, p. 162. The name "quadratrix" is due to the fact that the curve can also be used in the quadrature of the circle. On the application of this and other curves to the problems of geometry, see Volume II, Chapter V.

[4] Ἱπποκράτης. Born in Chios; fl. *c.* 460 B.C.

field of geometry. In his attempts at squaring the circle he discovered the first case of the quadrature of a curvilinear figure,[1] namely, the proof that the sum of the two shaded lunes here shown is equal to the shaded triangle. The proposition holds equally for any right-angled triangle, isosceles or not, although Hippocrates knew it only for the isosceles right-angled triangle. Proclus (*c.* 460) ascribes to him the method of reduction ('απαγωγή), the passing from one proposition to another that seems more simple, proving the latter, and then reversing the order. For example, Eratosthenes (*c.* 250 B.C.) tells us that Hippocrates showed that the duplication of a cube can be effected if two mean proportionals can be found between two given lines.[2]

Meton, Phaeinus, and Euctemon. That there was great interest in mathematical astronomy in Athens between the time of Pythagoras and that of Plato is seen in the work of the three astronomers Me'ton, Phaei'nus, and Eucte'mon.[3] It is not possible, however, to differentiate their contributions to the subject. The philosopher Theophras'tus[4] says that Phaeinus made astronomical observations on Lycabettus, at Athens, and that from these Meton constructed the cycle of 19 years, since known as the Metonic Cycle. The astronomer Ptolemy[5] says that Meton and Euctemon made observations at Athens and in other places. He adds that Meton made the length of the year to be $365\frac{1}{4}$ days + $\frac{1}{76}$ of a day, which is more than 30 minutes too long. Whether the 19-year cycle is really due to Meton, or was already known to Œnopides, or was obtained from Egyptian or other sources is, and is likely to remain, unknown.

[1] W. Lietzmann, *Der pythagoreische Lehrsatz* (Leipzig, 1912). For a further discussion see Volume II, Chapter V.

[2] That is, if $a : x = x : y = y : 2a$, then $x^2 = ay$, $x^4 = a^2y^2$, $y^2 = 2ax$, and hence $x^4 = 2a^3x$, or $x^3 = 2a^3$. F. Rudio, *Der Bericht des Simplicius über die Quadraturen des Antiphon und des Hippokrates*, Leipzig, 1907, with Greek and German text; P. Tannery, "Hippocrate de Chios et la quadrature des lunules," in the *Mémoires de Bordeaux* (1878); *La Géométrie Grecque*, p. 117 (Paris, 1887); "Le fragment d'Eudème sur la quadrature des lunules," *Mémoires scientifiques*, II, 46, 339 (Paris, 1912); Gow, *Greek Math.*, p. 164; M. Simon, *Archiv der Math.*, VIII (3), 269.

[3] Μέτων, Φαεινός, Εὐκτήμων. Fl. *c.* 432 B.C.

[4] Θεόφραστος. Fl. *c.* 350 B.C.

[5] Born *c.* 85; died *c.* 165.

The Method of Exhaustion. Iamblichus (*c.* 325) mentions Bry′son,[1] or Bryso, as one of the youths whom Pythagoras instructed in his old age. If this is true, Bryson must have been born about 520 B.C., but it is commonly believed that he flourished about 450 B.C. He was formerly thought to have contributed to what is known as the method of exhaustion, a crude approach to the integral calculus whereby the area between a curvilinear figure (say a circle) and a rectilinear figure (say an inscribed regular polygon) could be approximately exhausted by increasing the number of sides of the latter. There is, however, no reliable ancient authority for connecting his name with the theory.[2] The method was effectively used by later writers, notably by Eudoxus and Archimedes, and was extended to include the mensuration of solids.

Antiphon and the Method of Exhaustion. Aristotle mentions a Greek sophist named An′tiphon,[3] or Antipho, whose attempts at the quadrature of the circle led him into this phase of geometry. Antiphon inscribed a regular polygon in a circle, doubled the number of sides, and continued doubling until, as he seems to have believed, the sides finally coincided with the circle. Since he could construct a square equivalent to any polygon, he could then, as he thought, construct a square equivalent to the circle; that is, he could "square the circle," thus finding its area. We have here another phase of the method of exhaustion, the area between the polygon and the circle being exhausted as the process of doubling the number of sides proceeds. It is one of the first steps in the development of an infinitesimal calculus applied to integration,—a type of mathematics that had to wait two thousand years for serious consideration.

Archytas. In Plato's time Archy′tas,[4] a distinguished Pythagorean philosopher, achieved a high reputation as a mathematician, a general, a statesman, a philanthropist, and an educator, and Cicero (106–43 B.C.) speaks of him as a friend of the great

[1] Βρύσων. [2] F. Rudio, *Bibl. Math.*, VII (3), 378.

[3] Ἀντιφῶν. Fl. *c.* 430 B.C.

[4] Ἀρχύτας. Born at Tarentum, *c.* 428 B.C.; died *c.* 347. See Allman, *Greek Geom.*, p. 102, for an excellent summary of his work.

master himself. Horace (65–8 B.C.) refers to his death by shipwreck in the Adriatic, speaking of him in these words:

> The scanty present of a little dust
> Near the Matinian shore confines thee, O Archytas,
> Measurer of the sea, the earth, and the innumerable sand.[1]

Archytas lived in Magna Græcia, then much more tranquil than Greece itself, disturbed as the latter was by the Peloponnesian War. It was because of these wars that many Pythagoreans returned to Crotona and Tarentum, the result being that scholarship again flourished in this part of Italy. Vitruvius[2] says that Archytas solved the problem of the duplication of the cube by means of cylindric sections. He was the first to apply mathematics in any noteworthy way to mechanics, and he also applied the science to music and even to metaphysics.[3]

Eudemus (*c.* 335 B.C.), speaking of his work in geometry, tells us that he was one of those who "enriched the science with original theorems and gave it a sound arrangement," and from another statement we infer that he knew and doubtless proved the following propositions:

1. If a perpendicular is drawn to the hypotenuse from the vertex of the right angle of a right-angled triangle, each side is the mean proportional between the hypotenuse and its adjacent segment.
2. The perpendicular is the mean proportional between the segments of the hypotenuse.
3. If the perpendicular from the vertex of a triangle is the mean proportional between the segments of the opposite side, the angle at the vertex is a right angle.

[1] Te maris et terrae numeroque carentis harenae
Mensorem cohibent, Archyta,
Pulveris exigui prope litus parva Matinum
Munera.

Carmen i, 28

[2] *Praefatio* to his *De Architectura*, ix.

[3] For a list of fragments attributed to him see J. A. Fabricius, *Bibliotheca Graeca*, 14 vols., I, 833 (Hamburg, 1705–1728). There is a later edition, Hamburg, 1790–1809. See also Gow, *Greek Math.*, p. 157.

4. If two chords intersect, the rectangle of the segments of one is equivalent to the rectangle of the segments of the other.
5. Angles in the same segment of a circle are equal.
6. If two planes are perpendicular to a third plane, their line of intersection is perpendicular to that plane and also to their lines of intersection with that plane.[1]

Theodorus of Cyrene.[2] Among those who assisted in preparing the way for scientific mathematics as distinguished from the intuitive form, Theodo'rus of Cyrene deserves at least brief mention. He was a Pythagorean philosopher, and Proclus (*c.* 460) says that he was a little younger than Anaxagoras, who was born *c.* 499 B.C. According to Appuleius (*c.* 150) and Diogenes Laertius (2d century), Plato went to Cyrene to study geometry under Theodorus, possibly learning from him the theory of irrationality, which, as we know, had received attention in the school of Pythagoras.

Theæte'tus[3] **of Athens** was a pupil of Theodorus and of Socrates and is described by Plato as a man of unusual brilliancy.[4] Although his works are lost, there are references in the writings of the ancient historians to show that he discovered a considerable part of elementary geometry and wrote upon solids. Euclid seems to have been indebted to him and to Eudoxus for some of the material used in writing the *Elements*.

With Theætetus may be said to have closed the period which began with Pythagoras and which prepared the way for Plato and his school. Pythagoras made scientific study popular with the leisure class,—or at least he created an influential group of scholars. Without the work of his school, supplemented by the contributions of such schools as the one at Elea, the world would not have been ready for Plato. The period just closing supplied the raw material, and we shall find that Plato furnished the tools for making good use of this supply.

[1] Allman, *Greek Geom.*, p. 114.

[2] Θεόδωρος. Fl. *c.* 425 B.C. Cyrene (Κυρήνη) was a city on the north coast of Africa.

[3] Θεαίτητος. Fl. *c.* 375 B.C.; died 368 B.C. H. Vogt, *Bibl. Math.*, XIII (3), 200.

[4] Allman, *Greek Geom.*. p. 206.

5. Influence of Plato and Aristotle

Plato. To few men can the words of Carlyle be more appropriately applied than to Plato: "In every epoch of the world the great event, parent of all others, is it not the arrival of a Thinker in the world!" For never in all her early history was Greece so desperately in need of men of soul as she was when Plato[1] began his life work. While he was still a young man (404 B.C.) Athens fell before the Spartan forces. The century in which were first produced the great tragedies of Æschylus, Sophocles, and Euripides, which saw the Acropolis adorned with the masterpieces of Ictinus, Phidias, and Callicrates, and which knew Athens under the reign of Pericles the Magnificent,—this century had passed away, and with it had gone forever that glory of the city that appealed to the masses,—the glory of arms, of the drama, of architecture, and of sculpture. The new century was to see a new Athens, dead to the present but filled with intellectual ambitions for the future. Three great names

PLATO

A fanciful portrait. From a drawing by Raphael in the Accademia at Venice. Inserted to show this artist's conception of the philosopher

[1] Πλάτων (*Platon*). Born at Athens, *c.* 430 B.C.; died *c.* 349.

of Athenian citizens has that future preserved, and by them was it powerfully influenced,—the names of Plato, Aristotle, and Demosthenes.

Plato's Studies. Of the first of these three great leaders of men Cicero has this to say:

> It is reported of Plato that he came into Italy to make himself acquainted with the Pythagoreans, and that when there he made the acquaintance, among others, of Archytas and Timæus[1] and learned from them all the tenets of the Pythagoreans.[2]

It is also said that Plato visited Egypt, partly, no doubt, for purposes of trade, but chiefly that he might acquire knowledge. It may have been from the priests along the Nile, but more likely through the Pythagoreans, that he came to appreciate so highly the value of geometry. At any rate, in later years he is said to have placed above the entrance to his school of philosophy (the Academy) the words, "Let no one ignorant of geometry enter my doors,"[3]—the oldest recorded entrance requirement of a college,—and to have spoken of God as the great geometer.[4]

Plato studied under Socrates and also under a certain Eucleides (Euclid) of Meg'ara,[5] a philosopher who has often been confused with Euclid the geometer (*c.* 300 B.C.). He traveled extensively, visiting not only Egypt and lower Italy but also Sicily and possibly Asia. He thus came in contact with the mathematics and philosophy of these various countries and returned to Athens filled with enthusiasm for an era of splendid thought in place of the era of splendid action which had characterized the century that had just closed.

[1] Not the historian, but a native of Locri. Probably no works of his are extant, although there is one doubtfully attributed to him.

[2] *Tusculan Disputations*, I, 17.

[3] Μηδεὶς ἀγεωμέτρητος εἰσίτω μου τὴν στέγην.

[4] "God eternally geometrizes," Ἀεὶ θεὸς γεωμετρεῖ. This is not in Plato's works, but is stated by Plutarch as due to him. Plutarch, *Convivalium Disputationum libri novem*, viii, 2, ed. Didot (Paris, 1841).

[5] A Greek city.

Of his philosophy it is unnecessary to speak, since this has little bearing upon the problem in hand, but in the field of mathematics his great contribution was to the underlying principles of the science, including the method of analysis.

Plato's Interest in Arithmetic. In the study of numbers he, like all the ancient philosophers, was interested in arithmetic rather than logistic. In his *Republic* he says that the science has a double use, military and philosophical.

For the man of war must learn the art of numbers or he will not know how to array his troops;[1] and the philosopher also, because he has to arise out of the sea of change and lay hold of true being, and therefore he must be an arithmetician. . . . Arithmetic has a very great and elevating effect, compelling the mind to reason about abstract number.

Mysticism of Numbers. One thing that particularly interested Plato was the mysticism of numbers. In his *Republic* (Book VIII) he speaks in an obscure fashion of a certain mystic number, but does not make it clear what this number is. He calls it "the lord of better and worse births," and subsequent writers have often tried to find exactly what he meant. One theory is that 60^4, or 12,960,000, is the Platonic number. This number played an important part in the mysticism of the Hindus and the Babylonians, and it is possible that Pythagoras found it on the banks of the Euphrates, if he really studied there, and that he took it with him to Crotona, passing it on to his disciples, who, in turn, told it to Plato and his followers.

Although Plato esteemed the science of numbers highly,[2] he gives us no information concerning the way it was taught in his school or what it included. We are about as ignorant of the subject as presented by him, and of the ground it covered, as we are of the ancient logistic.[3]

[1]Evidently referring to the square, heteromecic, and triangular numbers described in Volume II, Chapter I.

[2]*Laws*, V.

[3]P. Tannery, "L'éducation platonicienne: L'arithmétique," in the *Revue scientifique*, XI (1881), 287.

Plato on Geometry. More than any of his predecessors Plato appreciated the scientific possibilities of geometry, of which more will be said in Volume II of this work. By his teaching he laid the foundations of the science, insisting upon accurate definitions, clear assumptions, and logical proof. His opposition to the materialists, who saw in geometry only what was immediately useful to the artisan and the mechanic, is made clear by Plutarch (1st century) in his *Life of Marcellus*. Speaking of the use of mechanical appliances in geometry, Plutarch remarks upon "Plato's indignation at it and his invections against it as the mere corruption and annihilation of the one good of geometry, which was thus shamefully turning its back upon the unembodied objects of pure intelligence." That Plato should hold the view here indicated is not a cause for surprise. The world's thinkers have always held it. No man ever created a mathematical theory for practical purposes alone. The applications of mathematics have generally been an afterthought.[1]

Immediate Followers of Plato. Among the followers of Plato was his nephew, Speusip′pus,[2] who accompanied him on his third journey to Syracuse and succeeded him as head of the Academy (347–339 B.C.). He wrote upon Pythagorean numbers,—integers like 3, 4, 5, which represent the sides and hypotenuse of a right-angled triangle. He also wrote upon proportion. We get some information concerning him from an anonymous work of uncertain date called the *Theologumena*, which is also the title of a lost work by Nicomachus (*c.* 100 A.D.). In this work it is related that Speusippus was the son of Potone, the sister of Plato, and that he "ceased not to study with diligence" the lessons of the Pythagoreans, and especially of Philolaus. The work also states that he treated with rare elegance the subjects of linear, polygonal, plane, and solid numbers.[3]

[1] On the general subject of mathematics in Plato's time see B. Rothlauf, *Die Mathematik zu Platons Zeit und Seine Beziehungen zu ihr* (Munich, 1878); Heath, *History*, I, 284.

[2] Σπεύσιππος.

[3] See Volume II, Chapter I.

Of the minor followers of Plato mention should be made of Leod'amas of Thasos,[1] who is referred to by Proclus (*c.* 460) and Diogenes Laertius (2d century) and is said to have made use of the analytic method of proof. There was also Philip'pus Medmæ'us,[2] an astronomer and geometer of Medma, or Mesma, in Magna Græcia, who, under the guidance of Plato, took up the study of mathematics.[3] Thymaridas, who devised a rule for solving simultaneous linear equations, seems to have lived about this time.

Eudox'us of Cnidus,[4] at one time a pupil of Plato, achieved eminence in astronomy, geometry, medicine, and law.[5] It is said that he introduced the study of spherics (mathematical astronomy) into Greece and made known the length of the year as he had found it given in Egypt. He was the first of the Greeks, so far as is known, to give a description of the constellations. Strabo asserts that the observatory of Eudoxus still existed at Cnidus in his time, that is, about the beginning of the Christian era. Seneca says that he brought from Egypt to Greece the theory of the motions of the planets; Aristotle records that he made separate spheres for the stars, sun, moon, and planets; and Archimedes says that he found the diameter of the sun to be nine times that of the earth and showed that a pyramid is one third of a prism of the same base and the same altitude, and similarly for a cone and cylinder. For the mensuration of the cone and cylinder he probably developed the method of exhaustion as a rigorous theory.[6] Vitruvius gives him credit for a new form of sundial called the spider's web,[7] which may, however, have been an astrolabe. Because of a note, possibly due to Proclus, he is often credited with having written a work on proportion which finally became Book V of Euclid, but for this statement there is no definite historical

[1] Λεωδάμας. Fl. *c.* 380 B.C. [2] Φίλιππος ὁ Μεδμαῖος. Born *c.* 375 B.C.

[3] He is also known as Opuntius. H. Vogt, *Bibl. Math.*, XIII (3), 193, 195.

[4] Εὔδοξος. Born *c.* 408 B.C.; died *c.* 355. See Allman, *Greek Geom.*, p. 128; Gow, *Greek Math.*, p. 183; Heath, *History*, I, 322.

[5] Diogenes Laertius, VIII, 86. [6] See Allman, *Greek Geom.*, pp. 96, 139.

[7] Ἀράχνη. Part of the astrolabe resembles a web.

sanction.[1] Our principal knowledge of Eudoxus and his work comes from an astronomical poem written by Ara'tus,[2] and from a commentary of Hipparchus upon it.[3]

Menæch'mus[4] was a pupil of Eudoxus and a friend of Plato,[5] and possibly it is to him that we owe the first treatment of conics. It is said that Alexander the Great was his pupil and that he asked that geometry be made more simple for him; whereupon Menæchmus replied: "O King, through the country there are private and royal roads, but in geometry there is only one road for all."[6] The conic sections which Proclus (*c.* 460) says were considered by him were probably the "Menæchmian Triads" of Eratosthenes (*c.* 230 B.C.). It is said that he obtained them by cutting cones by planes perpendicular to an element,—the parabola from a right-angled cone, the hyperbola from an obtuse-angled cone, and the ellipse from an acute-angled cone. A friend of his, Theudius of Magnesia, wrote a textbook on geometry.

Deinos'tratus, or Dinostratus,[7] was a brother of Menæchmus. He is known chiefly for his study of the quadratix, a curve already invented by one Hippias, very likely Hippias of Elis. This curve enabled him to square a circle.[8]

Xenoc'rates,[9] a native of Chalcedon, was a friend of Plato and Aristotle and was prominent both as a philosopher and as a diplomat. Besides various works on philosophy and government he wrote on physics, geometry, arithmetic, and astrology.

[1] For a discussion of the matter, see Allman, *Greek Geom.*, p. 136; Sir T. L. Heath, *The Thirteen Books of Euclid's Elements*, 3 vols., II, 112 (Cambridge, 1908) (hereafter referred to as Heath, *Euclid*).

[2] Ἄρατος. Fl. *c.* 270 B.C. The poem was the *Phaenomena* (Φαινόμενα), and certain fragments were preserved by Hipparchus.

[3] Ἵππαρχος. Fl. *c.* 150 B.C. J. B. J. Delambre, *Histoire de l'astronomie ancienne*, I, 106 (Paris, 1817). This poem of Aratus was first printed at Venice in 1499.

[4] Μέναιχμος. Fl. 365–350 B.C. See *Bibl. Math.*, XIII (3), 194.

[5] See Allman, *Greek Geom.*, p. 153; Max C. P. Schmidt, "Die Fragmente des Mathematikers Menaechmus," in *Philologus*, XLII (1884), p. 77; Gow, *Greek Math.*, p. 185; Heath, *History*, II, 110.

[6] The story is due to Stobæus, a late Greek writer, *c.* 500. It is also related of Euclid and King Ptolemy.

[7] Δεινόστρατος. Fl. *c.* 350 B.C.

[8] The details are considered in Volume II, Chapter V. See also Allman, *Greek Geom.*, p. 180; Gow, *Greek Math.*, p. 187.

[9] Ξενοκράτης. Born *c.* 396 B.C.; died 314.

Plutarch (1st century) tells us that he took the soul as a "self-moving number," and deified unity and duality,[1] speaking of the former as the first male existence, ruling in heaven, as father and Zeus, as uneven number and spirit; and duality as the first female, the mother of the gods, and the soul of the universe which reigns over the world,—all of which theory shows the Pythagorean influence. He also assumed the existence of indivisible lines and spoke of them as the elements of certain Platonic triangles, perhaps with some intuition of an infinitesimal calculus. He followed Speusippus as head of the Academy and wrote a history of geometry in five books, which, like his other works, is lost.

Ar'istotle[2] studied under Plato at Athens, and his diligence and brilliancy led the latter to call him the "intellect of the school."[3] He became one of the instructors of Alexander the Great, and later returned to Athens and founded the Peripatetic School of philosophy, probably so called from the place where he taught.[4] He was a voluminous writer, but although many of his works are extant the major part are lost. His interest in the mathematical sciences lay chiefly in their applications to physics. He speaks of mathematics as standing halfway between physics and metaphysics. He wrote two works of a mathematical nature, one on indivisible lines and the other on mechanical problems. Both have been edited and printed. We know that, contrary to the doctrines of the Pythagoreans, he advocated the separation of arithmetic and geometry. In his systematizing of logic he contributed indirectly to the great work of Euclid. To him, too, we owe the first known definition of continuity: "A thing is continuous when of any two successive parts the limits at which they touch are one and the same and are, as the word implies, held together."[5] Aristotle was also interested in the historical development of science, and this seems to have influenced the work of his disciples in

[1] Μονάς and δυάς.

[2] Ἀριστοτέλης. Born at Stageira (Stagira), the present Stavro, 384 B.C.; died at Athens, 322.

[3] Νοῦς τῆς διατριβῆς.

[4] Ὁ περίπατος.

[5] Gow, *Greek Math.*, p. 188.

gathering materials for the history of mathematics. Among those whose interests led them into this field was Theophras'tus,[1] a pupil of Plato as well as of Aristotle. He wrote on philosophy, oratory, poetry, botany, physics, politics, and mathematics, but his works are known chiefly from fragments.[2]

Eudemus. Eude'mus[3] of Rhodes, another disciple of Aristotle, who flourished *c.* 335 B.C., was also much interested in the history of mathematics. Most of his works are lost, but certain fragments remain and serve to throw considerable light upon the mathematics of the Aristotelian school. It seems to have been to his care that we are indebted for the preservation of certain works of Aristotle.

Dicæarchus. It is probable that Dicæar'chus[4] of Messina, a city just north of Syracuse, in Sicily, was also a disciple of Aristotle, although we know little of his life. He seems to have flourished *c.* 320 B.C. and to have died *c.* 285 B.C. His work in mathematics was connected chiefly with mensuration as applied to geography. There was another philosopher by the same name, a Pythagorean, whom Iamblichus (*c.* 325) quotes as having contributed to the history of mathematics, but his works are not extant.

Autolycus. Among the contemporaries of Aristotle should probably be included, although we are uncertain as to the date, the astronomer Autol'ycus.[5] Nothing is known of his personal history except that he wrote two treatises on astronomy, both of which are extant. These are the most ancient mathematical texts that have come down to us from the Greeks. The first is on the motion of the sphere, and the second is on the risings and settings of the fixed stars, and in each he shows considerable skill in geometry.

Aristæ'us,[6] known as Aristæus the Elder, is mentioned by Pappus, a mathematician of the 4th century, as one of the three geometers of the Greeks who were skilled in that branch of

[1] Θεόφραστος. Fl. *c.* 350 B.C.
[2] There are various editions of his works.
[3] Εὔδημος.
[4] Δικαίαρχος.
[5] Αὐτόλυκος. Fl. *c.* 330 B.C.
[6] Ἀρισταῖος. Fl. *c.* 320 B.C.

geometry which treats of analysis, the other two being Euclid and Apollonius. Pappus also relates that Aristæus wrote five books on solid loci,[1] supplementing five others on the elements of conics. Possibly these two works were the same. He also wrote on the five regular solids, and the 13th book of Euclid seems to have owed much to his skill. He was evidently one of those mathematicians of the 4th century B.C. who, inspired by Plato, helped to make possible the works of Euclid and Apollonius.

Callip'pus or Calippus,[2] an astronomer of Cyzicus, was a friend of both Eudoxus and Aristotle. Although not to be looked upon as a geometer, his astronomical observations deserve brief mention, being frequently referred to by Geminus and Ptolemy. The Callippic cycle of 76 years, 940 lunar months, or 27,759 days was such an improvement on the Metonic cycle of 19 years as to have been adopted by ancient astronomers. We have the testimony of Simplicius (6th century) that he was a pupil of Polemar'chus (4th century B.C.), who taught at Cyzicus, and that he lived for a time with Aristotle. Ptolemy tells us that he made astronomical observations on the shores of the Hellespont.

6. The Orient

Orient and Occident. The rise of mathematics in Greece, its remarkable development under the influence of such leaders as Thales, Pythagoras, and Plato, and its distinct characteristics, are such as to make it desirable to consider the Orient and the Occident separately from the time of Euclid until the two were joined by the new intellectual bond established by the Christian missions about the beginning of the 17th century. So little was accomplished in the Orient from 1000 B.C. to 300 B.C., however, that we may properly mention that little in the present chapter. Although each of these two great divisions of the world always influenced the other in developing a system of

[1] Τόποι στερεοί.

[2] Κάλλιππος or Κάλιππος. Fl. c. 325 B.C.

mathematics, the East has always been the East, and the West has always been the West. They have had many points in common, particularly in the application of mathematics to astronomy; but the development of a logical geometry, with all of its far-reaching results, is peculiar to the European peoples, while the less rigid and somewhat more poetic phases of mathematics have generally interested the Asiatic mind. Even the ancients recognized this difference, for Quintilian (*c.* 35–*c.* 96) remarks: "From of old there has been the famous division of Attic and Asiatic writers, the former being reckoned succinct and vigorous, the latter inflated and empty."[1]

China. It is an interesting fact that Egypt developed a worthy type of mathematics before 1000 B.C. and then stagnated, that Babylonia did the same, and that China followed a similar course. Was it that the world's vigor was concentrated in Greece? Had the older civilizations burned out? Or was there some subtle influence that subjected the original seats of mathematical thought to canonical expression instead of progressive action? Whatever the answer, between 1000 and 300 B.C. China produced no great classic in mathematics, unless possibly the *Nine Sections*[2] already mentioned, or the *Wu-ts'ao Suan-king* to be mentioned later, belongs to this period. It was rather in the impetus given to commercial calculation through the introduction of coins in the 7th century B.C., at about the same time as they appeared in Asia Minor, that China made her most noteworthy contribution to the progress of arithmetic. Knife money and spade money appeared *c.* 670 B.C., the coins representing such common articles of value as knives and spades. Circular coins were issued later and became the standard forms in the 3d century B.C. As to the methods used in calculating at this time, we are ignorant, but some mechanical means were probably employed in China as well as in other parts of the ancient world. About 542 B.C. the Chinese are known to have used in their calculations bamboo rods, in size and appearance somewhat like a new lead

[1] *Institutes of Oratory*, Bohn ed., XII, x, 16. [2] *K'iu-ch'ang Suan-shu.*

pencil. About 375 B.C. there appeared the earliest Chinese coins with weight or value inscribed upon them, and thus the monetary material for commercial arithmetic became fairly well perfected.

The Compass. As early as the 4th century B.C. there seems to have existed some kind of instrument for indicating the southern direction, probably the compass. In later literature the *chï-nan-kü*[1] (south-pointing chariot) is mentioned, but what it was is unknown.[2] In works of the 4th century it is ascribed to Huang-ti (2704–2595 B.C.) and is also mentioned as being in use in the reigns of Ki-li (1230–1185 B.C.) and his successor (1185–1135 B.C.)[3]

India. As already stated, we have no authentic records of India before the Mohammedan invasion (7th century), almost our only sources of information being the Vedic literature, the Buddhist sacred books, the heroic poems, such inscriptions as remain on monuments, and the metal land grants. Of these, the later Vedic literature, the heroic poems, and the Buddhist writings are all that give us any knowledge of the mathematics of the period from 1000 to 300 B.C. The Vedic writings probably extend down to about 800 B.C., although the *Vedāngas* ("Limbs for supporting the Veḍa") were written several centuries later. The dates of the *Śulvasūtra* period are unknown. Taking the opinions of various scholars and forming a rough estimate, we may put the ritualistic rules of the *Śulvasūtras* in the five centuries just preceding our era. The rules which have any mathematical interest relate indirectly to the proportions of altars in the temples. They include a statement about Pythagorean numbers, that is, numbers satisfying the relation $x^2 + y^2 = z^2$, and imply a statement of the Pythagorean Theorem itself. There is no reason for believing, however, that the Hindus had the slightest idea of the nature of a geometric proof. There is also evidence of a knowledge of irrationals and

[1] *Chï* means to point with the finger; *nan* means south; and *kü* means chariot.
[2] F. Hirth, *Ancient History of China*, p. 129.
[3] *Ibid.*, p. 135.

of an understanding of the uses of the gnomon.[1] The *Śulvasūtras* also state that the diagonal of a unit square is equal to

$$1 + \frac{1}{3} + \frac{1}{3 \cdot 4} + \frac{1}{3 \cdot 4 \cdot 34},$$

or 1.4142156. The area of the circle is asserted[2] to be

$$\left(\frac{1}{8} + \frac{1}{8 \cdot 29} - \frac{1}{8 \cdot 29 \cdot 6} + \frac{1}{8 \cdot 29 \cdot 6 \cdot 8}\right)^2 d^2.$$

Mathematics in the *Śulvasūtras*. The *Śulvasūtras* were changed more or less by such commentators as Āpastamba, Baudhāyana, and Kātyāyana. The following statements from the Baudhāyana edition show the style:[3]

"The chord stretched across a square produces an area of twice the size."[4]

"The diagonal of an oblong produces by itself both the areas which the two sides of the oblong produce separately."[5]

The *Lalitavistara*, one of the sacred books of the Hindus, speaks of the arithmetical prowess of the Buddha.[6] Sir Edwin Arnold has put the statement in verse in his *Light of Asia*.[7]

[1] L. von Schroeder, *Pythagoras und die Inder*, Leipzig, 1884; and *Indiens Literatur und Cultur*, Leipzig, 1887; H. Vogt, "Haben die alten Inder den Pythagoreischen Lehrsatz und das Irrationale gekannt?" *Bibl. Math.*, VII (3), 6; A. Bürk, "Das Āpastamba-Śulba-Sūtra," *Zeitschrift der deutschen Morgenländischen Gesellschaft*, LV, 543; LVI, 327; M. Cantor, "Ueber die älteste Indische Mathematik," *Archiv der Math. und Physik*, VIII (3), 63; B. Levi, "Osservazioni e congetture sopra la geometria degli Indiani," *Bibl. Math.*, IX (3), 97; Smith and Karpinski, *The Hindu-Arabic Numerals*, p. 13 and bibliographical notes throughout (hereafter referred to as Smith-Karpinski), Boston, 1911; G. R. Kaye, *Indian Mathematics*, Calcutta, 1915.

[2] Thibaut in the *Journal of the Royal Asiatic Soc. of Bengal*, XLV (1875), p. 227; Dutt, *History of Civ. in Anc. India*, I, 271.

[3] The translation is Dr. G. Thibaut's. See his memoirs "On the Śulvasūtras," *Journ. Royal Asiat. Soc. Bengal*, XLIV (1874); "The Baudhāyana Śulvasūtra," *The Pandit*, 1875; "The Kātyāyana Śulvasūtra," *The Pandit*, 1882; G. Milhaud, "La géométrie d'Apastamba," *Revue générale d. sci.*, XXI, 512.

[4] That is, the square on a diagonal of a square is twice the original square.

[5] That is, the square on a diagonal of a rectangle is equal to the sum of the squares on the two sides; essentially the Pythagorean Theorem.

[6] The date of the birth of Buddha is often placed at 543 B.C., but in Burma, Siam, and Ceylon it is usually given as 80 years earlier, that is, 623 B.C.

[7] It is given in Smith-Karpinski, p. 16. See also the translation of the Mahābhārata in E. Arnold, *Indian Poetry*, London, 6th ed., 1891.

In all this there is nothing that is definite, but there is enough to show that mathematics was not limited to the meager needs of trade and that it was related, as with all thoughtful peoples, to the higher life.

Mesopotamia. Just before the period of which we are speaking the Arameans[1] established a flourishing dominion to the north of the Hebrew territory of Palestine. Their mercantile interests extended into the ancient cities of Assyria, as is proved by such bronze weights as the one shown on page 39.

HEBREWS PAYING TRIBUTE TO SHALMANESER III, KING OF ASSYRIA

This was about 850 B.C. The original is now in the British Museum. From Breasted's *Ancient Times*

By 1000 B.C. they had developed a system of alphabetic writing, and their bills of exchange were known in Mesopotamia, Persia, and India, as those of Babylon had been known before them. All through this early period the records of taxes show that this form of applied arithmetic was ever present.

In the 8th century B.C. the Assyrians subdued Mesopotamia and much of the territory to the west and became the dominating power in Western Asia. They maintained the first great army equipped with weapons of iron and by this means held a large territory in subjection. Militarism, however, eventually proved a weakness, and they in turn succumbed to

[1] Or Syrians, as they are often called.

the power of the Kaldi,—Semitic nomads already mentioned, —who came from the South and who, known to us as the Chaldeans, finally became the ruling power in Mesopotamia.

Contributions of Babylonia and Assyria. In the midst of all these changes two steps in the history of mathematics deserve special mention: (1) the Arameans brought the arithmetic of commerce to a higher standard, and (2) the Babylonians and Chaldeans extended the earlier work in astronomy. The science of astrology had by this time developed as a potent force in civilization, and astronomy had become recognized as the science par excellence. Ptolemy the astronomer (*c.* 150) refers to a Chaldean record of a lunar eclipse of 721 B.C. and to the division of the circle into 360°. The recognition of a zodiac of twelve signs, and the study of the courses of the planets, about 600 B.C., are further evidences of the interest of the Chaldean astronomers in this phase of applied mathematics.

As to astrology,—that daughter of astronomy who nursed her own mother, as Kepler writes of it,—there are various tablets of this period which show in what high esteem it was held. In general they are reports of the following kind to the king: "Two or three times of late we have searched for Mars but have not been able to see him. If the king, my master, asks me if this invisibility presages anything, I reply that it does not."[1]

Science reached its highest point in the reign of Nebuchadnezzar, which closed in 561 B.C. It is true that we have lists of the planets dating from 523 B.C. and from other years,[2] statements of the irregular insertion of intercalary days at about the same period, and a definite recognition of the leap year between 383 and 351 B.C.,[3] but the mathematics of Mesopotamia practically ceased to exist with the decay of the Chaldean power.

[1] Bigourdan, *L'Astronomie*, p. 29; R. C. Thompson, *The Reports of the Magicians and Astrologers of Nineveh and Babylon in the British Museum*, London, 1900.

[2] F. Hommel, "Die Babylonisch-Assyrischen Planetenlisten," in the *Hilprecht Anniversary Volume*, p. 170.

[3] F. H. Weissbach, "Zum Babylonischen Kalender," in the Hilprecht volume above cited, p. 282.

TOPICS FOR DISCUSSION

1. Influences favorable to the development of mathematics among the Greeks from 1000 B.C. to 300 B.C.

2. The nature of logistic and of arithmetic and the reasons for their treatment as unrelated subjects.

3. The advantages of the Greek method of treating arithmetic from the geometric standpoint, particularly in relation to the nature of irrational numbers.

4. The influence of Thales upon the subsequent development of mathematics in Greece.

5. The influences which contributed to the making of the character of Pythagoras.

6. The influence of Pythagoras upon mathematics in general, and particularly upon geometry and the theory of irrationals.

7. Music as a branch of ancient mathematics.

8. Beginnings of a kind of infinitesimal calculus in Greece, particularly with respect to the method of exhaustion.

9. Types of geometric propositions that attracted special attention in this period, thus showing the nature of geometry before the time of Euclid.

10. The influence of Plato upon mathematics in general and upon geometry in particular.

11. The influence of astronomy upon mathematics in Greece, particularly with reference to geometry and a primitive trigonometry.

12. The early steps in the invention of conic sections.

13. The study of higher plane curves among the Greeks in the period under discussion.

14. The influence of Aristotle upon mathematics in general, and particularly upon its applications.

15. Nature of mathematics in the Orient in this period.

16. General distinction between the mathematics of Greece and that of the East.

17. Mysticism of numbers as found in the Orient, in Mesopotamia, and in the West.

18. Early studies in the history of mathematics among the Greeks.

19. The recognition of the sphericity of the earth by various leading Greek philosophers.

20. The nature of the mathematics of the *Śulvasūtras.*

CHAPTER IV

THE PERIOD FROM 300 B.C. TO 500 A.D.

1. THE SCHOOL OF ALEXANDRIA

Chronological and Geographical Considerations. The reason why the limitations of 300 B.C. and 500 A.D. are arbitrarily chosen for this chapter is that these dates mark approximately the period of influence of the greatest mathematical school of ancient times, the School of Alexandria. Moreover, the first of these dates is approximately that of Euclid, the world's greatest textbook writer, and the second is that of Boethius, whom Gibbon characterizes as "the last of the Romans whom Cato or Tully could have acknowledged for their countryman."

Within this period Greek civilization passed away, Rome rose and fell, and the ancient mathematics of the West descended from its most exalted to its most debased estate. We have, therefore, the most significant period of ancient mathematical history, at least in the matter of actual production, and we have the Mediterranean world, probably the most interesting of all ancient civilizations.

The School of Alexandria. The greatest mathematical center of ancient times was neither Crotona nor Athens, but Alexandria. Here it was, on the site of the ancient town of Rhacotis, in the Nile Delta, that Alexander the Great founded a city worthy to bear his name. Upon the death of the great Macedonian conqueror (323 B.C.) the vast domain which he had brought under his control was broken up. After the death of Antig'onus, his ablest general, the empire fell into three parts. Alexander's friend and counselor, and possibly his blood relative, Ptol'emy[1] So'ter (Ptolemy the Preserver), came into

[1] Πτολεμαῖος; Latin, *Ptolemaeus.*

possession of Egypt, Antigonus the younger laid claim to Macedonia, while Seleu'cus took for his part the provinces of Asia. Under Ptolemy's benevolent reign (323–283 B.C.) Alexandria became the center not only of the world's commerce but also of its literary and scientific activity.[1] Here was established the greatest of the world's ancient libraries and its first international university. Cardinal Newman, in speaking of these two features, says with poetic feeling that "as the first was the embalming of dead genius, so the second was the endowment of living." Here were trained more great mathematicians than in any other scientific center of the ancient world. With Alexandria are connected the names of Eu'clid, Archime'des, Apollo'nius, Eratos'thenes, Ptol'emy the astronomer, He'ron, Menela'us, Pap'pus, The'on, Hypa'tia, Diophan'tus, and, at least indirectly, Nicom'achus. Today, however, not the slightest trace remains of the famous library and museum, and even their exact locations are merely conjectural.

2. Euclid

Euclid.[2] Of all the great names connected with Alexandria, that of Euclid is the best known. He was the most successful textbook writer that the world has ever known, over one thousand editions of his geometry having appeared in print since 1482,[3] and manuscripts of this work having dominated

[1] For a summary of the causes of its rise and a description of its library see W. Kroll, *Geschichte der klassischen Philologie*, p. 12 (Leipzig, 1908); hereafter referred to as Kroll, *Geschichte*.

[2] Εὐκλείδης. Fl. *c.* 300 B.C. The leading work upon Euclid and his *Elements* is that of Sir T. L. Heath, *The Thirteen Books of Euclid's Elements*, 3 vols., Cambridge, 1908. The best Greek and Latin edition of Euclid's works is Heiberg and Menge, *Euclidis Elementa*, Leipzig, 1883–1916. Of the many works and articles on the life of Euclid the following may be consulted to advantage: A. De Morgan, "Eucleides," in Smith's *Dict. of Greek and Roman Biog.*; T. Smith, *Euclid, his life and system*, New York, 1902; W. B. Frankland, *The Story of Euclid*, London, 1902; G. B. Biadego, "Euclide e il suo Secolo," in Boncompagni's *Bullettino*, V, 1; P. Tannery, *Pour l'histoire de la science hellène*, Appendix II (Paris, 1887); Gow, *Greek Math.*, p. 195; M. C. P. Schmidt, *Realistische Chrestomathie aus der Litteratur des klass. Altertums*, I, 1 (Leipzig, 1900), hereafter referred to as Schmidt, *Chrestomathie*; Pauly-Wissowa's *Real-Encyclopädie*, Vol. VI (Stuttgart, 1909), has an extensive article on Euclid.

[3] P. Riccardi, *Saggio di una Bibliografia Euclidea*, p. 4. Bologna, 1887.

the teaching of the subject for eighteen hundred years preceding that time. He is the only man to whom there ever

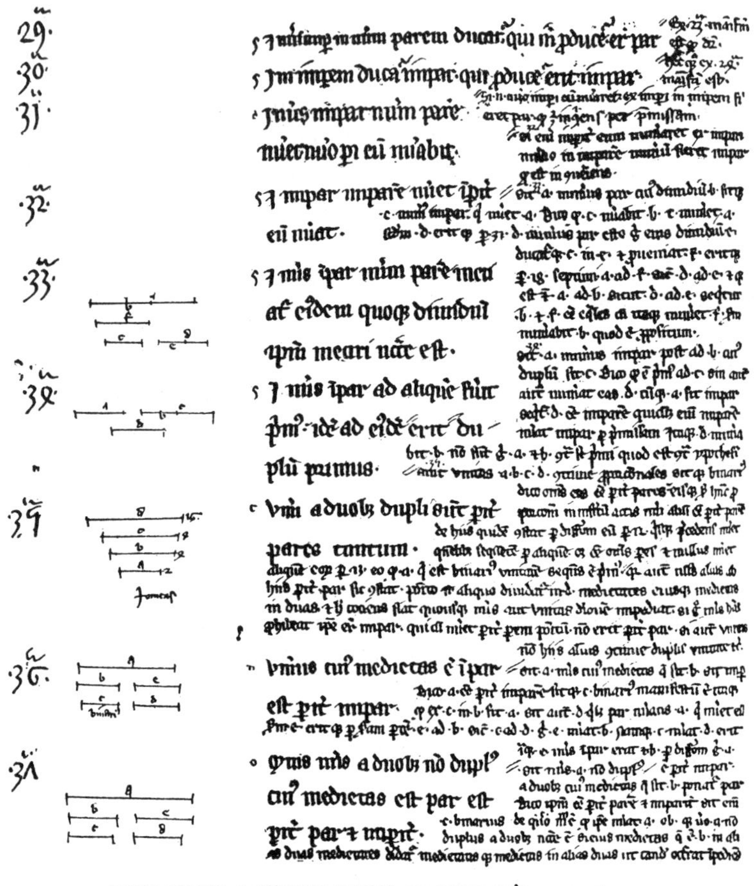

PAGE FROM A TRANSLATION OF EUCLID'S ELEMENTS

This manuscript was written *c.* 1294. The page relates to the propositions on the theory of numbers as given in Book IX of the *Elements*. The first line gives Proposition 28 as usually numbered in modern editions

came or ever can come again the glory of having successfully incorporated in his own writings all the essential parts of the accumulated mathematical knowledge of his time.

Of the life of Euclid nothing definite is known. A recent writer expresses the belief that the evidence indicates that he was born as early as 365 B.C. and that he wrote the *Elements* when he was about forty years old,[1] but we have no precise information as to his birthplace, the dates of his birth and death, or even his nationality. It was formerly asserted that he was born at Meg'ara, a Greek city, but it is now known that Euclid of Megara was a philosopher who lived a century before Euclid of Alexandria. The Euclid in whom we are interested may have been a Greek or he may have been an Egyptian who came to the Greek colonial city of Alexandria to learn and to teach. There is some reason for believing that he studied in Athens, but in the way of exact information we have nothing concerning him. Under any circumstances, the period of his real influence upon mathematics begins about 300 B.C.

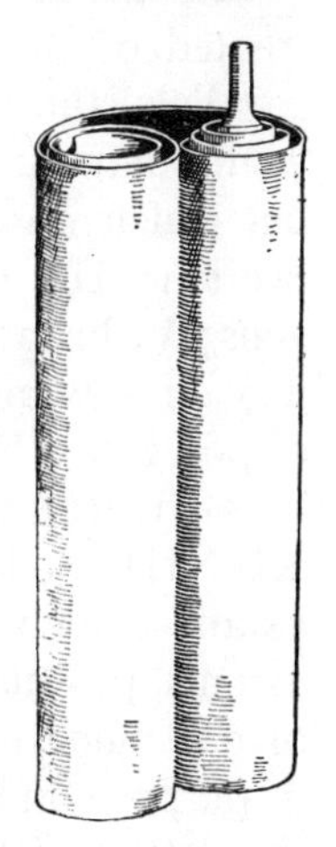

PARCHMENT ROLL

Upon such a roll a "book" of Euclid was written

The Books of Euclid. As was the custom in the days when all treatises were written on long strips of parchment or papyrus, the separate parts of his work were rolled up and called volumes, from a Latin root meaning to roll. Because of the difficulty of handling large rolls, they were cut into smaller rolls known as *biblia* (*βιβλία*, bibles), a word meaning books. Hence we have the books of Homer, the books of geometry, the books of the Bible, and so on. Euclid's greatest work is known as the *Elements*,[2] and in the books relating to geometry there was arranged that mass of material treating of circles, rectilinear figures, and ratios which had accumulated during the two centuries following the death of Pythagoras. No doubt there were also many propositions that were original with Euclid; but the feature which made his treatise famous, and which

[1] H. Vogt, *Bibl. Math.*, XIII (3), 193; but see Heath, *History*, I, 354.

[2] *Στοιχεῖα*. So great was Euclid's fame that he was known to the Greeks as *ὁ στοιχειωτής*, "teacher of the *Elements*."

accounts for the fact that it is the oldest scientific textbook still in actual use, is found in its simple but logical sequence of theorems and problems. It has been said of Shakespeare that he "took the stillborn children of lesser men's brains and breathed on them the breath of life,"—and so it was with Euclid.

Contents of the *Elements*. The various books of the *Elements* treated of the following topics, respectively: I, Congruence, parallels, the Pythagorean Theorem; II, Identities which we would now treat algebraically, like $(a+b)^2=a^2+2ab+b^2$, but which were then treated geometrically; areas; the Golden Section; III, Circles; IV, Inscribed and circumscribed polygons; V, Proportion treated geometrically; in part, a geometric way of solving fractional algebraic equations; VI, Similarity of polygons; VII–IX, Arithmetic (the ancient theory of numbers) treated geometrically; X, Incommensurable magnitudes; XI–XIII, Solid geometry. We have in this text the earliest extant evidence of a systematic arrangement of definitions, axioms, postulates, and propositions. Euclid differs from most of our modern writers on geometry in his greater seriousness of purpose, in his desire to be more rigorous, and in the following details of treatment: He has no intuitive geometry as an introduction to the logical; he uses no algebra as such; he demonstrates the correctness of his constructions before using them, whereas we commonly assume the possibility of constructing figures and postpone our proofs relating to constructions until we have a fair body of theorems; he does not fear to treat of incommensurable magnitudes in a perfectly logical manner; and he has no exercises of any kind.

Euclid's Other Works. Euclid wrote a number of other works. Among them are the *Phænomena*,[1] dealing with the celestial sphere and containing twenty-five geometric propositions; the *Data*; possibly a treatise on music;[2] and works on optics,[3]

[1] Φαινόμενα.

[2] Εἰσαγωγὴ 'Αρμονική or else the Κατατομὴ Κανόνος, both doubtful. See Heath, *Euclid*, I, 17.

[3] 'Οπτικά.

porisms,[1] and catoptrics.[2] He also wrote a work on divisions of figures, touching upon questions which arise, for example, in surveying.[3]

Immediate Effect of Euclid's Work. The natural effect of Euclid's work on geometry was to give rise to the feeling that elementary geometry had attained to perfection and that the next step in the progress of mathematics must be in the direction of some kind of higher geometry or else in the field of mensuration. As a result, mathematics pursued both courses, at first with little effect, as was the case with the predecessors of Euclid, and then, when another genius appeared, with great rapidity.

Minor Writers. There were, for example, at first such minor writers as Co'non[4] of Samos, who, influenced by his observations of the coiled basketry work of the Egyptians, may have invented the spiral of which Archimedes developed the properties. He is also mentioned by Apollonius (*c.* 225 B.C.) as having studied the number of points of intersection of two conics. There was also Nicot'eles[5] of Cyrene, possibly a student in Alexandria, of whom Apollonius speaks as his predecessor in the study of conics. Still another writer of influence appeared in the person of the astronomer Aristar'chus,[6] a native of Samos but a teacher at Alexandria. It was he who first showed how to find, by means of the Pythagorean triangle, the relative distances of the sun and the moon from the earth, and for nearly two thousand years no better plan was known. His instruments of observation were such as to make his result far from being even approximately correct.[7] His greatest glory, however, lies

1 Relating to methods of solution. See M. Breton, in *Journal de Math. pures et appliquées*, XX (1), III (2).

2 Κατοπτρικά.

3 Περὶ Διαιρέσεων βιβλίον. R. C. Archibald, *Euclid's Book on Divisions of Figures*, Cambridge, 1915.

4 Κόνων. Fl. *c.* 260 B.C.

5 Νεικοτέλης or Νικοτέλης. Fl. *c.* 250 B.C.

6 Ἀρίσταρχος. Born *c.* 310 B.C.; died *c.* 230. See Sir T. L. Heath, *Aristarchus of Samos*, Oxford, 1913; *The Copernicus of Antiquity*, London, 1920.

7 Carl Snyder, *The World Machine*, chap. vii, "Aristarchus and the distance and grandeur of the sun" (London, 1907); Bigourdan, *L'Astronomie*, 252.

in the fact that he was the first to place the sun in the center of the universe, asserting that the earth and the other planets revolved about it, thus anticipating Copernicus by seventeen centuries. In the field of arithmetic he found $\sqrt{2}$, possibly by a method analogous to that of continued fractions.[1]

There are also extant various papyri of the Ptolemaic period containing information about the financial problems of Egypt. These problems relate chiefly to taxes and the cost of various commodities, but they add nothing to our information as to methods of calculation in ancient times.[2]

Men like Conon were merely the usual heralds calling out the approach of genius. The three men whose advent they heralded were Archimedes, Apollonius, and Heron. Before speaking of Archimedes, however, reference should be made to a scholar whose interests were so scattered as to make his contributions to pure mathematics of relatively little importance. This man was the poet, librarian, arithmetician, and geographer Eratos'thenes.

3. Eratosthenes and Archimedes

Eratosthenes[3] lived some years after Euclid, and was one of the greatest scholars of Alexandria.[4] His admirers exaggerated his attainments by calling him "the second Plato,"[5] and some have thought that his nickname "Beta" signified that he was the second of the wise men of antiquity, the Greek letter *beta* standing for two. Others have said that he was called by this name because his room in the university bore the number two. But whether or not his followers ranked him second among the wise men of Greece, we are justified in calling him the first prominent geographer of antiquity. He was educated at Athens

[1] P. Tannery, *Mémoires de Bordeaux*, V (2), 237; IV (3), 79.

[2] For a bibliography, a list of the papyri, and a summary of the information available, see H. Maspero, *Les Finances de l'Égypte sous les Lagides*, Paris, 1905.

[3] Ἐρατοσθένης. Born at Cyrene, *c.* 274 B.C.; died *c.* 194.

[4] Schmidt, *Chrestomathie*, I, 29, 114. See also *Bibl. Math.*, XIII (3), 193.

[5] B. Baldi, *Cronica de Matematici*, p. 29 (Urbino, 1707), hereafter referred to as Baldi *Cronica*; Heath, *History*, II, 104.

and is known to have taught at Alexandria after *c.* 240 B.C., to have been librarian of the university, and to have been a poet of some merit. His contribution to arithmetic was his sieve,[1] a method of sifting out the composite numbers in the natural series, leaving only primes. This he did by writing all the odd numbers and then canceling the successive multiples of each, one after the other, thus: 3, 5, 7, ~~9~~, 11, 13, ~~15~~, 17, 19, ~~21~~, 23, ~~25~~, ~~27~~, 29, 31, ~~33~~, ~~35~~, 37, ~~39~~, . . .[2] Prime numbers

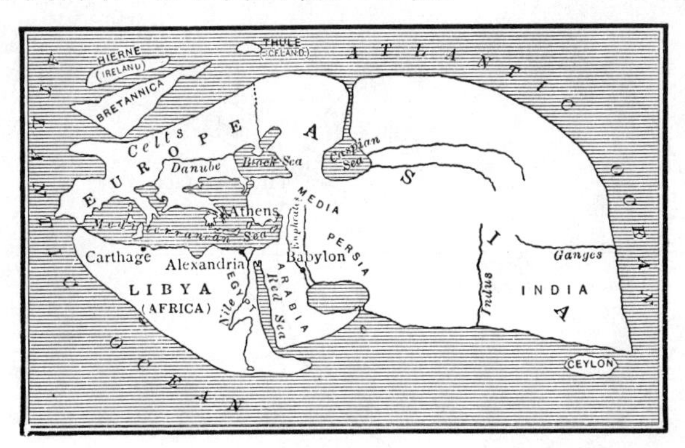

MAP OF THE WORLD ACCORDING TO ERATOSTHENES

This shows the knowledge of the geography of the world in the 3d century B.C. and should be compared with the map of Hecatæus (517 B.C.) shown on page 77. From Breasted's *Ancient Times*

have been studied from that time until the present, but no general formula is yet known for detecting all of them. For example, we do not yet know whether there are an infinite number of primes of the form $x^2 + 1$, whether $2 = x - y$ has an infinite number of prime solutions, or whether a prime number can always be found between n^2 and $(n + 1)^2$.

Earth Measure. Of the mathematical achievements of the Greek astronomers none is more interesting than the measurement of the circumference and diameter of the earth by Eratosthenes, the first noteworthy step in the science of

[1] Κόσκινον. It was called the *Cribrum Arithmeticum* by Latin writers.

[2] M. C. P. Schmidt, *loc. cit.*, I, 114 (Greek text).

geodesy.[1] Learning that the sun at noonday was exactly in the zenith at Syene[2] when it was 7° 12′ south of the zenith at Alexandria, he decided that Alexandria was 7° 12′ north of Syene on the earth's surface. Since the distance was known to be 5000 stadia, and since 7° 12′ = $\frac{1}{50}$ of 360°, he judged the circumference of the earth to be 50 × 5000 stadia, or 250,000 stadia. This result he altered to 252,000 stadia so as to have

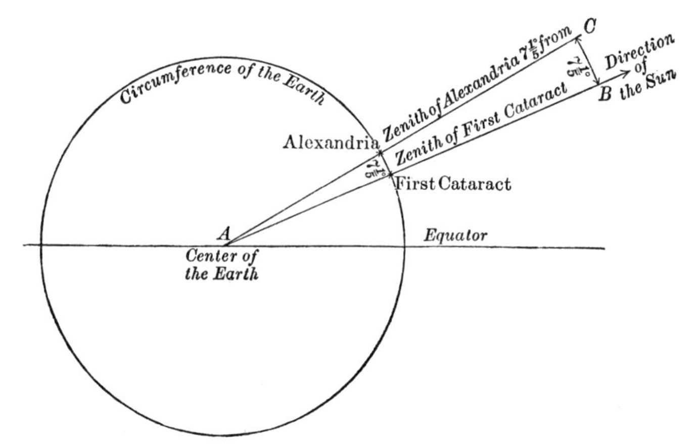

DIAGRAM SHOWING METHOD USED BY ERATOSTHENES IN MEASURING THE EARTH

Eratosthenes found that, when the sun was directly over Syene, at the First Cataract, it was 7° 12′ south of Alexandria. From this he computed the circumference of the earth. From Breasted's *Ancient Times*

700 stadia, a more convenient number, to a degree, and from this he computed the diameter to be the equivalent of 7850 miles, in our system of measure, which is only 50 miles less than the polar diameter as we know it.[3]

Eratosthenes also stated that the distance between the tropics is $\frac{11}{83}$ of the circumference, which makes the obliquity of the

[1] Carl Snyder, *The World Machine*, chap. vi, "Eratosthenes and the earliest measures of the earth" (London, 1907); Schmidt, "Erdmessung des Eratosthenes," Greek text, *Chrestomathie*, II, 105; see also I, 29.

[2] Sy e'ne, Συήνη, the modern Assouan (Arabic from *al* + *Syene*) at the first cataract of the Nile.

[3] The problem of earth measure is more fully treated in Volume II, Chapter V.

ecliptic[1] 23° 51′ 20″. Plutarch tells us that he found the sun to be 804,000,000 stadia from the earth, and the moon to be 780,000 stadia,—results which are remarkably close when we consider the instruments then in use. That the knowledge of geography had increased in the preceding 250 years may be seen by comparing his map with that of Hecatæus (517 B.C.).[2]

In one of his letters Eratosthenes also discussed the problem of the duplication of the cube.

Archime'des[3] was a friend of Eratosthenes and, if the testimony of Plutarch is accepted, was related to King Hiero. Leibniz praised his genius by saying that those who knew his works and those of Apollonius marveled less at the discoveries of the greatest modern scholars.[4] These words are justified, for Archimedes anticipated by nearly two thousand years some of the ideas of Newton and his contemporaries, and in the application of mathematics to mechanics he had no equal in ancient times. One of the Italian historians of mathematics[5] uses the happy phrase that he had "a genius more divine than human," and Pliny calls him "the god of mathematics," a phrase which one of his French translators felicitously renders as "the Homer of geometry."

It is related that Archimedes set fire to the besieging ships in the harbor of Syracuse by the aid of burning mirrors, and there is nothing improbable in the idea that he may have

[1] This term seems first to have been used by Ambrosius Aurelius Theodosius Macrobius, a grammarian of *c.* 400.

[2] On the general subject of the history of mathematical geography consult S. Günther, *Studien zur Geschichte der math. und physik. Geographie*, with extensive series of bibliographies (Halle a. S., 1879).

[3] Ἀρχιμήδης. Born at Syracuse, the modern Siracusa, Sicily, 287 B.C.; died at Syracuse, 212 B.C. Sir T. L. Heath, *Archimedes*, Cambridge, 1897 (hereafter referred to as Heath, *Archimedes*); German translation by Kliem, Berlin, 1914; P. Midolo, *Archimede e il suo tempo*, Syracuse, 1912; C. Snyder, *The World Machine*, chap. x, "Archimedes and the first ideas of gravitation," London, 1907; Schmidt, *Chrestomathie*, III, 64, 100; Heath, *History*, II, 16.

[4] "Qui Archimedem et Apollonium intelligit, recentiorum summorum virorum inventa parcius mirabitur," *Archimedis Opera* (Geneva, 1768), V, 460. The definitive edition is that of Heiberg, Leipzig, 1880–1915.

[5] Baldi, *Cronica*, p. 26: "Hebbe ingegno più divino, che humano."

made them at least untenable by the soldiers.[1] A glance at the map, and particularly at the lesser harbor then in use, and a consideration of the fact that the ships were then hardly larger than our pleasure yachts of today, and that they were all anchored close to the rocky shore, will show that the task was not so great as one might at first suppose. At a time when the breeze was blowing in from the sea, escape would have been difficult, even with oars.

ARCHIMEDES

Conjectural portrait bas-relief in the Capitoline Museum at Rome. Date uncertain

Archimedes and Mechanics. Plutarch, in his life of Marcellus, relates this incident to illustrate the genius of Archimedes in mechanics:

> Archimedes . . . had stated that, given the force, any given weight could be moved; and even boasted . . . that, if there were another earth, by going into it he could remove this one. Hiero being struck with amazement at this, . . . [Archimedes] fixed accordingly upon a ship of burden . . . which could not be drawn out of the dock without great labor and many men; and loading her with many passengers and a full freight, sitting himself the while far off with no great endeavor, but holding the head of the pulley in his hand and drawing the cords by degrees, he drew the ship in a straight line as smoothly and evenly as if she had been in the sea.

[1] On the subject of burning mirrors in Greek literature see Sir T. L. Heath, in *Bibl. Math.*, VII (3), 225.

The Sand Reckoner. Archimedes saw the defects of the Greek number system, and in his *Sand Reckoner*[1] he suggested an elaborate scheme of numeration, arranging the numbers in octads, or the eighth powers of ten. In this work he

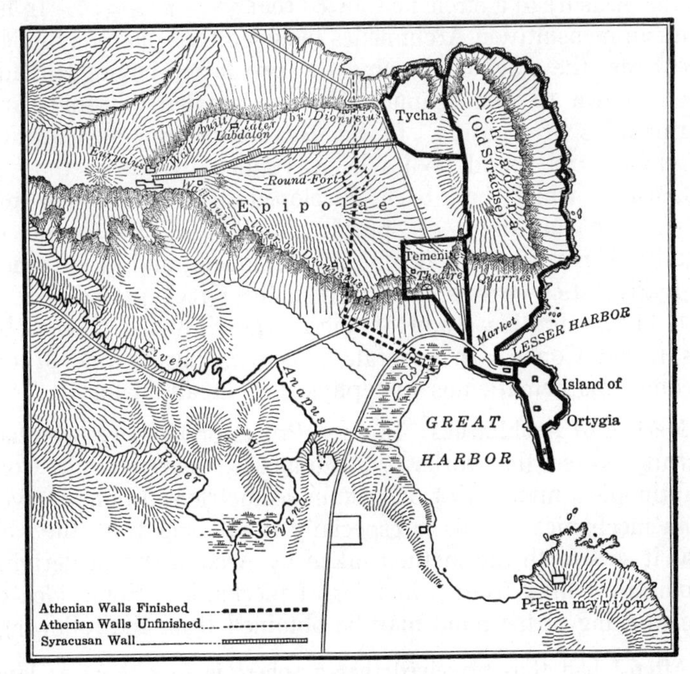

MAP OF ANCIENT SYRACUSE

Showing the general situation at the time of the Third Peloponnesian War and continuing until the time of Archimedes. On the river Anapus here shown papyrus still grows luxuriantly. From Breasted's *Ancient Times*

recognized, in substance, that $a^m a^n = a^{m+n}$, a law that is the basis of our present operations by logarithms.

Other Mathematical Activities. Among his many activities was the summation $\sum_1^n n^2$, the first example of the systematic treatment of higher series of any kind. By the intersection of

[1] Ψαμμίτης (*psammites*); Latin, *arenarius* or *harenarius*.

conics he was able to solve cubic equations which we should now write in the form $x^3 \mp ax^2 \pm b^2c = 0$. He also succeeded in squaring a parabola,[1] that is, in finding the area of a segment, showing that it is two thirds of a circumscribed parallelogram. In the measure of a circle he showed that $3\frac{1}{7} > \pi > 3\frac{10}{71}$. In his work in mensuration Archimedes included the sphere, cylinder, and cone, the rules concerning the two latter having already been known to Menæchmus. He also studied ellipsoids and paraboloids of rotation. In his treatise on the mensuration of circles and round bodies he was aided by the method of exhaustion which had been developed by Menæchmus and others. In the study of specific gravity and the center of gravity of planes and solids he was a pioneer, and in the study of hydrostatics he was unequaled in the Greek period. He is also known for his study of spirals, possibly led thereto by his friend Conon. In general, he stands out as one of the greatest mathematicians and physicists in all history.

Method of Archimedes. In 1906 Professor Heiberg, who had already edited the works of Archimedes, discovered in Constantinople a manuscript on certain geometric solutions derived from mechanics.[2] This is especially interesting from the fact that it sets forth the method taken by Archimedes in deriving geometric truths from principles of mechanics. Some idea of the working of his mind may be obtained from the following:

After I had thus perceived that a sphere is four times as large as the cone whose base is the largest circle of the sphere and whose altitude is equal to the radius, it occurred to me that the surface of a sphere is four times as great as its largest circle, in which I proceeded from the idea that just as a circle is equal to a triangle whose base is the periphery of the circle and whose altitude is equal to the radius, so a sphere is equal to a cone whose base is the same as the surface of the sphere and whose altitude is equal to the radius of the sphere.[3]

[1] See Volume II, Chapter X.

[2] Translated by Lydia G. Robinson (Chicago, 1909) and by Sir T. L. Heath (Cambridge, 1912).

[3] For his method with respect to the parabola, see Volume II, Chapter X.

Death of Archimedes. Of the death of Archimedes at the siege of Syracuse under Marcellus (212 B.C.), Plutarch has this interesting record:

Nothing afflicted Marcellus so much as the death of Archimedes who was then, as fate would have it, intent upon working out some problem by a diagram, and having fixed both his mind and his eyes upon the subject of his speculation, he did not notice the entry of the Romans nor that the city was taken. In this transport of study and contemplation a soldier unexpectedly came up to him and commanded him to go to Marcellus. When he declined to do this before he had completed his problem, the enraged soldier drew his sword and ran him through. Others write that a Roman soldier ran towards him with a drawn sword and threatened to kill him, whereupon Archimedes . . . earnestly besought him to stay his hand that he might not leave his work incomplete; but the soldier, unmoved by his entreaty, instantly slew him. Others again relate that Archimedes was carrying to Marcellus some mathematical instruments, dials, spheres, and angles, by which the size of the sun might be measured . . ., and some soldiers . . . thinking that he carried gold in a vessel, slew him. Certain it is that his death brought great affliction to Marcellus; that he ever after regarded the one who killed him as a murderer; and that he sought for the kindred of Archimedes and honored them with signal favors.

Discovery of the Tomb of Archimedes. In his *Tusculan Disputations* (V, 23) Cicero relates that he himself discovered the tomb of Archimedes "when the Syracusans knew nothing of it and even denied that there was any such thing remaining." He relates the incident as follows:

I remembered some verses which I had been informed were engraved on his monument, and these set forth that on the top of the tomb there was placed a sphere with a cylinder. When I had carefully examined all the monuments . . . I observed a small column standing out a little above the briers, with the figure of a sphere and a cylinder upon it. . . . When we could get at it and were come near to the front of the pedestal, I found the inscription, though the latter parts of all the verses were effaced almost half away. Thus one of the noblest cities of Greece, and one which at one time likewise had been very celebrated for learning, had known nothing of the monument of its greatest genius, if it had not been discovered to them by a native of Arpinum.[1]

Of his works that have come down to us, those which are of chief interest in the history of mathematics are the ones on the quadrature of the parabola, on the sphere and the cylinder, on the measure of a circle, on spirals, conoids, and spheroids, and on notation. Archimedes seems also to have been interested in astronomy,[2] although no work of his upon this subject is extant.

4. Apollonius and his Successors

Apollo'nius of Per'ga[3] was known as "the great geometer" because of his work on conic sections. He was educated in Alexandria, and since he died under Ptolemy IV (Philop'ator, reigned 222–205 B.C.) he very likely knew Eratosthenes. He improved on the numeration system of Archimedes by using 10^4 as the base. This number, the myriad,[4] had long been in use in the Orient, and was the base of all great systems of numeration in the East as well as in Europe for many centuries. His chief work was on the conic sections, to which he gave the names ellipse, parabola, and hyperbola.[5]

This work[6] consisted of eight books, the first four of which have come down to us in Greek and the next three in Arabic, the last book being lost. In the first book Apollonius shows how the three conics are produced from the same cone. He uses a kind of coordinate system, the diameter serving for what we call the x-axis, and the perpendicular at the vertex serving for the y-axis. Books I–IV probably contain little that was

[1] Translated by C. D. Yonge. London, 1891.

[2] Livy (XXIV, 34) speaks of him as "unicus spectator coeli siderumque."

[3] Apollonius Pergæus, Ἀπολλώνιος. Fl. *c.* 225 B.C.; born at Perga, in Pamphylia, on the south coast of Asia Minor; Heath, *History*, II, 126.

[4] Μύριοι, ten thousand.

[5] Ἔλλειψις, παραβολή, ὑπερβολή.

[6] Sir T. L. Heath, *Apollonius of Perga* (Cambridge, 1896); J. L. Heiberg, edition of his works (Leipzig, 1891); E. Halley, *Apollonii Pergaei Conicorum libri octo* . . . (Oxford, 1710), and his *De Sectione Rationis libri duo* (Oxford, 1706); G. Eneström, *Bibl. Math.*, XI (3), 7.

not already known, but he arranged the material anew, as Euclid had arranged systematically many propositions that his own predecessors had known. Books V–VII seem to contain the discoveries which he himself had made. Book V treats of normals to a curve; Book VI, of the equality and similarity of conics; and Book VII, of diameters and rectilinear figures described upon these diameters. In general, his propositions are those which we now treat by analytic geometry, his method being synthetic and analogous to that of Euclid with respect to the circle and rectilinear figures.

Apollonius wrote various other works on geometry, including one on plane loci.[1] Ptolemy speaks of him as having been also a contributor to astronomy, but he probably confused him with another Apollonius who lived a little earlier.

In the works of Apollonius Greek mathematics reached its culminating point. Without Euclid as a guide, Apollonius could never have reached the summit; together, they dominated geometry for two thousand years.

Minor Writers. After the death of Apollonius no great writers on mathematics appeared for about two centuries. Greek civilization was receding. War was taking its toll. Arithmetic seemed for a time to sink into a comatose state after the slight attempts of Eratosthenes, and elementary geometry seemed to die with Euclid and Apollonius.

There was some indication at this time of the coming birth of a geometry of higher plane curves, just as the centuries immediately preceding Euclid and Apollonius foretold the appearance of these masters. The Greek civilization, however, had not strength to fulfill the momentary promise. Many generations had to come and go before another people, living far to the north, speaking a new language and making use of new symbols and of a new method, brought to light the theory.

[1] For his geometric works see a convenient list in Gow, *Greek Math.*, pp. 246, 261. See also various restorations of his lost works, such as Woepcke's in the *Mémoires présentés par divers savants à l'Académie des Sciences*, XIV; reprint (Paris, 1856); G. Eneström, *Bibl. Math.*, XI (3), 7, 8. Since we are concerned at present with the elementary field, the reader who wishes to consider the general history of conics should consult the *Encyklopädie*, III (ii), 1.

Perseus. Of those who treated of these curves, one of the first was Per′seus,[1] who lived *c.* 150 B.C. He wrote on sections of the anchor ring,[2] but his works are known only by references of later writers.

Nicomedes. Among the minor geometers of this period one of the best known was Nicome′des,[3] who flourished *c.* 180 B.C. and who invented a curve called the conchoid (mussel-shaped), by which the trisection of an angle is easily effected.[4]

Diocles. A probable contemporary of his, Di′ocles,[5] invented the cissoid,[6] by which the duplication of a cube can be accomplished.[7] He also studied[8] the problem of Archimedes, to cut a sphere by a plane in such a way that the volumes of the segments shall have a given ratio.

At about the same time, say 180 B.C., Zenodo′rus[9] wrote upon isoperimetry,[10] but most of his writings are lost.

A little later, Poseido′nius[11] taught in Rhodes, acquired a high reputation as a cosmographer and geometer, and had the honor of claiming Cicero and Pompey as his pupils. His measurements of the distance to the sun and of the circumference of the earth, known to us through the works of Cleome′des (*c.* 40 B.C.), were far from being as accurate as those of Eratosthenes, but his results seem to have been more generally accepted by ancient geographers.

Contributions of the Astronomers. Owing largely to the influence of the Egyptian and Chaldean priest-astronomers, whose achievements had attracted more and more attention on the part of the Greeks as intercourse became more free,

[1] Περσεύς. Heath, *History*, II, 203.

[2] Σπεῖρα, a torus, or ring-shaped solid of revolution (in a special case, an anchor ring), a solid already studied by Eudoxus.

[3] Νικομήδης. His birthplace is unknown. Heath, *History*, II, 199.

[4] This method and the use of other important curves are considered in Volume II, Chapter V.

[5] Διοκλῆς.

[6] "Ivy-shaped" curve.

[7] Such cases are more fully discussed in Volume II, Chapter V.

[8] In his Περὶ πυρείων.

[9] Ζηνόδωρος.

[10] Fourteen of his propositions have been preserved by Pappus (V, Pt. I) and Theon of Alexandria (*Comment. Almagest.*).

[11] Ποσειδώνιος, born at Apameia, in Syria, *c.* 135 B.C.; died *c.* 44. Sometimes called the Apamean. The name is also spelled Posidonius.

the 2d century B.C. was noteworthy for its advance in the study of the stars. In this century two names stand out prominently, not merely for their work as observers but because of their mathematical attainments.

Hypsicles. The first of these astronomers was Hyp'sicles[1] of Alexandria, who may have written the so-called fourteenth book of Euclid's *Elements*, containing seven propositions on regular polyhedrons. He was also interested in polygonal numbers,[2] in progressions, and in certain indeterminate equations. His prime interest, however, was in astronomy, and about his time there begins, among the Greeks, the division of the circle into 360° and the definite, scientific use of sexagesimal fractions, which the Babylonians had already suggested.

Hipparchus. About this time Hippar'chus,[3] working chiefly in Rhodes, wrote a famous work on astronomy in which were set forth the basic principles of the science. For this work he needed to measure angles and distances on a sphere, and hence he developed a kind of spherical trigonometry. Plane trigonometry had as yet taken only rudimentary form, and, so far as we know, there were no tables of functions. Hipparchus worked out a table of chords, that is, of double sines of half the angle, and thus was definitely begun the science of trigonometry. With him also began the theory of stereographic projection, a phase of geometry which Agatharchus (470 B.C.) had already put in practice. Hipparchus used it for the purpose of representing the projection of the celestial sphere upon the plane of the equator. He left a catalogue of 850 fixed stars, a number which Ptolemy (*c.* 150) increased to 1022 and which was not further materially increased until modern times.[4]

[1] Ὑψικλῆς. Fl. *c.* 180 B.C.. De Morgan places him *c.* 160 A.D. on general rumor, but asserts that he could not have lived before 550 A.D. and places Diophantus even later! The Arab writers say that he was born in Ascalon. See Smith's *Dict. of Greek and Roman Biog.*, II, 541. [2] See Volume II, Chapter I.

[3] Ἵππαρχος. Born at Nicæa, in Bithynia, Asia Minor, *c.* 180 B.C.; died *c.* 125.

[4] F. Boll, "Die Sternkataloge des Hipparch und des Ptolemaios," *Bibl. Math.* II (3), 185. This article disputes the usual assertion that Hipparchus listed 1022 stars, and asserts that he knew only about 850, the rest being catalogued by Ptolemy. See also Heath, *History*, II, 255.

Mathematics of Rome. The next two or three centuries witnessed the rise of the Roman military power and the consequent suppression of intellectual ideals. Art, philosophy, science, politics, ethics, and mathematics had all sunk to a low level. In literature, however, Rome made progress, although Vergil took Homer as his model, and Cicero followed in the footsteps of Demosthenes. Even as early as the 6th century B.C. Etruscan art had become wholly Greek in its technique and in its use of Greek mythology and customs. Rome simply followed in the same lines, not merely in art but in letters and science as well. In mathematics she showed no originality and possessed no high ideals. The science was worth to her precisely what it would fetch in the coin of the realm and no more. Rome created a goddess Numeraria, but she favored the acquisition of wealth rather than the creation of men. Money is everything only when man is nothing.[1] Whenever a genius like Heron of Alexandria, for example, arose in Greco-Roman territory, his interests were usually in the applications of the science as already developed, not in extending the boundaries. As to Rome herself, it is noteworthy how many of her scholars and literary men were born outside of Italy. Spain furnished the two Senecas, Lucanus, Martial, Quintilian, and probably Hyginus; France, Favorinus and Domitius Afer; Palestine, Josephus; Egypt, Philo; and Greece, Plutarch and Epictetus.[2] If Pythagoras and Archimedes may be ranked as dwellers in Italy, they were essentially Greek, and after the death of the latter, exact science may be said to have taken her departure. Cicero lamented this attitude of the Latin mind, contrasting the high honor in which geometry was held among the Greeks with the lack of appreciation on the part of the Romans.[3]

In this period there flourished Marcus Terentius Var'ro (116–28 B.C.), whom Quintilian called the most learned of the

[1] "L'argent n'est tout que dans les siècles où les hommes ne sont rien." Libri, *Histoire*, I, p. xiv.

[2] Libri, *Histoire*, I, p. 53.

[3] "In summo honore apud Graecos geometria fuit; itaque nihil mathematicis illustrius; at nos ratiocinandi metiendique utilitate huius artis terminavimus modum." *Tusculanarum Disputationum Libri V*, I, 2.

Romans. St. Augustine said of him that he had read so much that we wonder that he had time to write anything, and that he had written so much that we can scarcely believe that anyone could find time to read it all. Of such a dilettante nothing very scientific could be expected, and his one extant work[1] certainly has no great merit. In his *Disciplinarum Libri* he treated of arithmetic, and he wrote a *Mensuralia* or *De Mensuris* which related to practical mensuration, but, so far as we know, his works were mere compilations.[2] He is the one of the few pre-Christian mathematicians of whom we have a contemporary portrait, his profile appearing on a coin struck when he was the proquæstor of Pompey.

Geminus. Among those who showed any interest in the history of mathematics at this time the best known is Gem′inus,[3] who was a native of Rhodes but may have written in Rome. He is said to have divided mathematics into two groups, the pure group, including arithmetic (in the ancient sense) and geometry, and the applied group, including mechanics, astronomy, optics, geodesy, canonics, and logistic. Proclus, who lived in the 5th century A.D., tells us that he wrote a geometry which treated of spirals, conchoids, and cissoids. Only one of his works is extant, the *Phænomena*,[4] a treatise on astronomy. Proclus has numerous historical notes based upon the works of Geminus, these notes being found mostly in fragments that remain of the latter's *Arrangement of Mathematics.*

Minor Sources. Another source of information in the history of mathematics was written by a Sicilian who flourished a little later than this, having apparently been living in the year 8 B.C. Diodo′rus, usually called Diodorus Siculus, was born in Agyrium on the island of Sicily. He wrote forty books on history, and while his style is not good and his facts are ill-sorted, his

[1] *De Re Rustica Libri III.*

[2] Montucla, *Histoire*, I (2), 488, tells us that a MS. of his arithmetic was extant as late as the close of the 16th century, but it is now lost.

[3] Γεμῖνος or Γεμεῖνος. Fl. *c.* 77 B.C. M. C. P. Schmidt, "Was schrieb Geminos?" *Philologus*, XLV, 63; *Chrestomathie*, I, 45; C. Tittel, *De Gemini Studiis Mathematicis*, Leipzig, 1895; Heath, *History*, II, 222.

[4] Εἰσαγωγὴ εἰς τὰ Φαινόμενα. It was first printed, in Greek and Latin, in 1590.

works give us considerable information on the nature of the mathematics studied in the classical period, particularly in the schools of Egypt.

Still another writer to whom we are indebted for numerous bits of knowledge relating to the ancient mathematics is Strabo[1] the geographer. His second book deals with mathematical geography and passes certain criticisms on the map of the world prepared by Eratosthenes.

A little later than Geminus there lived P. Nigid'ius Fig'ulus,[2] a Pythagorean philosopher, who was highly esteemed in his time. He was known to his contemporaries as a philosopher, statesman, mathematician, and astrologer, but his contributions had little influence. In the field of mathematical astronomy he wrote *De Sphaera Barbarica et Graecanica*, but only fragments of his works have come down to us.

Probably contemporary with Figulus (but we are not sure of the dates) there was a certain geometer named Dionysodo'-rus,[3] who lived in Asia Minor, probably in Ami'sus.[4] He is known for a solution of the problem proposed by Archimedes and already discussed by Diocles,—to cut a sphere by a plane in a given ratio.[5] He also invented a new type of conic sundial.

Cæsar the Mathematician. It is also proper to speak of the contributions of Julius Cæsar (100–44 B.C.) to the reform of the calendar (46 B.C.), a work undertaken with the help of Sosig'enes[6] of Alexandria, an astronomer of whom almost nothing further is known. Cæsar himself was well versed in astronomy and wrote a poem on the subject and a work *De Astris*, neither of which is extant. He also planned extensive surveys of the empire.

About 40 B.C. the Greek astronomer Cleome'des[7] seems to have flourished and to have composed a treatise on the circular

[1] Born *c.* 66 B.C.; died *c.* 24 A.D.

[2] Fl. *c.* 60 B.C.; died in exile 44 B.C.

[3] Διονυσόδωρος. Fl. *c.* 50 B.C.

[4] Αμισός.

[5] It is preserved in Eutocius's commentary (*c.* 560) on II, 5, of the work of Archimedes on the sphere and cylinder. The method employed is that of the intersection of a parabola and a hyperbola.

[6] Σωσιγένης.

[7] Κλεομήδης.

theory of the heavenly bodies.[1] It was said three centuries ago that manuscripts of his treatises on arithmetic and the sphere were still in existence,[2] but they have since been lost.

Vitruvius. Of the Romans who made extensive practical use of mathematics, none is more prominent than Marcus Vitru'vius Pollio, commonly known as Vitruvius. Although the dates are uncertain, it is thought that his great work on architecture[3] was written between 20 and 14 B.C. In Book IX he treats of various types of sundials, and throughout the work he shows his early training as an engineer. He also has something to say on perspective, the ancient science of optics.

Referring to the same general line of applied mathematics, Lucius Junius Moderatus Columel'la (*c.* 25 A.D.) of Gades (Cadiz) wrote on agriculture[4] and included in his treatise a certain amount of information on astronomy, the calendar, and the art of surveying.

Although the name of Gaius Plin'ius Secundus,[5] commonly known as Pliny, is connected chiefly with his *Natural History*, a work in thirty-seven books, it should be recalled that he incorporated a certain amount of mathematics in his treatise. Book II contains a brief account of astronomy and is particularly valuable because of its historical information. Our knowledge of the practical use of the Roman numerals is enriched by his frequent reference to them in this work.

Frontinus. Next to Vitruvius, the most prominent of the Roman writers who made any practical use of mathematics was Sextus Julius Fronti'nus (*c.* 40–106), general, superintendent of water supply, and author of a work on war[6] and of one on

[1] Κυκλικῆς θεωρίας μετεώρων βιβλία δύο, first printed in Latin at Venice, 1498; in Greek, at Paris, 1539.

[2] Baldi, *Cronica*, p. 43. The date of Cleomedes is often given as a century later, but since he mentions no writer later than Poseidonius (died *c.* 44 B.C.), it is probable that he lived in the 1st century B.C.

[3] *De Architectura Libri X* (first printed at Rome *c.* 1486); *The Ten Books on Architecture*, translated by M. H. Morgan, Cambridge, Massachusetts, 1914.

[4] *De Re Rustica.*

[5] Born at Como, 23; died at the destruction of Pompeii, 79. The first name often appears as Caius.

[6] *Strategematicon Libri IV*. There is an edition by Gundermann, Leipzig, 1888.

aqueducts.[1] Some appreciation of the engineering works of this period may be formed from a consideration of the aqueduct of Claudius, which was constructed in the 1st century A.D. There are also preserved certain other books, generally believed to have been written by Frontinus, setting forth the principles of land surveying as commonly practiced by the Romans.[2]

THE AQUEDUCT OF CLAUDIUS

Constructed about the time of Frontinus, or probably just before his period of activity. From Breasted's *Ancient Times*

Hyginus. Among those who made use of mathematics in the work of surveying, Hygi'nus[3] (*c.* 120), known as Gromaticus (the surveyor), is one of the most prominent. The *gromatici* were those who used the *groma*,[4] an instrument employed in measuring and laying out land, and Hyginus was well known as a writer on the subject, although the fragments of his works

[1] *De Aquaeductibus urbis Romae Libri II.* First printed in Rome *c.* 1490; there is a recent edition in English by Herschel, London, 2d ed., 1913.

[2] His writings are collected in the so-called *Codex Arcerianus*. See K. Lachmann and A. Rudorff, *Gromatici Veteres*, being Vol. I of F. Blume, K. Lachmann, and A. Rudorff, *Die Schriften der Römischen Feldmesser*, 2 vols. (Berlin, 1848).

[3] Also spelled Hygenus and Higinus.

[4] Also spelled *gruma*. It is from γνώμων (*gnomon*), the shaft set up for the ancient shadow-reckoning, for sundials, and for general astronomical purposes. See Lachmann and Rudorff, *loc. cit.*, I, 108.

extant show no mathematical contributions to the science. There was an earlier Hyginus,[1] who wrote a work of no merit on astronomy,[2] and who is sometimes confused with his more prominent namesake, the surveyor.

The Roman surveyor Balbus (*c.* 100) was very likely contemporary with Hyginus Gromaticus, but his contributions[3] were unimportant.

Theodosius. There lived about this time, and certainly in the reign of Trajan (98–117), the mathematician and astronomer Theodo′sius.[4] He seems, on the testimony of Suidas, to have been a native of Tripoli, on the Phœnician coast. He wrote several works, the most important being his treatise on the sphere.[5] While this work possessed but little merit, it was translated into the Arabic along with most of the other Greek works on astronomy, and its brevity gave it considerable standing in Arabian schools. He is often confused with a Theodosius of Bithynia, who lived *c.* 50 B.C. and wrote on the sundial.

5. Period of Menelaus

Heron, or He′ro, of Alexandria[6] represented the applications of mathematics more completely than any other writer of about the beginning of our era. He seems to have been an Egyptian,

[1] Gaius Julius Hyginus, a friend of Ovid and therefore living in the 1st century B.C.

[2] *Poeticon astronomicon Libri IIII.*

[3] *Expositio et ratio omnivm formarvm.* See Lachmann and Rudorff, *loc. cit.*, I, 91.

[4] Θεοδόσιος. Fl. *c.* 100.

[5] Σφαιρικὰ ἐν βιβλίοις τρισίν. It was first printed, in Latin, at Paris, in 1529. See A. A. Björnbo, "Wann lebte Theodosios?" *Abhandlungen, zur Geschichte der Mathematik*, hereafter referred to as *Abhandlungen*, Leipzig, *v. d.*, XIV, 64.

[6] Ἥρων. Fl. *c.* 50 A.D. This date is based upon the careful researches of Wilhelm Schmidt, *Heronis Alexandrini Opera quae supersunt omnia*, Leipzig, 1899–1914. He places Heron in the 1st century A.D. It was formerly thought that he lived under the Ptolemies Philadel′phus and Euer′getes (283–222 B.C.), and it was also asserted that he flourished *c.* 100 B.C. See also R. Meier, *De Heronis aetate*, Leipzig, 1905; *Abhandlungen*, VIII, 195; T. H. Martin, "Recherches sur la vie et les ouvrages d'Héron d'Alexandrie" in *Mémoires présentés par divers savants à l'Académie des Inscriptions*, IV (1) (Paris, 1854); F. Hultsch, *Heronis Alexandrini geometricorum et stereometricorum reliquiae*, Berlin, 1864. Heath, *History*, II, 298, states that the evidence at present favors the 3d century A.D., but at best the date is very uncertain.

his style not being that of a Greek. He invented the pneumatic device commonly known as Heron's Fountain, a simple form of the steam engine, and various other machines, showing much ingenuity in all his numerous activities. He wrote on pneumatics, dioptrics, and mechanics, but from the standpoint of mathematics his work on mensuration is the most interesting. In this he treats of land surveying, probably summarizing the methods in use by the Egyptians. As is the case with many of the Greek scholars, some of his works are lost. His formula for the area of a triangle, $A = \sqrt{s(s-a)(s-b)(s-c)}$, is well known. It appears in the geodesy,[1] which is contained in his metrics,[2] but the proof is given (possibly an interpolation) in his dioptrics.[3] In his geometry may be found the first definite use of the trigonometric rule which we express by the formula $c = \frac{n}{4} \cdot \cot \frac{180^\circ}{n}$, where n is the number of sides of a regular polygon of area A and side s, and where $c = A/s^2$. He computed c for $n = 3, 4, \cdots, 12$, but his method is unknown. He was able to solve the equation which we write in the form $ax^2 + bx = c$, so that the general quadratic as we know it today was thus fully mastered by the Greek mathematicians.

About this time there lived Sere'nus of Antinoop'olis.[4] He was the author of a treatise on the *Section of the Cylinder*, containing thirty-three propositions, and of one on the *Section of the Cone*, with sixty-nine propositions. The latter has considerable work on maxima and minima. He also employed the principle of a harmonic pencil of rays.

Menelaus. Of those who, in the period of decay of Greek mathematics, showed any evidence of genius, Menela'us[5] was one of the most prominent. He was a native of Alexandria and wrote

[1] Γεωδαισία. [2] Μετρικά.

[3] Περὶ διόπτρας. On the formula in the Middle Ages, see G. Eneström, *Bibl. Math.*, V (3), 311.

[4] Antin'oe, Ἀντινόεια, a city on the eastern bank of the Nile. He is often called Serenus of Antissa. See Cantor, *Geschichte*, I, chap. 20. The date of Serenus is quite uncertain. J. L. Heiberg, who edited his *Opuscula* (Leipzig, 1895), is inclined (p. xvii) to place him in the 4th century.

[5] Μενέλαος. Fl. *c.* 100. Heath, *History*, II, 260.

a treatise on the sphere,[1] particularly with respect to the geometric properties of spherical triangles. He is known to have made astronomical observations in Rome in the year 98. Besides his treatise on the sphere he also wrote six books on the calculation of chords. One of his most important theorems states that if the three lines forming a triangle are cut by a transversal, the product of the lengths of three segments which have no common extremity is equal to the products of the other three. This appears as a lemma to a similar proposition relating to spherical triangles, "the chords of three segments doubled" replacing "three segments." The proposition was often known in the Middle Ages as the *regula sex quantitatum* because of the six segments involved. He also knew the invariant property of the anharmonic ratio of the line segments formed by a transversal cutting four concurrent lines,—a property the discovery of which was formerly attributed to Pappus,[2] who flourished about two centuries later.

Nicomachus. The best known of the Greek writers on arithmetic, although not the greatest arithmetician, was Nicom'achus[3] of Gerasa, his birthplace being probably the modern Jerash, a town situated about fifty-six miles northeast of Jerusalem. Since he mentions Thrasyl'lus,[4] who lived under Tiberius (reigned 14–37), but says nothing of the work of Theon of Smyrna, who lived under Hadrian (reigned 117–138), and since his work was translated from Greek into Latin by Appuleius, who lived in the time of Antoninus Pius (reigned 138–161), we are safe in asserting that he lived about the close of the first century.

Nicomachus wrote a treatise on music and a work in two books on arithmetic.[5] The arithmetic as it has come down to us may be only a compendium of a larger work which has

[1] The Latin title, by which it is best known, is *Sphaericorum Libri III*. There are editions by Maurolycus (1558), Mersenne (1644), and later writers.

[2] See the *Abhandlungen*, XIV, 96, 99.

[3] Νικόμαχος Γερασηνός, or Γερασινός. Fl. *c.* 100. Heath, *History*, I, 97.

[4] Probably Thrasyllus of Rhodes, died *c.* 36.

[5] This was first printed in 1538, at Paris. The best edition is that of Hoche, Leipzig, 1866.

long since been lost. Some such work seems to have been known to Boethius (*c.* 510) and to have been used by him in compiling his own treatise on the subject.

The Works of Nicomachus. Nicomachus belonged to the Neopythagoreans, a sect of philosophers then flourishing in Alexandria and trying to revive the teachings of Pythagoras. It is therefore quite possible that Nicomachus made the journey from Gerasa to Alexandria to study their doctrines.[1] At any rate there is a considerable amount of the Pythagorean theory of numbers in the tiresome treatment that he accords to arithmetic. The period was one of intellectual decadence, and had he not happened to summarize the ancient teachings in a field that had not been entered by writers of the first rank, we should never have heard of him.[2] His arithmetic[3] was rather an introduction to the philosophy of the subject than a scholarly treatment of the science itself. For lack of anything better it was adopted as a textbook in the few remaining schools of philosophy, and Boethius did much to perpetuate its influence.[4] In the *Philop'atris*,[5] probably a spurious dialogue inserted among the genuine works of Lucia'nus,[6] perhaps as late as the 10th century,[7] it is said of a certain man that "he reckons like Nicomachus of Gerasa."[8] The remark is ludicrous, and very likely was so intended, because there is no evidence that Nicomachus could reckon with any skill whatever, his interest being rather in the theory of numbers, which, as we have seen, was quite distinct from logistic.

[1] On the rise of other intellectual centers, however, as Alexandria began to lose prestige, see Kroll, *Geschichte*, p. 32.

[2] P. Tannery, *Revue philosophique*, XI, 289.

[3] *Introductionis Arithmeticae Libri duo*; Greek, Ἀριθμητικῆς εἰσαγωγῆς βιβλία β. There are various editions in Latin and Greek. For a summary of the work in English see G. Johnson, *The Arithmetical Philosophy of Nicomachus of Gerasa* (Lancaster, Pennsylvania, 1916); Heath, *History*, I, 97, and II, 238.

[4] It was also known in Hebrew, at least in paraphrase, in 1317. See M. Steinschneider, "Die Mathematik bei den Juden," *Bibl. Math.*, XI (2), 79.

[5] Φιλόπατρις.

[6] Λουκιανός. A humorous Greek writer of the 2d century.

[7] For discussion, see M. C. P. Schmidt, *Chrestomathie*, III, 19.

[8] Καὶ γὰρ ἀριθμέεις ὡς Νικόμαχος ὁ Γερασηνός.

Nicomachus mentions the sieve of Eratosthenes and often cites the Pythagorean doctrines. He gives an extended treatment of figurate numbers, and in his work appears an early form of the Greek multiplication table. Extensive multiplication tables are found in the Babylonian tablets, but no earlier Greek example is known, unless it be the one on the ancient wax tablet mentioned on page 58. The medieval name, *mensa Pythagorica*, may mean that a certain form of the multiplication table, mentioned in Volume II, came from the Neopythagoreans.

Another work of Nicomachus, the *Theologumena*,[1] has been lost, the extant work by that name being a later compilation.

The'on[2] of Smyrna, so called to distinguish him from Theon of Alexandria, who is mentioned later, lived in the time of Hadrian (reigned 117–138). He was interested in arithmetic and astronomy, and was the author of a work[3] which is commonly known in the Latin translation as the *Expositio*. Of this work, which set forth the mathematics necessary for the reading of Plato, two books are extant, one on arithmetic and one on astronomy, and very likely these are all that he wrote. The former resembles the work of Nicomachus but is less systematic.[4]

Marinus of Tyre. Mari'nus[5] of Tyre, a Greek scientist, who lived *c.* 150, may properly be called the founder of ancient mathematical geography. Apparently with greater success than Hipparchus (*c.* 150 B.C.) he definitely located places by reference to two coordinates, namely, latitude and longitude, and his maps set a new standard which the astronomer Ptolemy recognized a little later. The maps themselves, however, have not come down to us. He established the prime meridian

[1] Θεολογούμενα ἀριθμητικά.

[2] Θέων. Fl. *c.* 125. Heath, *History*, II, 238.

[3] Τῶν κατὰ τὸ μαθηματικὸν χρησίμων εἰς τὴν τοῦ Πλάτωνος ἀνάγνωσιν (βιβλία). The best Greek edition is E. Hiller, *Theonis Smyrnaei Philosophi Platonici Expositio* . . . (Leipzig, 1878). There is a French translation by J. Dupuis (Paris, 1892).

[4] On his astronomy see the edition by T. H. Martin, *Theonis Smyrnaei Platonici Liber de Astronomia* (Paris, 1849); J. B. Biot, review in the *Journal des Savants* (April, 1850).

[5] Μαρῖνος.

through the *Fortunatae Insulae*,[1] and this meridian was adopted by Ptolemy. At a later date the meridian was more definitely located through Ferro,[2] one of the Canary Islands, and this position was recognized until modern times.

6. Ptolemy and his Successors

Ptol'emy, or Claudius Ptolemæus,[3] whose period of greatest activity was *c.* 140–160, did for astronomy what Euclid did for plane geometry, Apollonius for conics, and Nicomachus for

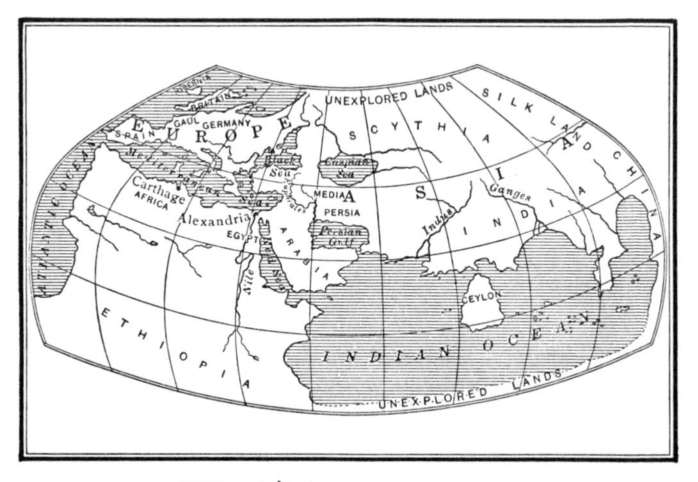

PTOLEMY'S MAP OF THE WORLD

This shows the great growth in the knowledge of geography from the time of Eratosthenes. See page 109. From Breasted's *Ancient Times*

arithmetic. He brought together in a single treatise the discoveries of his predecessors, arranging the material systematically, and, like the first two mentioned, was possessed of such genius as to make his work a standard of excellence for many

[1] Αἱ τῶν Μακάρων νῆσοι, Islands of the Blessed, probably including the Canary, Madeira, and Azores groups. It was here, in what Milton calls the "thrice happy isles," that Hesiod and Pindar placed Elysium.

[2] Ancient *Pluvialia*, the Πλουϊτάλα of Ptolemy.

[3] Πτολεμαῖος Κλαύδιος. Born *c.* 85; died *c.* 165. Heath, *History*, II, 273.

centuries. As to his life we know only that he taught in Athens and Alexandria. His greatest work, commonly known as the *Almagest*,[1] contains much information on the history of ancient astronomy. He also wrote on the planisphere, on music, and on applied mathematics. There is a question as to the genuineness of a work on optics that is often attributed to him. In the *Almagest* there is a summary of the computations of Eratosthenes, Poseidonius, and others as to the size of the earth, the position of certain places, and the size of islands and countries. In the application of mathematics to astronomy and geography Ptolemy stands preëminent among Greek scholars. He extended the use of sexagesimal fractions and elaborated the table of chords already used by Hipparchus. He also wrote a treatise on the postulate of parallels and a work of an astrological nature which is generally known in English as the *Tetrabiblos*.[2]

Minor Writers. Among the minor writers who came after Ptolemy there was the jurist Domi'tius Ulpia'nus (*c.* 170–228), a prolific contributor to the law and the compiler of the first table of mortality of which we have any knowledge.

Probably in the same period (*c.* 180) there lived the Roman surveyor Marcus Junius Nip'sus, but his contributions to the science relate chiefly to mensuration and are unimportant.[3] At

[1] The original title is usually given as Μαθηματικῆς συντάξεως βιβλίον πρῶτον; but on this question see J. L. Heiberg, *Ptolemaei Opera*, II, p. cxl (Leipzig, 1898–1907). Since he wrote another σύνταξις, the Arabs seem to have called the greater work *al* μεγάλη, and afterwards *al* μεγίστη (Smith's *Dict. of Greek and Roman Biog.*, III, 570). From μεγίστη, with the Arabic *al* (the), the Arabs made the word which has come to us as Almagest, so that to speak of "the almagest" is like speaking of "the the-greatest." The work was first printed, in an abridged form prepared by Regiomontanus, at Venice in 1496; the first complete edition appeared in Venice in 1515. For the latest work on the subject see C. H. F. Peters and E. B. Knobel, *Ptolemy's Catalogue of the Stars, a Revision of the Almagest*, Washington, 1915.

[2] Τετράβιβλος σύνταξις. The first printed edition appeared at Venice, 1484; first Greek edition, Nürnberg, 1535. It is also known by the Latin name, *Quadripartitum*.

[3] They are given in the *Codex Arcerianus* under the following titles: *fluminis uaratio, limitis repositio, uarationis repositio, lapides etc., podismus.* See Lachmann and Rudorff, *Gromatici Veteres*, I, 285.

about the same time (*c.* 200) there flourished another Roman surveyor named Epaphrodi'tus, who wrote not only on surveying but also on the theory of numbers.[1] He showed that if r is the radius of the circle inscribed in a right-angled triangle of sides a, b, and hypotenuse c, then $2r = a + b - c$. It is proper to refer, chiefly for the sake of showing the low estate to which learning had fallen, to the *Chronicon* of Sextus Julius Africa'nus (*c.* 220), a considerable part of which work is lost, but the extant portion of which contains information of value on the history of the calendar, and also to another work attributed to him, in which some notes appear on the history of other branches of mathematics.

Among the lesser Roman geometers and astronomers there was Censori'nus (*c.* 235), who wrote a book (238) entitled *De die natali*, a work primarily on astrology but containing a limited treatment of chronology, astronomy, and computation. It has been stated that he also wrote a geometry, although the work, if it ever existed, is lost.

We are also told by early writers of the interest taken in mathematics by the wealthy Roman dilettante Quintus Sammonicus Sere'nus (died 212). He was a prolific writer and his works include medicine, mathematics, and other sciences, but in general they merely show the debased state of learning. He is not to be confused with Serenus of Antinoopolis, already mentioned.

A little later (*c.* 275) Spo'rus[2] of Nicæ'a wrote a work from which we derive certain information relating to the history of early mathematics, particularly with reference to duplicating the cube and squaring the circle. He may have been the teacher of Pappus, who is usually put a century later.[3]

[1] V. Mortet, "Un Nouveau Texte des Traités d'Arpentage et de Géométrie d'Epaphroditus et de Vitruvius Rufus," *Notices et Extraits des Manuscrits de la Bibl. Nat.*, XXXV (1896), p. 510.

[2] Probably the same as Porus of Nicæa. The date is very uncertain; it is often given as of the 2d century.

[3] P. Tannery, *Mémoires de Bordeaux*, V (2), 211, and *Mémoires scientifiques*, Paris, 1912, I, 178, thinks he was the teacher of Pappus, or possibly one of his older pupils. The dates are so uncertain as to allow of either possibility.

It is possible that Metrodo'rus,[1] the compiler of the arithmetical epigrams in the Greek Anthology,[2] flourished about 325, but the date *c.* 500 is more probable. These epigrams were puzzle problems, like the one about the pipes filling the cistern, which we should now solve by algebra. For a long time such problems have interested students of arithmetic and algebra, and will doubtless continue to do so for all time to come. Sir Thomas Heath believes that their use dates back at least to the 5th century B.C.

7. Diophantus and his Successors

Diophan'tus[3] of Alexandria was one of the greatest mathematicians of the Greek civilization. That he flourished about the middle of the 3d century seems now fairly certain, although various other dates have from time to time been given. Psellus (11th century) says that Diophantus and Anato'lius[4] wrote on Egyptian computation and that "the very learned Anatolius collected the most essential parts of the doctrine . . ., dedicating his work to Diophantus." Very likely, therefore, Anatolius may have studied under Diophantus. Since he became bishop of Laodicea *c.* 280, he doubtless wrote this work some time before that date, and so Diophantus, who seems to have been the elder, probably flourished *c.* 250–275.[5]

[1] Μητρόδωρος. [2] English translation by W. R. Paton, London, 1918, p. 25.

[3] Διόφαντος. Also written Diophantes, Diophantis, and Diophantos. Fl. *c.* 250–275. There were several writers by this name. Sir T. L. Heath, *Diophantus of Alexandria*, 2d ed. (Cambridge, 1910). On the text see also Tannery's edition of his *Opera Omnia* (Leipzig, 1893, 1895). The first Latin edition of his works was that of Xylander (Wilhelm Holzmann), Basel, 1575; the second, that of Bachet (Paris, 1621), contained the Greek text; the third was that of Bachet with Fermat's notes, Toulouse, 1670. Stevin published a French translation of the first four books in his *Arithmétique*, Leyden, 1585, with editions in 1625 and 1634.

[4] 'Ανατόλιος. *Bibl. Math.*, IV (3), 396. Some fragments of his works are given in J. A. Fabricius, *Bibliotheca Graeca*, III, 275 (Hamburg, 1716). His computus was published by J. P. Migne, *Patrologia Graeca*, Vol. X (Paris, 1857).

[5] Heath, *Diophantus*, 2d. ed., p. 1. Tannery confirms this by an ingenious study of the price of wine at this time, finding that it conforms to that which Diophantus gives. See his *Mémoires scientifiques*, I, 62 (Paris, 1912).

All that is known of his life is given in a curious problem in the Greek Anthology, probably dating from the 5th century. The problem states that his boyhood lasted $\frac{1}{6}$ of his life, his beard grew after $\frac{1}{12}$ more, after $\frac{1}{7}$ more he married, 5 years later his son was born, the son lived to half his father's age, and the father died 4 years after his son. While the statement is obscure at one point, it is generally thought to mean that Diophantus married at 33 and died at 84.

Works of Diophantus. Diophantus wrote three works: (1) *Arithmetica*, originally in thirteen books, of which six are extant[1]; (2) a tract *De polygonis numeris*[2] of which a portion is extant; (3) a number of propositions under the title of *porisms*. Of these, the work of greatest importance is the *Arithmetica*. This work relates, as the title indicates, to the theory of numbers as distinct from computation, and covers much that is now included in algebra. The equations of the first degree are determinate and are so framed as to give positive values for the unknowns. In solving determinate quadratic equations Diophantus used only one root, even where both are positive. He solved a single special case of a cubic equation, but it is thought that further work on such equations may have been given in the lost books. His indeterminate quadratic equations are generally of the types $Ax^2 + C = y^2$ and $Bx + C = y^2$. His simultaneous quadratics relate only to special cases.[3]

Diophantus introduced a better algebraic symbolism than had been known before his time. In general he anticipated by several centuries the progress of algebra, as this progress appears in the works of other writers; and his work, while known to the Arabs, was not really appreciated until its discovery in Europe in the 16th century. He stands out in the history of science as one of the great unexplained geniuses. We do not know what teachers inspired him, we do not know

[1] Heath, p. 16, lists altogether twenty-five MSS., each containing more or less of the works of Diophantus. See also the Tannery edition, I, xxii.

[2] Περὶ πολυγώνων ἀριθμῶν. See the Tannery edition, I, 450; Heath, *Diophantus*, 2d ed., p. 247.

[3] Heath, *Diophantus*, 2d ed., p. 93.

the books he read, and we cannot explain how it happened that he appeared like a giant in a century of pigmies. Perhaps Seneca's statement that "no age is shut against great genius"[1] is the only explanation to be expected.

Lesser Writers. Not far from this time there also flourished the Neoplatonist Porphy'rius,[2] originally known as Malchus[3] the Tyrian and commonly spoken of as Porphyry. He wrote on the life of Pythagoras[4] and a work on the music of Ptolemy. He resided in Athens and Rome, spent some time in Sicily, and is known chiefly for his philosophical works and his antagonism to Christianity. His tomb, or one traditionally designated as his, is still pointed out in Constantinople.

One of the pupils of Anatolius and Porphyrius was Iam'blichus,[5] the author of several works, including one on arithmetic. He wrote a commentary on Nicomachus,[6] and we are indebted to him for considerable information relating to the latter, to Pythagoras,[7] and to other Greek writers. To him is due the theorem that if a number equal to the sum of the three integers $3n$, $3n-1$, $3n-2$ is taken, and if the separate digits of this number are added, and the digits of this result, and so on, the final sum is 6.

About 340 Julius Fir'micus Mater'nus, a Sicilian, wrote a work entitled *Eight Books on Mathematics*[8] but concerned

[1] "Nullum saeculum magnis ingeniis clausum est."

[2] Πορφύριος. Born in Syria, 232 or 233; died *c.* 300.

[3] From *Melekh*, the Hebrew for "king"; in the Greek of that period, Μάλχος. The name was changed, according to tradition, to Porphyrius (wearer of the purple).

[4] Πυθαγόρου βίος, possibly a fragment of his history of the philosophers.

[5] Ἰάμβλιχος. Born at Chalcis, Cœlesyria, *c.* 283; died *c.* 330.

[6] It appeared in various editions in the 16th century. The title-page of the 1668 edition begins: *Jamblichus Chalcidensis ex Coele-Syria in Nicomachi Geraseni Arithmeticam introductionem* (Arnheim, 1668).

[7] Περὶ Πυθαγόρου αἱρέσεως, of which four books are extant, the first containing the life of Pythagoras. The latter was published in Greek and Latin, at Franeker, in 1598. There have been other editions. See *Bibl. Math.*, VIII (3), 309.

[8] *Iulii Firmici Materni Junioris Siculi V. C. Matheseos Libri VIII.* It was first printed at Venice in 1497. The definitive edition is that of Kroll and Skutsch, Leipzig, 1897–1913. L. Thorndike, "A Roman Astrologer as a Historical Source: Julius Firmicus Maternus," *Classical Philology*, VIII, 415.

exclusively with judicial astrology according to the precepts of the Babylonians and Egyptians. Such works have little place in a history of mathematics except as they show from time to time the tendencies of the devotees of the science.

There are also various other isolated cases of mathematical interest in this period of general decay of scholarship, as in the constructing of an astrolabe by Syne'sius of Cyrene (*c.* 378–*c.* 430), the poet and orator, a pupil of Hypa'tia. He became bishop of Ptolemais in 410.

About 390 The'on of Alexandria, known as Theon the Younger, father of the learned Hypatia, edited Euclid's *Elements* and the great work of Ptolemy, wrote various scientific treatises, and set forth a method for finding square roots by the aid of sexagesimal fractions. Manuscripts of his edition of Euclid have been helpful to modern writers in determining the accurate text of the *Elements*.

A little later (*c.* 450) Domni'nus[1] of Larissa, in Syria, wrote on arithmetic, philosophy, and optics. He followed the geometric, deductive method of Euclid rather than the inductive method of Nicomachus, and seems to have had access to some important work that is now lost on the theory of numbers.

Pap'pus[2] of Alexandria, a late Greek geometer, flourished probably in the 3d century, although the date is uncertain. Suidas (*c.* 10th century), not a very careful writer, however, places him in the reign of Theodosius (379–395), but others believe him to have lived two centuries earlier. Of his greatest work, the *Mathematical Collections*,[3] only the last six of the eight books that it originally contained have come down to us. The third book treats of proportion, inscribed solids, and the duplication of the cube; the fourth, of spirals and of such other higher plane curves as the quadratrix; the fifth, of maximum and isoperimetric figures; the sixth, of the sphere; the

[1] Δομνῖνος. P. Tannery, Darboux *Bulletin*, VIII (2), 288.

[2] Πάππος. Fl. *c.* 300. Heath, *History*, II, 355.

[3] Μαθηματικῶν συναγωγῶν βιβλία. The text of this work in Greek and Latin was published with notes by Hultsch, Berlin, 3 vols., in 1876–1878. There was a Latin edition published at Pesaro in 1588, reprinted without change at Venice in 1589 and at Pesaro in 1602. See also *Bibl. Math.*, XII (3), 252.

seventh, of analysis and its history among the Greeks; and the eighth, of mechanics. Two well-known theorems bear his name, one on the generation of a solid by the revolution of a plane figure about an axis, later known as Guldin's Theorem, and the other a generalization of the Pythagorean Theorem. He also knew the doctrine of the involution of points and the constancy of anharmonic ratios in the case of a transversal cutting a pencil, the latter having already been known to Menelaus.

Hypa'tia[1] of Alexandria was the first woman who took any noteworthy position in mathematics, and on this account and because of her martyrdom she has occupied an unduly exalted place in history. She was the daughter and pupil of Theon, and such were her attainments that she was called upon, so tradition says, to preside over the Neoplatonic School at Alexandria. Much that passes for history in her case seems to be fiction, as the statement of Suidas (*c.* 10th century) that she married Isidorus of Gaza, the Neoplatonist. It seems certain, however, that she was slain in one of the city brawls between followers of rival sects. Suidas says that she wrote a commentary on an astronomical table of a certain Diophantus, possibly the algebraist, and one on the conics of Apollonius. Her works, however, are all lost.[2]

Pro'clus, surnamed the Successor[3] because he was looked upon as the successor of Plato in the field of philosophy,[4]

[1] Ὑπατία. Born at Alexandria, *c.* 370; died at Alexandria, 415.

[2] For the romantic side of her life, see J. Toland, *Hypatia, or the history of a most beautiful, most vertuous, most learned . . . lady*, London, 1720; C. Kingsley, *Hypatia*, London, 1853; F. Mauthner, *Hypatia, Roman aus dem Altertum*, 2d ed., Stuttgart, 1892. For a critical study, see R. Hoche, "Hypatia, die Tochter Theons," in *Philologus*, XV (1860), 435; S. Wolf, *Hypatia, die Philosophin von Alexandria*, Vienna, 1879; W. A. Meyer, *Hypatia von Alexandria*, Heidelberg, 1886. See also Heath, *History*, II, 528.

[3] Πρόκλος Διάδοχος. Born at Byzantium, *c.* 412; died 485. A certain Marinus, not to be confused with Marinus of Tyre, gives his birth as February 8, 412. The name also appears as Proculus. The best of the partial editions of his works is that of Cousin, *Procli Opera*, 6 vols., Paris, 1820–1827; 2d ed., 1864. The best edition of his commentary on Euclid I is that of G. Friedlein, *Procli Diadochi in primum Euclidis Elementorum librum*, Leipzig, 1873. His *Institutio Physica*, edited by A. Ritzenstein, was published at Leipzig in 1912.

[4] Or because he succeeded Syrianus, the philosopher, at Alexandria.

studied at Alexandria and taught at Athens. He was a prolific writer and his works include a paraphrase of difficult passages from Ptolemy, a work on astronomy, a commentary on Euclid I, and a brief treatise on astrology. He also shows evidence of a study of certain higher plane curves. His works are valuable sources of information on the history of Greek geometry. For information concerning his life we are indebted to Mari'nus,[1] of Flavia Neapolis in Palestine (the old Sichem), who succeeded him in 485.[2] This Marinus, very likely a Jewish scholar,[3] also wrote an introduction to the *Data* of Euclid.

At about this time Victo'rius[4] of Aquitania (457) wrote a *Canon Paschalis*, one of the first of the *Computi*,—books on the finding of the date for Easter. He suggested beginning our era at the time of the first full moon after the death of Christ. He also wrote a *calculus*, that is, a practical arithmetic. In this he gave considerable attention to fractions and to tables for the multiplication of large numbers.

The name of Capella might, for chronological reasons, be included in this chapter, but on account of the relation of his work to that of writers of the 6th century it is considered in Chapter V.

8. The Orient

China. The period from 300 B.C. to 500 A.D. was one of mathematical activity in China, and some slight but noteworthy trace remains of an interest in numbers in Japan.[5] At the beginning of this period the event of greatest concern in the history of Chinese mathematics was the burning of all books[6] (213 B.C.), as already mentioned in Chapter II, by order of the emperor Shï Huang-ti,[7] founder of the Ch'in (Ts'in)

[1] Μαρῖνος. As stated above, he must not be confused with the astronomer already mentioned.

[2] His life of Proclus was first printed at Zürich in 1559.

[3] S. Krauss, *Jewish Quarterly Review*, 1897, p. 518.

[4] Often written Victorinus. It is thought that he was born in Limoges.

[5] Smith and Mikami, *History of Japanese Mathematics*, chap. i (Chicago, 1914); hereafter referred to as Smith-Mikami.

[6] An exception was made of books on medicine, agriculture, and divination.

[7] She Huang-ti, "the First Emperor," born 259 B.C.; died 210 or 211.

Dynasty (221 B.C.), who wished to appear in the eyes of posterity as the creator of a new era of learning. The penalty for not burning the books was branding and four years' service on the Great Wall. The records say that four hundred and sixty scholars protested against this odious law and were buried alive as an example to others. How many of the ancient classics survived, or how many were faithfully transmitted by means of copies made from memory, we do not know, but it is probable that Chinese scholars will in due time apply the methods of textual criticism to the determination of this point.

About this time, and probably just after the burning of the books, there lived the learned Ch'ang Ts'ang (*c.* 250–152 B.C.), a statesman of highest rank, who wrote (176 B.C.) a new *K'iu-ch'ang Suan-shu* (*Arithmetic in Nine Sections*),[1] basing it upon fragments of the earlier work of the same name. The nine chapters or sections have already been given (page 32).

Ch'ang Ts'ang gave the area of a segment of a circle as $\frac{1}{2}(c+a)a$, where c is the chord and a is the altitude of the segment. Among his problems is that of finding the height of the trunk of a tree, the upper part of which was 10 feet high but has fallen over and reaches the ground 3 feet from the base. The rule for the area of the segment of a circle is later found in the work of the Hindu Mahāvīra (*c.* 850), and the problem about the tree is found in various Hindu mathematical works after the time of Āryabhaṭa (*c.* 510).

Minor Chinese Writers and Events. The period following the burning of the books was, as might have been expected from the need thus created, one of considerable intellectual activity. In this respect, but from a wholly different cause, it was not unlike the century following the impetus given to learning by Plato. Ch'êng Kiang Chen (also known as Chŭn Shuen), who died in 200 B.C., wrote on knotted cords which perhaps, like the Peruvian *quipu*, were for keeping accounts.

[1] K. L. Biernatzki, "Die Arithmetik der Chinesen," Crelle's *Journal*, LII, 11; A. Wylie, *Chinese Researches*, Part III (Shanghai, 1897). These writers put the date *c.* 100 B.C., but Ch'ang Ts'ang appears to have died in 152 B.C., upwards of 100 years old, and to have written the work in 176 B.C. See Mikami, *China*, p. 9.

Then as always in Chinese history the regulation of the calendar occupied the attention of scholars. Thus it is recorded that *c.* 104 B.C. the emperor reëstablished official astronomy and a new calendar was devised.[1] It is also worthy of note, as bearing upon the arithmetic of commerce, that about this time (135 B.C.) coinage became a government prerogative.[2] The Chinese annals of this period also speak of the efforts of the emperor[3] to open up communication with the region about the river Oxus, all such efforts having relation to the unsolved problem of the transmission of mathematical knowledge between the East and the West. The famous Chinese general Ch'ang K'ién went to the countries of the Jaxartes and the Oxus in the 2d century B.C., and about 100 B.C. an envoy was sent as far west as Lake Baikal.[4] This intercourse between the East and the West was maintained for several centuries. For example, an Aramaic manuscript of the 1st century (*c.* 1–20), the earliest known specimen of rag paper, has been found on the Chinese border.[5] That China had intercourse with India is evident from the fact that the records show such relations as early as 218 B.C. and that the name Sin-du appears in the Chinese annals of about 120 B.C. It is also well established that China was known in the West at this period. Ptolemy the astronomer (*c.* 150) speaks of the country under the name of Thin, and in 166 Marcus Aurelius sent an embassy to the emperor's court.

[1] J. B. Biot, *Études sur l'astronomie Indienne et sur l'astronomie Chinoise*, p. 299. Paris, 1862.

[2] H. B. Morse, "Currency in China," from the *Journal of the North-China Branch of the Royal Asiatic Society*, XXXVIII; reprint, p. 2.

[3] Wu-tí (140–87 B.C.). On the general subject of the relations of China with the West see S. W. Williams, *A History of China*, p. 58 (New York, 1897); F. Hirth, "The Story of Chang K'ién," *Journal of the Amer. Oriental Soc.*, XXXVII, 89, 185, 186; T. W. Kingsmill, "The Intercourse of China with Central and Western Asia in the 2d Century B.C.," *Journal of the China Branch of the Royal Asiat. Soc.*, XIV (N.S.), 1; Hirth and Rockhill, *Chau Ju-Kua: His Work on the Chinese and Arab Trade in the twelfth and thirteenth centuries* (Petrograd, 1911), the preface to which considers the whole question from earliest times to the 13th century.

[4] E. Bretschneider, *Mediaeval Researches*, I, 32. London, 1910.

[5] M. A. Stein, *Ruins of the Desert of Cathay*, II, 114. London, 1912.

It is probable that this continued interchange of thought is one of the causes of the frequent changes in the calendar and of the study of the related geometric figure of the circle. About 25 A.D. there lived a well-known philosopher and astronomer named Liu Hsiao, who was of the Imperial house of the Han Dynasty.[1] He was one of the most prominent of the "circle squarers" of his day. His son, Liu Hsing,[2] devised a new calendar,[3] thus using his time to better advantage than the father. A few years later (*c.* 75 A.D.) Pan Ku wrote a work[4] in which the use of the bamboo rods, a primitive form of abacus, is mentioned. At about this time Ch'ang Höng (78–139), chief astrologer and minister under the emperor An-tí, constructed an armillary sphere and wrote on astronomy and geometry. He gave $\sqrt{10}$ as the value of π, this being one of the earliest uses of this approximation.[5] Perhaps contemporary with him, although we are uncertain, there lived Ch'ang ch'un-ch'ing, who wrote a commentary on the *Chóu-peï*. About 190 there flourished Ts'ai Yung,[6] one of the numerous experts on the calendar, but his works are lost. He was sentenced to death for political reasons, but the sentence was commuted to having his hair pulled out. His convivial habits gave him the name of Drunken Dragon.

Wu-ts'ao Suan-king. Possibly about the beginning of the Christian era, for the date is so uncertain[7] that we are not safe in fixing the time even within the limits of several centuries, there was written one of the best-known but least worthy Chinese classics on mathematics, the *Wu-ts'ao Suan-king*.[8] The author seems to have been Sun-tzï,[9] but

[1] This dynasty lasted from 206 B.C. to 25 A.D.

[2] Biot (p. 305) transliterates the name as Lieou-hin.

[3] The *San-t'ung* calendar, devised in the year 66.

[4] The *Han Shu*. Pan Ku died in 92.

[5] On account of the unreliability of early Chinese texts, all such statements are open to some doubt.

[6] Born 133; died 192.

[7] Mikami, *loc. cit.*, p. 37, says in the former (beginning *c.* 206 B.C.) or later (*c.* 25–220 A.D.) Han Dynasty.

[8] *Arithmetic Classic in Five Books.*

[9] Also given as Sun Tsze, Sun Tsŭ, Suentse, Sun Wu tsze, and Sun Tsu Yeh Ch'i-sun. The work is also known as the *Sun-tzï Suan-king*. Père Vanhée puts the date as probably the 1st century A.D., while Biernatzki (p. 21) says that Sun-tzï may have lived 220 B.C.

even as to this we are uncertain. The work is obscurely written and is not so accurate in its statements as the *Nine Sections.* It relates chiefly to the mensuration of areas. A single problem will serve to show its nature:

"There is a quadrangular field of which the eastern side is 35 paces, the western side 45 paces, the southern side 25 paces, and the northern side 15 paces. Required the area of the field."

Evidently a solution is impossible through lack of sufficient data; but the author assumes that he may take one fourth the product of the sums of the pairs of opposite sides,[1] such approximations as this being not uncommon all through the East in these early times.

Liu Hui. The best-known Chinese mathematician of the 3d century was Liu Hui.[2] In 263 he wrote the *Sea Island Arithmetic Classic*,[3] a work which probably took its name from the first problem that it contains, this problem beginning with the statement, "There is a sea island that is to be measured." The work is concerned with the mensuration of heights and distances, the rules seeming to show some familiarity with the manipulation of algebraic formulas.

Liu Hui also wrote a commentary on the *Nine Sections*, and it seems to have been in the performing of this task that he accumulated the materials for his "Sea Island" work.

Minor Chinese Writers from 200 to 500. Of the minor writers of the 3d century mention may properly be made of Wang Pi (*c.* 225–249), the leading authority on the mysticism of the *I-king*[4]; of Wang Fan (229–267), the astronomer, who asserted that $\pi = \frac{142}{45}$; of Siu Yo (*c.* 250), who wrote the *Omissions noted in the Art of Numbers*[5]; of Li Ping, the great

[1] Mikami, *China*, p. 38.

[2] Also transliterated Lew Hui, Lew Hwuy, and Lieou Hoei.

[3] *Hai-tau Suan-king.* Wylie says that this title first appeared in an edition prepared in the 8th century.

[4] See page 25.

[5] *Shu-shu-ki-yi*, or *Chou-chou-ki-yi.* There are many commentaries on this work. See A. Vissière, *Recherches sur l'origine de l'abaque Chinois* . . ., p. 22 (Paris, 1892).

irrigation engineer of the 3d century;[1] of Liu Chih[2] (*c.* 289), who is possibly the one who gave the so-called "Chih's value of π," that is, $\pi = 3\frac{1}{8}$; and Hsü Yüeh, who wrote a commentary on Siu Yo's work above mentioned.

The 5th century is more interesting because of the evidence that we have of intercourse between China and the rest of the world than because of any definite contributions to mathematics. A few names of mathematicians are known,[3] but it was the visit of the Buddhist missionaries and pilgrims from India that is significant. The result of this visit was the translation of an arithmetic and of various astronomical works of the Brahmans, which stimulated the activity of Chinese scholars in these fields. This interchange of thought was not new, for Buddhism was transmitted from India to China at least as early as the year 65. In 399 a Chinese Buddhist, Fa-hién, went to India, and after his return in 414 he devoted his life to the translation of Hindu works. Since religion was closely related to astronomy, and astronomy to mathematics, the influence of this interchange of religious thought must have been stimulating to the science of China. Moreover, after about the year 450 there are many references in the Chinese annals to the people of Po-ssï (Persia), and thereafter many embassies passed between the two countries.

Among the mathematicians of this period whose names have come down to us is P'i Yen-tsung (*c.* 400–*c.* 450), who is said to have computed a noteworthy value of π which has since been lost. There is also Tsu Ch'ung-chih (430–501), an expert in mechanics, who revived the knowledge of the "south-pointing vehicle" and constructed a motor boat, all details of which are lost. He gave $\frac{22}{7}$ as an "inaccurate value" of π, and $\frac{355}{113}$ as the "accurate value," and he also showed that π lies between our present decimal forms 3.1415926 and 3.1415927. About the year 450 a new calendar was devised by Ho' Ch'êng-t'ien,

[1] H. K. Richardson, *Asia*, XIX, 441.

[2] There was another mathematician of the same name (1311–1375), who devised a new official calendar.

[3] For example, Tun Ch'üan (*c.* 425), who wrote the *San-tŏng-shu*, and Wang Jong, an arithmetician.

and at about the same time one Wu, a geometer, gave the equivalent of 3.1432 + as the value of π. These details have little significance except as they show the nature of the scientific interests of China during this long period.

Japan in Earliest Times. Prior to the year 500 Japan seems to have made no progress either in literature or in science. There is a tradition that Chinese ideograms made their way through Korea and into Japan in the year 284. There is also reference to the *Jindai monji*, or "letters of the era of the gods," in early times, possibly a kind of system of cabala with numerical values assigned to the letters, but nothing is definitely known upon the subject. A tradition also exists that in 660 B.C. the Japanese had a system of numeration extending to very high powers of ten. In this system the special name *yorozu* was used for 10,000, corresponding to the Greek myriad already mentioned, and this may possibly be some slight evidence of the early interrelations between the East and the West.[1]

Of the rest of Japanese mathematics in the early periods we know only that there was a system of measures and that, as among all other ancient peoples of any intellectual standing, a calendar existed.

India. The noteworthy contribution of India in this period was probably the Hindu numeral system, which will be discussed later.[2] A second event of importance in the history of mathematics in India, and one which chronologically precedes the writing of the numerals, was the invasion of this country by the army of Alexander the Great (327 B.C.) and the sending of Greek ambassadors to reside in Indian courts. How much influence this event had upon the science and particularly upon the astronomy of the Hindus it is difficult at present to say. It is worthy of note, however, that the later Hindu writers used such Greek adaptations as *jāmitra* (from the Greek διάμετρος), *kendra* (κέντρον), and *dramma* (δραχμή).[3]

[1] For discussion and bibliography see Smith-Mikami, p. 4.

[2] See Volume II, Chapter II.

[3] G. R. Kaye, *Indian Mathematics*, p. 26 (Calcutta, 1915) (hereafter referred to as Kaye, *Indian Math.*); H. T. Colebrooke, *Algebra with Arithmetic and*

Just before the beginning of the Christian era there were numerous invasions from the north that interfered seriously with the spread of Greek science, and in the 4th century A.D. there appeared at least one work which definitely sought to replace the astronomy of Greece by the ancient science of India.

The first important work on astronomy produced in India, so far as now known, was the *Sūrya Siddhānta*,[1] probably written about the beginning of the 5th century, although known to us only in later manuscripts. The ritualistic mathematical formulas of the *Śulvasūtras* now gave place to the mathematics of the stars. This change was possibly due to the influence of Greek scholars whose works might still have been appreciated by the descendants of the ancient Greeks who settled in India after Alexander's time. Varāhamihira, who will be mentioned later, speaks of five *Siddhāntas*, but places the *Sūrya Siddhānta* at the head. Among the five is the *Paulisa Siddhānta*, probably of about the same period. This contains an excellent summary of early Hindu trigonometry, the rules, expressed in modern symbolism, being as follows:

$$\sin 30^\circ = \tfrac{1}{2}, \qquad \pi = \sqrt{10},$$

$$\sin 60^\circ = \sqrt{1 - \tfrac{1}{4}}, \quad \sin^2\phi = \left(\frac{\sin 2\phi}{2}\right)^2 + \left(\frac{1 - \sin(90^\circ - 2\phi)}{2}\right)^2.$$

There is also included in this work a table of sines which was apparently derived from Ptolemy's table of chords.

The absence of an authentic Hindu chronology and of a careful study of the effect of the Greek civilization upon the sciences in India renders difficult a satisfactory assessment of her mathematical achievements in this period.

Mensuration, from the Sanscrĭt, p. lxxx (London, 1817) (hereafter referred to as Colebrooke, *Āryabhaṭa*, or *Brahmagupta*, or *Bhāskara*, according to the part of the work considered, and with the modern spellings as here).

[1] E. Burgess, "The Sūrya Siddhānta," in the *Journ. of the Am. Oriental Soc.*, VI (New Haven, 1860); G. R. Kaye, "Ancient Hindu Spherical Astronomy," *Journ. and Proc. of the Asiatic Soc. of Bengal*, Vol. XV; Bapu Deva Sastri and L. Wilkinson, *The Sūrya Siddhānta and the Siddhānta Śiromani* (Calcutta, 1861). Albêrûnî, the Arab writer on India (*c.* 1000), speaks of the work as "the Siddhānta of the sun, composed by Lāṭa."

Decay of Civilization in Mesopotamia. For about two thousand five hundred years before the period now under consideration Mesopotamia had maintained a high civilization. Assyria, Sumeria, Babylonia, and Chaldea had contributed in a large way to the world's commercial machinery, to its science, to its laws, and to its art. Mathematics, medicine, religion, sculpture, architecture, literature, and the science of government are all indebted to the genius of those who dwelt in the lands bordering upon or in the vicinity of the Two Rivers.

With the close of the 6th century B.C., however, there came a change that was disastrous to the native civilization of this region. The Persian conquest of 539 B.C. and the subsequent coming of the Parthians, the Greeks, and the Romans, each of whom held in subjection some or all of the territory of Mesopotamia, left little of her ancient glory. Trajan, hoping to repeat the conquests of Alexander, visited Babylon early in the 2d century A.D., and "saw nothing worthy of such fame, but only heaps of rubbish, stones, and ruins," and this was symbolic of the decay of a civilization which had perhaps exerted a greater influence upon the world than any that had existed prior to the rise of Greece.

Astrology continued to retain its power over the mass of people, as it does in a large part of Asia today. This is shown by tablets of the 2d century B.C., in which reports are made to the king with respect to predictions as to the positions of the planets. If superstition affected the court, much more would it have affected the people at large.

In all the records of this region only a single name stands out that is worthy of mention in the history of the mathematics of this period, and this only in connection with a sister science. About 250 B.C. Berosus (probably Bar Oseas, that is, the son of Oseas), a Chaldean, founded a school on the island of Cos, and introduced into Greece the astronomy and the astrological beliefs of his people, constructing a sundial and probably other instruments.[1]

[1] A. Wittstein, "Bemerkung zu einer Stelle im Almagest," *Zeitschrift für Math.*, XXXII (Hl. Abt.) (Leipzig, 1887), 201.

TOPICS FOR DISCUSSION

1. The School of Alexandria, its rise, its influence, the great scholars connected with it, and its decay.

2. Euclid, his life, his works, and his influence.

3. The work of Eratosthenes, particularly with respect to geodesy.

4. The life, the works, and the influence of Archimedes.

5. Apollonius and his contribution to the study of conics.

6. The mathematical contributions of the Greek astronomers.

7. Mathematics in the Roman civilization. Causes of the disregard for the science.

8. The life and works of Heron. His influence upon the development of applied mathematics as compared with that of Archimedes.

9. The work of Nicomachus compared with the works of Euclid and Apollonius.

10. The work of Claudius Ptolemæus, or Ptolemy.

11. The life and works of Diophantus.

12. The decay of Greek geometry, with a special consideration of the work of Menelaus, Hypatia, Proclus, and Pappus.

13. Causes and probable effects of the burning of the books in China in 213 B.C.

14. The period in which the *Nine Sections* was written and the general nature of this work.

15. The knotted cords of China and the general subject of knotted cords in the keeping of records and in religious ceremonial.

16. Efforts at opening communications between the East and West at this period, and the probable effect of these efforts on science in general and mathematics in particular.

17. The periods and nature of the *Arithmetic Classic in Five Books* and the *Sea Island Classic*.

18. Influx of Hindu learning into China in this period and the probable effect of this intercourse on the mathematics of both China and India.

19. The invasion of India by Alexander the Great and its effect upon the mathematics of the East.

20. The nature of the *Sūrya Siddhānta* and the bearing of this work upon the mathematics of India.

21. Causes of the decay of mathematics in Mesopotamia in the five centuries after the time of Alexander.

CHAPTER V

THE PERIOD FROM 500 TO 1000

1. China

Intercourse with India and the West. The five centuries extending from 500 to 1000 saw the general trend of mathematics to the West rather than in the opposite direction. Europe was intellectually dormant, drugged with a new narcotic, while most of the East was, as always, superstitious but inquisitive. On this account it is proper to consider first the work of this period as it appears in the Orient. Even in the Dark Ages, however, the West influenced the East, passing traces of the later Greek culture on to the intellectual centers of China and probably to those of India.

In so far as this intercourse was commercial it influenced the art of calculation, while the travel of pilgrims and the movements of armies resulted in the exchange of a knowledge of both astronomical and abstract mathematics. Moreover, the priest, whose leisure allowed time for the study of mathematics, was often an astronomer, and he or the professional astrologer was looked upon as a natural attendant at court or a necessary adjunct to the general's staff. Where the army went, there went a knowledge of mathematics. Astrologers of one country thus consulted with those of another. The itinerant tradesman, the pilgrim, and the army were the means of the exchange of ideas in all ancient times, just as books and periodicals are the corresponding media in our day.

Evidences of this Intercourse. Of the many evidences of intercourse that we have in this period, a few may be mentioned simply as typical. In 518 Hui-sing, a Buddhist pilgrim, visited India; sometime in the 7th century a Sanskrit calendar[1] was translated into Chinese; in 615 an Arab[2] embassy visited China; in 618 a Hindu astronomer[3] was employed by the Chinese Bureau of Astronomy to devise a new calendar; in 629 Hüan-tsang[4] went to India and after his return in 645 he devoted his life to the translation of Hindu works, of which he had brought no less than 657 from India; in 636, so the Chinese records assert, a Roman priest whom these records speak of as A-lo-pen came to the capital of China; and at the end of the 7th century Buddhist pilgrims sailed from Canton to Java and Sumatra. In the 8th century Arab ambassadors visited China several times, in particular in 713, 726, 756, and later; in 719 an ambassador[5] was sent from Rome to the Chinese court; between 713 and 825, foreign ships of large tonnage visited Canton, and an important customhouse is known to have existed there at that time; and about 775 the geographer Kia Tan (730–805) wrote[6] the itinerary of a voyage by sea from Canton to Persia. About 800, when Bagdad was rapidly becoming the center of the mathematical world, the Chinese received an embassy from A-lun (Harun al-Rashid). In records of the Tang Dynasty (618–907) there are numerous references to the Arabs (Ta-shi), and until the 12th century the intercourse between the Chinese and these people is frequently mentioned. Mas'ûdî (died at Cairo, 956), the famous Arab geographer and historian, visited India, Ceylon, and China in 915, and his *Meadows of Gold*, in which he mentions these countries, is well known. With such evidences as these we have a simple answer to the question as to whether it is probable that China knew of the status of Western mathematics before her own period of remarkable activity, and whether, on the other hand, the West could have known anything of Oriental progress. The answer is that it would have been very strange if each had not been the case.

[1] The *Chiu-chï-li*, as it was called in Chinese. The translator was Chü-t'an Hsi-ta.

[2] That is, if the name *Ta-shi* is taken, as usual, to mean "Arab."

[3] In Chinese, Chü-t'an Chüan.

[4] Original name was Ch'ön I. See Giles, *Biog. Dict.*, No. 801.

[5] Called in Chinese by the name T'u-huo-lo.

[6] In the *T'ang shu*.

The Sixth Century. The 6th century is an important one in the history of Chinese mathematics, owing to the appearance of several works of considerable merit. The earliest of the prominent writers was probably the learned Buddhist Ch'ön Luan,[1] who seems to have been living in 535, but who devised a calendar in the second half of the century.[2] He wrote the *Arithmetic in the five classics*,[3] in which he included various problems of the standard type that had appeared in earlier works. He also wrote commentaries on several of the earlier treatises.[4]

Probably about the same time as Ch'ön Luan there lived Ch'ang K'iu-kien[5] (*c.* 575), whose arithmetic[6] in three books is nearly all extant. The work is devoted chiefly to fractions, and it seems quite clear that the author knew the modern rule of division by multiplying by the reciprocal of the divisor. It also treats of arithmetic progression, the Rule of Three, mensuration, and indeterminate linear equations.

Another contemporary of Ch'ön Luan was the arithmetician Hsia-hou Yang[7] (*c.* 550), the author of a treatise that is still extant.[8] This work includes, as was the custom in most cases of the kind, some problems in mensuration as well as a treatment of certain processes of arithmetic. The arithmetic problems all relate to multiplication, division, and percentage.

In this century there also flourished a geometer by the name of Men (*c.* 575), of whom little is known, but who is said to have given 3.14 as the value of π.

Seventh to the Tenth Century. The most prominent Chinese mathematician of the 7th century was Wang Hs'iao-t'ung,[9]

[1] Given by Père Vanhée as Tsen Loan and by Biernatzki (p. 12), who puts him early in the 7th century, as Tschin Lwan. On all these names Mikami's work has been freely used.

[2] In the reign of Wu-tí, of the Chou monarchy (557–581), in the Chin Dynasty.

[3] *Wu-king Suan-shu.*

[4] For example, on the *Chóu-peï* and the *K'iu-ch'ang Suan-shu.*

[5] Biernatzki (p. 12) transliterates the name as Tschang Kiu Kihn and gives the date as early in the 7th century.

[6] *Ch'ang K'iu-kien Suan-king.*

[7] Biernatzki gives the name as Hea Hau yang. The date is uncertain, but he probably lived in the period from *c.* 550 to *c.* 600.

[8] The *Hsia-hou Yang Suan-king* (*Arithmetic Classic of Hsia-hou Yang*).

[9] Also written Wang Hiao-t'ong and Wang Heau tung.

known to have been living in 623 and in 626. He was an expert on the calendar and was one of the first of the Chinese to write on cubic equations. His work,[1] most of which is extant, contains twenty problems on mensuration, and in some of these problems the cubic equation enters. No method of solving such equations, however, is given.

The 8th century saw no work of importance in mathematics. In 727 I-hsing devised a new calendar,[2] and two centuries later (*c.* 925) there appeared an astrological treatise of some merit,[3] but neither contained any mathematics beyond such as was needed in the work on the calendar. The Dark Ages of the West had spread over the East as well.

SHŌTOKU TAISHI, *c.* 600

From a bronze of the 18th century, showing the prince with a *soroban*, a chronological impossibility

2. Japan

Beginnings of Japanese Mathematics.[4] Although Chinese influence had begun to show itself in the intellectual development of Japan before 500, it was not until the Buddhist missionaries began to appear, in 522,[5] that any very pronounced results were noticed. Indeed, it was not until 552 that Buddhism was really introduced, and not until two years later that two scholars, learned in matters pertaining to the calendar,[6] crossed over from Korea and brought to Japan the Chinese system of chronology. Not far from the year 600 a Korean priest, Kanroku, presented to the empress a set of books on astrology and the calendar, and Prince Shōtoku Taishi showed so much interest in calculation that tradition thereafter made him the father of Japanese arithmetic.

[1] *Ch'i-ku Suan-king.*

[2] The *T'ai-yen* calendar.

[3] The *K'ai-yüan Chan-king.*

[4] See Smith-Mikami.

[5] The first to come was Szŭ-ma Ta, known in Japanese as Shiba Tatsu.

[6] These were Wang Pao-san and Wang Pao-liang. See Smith-Mikami, p. 8.

Chinese Influence in Japan. From now on for many generations Japan came completely under Chinese influence in all her intellectual life. The Chinese system of measures was adopted, a school of arithmetic was founded (*c.* 670), an observatory was established at about the same time, and in 701 a university system was inaugurated. Nine Chinese works were specified for students of mathematics,[1] and these seem to have been the classics which influenced the Japanese study of mathematics for several centuries.

TENJIN, PATRON OF MATHEMATICS, *c.* 890

From a bronze. The portrait is found also in early paintings of the Japanese

Aside from Shōtoku Taishi the man whose name stands out most prominently in the history of Japanese mathematics in this period is Tenjin,[2] counselor and teacher at the imperial court (*c.* 890) and a great patron of science and letters.

Altogether the era was one of preparation, contributing nothing new to what China had already developed. Indeed it was not until the 17th century that Japan really awoke to her possibilities in the field of mathematics.

3. India

General Nature of the Work. In the period from 500 to 1000 there were four or five mathematicians of prominence in India. These were the two Āryabhaṭas,[3] Varāhamihira the astronomer, Brahmagupta, and Mahāvīrācārya. In the works of all these writers there is such a mixture of the brilliant and the

[1] These were (1) *Chóu-peï Suan-king*, (2) *Sun-tzï Suan-king*, (3) *Liu-chang*, (4) *San-k'ai Chung-ch'a*, (5) *Wu-ts'ao Suan-shu*, (6) *Hai-tau Suan-shu*, (7) *Kiu-szu*, (8) *Kiu-ch'ang*, (9) *Kiu-shu*, of which the third, fourth, and seventh are lost.

[2] His name was Michizane, but after his death he was canonized as Tenjin, "Heaven man."

[3] For rules for pronouncing Hindu names, see page xxi.

commonplace as to make a judgment of their qualities depend largely upon the personal sympathies of the student. Albêrûnî (*c.* 1000), the Arab historian, speaks of this peculiarity of their writings in these words:

I can only compare their mathematical and astronomical literature . . . to a mixture of pearl shells and sour dates, or of pearls and dung, or of costly crystals and common pebbles. Both kinds of things are equal in their eyes, since they cannot raise themselves to the methods of a strictly scientific deduction.[1]

Āryabhaṭa.[2] The first of the great writers whose name has come down to us is the elder Āryabhaṭa,[3] born at Kusumapura (Kousambhipura), the City of Flowers,[4] a small town on the Jumma just above its confluence with the Ganges.[5] The place is not far from the present Patna (Patnā), called by the Mohammedans Azimabad, by the ancient Buddhists Pāṭaliputra (Pātoliputra), and by Megasthenes, the Syrian ambassador,

[1] Albêrûnî's *India*, translated by E. C. Sachau, 2 vols., I, 25 (London, 1910); hereafter referred to as Albêrûnî's *India.* On the relation of Greek and Hindu arithmetic see H. G. Zeuthen, *Bibl. Math.*, V (3), 97. On the relation of India to the West in general, see H. G. Rawlinson, *Intercourse between India and the Western World from the Earliest Times to the Fall of Rome* (Cambridge, 1916). For extreme Hindu claims, see Benoy Kumar Sarkar, *Hindu Achievements in Exact Science* (New York, 1918).

[2] Born 475 or 476; died *c.* 550.

[3] Sir M. Monier-Williams, *Indian Wisdom*, 4th ed., p. 175 (London, 1893) (hereafter referred to as Monier-Williams, *Indian Wisdom*); J. Garrett, *Classical Dictionary of India*, p. 767 (Madras, 1871); C. M. Whish, "On the Alphabetical Notation of the Hindus," *Trans. of the Literary Society of Madras* (London, 1827); L. Rodet, "Leçons de Calcul d'Âryabhaṭa," *Journal Asiatique*, XIII (7), 393; L. Rodet, "Sur la véritable signification de la notation numérique inventée par Âryabhaṭa," *ibid.*, XVI, p. 440. Certain fragments of his works were published by H. Kern in the *Journal of the Royal Asiatic Society*, XX (1863), 371. See also G. R. Kaye, *Indian Math.*, p. 11, and "Āryabhaṭa," *Journ. and Proc. of the Asiatic Soc. of Bengal*, IV (N. S.), p. 111 (hereafter referred to as Kaye, *Āryabhaṭa*); an article on "Ancient Hindu Spherical Astronomy," *ibid.*, XV; and an article in *Scientia*, XXV, 1, all claiming Greek origin for most of the Hindu work.

[4] The term is also applied to Pāṭaliputra. If we may trust to the rather obscure statements of Albêrûnî, it was the younger Āryabhaṭa, however, who was born at Kusumapura. See the mention of him later.

[5] Sir E. Clive Bayley, *Journal of the Royal Asiatic Society*, XV (N.S.), 21.

Palibothra.[1] Because of this geographic proximity Āryabhaṭa is often said to have been born at Pātaliputra. A tradition says that the city was originally called Pāṭaliputraka, being founded by Putraka, the knight of the magic cup and staff and slippers, who married the princess Pāṭali.[2] The tradition further asserts that Buddha, toward the close of his life, crossed the Ganges at this point and prophesied the future greatness of the city.[3] By the beginning of the 5th century, and nearly a century before the birth of Āryabhata, it had lost some of its ancient prestige, for Fa-hién, the Chinese traveler already mentioned, describes (*c.* 400) the ruins of the royal palace which Aśoka commissioned the genii to build, although he speaks of the remarkable hospitals and other institutions still to be found there.[4] Āryabhaṭa evidently wrote there or at

MATHEMATICAL-HISTORICAL MAP OF INDIA

At Delhi, Jaipur, and Benares are interesting relics of native observatories; Patna is approximately the birthplace of Āryabhata (*c.* 475); Ujjain was the leading mathematical center of ancient India and is known particularly for Varāhamihira (*c.* 505), Brahmagupta (*c.* 628), and Bhāskara (*c.* 1150); about 75 miles from Poona are the Nānā Ghāt inscriptions with early numerals; it was at Mysore that Mahāvīra (*c.* 850) lived

[1] E. Reclus, *Asia*, American ed., III, 222. [2] J. Garrett, *loc. cit.*, p. 779.

[3] R. W. Fraser, *A Literary History of India*, p. 143 (N. Y., 1898); E. W. Hopkins, *Religions of India*, pp. 5, 311 (Boston, 1898).

[4] Dutt, *Hist. of Civ. in Anc. India*, II, 58 (London, 1893). On the sojourn of Megasthenes, 306–298 B.C., see Fraser, *loc. cit.*, p. 175. On its importance at about the time of Āryabhaṭa, see the inscriptions of Chandragupta II in

Kusumapura, for he says in one of his works: "Having paid homage to Brahma, to Earth, to the Moon, to Mercury, to Venus, to the Sun, to Mars, to Jupiter, to Saturn, and to the constellations, Āryabhaṭa, in the City of Flowers, sets forth the science venerable."[1]

It was probably because Āryabhaṭa lived so far from Ujjain, the ancient center of mathematics and astronomy, that his works were so little known among Hindu scholars of the centuries immediately following.

Āryabhaṭa's Work. His work, often called the *Āryabhaṭīya*[2] or *Āryabhaṭīyam*, consists of the *Gītikā* or *Daśagītikā*, a collection of astronomical tables, and the *Āryāṣṭasata*, which includes the *Ganita*, a treatise on arithmetic[3]; the *Kālakriyā*, on time and its measure; and the *Gola*, on the sphere.

The arithmetic carries numeration by tens as far as 10^8, treats of plane and solid numbers, and gives a rule for square root. It contains a rule for summing an arithmetic series after the pth term, which may be expressed in modern symbols thus:

$$s = n\left[a + \left(\frac{n-1}{2} + p\right)d\right].$$

It also has a rule which we express by the formula

$$n = \frac{1}{2}\left(1 + \frac{-2a \pm \sqrt{(d-2a)^2 + 8sd}}{d}\right).$$

The rest of the work shows a knowledge of the quadratic equation and of the indeterminate linear equation.

J. F. Fleet, *Corpus Inscriptionum Indicarum*, III, pl. iv, A, B (London, 1888). A century and a half later (629–645) the Chinese pilgrim Hüan-tsang remarked, "Although it has long been deserted, its foundation walls still survive." See Fraser, *loc. cit.*, p. 248.

[1] Rodet, *loc. cit.*, p. 396. For a slightly different translation see Kaye, *Āryabhaṭa*, p. 116.

[2] Monier-Williams, *Indian Wisdom*, p. 175; Mrs. Manning, *Ancient and Mediaeval India*, 2 vols. (London, 1869), largely from the *Journal of the Royal Asiatic Soc.*, I (N.S.), 392, and XX, 371.

[3] Rodet, "Leçons," *loc. cit.*, p. 395. He translates the second part, p. 396. See also Kaye, *Āryabhaṭa*, p. 111; *Bibl. Math.*, XIII (3), 203.

Among the rules relating to areas is one for the isosceles triangle, and this will serve to show the imperfect form of statement used by Āryabhaṭa: "The area produced by a trilateral is the product of the perpendicular that bisects the base, and half the base." The formula for the volume of a sphere is very inaccurate, being $\pi r^2\sqrt{\pi r^2}$, which would make π equal to $\frac{16}{9}$, possibly an error for the $(\frac{16}{9})^2$ of Ahmes.

The rule for finding the value of π is given as follows: "Add four to one hundred, multiply by eight, and add again sixty-two thousand; the result is the approximate value of the circumference when the diameter is twenty thousand." This makes π equal to $\frac{62832}{20000}$, or 3.1416.[1] Āryabhaṭa also gives a rule for finding sines, and the *Gītikā* has a brief table of these functions.

His work is also noteworthy as containing one of the earliest attempts at a general solution of a linear indeterminate equation by the method of continued fractions.[2]

As stated above, the Āryabhaṭa here mentioned is known as the elder of the two mathematicians of the same name. This fact appears in the work of Albêrûnî[3] and has been the subject of comment by recent writers.[4] The date of the younger Āryabhata is unknown, nor is it possible as yet to differentiate clearly between the works of the two. He seems, from the meager authorities now known, to have been born at Kusumapura.

Varāhamihira (*c.* 505). Among the astronomers of India[5] two appear with the name of Varāhamihira, one living *c.* 200 and the other *c.* 505.[6] The latter of these scholars is the most celebrated of all the writers on astronomy in early India. He wrote several works, of which the *Pañca Siddhāntikā*, treating of astrology and astronomy, is the best known. It includes the computation necessary for finding the position of

[1] Kaye, in his *Āryabhaṭa*, questions whether this is from the works of the elder Āryabhaṭa, the one of whom we are speaking. He thinks it is due to the younger one mentioned below.

[2] Kaye, *Indian Math.*, p. 12.

[3] *India*, II, 305, 327.

[4] See summary by Kaye, *Āryabhaṭa*, p. 113.

[5] A list with dates is given in Colebrooke, *loc. cit.*, p. xxxiii.

[6] The date is quite uncertain. Varāhamihira is said by some Oriental authorities to have died *c.* 587.

a planet, shows an advanced state of mathematical astronomy, but is chiefly valuable in the history of mathematics because of the description that it gives of the five *Siddhāntas* which had been written just before this time.[1] He urged his people to appreciate the work of the Greeks, saying: "The Greeks, though impure, must be honored, since they were trained in the sciences and therein excelled others. What then, are we to say of a Brahman if he combines with his purity the height of science?"[2]

Varāhamihira taught the sphericity of the earth, and in this respect he was followed by most of the other Hindu astronomers of the Middle Ages.[3] Two of his works were translated into Arabic by Albêrûnî (*c.* 1000).[4]

Brahmagupta. The most prominent of the Hindu mathematicians of the 7th century was Brahmagupta,[5] whose period of activity has been fixed as *c.* 628, both from astronomical data and from the testimony of various Hindu writers.[6] He lived and worked in the great astronomical center of Hindu science, Ujjain or Ujjayinī, a town in the state of Gwalior, Central India, said to have been the viceregal seat of Aśoka during his father's reign at Patna. Varāhamihira also carried on his work at the observatory in Ujjain.

When he was only thirty years old, Brahmagupta wrote an astronomical work in twenty-one chapters entitled *Brahmasiddhānta,*[7] which includes as special chapters the *Gaṇitād'hāya*[8]

[1] G. Thibaut and Sudharkar Dvivedi, *The Pañcha-siddhāntikā of Varaha Mihira.* Benares, 1889.

[2] Albêrûnî's *India,* I, 23.

[3] Albêrûnî, *loc. cit.*, I, 266.

[4] For a list of his works, see Albêrûnî, *loc. cit.*, I, xxxix. For the influence of the Greeks upon his work and upon Hindu astronomy in general, see Colebrooke, *loc. cit.*, p. lxxx.

[5] Colebrook, *loc. cit.* Albêrûnî (*c.* 1000) speaks of him as "the son of Jishṇu, from the town of Bhillamāla." Sūryadāsa, a commentator on Bhāskara, also speaks of him as the son of Jishṇu.

[6] Colebrooke (*loc. cit.*, p. xxxv) makes the date 581 or 582, from Brahmagupta's reference to the position of the star *Chitrá* (*Spica Virginis*). The Hindu astronomers make it *c.* 628. He seems to have been born *c.* 598.

[7] Also called the *Brāhma-sphuṭa-sidd'hānta,* "Brahma correct system," possibly a revision. Albêrûnî (*c.* 1000) gives twenty-four chapters, with the title of each. See his *India,* I, 154; II, 303.

[8] *Lectures on Arithmetic.*

and the *Kuṭakhādyaka*.[1] The former begins by a definition of a *gaṇaca*, that is, a calculator who is competent to study astronomy: "He who distinctly and severally knows addition and the rest of the twenty logistics and the eight determinations, including measurement by shadow, is a *gaṇaca*."[2]

Nature of Brahmagupta's Arithmetic. The arithmetic includes work with integers and fractions, progressions, barter, Rule of Three, simple interest, the mensuration of plane figures, and problems on volumes and on shadow reckoning (a primitive plane trigonometry applied by him to the sundial). The mensuration is often faulty, as where Brahmagupta states a rule which would give the area of an equilateral triangle of side 12 as 6×12, or 72; that of the isosceles triangle 10, 13, 13 as 5×13, or 65; and that of the triangle 13, 14, 15 as $7 \times \frac{1}{2} \times (13 + 15)$, or 98. He also states that the area of any quadrilateral whose sides are a, b, c, d is

$$\sqrt{(s-a)(s-b)(s-c)(s-d)},$$

where $s = \frac{1}{2}(a+b+c+d)$, a formula that is true only for cyclic quadrilaterals. His rule for the quadrilateral is as follows: "Half the sum of the sides set down four times, and severally lessened by the sides, being multiplied together, the square root of the product is the exact area."[3] He uses 3 as the "practical value" of π and $\sqrt{10}$ as the "neat value."

Brahmagupta's Algebra. The *Kuṭakhādyaka* applies algebra to astronomical calculations. For example, "One who tells, when given positions of the planets, which occur on certain lunar days or on days of other denomination of measure, will recur on a given day of the week, is versed in the pulverizer."[4]

[1] *Lectures on Indeterminate Equations.* The *kuṭaka* (*kuṭṭaka*, *cuṭacā*) is defined by Colebrooke (*loc. cit.*, p. vii) as "a problem subservient to the general method of resolution of indeterminate problems of the first degree." The word means "pulverizer" and is used as a name for algebra. *Ibid.*, p. 325; J. Taylor, *Lilawati*, p. 129 (Bombay, 1816); hereafter referred to as Taylor, *Lilawati*, with this spelling. The word *khādyaka* means "sweetmeat," such fanciful names being common in the East.

[2] Colebrooke, *loc. cit.*, p. 277.

[3] *Ibid.*, p. 295.

[4] Question 7 of the Colebrooke translation.

In his chapter on computation[1] Brahmagupta gives the usual rules for negative numbers. He also has a chapter on quadratic equations, the rule[2] for solving an equation of the type $x^2 + px - q = 0$ being substantially a statement of the formula

$$x = \frac{\sqrt{p^2 + 4q} - p}{2},$$

which evidently gives one root correctly.

In the case of simultaneous equations of the first degree the unknowns are spoken of as "colors," and the problems are chiefly astronomical. Indeed, Brahmagupta was the first Indian writer, so far as we know, who applied algebra to astronomy to any great extent. While the fanciful problems so often found in Indian works are generally wanting, a commentator has supplied various examples to illustrate certain of his rules. Two such problems are as follows:

On the top of a certain hill live two ascetics. One of them, being a wizard, travels through the air. Springing from the summit of the mountain he ascends to a certain elevation and proceeds by an oblique descent diagonally to a neighboring town. The other, walking down the hill, goes by land to the same town. Their journeys are equal. I desire to know the distance of the town from the hill, and how high the wizard rose.

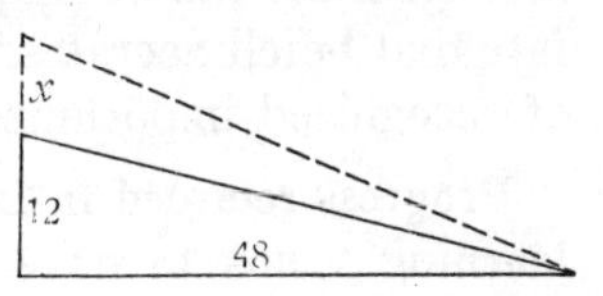

The commentator takes the case here shown, and finds x to be 8.

A bamboo 18 cubits high was broken by the wind. Its tip touched the ground 6 cubits from the root. Tell the lengths of the segments of the bamboo.[3]

Indeterminate Equations. It is indicative of the state of algebra at this time that Brahmagupta was interested in the solution of indeterminate equations. Āryabhaṭa had already

[1] *Shaṭ-trinṣat-paricarman.* [2] Page 346 of the Colebrooke translation.
[3] From his arithmetic. For an early Chinese version, see page 139.

considered the question of the integral solutions of $ax \pm by = c$, but Brahmagupta actually gave as the results

$$x = \pm cq - bt,$$
$$y = \mp cp + at,$$

where t is zero or any integer and p/q is the penultimate convergent of a/b.[1] He also considered the so-called Pell Equation of the form

$$Du^2 + 1 = t^2,$$

but the solution was first effected, so far as we know, by Bhāskara in the 12th century.

For the sides of the right-angled triangle Brahmagupta gave the two sets of values

$$2\,mn, \quad m^2 - n^2, \quad m^2 + n^2,$$

and $$\sqrt{m}, \quad \frac{1}{2}\left(\frac{m}{n} - n\right), \quad \frac{1}{2}\left(\frac{m}{n} + n\right),$$

values which he probably obtained from Greek sources.

Brahmagupta was accused of propagating falsehoods relating to science for the purpose of pleasing the bigoted priests and ignorant rabble of his country, hoping thus to avoid the fate that befell Socrates,[2] all of which shows that he was a man of recognized importance in his day.

Progress retarded in India. From this time to the year 1000 learning seems to have made but little progress in northern India. In the 8th century the Rajput dynasty succeeded the high-minded Valabhis, and for two hundred years the history of this part of India is a blank. Not a piece of literature of any value remains, nor any work of art or of industry.[3] The abode of mathematics now moved northward and is found for two or three centuries in Persia and in the other lands which had been brought under Moslem rule. In southern India, however, there must have been some encouragement of mathematics, as will be seen from the great work of Mahāvīra.

[1] Kaye, *Indian Math.*, p. 16.
[2] See Sachau's note in his translation of Alberûnî's *India*, II, 304.
[3] Dutt, *History of Civ. in Anc. India*, II, 162.

Mahāvīra. The third of the great Hindu writers of this period is Mahāvīrācārya, Mahāvīra the Learned, who wrote the *Gaṇita-Sāra-Sangraha.*[1] This writer probably lived at the court of one of the old Rāshṭrakūta monarchs who ruled over what is now the kingdom of Mysore, and whose name is given as Amoghavarsha Nirpatuṅga. This king ascended the throne in the first half of the 9th century, so that we may roughly fix the date of the treatise in question as *c.* 850, or between the dates of Brahmagupta and Bhāskara,[2] though nearer to the former.

The work begins, as is not unusual with Oriental treatises, with a salutation of a religious nature. In this case the words are addressed to the author's patron saint, the founder of the religious sect of the Jainas (Jinas), a contemporary of Buddha:

> Salutation to Mahāvīra, the Lord of the Jinas, the protector [of the faithful], whose four infinite attributes, worthy to be esteemed in [all] the three worlds are unsurpassable [in excellence].
>
> I bow to that highly glorious Lord of the Jinas, by whom, as forming the shining lamp of the knowledge of numbers, the whole of the universe has been made to shine.

Mahāvīra's Sources. In general it may be said that Mahāvīra seems to have known the work of Brahmagupta. It would have been strange if this had not been so, for the *Brāhma-sphuṭa-siddhānta* was probably recognized in his time as one of the standard authorities. Mahāvīra seems to have made the effort to improve upon the work of his predecessor, and certainly did so in his classification of the operations, in the statement of rules, and in the nature and number of problems. As a result his work became well known in southern India, although there is no definite proof that Bhāskara (*c.* 1150), living in Ujjain, far to the north, was familiar with it.

Mahāvīra's Work. The work itself consists of nine chapters. The first is introductory and relates chiefly to the measures

[1] M. Raṅgācārya, *The Gaṇita-Sāra-Sangraha of Mahāvīrācārya*, Sanskrit and English, Madras, 1912; hereafter referred to as *Mahāvīra. Gaṇita-Sāra* means "Compendium of Calculation."

[2] He lived *c.* 1150. See page 275.

used, the names of the operations, numeration, negatives, and zero. Eight operations with numbers are given, addition (except in series) and subtraction (even with fractions) being omitted as if presupposed. One interesting feature is the law relating to zero, which is stated thus: "A number multiplied by zero is zero, and that [number] remains unchanged when it is divided by, combined with, [or] diminished by zero." That is, the law given by Bhāskara for dividing by zero is not here recognized, division by zero being looked upon as of no effect. The law of multiplication by negative numbers is stated, and the imaginary number is thus disposed of: "As in the nature of things a negative [quantity] is not a square [quantity], it has therefore no square root."

In his arithmetic operations he first treats of multiplication. He then considers in order the topics of division, squaring, square root, cubing, cube root, and the summation of series. In his work in series he includes some treatment of arithmetic and geometric progressions and of *Vyutkalita*, that is, the summation of a series after a certain number of initial terms (*ista*) have been cut off, a theory which, as we have seen (p. 155), occupied the attention of Āryabhaṭa.

The most noteworthy feature in his treatment of fractions is that relating to the inverted divisor, the rule being set forth as follows: "After making the denominator of the divisor its numerator [and *vice versa*], the operation to be conducted then is as in the multiplication [of fractions]." It is curious that this device, which from another source we know to have been used in the East, became a lost art until again adopted in Europe in the 16th century.

His method of approach to the subject of quadratic and radical equations is through fanciful problems of which the following is a type:

One fourth of a herd of camels was seen in the forest; twice the square root [of that herd] had gone on to mountain slopes; and three times five camels [were] however, [found] to remain on the bank of a river. What is the [numerical] measure of that herd of camels?

This evidently requires the finding of the positive root of the equation $\frac{1}{4}x + 2\sqrt{x} + 15 = x$, or, in general, the solution of an equation of the type $x-(bx+c\sqrt{x}+a)=0$, the rule for which is given. The chapter also contains various other types of equations involving some knowledge of radical quantities.

A single example will also suffice to show the nature of his indeterminate problems:

Into the bright and refreshing outskirts of a forest, which were full of numerous trees with their branches bent down with the weight of flowers and fruits, trees such as jambu trees, lime trees, plantains, areca palms, jack trees, date palms, hintala trees, palmyras, punnāga trees, and mango trees—[into the outskirts], the various quarters whereof were filled with the many sounds of crowds of parrots and cuckoos found near springs containing lotuses with bees roaming about them—[into such forest outskirts] a number of weary travelers entered with joy. [There were] sixty-three [numerically equal] heaps of plantain fruits put together and combined with seven [more] of those same fruits, and these were equally distributed among twenty-three travelers so as to have no remainder. You tell me now the numerical measure of a heap of plantains.

Mahāvīra's Treatment of Areas. His work in the measurement of areas is somewhat like the corresponding chapter in Brahmagupta's treatise, although it is distinctly in advance of the latter. Mahāvīra makes the same mistake as Brahmagupta with respect to the formula for the area of a trapezium (trapezoid) in that he does not limit it to a cyclic figure. The same error enters into his formula for the diagonal of a quadrilateral, which he gives as

$$\sqrt{\frac{(ac+bd)(ab+cd)}{ad+bc}} \quad \text{or} \quad \sqrt{\frac{(ac+bd)(ad+bc)}{ab+cd}}.$$

For the Pythagorean triangle Mahāvīra gives rules similar to those of Brahmagupta. For π he uses $\sqrt{10}$, a common value all through the East and also in medieval Europe. He was the only Hindu scholar of the native school who made any serious attempt to treat of the ellipse, but his work was inaccurate.

His rule for the sphere is interesting, the approximate value being given as $\frac{9}{2}(\frac{1}{2}d)^3$, and the accurate value as $\frac{9}{10}\cdot\frac{9}{2}(\frac{1}{2}d)^3$, which means that π must be taken as $3.03\frac{3}{4}$.

All things considered, the work of Mahāvīra is perhaps the most noteworthy of the Hindu contributions to mathematics, possibly excepting that of Bhāskara, who lived three centuries later. Mahāvīra may have known the works of Chinese scholars, for the value that he gives for the area of the segment of a circle, $\frac{1}{2}(c+a)a$, was given six centuries earlier by Ch'ang Ts'ang, but in any case he was a man of scientific attainments.

Bakhshālī Manuscript. Another work that stands out with some prominence in this period is the Bakhshālī manuscript.[1] This work, of uncertain origin and date, contains material relating to both arithmetic and algebra. It was formerly referred to the early part of our era and then to the 8th or 9th century, but it gives evidence of having been written even after the latter period, and possibly it is not even of Hindu origin. The nature of the work may be inferred from a single problem:

> A merchant pays duty on certain goods at three different places. At the first he gives $\frac{1}{3}$ of the goods, at the second $\frac{1}{4}$ [of the remainder] and at the third $\frac{1}{5}$ [of the remainder]. The total duty is 24. What was the original amount of the goods?[2]

4. Persia and Arabia

Persia. We are apt to think that the rise of learning in the lands conquered by the Mohammedans was due solely to Arab influence, but this is not the case. In Persia, for example, Khosrú the Holy,[3] a generous patron of science, invited to his

[1] R. Hoernle, "The Bakhshālī Manuscript," *Indian Antiquary*, Vol. XVIII (1888); G. R. Kaye, "Notes on Indian Mathematics," in *Journ. and Proc. of the Asiatic Soc. of Bengal*, III (2), 501 (hereafter referred to as Kaye, *Notes*); and "The Bakhshālī Manuscript," *ibid.*, VIII (2), 349 (hereafter referred to as Kaye, *Bakhshālī*).

[2] The answer is 40, which necessitates the bracketed words.

[3] Khosrú I, Anôschirvân. He was a contemporary of Justinian, who was crowned emperor in Constantinople in 527. See W. S. W. Vaux, *Persia*, p. 169 (London, 1875); T. Nöldeke, *Aufsätze zur persischen Geschichte*, p. 113 (Leipzig, 1887).

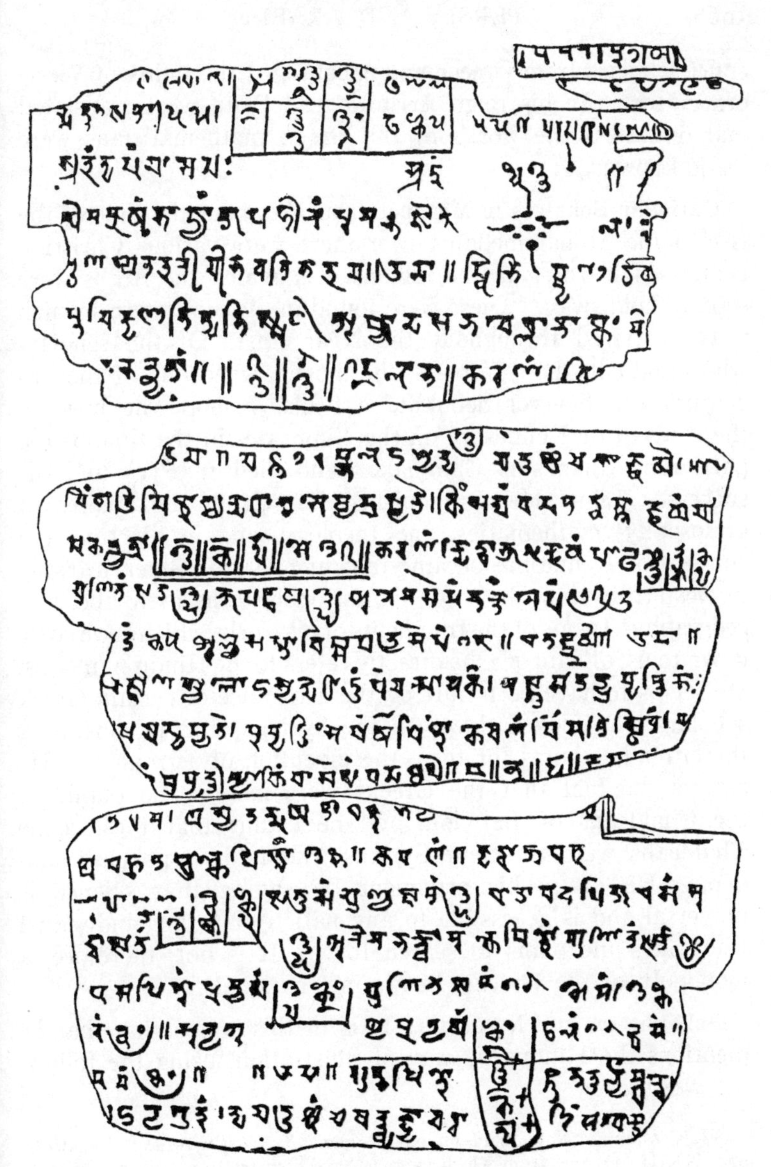

BAKHSHĀLĪ MANUSCRIPT

A portion of this manuscript of which the date is still unsettled. It may be of the 10th century. From Kaye's *Indian Mathematics*

court scholars from Greece and encouraged the influx of Western culture. In his reign Aristotle and Plato were translated and doubtless the works of the Greek mathematicians were made known.

Christian Scholars in Mesopotamia. At about the time of the rise of the Mohammedan power there were various Christian centers of learning in the regions over which the Arabs were soon to hold sway. These were found in the monasteries which were scattered throughout the Near East. Of the scholars who taught in these retreats, the most learned one of the 7th century was Severus Sebokht,[1] a titular bishop who lived in the convent of Kenneshre on the Euphrates in the time of the patriarch Athanasius Gammala (who died in 631) and his successor John. He distinguished himself in the studies of philosophy, mathematics, and theology, and in his time the convent of Kenneshre became the chief seat of Greek learning in western Syria. He wrote on astronomy, the astrolabe, and geography. In one of the fragments of his works which have come down to us, of date 662, he directly refers to the Hindu numerals. He seems to have been hurt by the arrogance of certain Greek scholars who looked down on the Syrians, and in defending the latter he claims for them the invention of astronomy. He asserts the fact that the Greeks were merely the pupils of the Chaldeans of Babylon, and he claims that these same Chaldeans were the very Syrians whom his opponents condemn. He closes his argument by saying that science is universal and is accessible to any nation or to any individual who takes the pains to search for it. It is not, therefore, a monopoly of the Greeks, but is international.

Sebokht on our Numerals. It is in this connection that he mentions the Hindus by way of illustration, using the following words:

[1] J. Ginsburg, "New Light on our Numerals," *Bulletin of the Am. Math. Soc.*, XXIII (2), 366, from which extracts have been freely made. Attention of English readers was first called to this writer's mathematical works by Professor Karpinski, *Science* (U.S.), June, 1912. See also E. R. Turner, *Popular Sci. Mo.*, December, 1912.

I will omit all discussion of the science of the Hindus, a people not the same as the Syrians; their subtle discoveries in this science of astronomy, discoveries that are more ingenious than those of the Greeks and the Babylonians; their valuable methods of calculation; and their computing that surpasses description. I wish only to say that this computation is done by means of nine signs. If those who believe, because they speak Greek, that they have reached the limits of science should know these things they would be convinced that there are also others who know something.

Bagdad. It was at Bagdad, on the Tigris River, that mathematics had its greatest encouragement under the Mohammedan ascendancy. Built upon the ruins of an ancient town by the caliph[1] al-Mansûr (712–774/5), one of the Abbasides,[2] Bagdad[3] became the intellectual center of the Mohammedan world, a second Alexandria in its fostering of learning. In al-Mansûr's reign (*c.* 766) a work mentioned as the *Sindhind* is said to have been brought to his court by an Indian scholar named Kankah (Mankah?), the Hindu astronomy and mathematics being thus made known to the scholars of Bagdad. This work may have been the *Sūrya Siddhānta* or it may have been some other work bearing the title *Siddhānta*, this name being nearer to *Sindhind* than any other Sanskrit word likely to be meant. It is generally believed, however, that it was the *Brahmasiddhānta* of Brahmagupta, whose works are known to have been brought to Bagdad at this time.[4]

To the court of the Caliphs there also came, so the story goes, a Persian by the name of Ya'qûb ibn Ṭâriq (died 796). He is said to have written (775) on the sphere (mathematical astronomy) and the calendar, and to have edited, and probably to have assisted in translating, the works of Brahmagupta

[1] Calif, from *Khalīfah*, successor (of the Prophet).

[2] Ab bäs'īdes or Ab'ba sīdes, the (at least pretended) descendants of Abbâs, uncle and adviser of Mohammed. Al-Mansûr reigned from 753/4 to 774/5. For rules for pronouncing Arabic names, see page xx.

[3] Persian *Bagadata*, "God-given"; in Arabic, Dār al-Salām, "Abode of Peace." Also spelled Baghdad.

[4] Sachau's preface to his translation of Albêrûnî's *India*, I, xxxi. As to the absence of Arabic records to prove that any embassy came from India at this time, see *ibid.*, II, 313.

above mentioned. To the same court there came (and the records of this fact are somewhat more trustworthy) the astronomer Abû Yaḥyâ,[1] and there he translated the *Tetrabiblos* of Ptolemy, thus assisting to begin the great movement that led to the introduction of the classics of Greek mathematics into the court of the Caliphs.

About the same time al-Fazârî[2] (died 777), working also at Bagdad, wrote on astrology and the calendar. He was the first Moslem, so far as is known, to construct astrolabes and to write on mathematical instruments. His famous contemporary, Jeber,[3] the greatest alchemist of the Arabs, also wrote on the astrolabe and possibly on mathematics.[4]

It was in this reign that another al-Fazârî,[5] son of the one already mentioned, a man of unusual scholarship, particularly in the field of astronomy, was asked by the caliph to translate the *Siddhānta* brought to Bagdad by Kankah. It was on this translation that Mohammed ibn Mûsâ al-Khowârizmî (*c.* 825) based his astronomical tables.

Harun al-Rashid. Harun al-Rashid,[6] well known to us from the *Arabian Nights Tales*, was a great patron of learning. Under his influence several of the Greek classics in science, including part of Euclid's works, were translated into Arabic. Indeed, it is to the Arabic versions that medieval Europe was indebted for its first knowledge of Euclid's *Elements*. In his reign there was a second influx of Hindu learning into Bagdad, especially in the line of medicine and astrology.

[1] Abû Yaḥyâ al-Baṭrîq, who died about 796–806.

[2] Ibrâhîm ibn Ḥabîb ibn Soleimân ibn Samora ibn Jundab, Abû Isḥâq al-Fazârî.

[3] Jâbir ibn Ḥaiyân al-Ṣûfî, Abû 'Abdallâh (died *c.* 777), one of two prominent scholars known by the name of Geber in the Middle Ages.

[4] The matter is discussed briefly in H. Suter, "Die Mathematiker und Astronomen der Araber und ihre Werke," in Volume X of the *Abhandlungen*. The transliteration of Arabic names is taken from Suter's list, with the change of *el* to *al*, of *ǧ* to *j*, of *š* to *sh*, of *ch* to *kh*, of *w* to *v*, and of *j* to *y* as in English. While this is not always desirable in the case of *sh*, as in Isḥâq, it is much simpler for the general reader. With respect to all these names the student should consult Suter's work.

[5] Mohammed ibn Ibrâhîm ibn Ḥabîb, Abû 'Abdallâh al-Fazârî. He died between 796 and 806.

[6] Hârûn al-Rashîd, Aaron the Just. He reigned from 786 to 808/9.

Al-Mâmûn. Harun al-Rashid's son, al-Mâmûn (reigned 809–833), was also a great patron of learning; indeed, he was more than a mere patron, for he erected an observatory at Bagdad and himself took observations there. He is also credited with supervising two geodetic surveys in Mesopotamia for the purpose of determining the length of a degree of the meridian. Under his direction the translation of the Greek classics continued, the *Almagest* of Ptolemy being put into Arabic and the translation of the *Elements* of Euclid being completed. In order to show the great activity among the Arabs in the field of mathematics, and the general nature of the work accomplished, a brief list of names will be given, with notes that are necessarily condensed, although it is apparent that most of the names are unfamiliar and most of the details will pass from the reader's mind.

It is evident that astronomy was the science of this period that did most to bring mathematics into a favorable light at court. Linked up with astrology on the one hand and with mathematics on the other, it introduced just enough of superstition through the former to help establish the latter science.

Writers in al-Mâmûn's Reign. Among those who, in al-Mâmûn's remarkable reign, wrote upon mathematical astronomy, thus assisting to advance the study of trigonometry, the following scholars deserve special mention, not so much for their genius as for their spirit: al-Tabarî,[1] who wrote a commentary on Ptolemy's *Tetrabiblos*; al-Nehâvendî,[2] who prepared a set of astronomical tables; al-Mervarrûḏî,[3] who made astronomical observations at Damascus and Bagdad (*c.* 830); al-Aṣṭorlâbî,[4] who lived in Bagdad (*c.* 830), wrote on astronomy and geodesy, and was celebrated as a maker of astrolabes and other astronomical instruments; Messahala,[5] a Jewish

[1] Omar ibn al-Farrukhân, Abû Ḥafṣ al-Tabarî, died *c.* 815.

[2] Aḥmed ibn Mohammed al-Nehâvendî, died *c.* 835–845.

[3] Khâlid ibn 'Abdelmelik al-Mervarrûḏî.

[4] 'Alî ibn 'Isâ al-Aṣṭorlâbî.

[5] Mâ-shâ'-allâh ibn Aṯarî. The spelling in the text is the one commonly used in the West. The text of one of his MSS. was published by W. W. Skeat in his edition of Chaucer's *Astrolabe*, London, 1872. His chief work was done just before al-Mâmûn's reign.

astrologer, who wrote (*c.* 800) a treatise on the astrolabe that seems to have influenced the later works of Rabbi ben Ezra (*c.* 1150) and Chaucer (*c.* 1400); and Alfraganus[1] (*c.* 833), to use his European name, who wrote on sundials, astronomy, and the *Almagest.*

Mohammed ibn Mûsâ al-Khowârizmî. The greatest mathematician at the court of al-Mâmûn was Mohammed ibn Mûsâ al-Khowârizmî,[2] Abû 'Abdallâh (died between 835 and 845), a native of Khwarezm, the country in which is now the city of Khiva. Although an astronomer and the author of several astronomical tables and of works on dials, the astrolabe, and chronology, he is best known for having written the first work bearing the name "algebra," a treatise based upon Greek models.[3] He also wrote on arithmetic, this work being translated into Latin by Robert of Chester or by Adelard of Bath under the title *Algoritmi de numero Indorum*, whence such words as *algorism* and *augrim*,[4] derived from al-Khowârizmî. The title of the algebra was *'ilm al-jabr wa'l muqabalah*, "the science of reduction and cancellation."[5] After al-Mâmûn's death mathematics continued to flourish in Bagdad for about a century and a half, although, as might be expected, with somewhat less encouragement.[6]

[1] Mohammed ibn Ketîr al-Fargânî. The European translators also used such forms as Alfergani and Alfragan. Johannes Hispalensis translated his version of the *Almagest* into Latin, and it was printed at Ferrara in 1493 and again, with a preface by Melanchthon, at Nürnberg in 1537.

[2] This transliteration is more familiar to English readers than is Suter's el-Chowârezmî or Chwârezmî. The name means Mohammed son of Moses, the Khwarezmite. C. Huart, *History of Arab Literature*, pp. 131, 292, 297 (London, 1903), says that there were two others by the name of al-Khowârizmî, one (the geographer) of 935–993 (or 1002), and the other of *c.* 1036, but Huart seems to have been confused in this matter. The name also appears as al-Khowârazmî and as al-Khowâruzmî.

[3] L. C. Karpinski, *Robert of Chester's Latin Translation of the Algebra of al-Khowarizmi*, New York, 1915.

[4] So Chaucer speaks, in the *Canterbury Tales*, of "augrim stones."

[5] See Volume II, Chapter VI.

[6] Another Mohammed of Bagdad wrote a work on the division of surfaces. On the relation of this work to Euclid's book on the divisions of figures, see the careful study by Professor R. C. Archibald in his *Euclid's Book on Divisions of Figures*, pp. 1–8 (Cambridge, 1915).

Other Scholars of Bagdad. Almâhânî[1] (*c.* 860), as he is commonly called, an astronomer of high standing, is perhaps best known for having written upon the familiar problem of Archimedes relating to the cutting of a sphere into segments having a given ratio of volume. In his stereometric solution of the cubic equation involved in this problem he made use of the sine of a trihedral angle. He also wrote commentaries on Books V and X of Euclid's *Elements* and on the work of Archimedes on the sphere and cylinder.

Alchindi[2] (*c.* 860), to use the name by which he was generally known in medieval Europe, was commonly called "the philosopher of the Arabs." He wrote on a large variety of topics, including astronomy, astrology, optics, and number. Gherardo of Cremona (*c.* 1150) translated his work on optics into Latin.

About 870 there lived in Bagdad three scholars known as the Benî Mûsâ (sons of Moses) or the Three Brothers.[3] They were the sons of Mûsâ ibn Shâkir, a reformed robber who had finally devoted himself to geometry and astronomy in al-Mâmûn's court. Of these brothers, Mohammed, Aḥmed, and al-Ḥasan, the first-named was the most celebrated, but all three gave attention to securing the best scientific works of the Greeks and to having them translated. They wrote on medicine, conics, geometry, mensuration, the trisection of an angle, and other scientific subjects. They used the conchoid in the trisection problem and the string fastened to the foci in the construction of an ellipse.

At this period there worked for a time in Bagdad the celebrated Tâbit ibn Qorra[4] (826–901), a physician of prominence, but better known for his work in philosophy and mathematics, and particularly for the claim that he was successful in applying

[1]Mohammed ibn 'Îsâ, Abû 'Abdallâh al-Mâhânî, of Bagdad, died probably between 874 and 884.

[2]Ya'qûb ibn Isḥâq ibn al-Ṣabbâḥ al-Kindî, Abû Yûsuf, died *c.* 873/4.

[3]M. Curtze, "Liber Trium Fratrum de Geometria," in *Nova Acta der K. Leop.-Carol. Deutschen Akad. der Naturforscher*, XLIX, No. 2 (Halle, 1885).

[4]Ṯabit ibn Qorra ibn Mervân, Abû-Ḥasan, al-Ḥarrânî, a native of Ḥarrân in Mesopotamia, where he also spent some of his later years.

algebra to geometry. He revised the translation of Euclid's *Elements* made by Isḥâq ibn Ḥonein, a renowned physician (died 910), and the translation of the so-called "middle books," that is, of those books written between the time of Euclid and that of Ptolemy.[1] He also wrote extensively on astronomy, the *Almagest*, conics, elementary geometry, Euclid, magic squares, amicable numbers, and astrology. Gherardo of Cremona (*c.* 1150) and Johannes Hispalensis (*c.* 1140) translated certain of his works. He had a son,[2] a physician, who also followed in his father's steps, writing on astronomy and geometry, and revising one of the translations of Archimedes from the Syriac into Arabic.

At about this time an Egyptian—Aḥmed ibn Yûsuf[3]—wrote on proportion and astronomy and discussed the *figura cata*, that is, the proposition of Menelaus relating to the segments of the sides of a triangle cut by a transversal.

Christian and Jewish Scholars in Bagdad. To Bagdad there also came at this time various Jewish and Christian writers, their names being commonly given in Arabic form. Among these were Sahl ibn Bishr,[4] an astrologer, who had already gained considerable reputation in Khorâsân. He wrote a work on algebra. Part of his writings appeared in print in Venice (1493) and part in Basel (1533). There was also Abû'l-Ṯaiyib,[5] who gave up his Jewish religion and adopted the faith of Islam. He compiled a set of astronomical tables and seems to have written on trigonometry. Among the Christians there was Qosṭâ ibn Lûqâ al-Ba'albekî[6] (died *c.* 912/3), a

[1] L. M. L. Nix, *Das fünfte Buch der Conica des Apollonius von Perga*, with Arabic text and German translation. Leipzig, 1889.

[2] Sinân ibn Ṯâbit ibn Qorra, Abû Sa'îd, died 943. See Suter's list, *Abhandlungen*, X, 51.

[3] Aḥmed ibn Yûsuf ibn Ibrâhîm, Abû Ja'far, al-Miṣrî (died *c.* 912/3). Al-Miṣrî means the Egyptian, and the name is applied to other writers as well. There is some doubt as to his works. He was the son of Yûsuf ibn Ibrâhîm ibn al-Dâya, who was known as "the Arithmetician" and lived in Damascus, Bagdad, and Egypt.

[4] Sahl ibn Bishr ibn Ḥabîb ibn Hânî (or Hâyâ), Abû 'Oṯmân (*c.* 850).

[5] Sind ibn 'Alî, Abû'l-Ṯaiyib (*c.* 850).

[6] Kosta, son of Luke, from Baalbek, known to early Europeans as Kusta ben Luca.

PYTHAGOREAN THEOREM IN T̲ÂBIT IBN QORRA'S TRANSLATION OF EUCLID

The translation was made by Isḥâq ibn Ḥonein (died 910) but was revised by T̲âbit ibn Qorra, *c.* 890. This manuscript was written in 1350

physician, who translated the *Spherics* of Theodosius and parts of Aristarchus, Autolycus,[1] Hypsicles, Heron, and Diophantus, and who wrote a geometry in catechism form. There was also a Greek Christian, Nazîf ibn Jumn (or Jemen), known as al-Qass (the priest), who translated Euclid X; and another of the same faith, al-Jorjânî,[2] a physician, who wrote a compendium of the *Almagest*.

It is possible that it was about this time and in this region that the anonymous Hebrew work entitled *Mishnath ha-Middoth* (*Theory of Measures*) was written, but the place and date are quite unknown. It is primarily on the measurement of geometric solids, and some of its features recall the work of al-Khowârizmî on mensuration.[3]

Later Writers. After the reigns of the first three caliphs of Bagdad the science of astronomy still continued to be the antechamber of mathematics. Thus we find such writers in this field as al-Mervazî,[4] who wrote extensively on astronomy and astronomical instruments; Albumasar[5] (died 886), the most celebrated of the Arab writers on astrology, who was led by this science to the study of astronomy; Aḥmed ibn al-Ṯaiyib[6] (*c.* 890), of Persian origin, a pupil of Alchindi's, who wrote on algebra and arithmetic as well as on astrology and music; and al-Dînavarî,[7] who wrote on algebra, astronomy, and the Hindu methods of computation. There was also the

[1] A Greek astronomer who lived *c.* 360 B.C. The others have already been mentioned.

[2] 'Îsâ ibn Yaḥyâ al-Masîḥî, Abû Sahl, al-Jorjânî, died *c.* 1009/10. Al-Masîḥî means a believer in the Messiah, a Christian. The Suter list does not give the place where either of the last two lived.

[3] M. Steinschneider, *Festschrift Zunz* (Berlin, 1864); H. Shapiro, *Abhandlungen*, with translation and commentary, III, 3; F. Rosen, *The Algebra of Mohammed ben Musa*, p. 70 (London, 1831).

[4] Aḥmed ibn 'Abdallâh al-Mervazî, a native of Merv (probably died between 864 and 874), known as Ḥabash al-Ḥâsib ("Ḥabash the computer").

[5] As he was commonly known in medieval Europe. His name was Ja'far ibn Mohammed ibn 'Omar al-Balkhî (from Balkh, in Khorâsân), Abû Ma'shar.

[6] Aḥmed ibn Mohammed ibn Mervân, Abû'l-'Abbâs, al-Sarakhsî, known as Aḥmed ibn al-Ṯaiyib.

[7] Aḥmed ibn Dâ'ûd, Abû Ḥanîfa, al-Dînavarî (died 895). He lived most of the time in Dînavar, his native place.

well-known scholar Albategnius[1] (died 929), as he was called in Europe, who was justly esteemed for his astronomical writings[2] and tables. Among the many other scholars of this period there may be mentioned Rhases (died 932),[3] to use his European name, a celebrated physician who wrote on geometry and astronomy; a grandson of Ṯâbit ibn Qorra,[4] also a physician, who wrote on conics, dialing, and elementary geometry; al-Fârrâbî,[5] a native of Fârâb in Turkestan, who wrote a commentary on Euclid and was a philosopher of high standing; Ibn Yûnis,[6] who, next to al-Battânî, was the most celebrated astronomer among the Arabs; and al-Ḥarrânî,[7] who wrote a commentary on Euclid.

The 10th century saw several writers of somewhat higher attainments, among whom the best-known was Abû'l-Wefâ[8] (940–998), celebrated for his improvements in trigonometry, his introduction of the tangent (*umbra versa*), and his computation of tables of sines and tangents for every 10′; it is also very likely that he is entitled to credit for the use of secants and cosecants. He was also prominent as a writer on arithmetic, algebra, geometry, and astronomy.

Among the other writers of this period who are worthy of special mention were al-Ḥaiṯam of Baṣra,[9] who wrote on algebra, astronomy, geometry, gnomonics, and optics; Abû Ja'far al-Khâzin (died between 961 and 971), who attempted

[1] Mohammed ibn Jâbir ibn Sinân, Abû 'Abdallâh, al-Battânî, a native of Battan, in Mesopotamia. He is also known as al-Raqqî, from the fact that he made his observations at Raqqa on the Euphrates.

[2] Translated by Robert of Chester (*c.* 1140) or Robertus Retinensis, referred to later. The work was printed in 1537.

[3] Mohammed ibn Zakarîyâ al-Râzî, Abû Bekr.

[4] Ibrâhîm ibn Sinân Ṯâbit ibn Qorra, Abû Isḥâq, son of the Sinân already mentioned. Born 908/9; died 946.

[5] Mohammed ibn Mohammed ibn Ṭarkhân ibn Auzlag, Abû Naṣr, al-Fârrâbî; died at Damascus, 950/1.

[6] 'Alî ibn Abî Sa'îd 'Abderraḥmân ibn Aḥmed ibn Yûnis (or Yûnos), Abû'l-Ḥasan, al-Ṣadafî; died 1009.

[7] Ibrâhîm ibn Hilâl ibn Ibrâhîm ibn Zahrûn, Abû Isḥâq, al-Ḥarrânî. Born 923; died at Bagdad, 994.

[8] Mohammed ibn Mohammed ibn Yahyâ ibn Ismâ'îl ibn al-'Abbas, Abû'l-Wefâ al-Bûzjânî.

[9] Al-Ḥasan ibn al-Ḥasan ibn al-Ḥaiṯam, Abû 'Alî, *c.* 965–1039.

the solution of the cubic equation by the aid of conics and who wrote on Euclid and astronomy; and Kûshyâr ibn Lebbân,[1] who wrote on arithmetic, trigonometry, and astronomy.

Al-Nairîzî[2] (died *c.* 922/3) was one of the notable 10th century writers on Euclid. He was interested in astronomy and geometry, writing commentaries on both Ptolemy and Euclid, but it is the commentary on the *Elements*, translated into Latin by Gherardo of Cremona, that is best known.

As a type of the lesser commentators on Euclid in the 10th century there may be mentioned al-Ḥasan ibn 'Obeidallâh,[3] who wrote a commentary on the difficult parts of the *Elements*.

Translators into Arabic. Among the noteworthy translators of this period were al-Hajjâj,[4] who made two translations of at least six books of Euclid's *Elements*, and also translated Ptolemy's *Almagest*; al-Jauharî,[5] who made astronomical observations at Bagdad and Damascus (*c.* 830) and wrote a commentary on the *Elements* of Euclid; Ḥonein ibn Isḥâq,[6] who translated various Greek works, possibly including Ptolemy's *Tetrabiblos*, and who wrote on astronomy, but was more celebrated as a physician and a philosopher; and his son Isḥâq,[7] who was a physician and translated Euclid's *Elements* and *Data*, the *Almagest*, Archimedes on the sphere and cylinder, and probably the *Spherics* of Menelaus. Somewhat less well known, but worthy of mention, are al-Arjânî,[8] who wrote a commentary (*c.* 850) on Euclid X; al-Ḥimṣî,[9] who translated the first four books of Apollonius; and Sa'îd ibn Ya'qûb,[10] a physician, who translated parts of Euclid and of Pappus.

[1] Kûshyâr ibn Lebbân ibn Bâshahrî al-Jîlî, Abû'l-Ḥasan, *c.* 971–*c.* 1029.

[2] Al-Faḍl ibn Ḥâtim al-Nairîzî, Abû'l-'Abbâs.

[3] Al-Ḥasan ibn 'Obeidallâh ibn Soleimân ibn Vahb, Abû Mohammed, *c.* 925.

[4] Al-Hajjâj ibn Yûsuf ibn Maṭar, *c.* 786–*c.* 835.

[5] Al-'Abbâs ibn Sa'îd al-Jauharî.

[6] Ḥonein ibn Isḥâq, al-'Ibâdî Abû Zeid; born 809/10; died at Bagdad, 873.

[7] Isḥâq ibn Ḥonein ibn Isḥâq al-'Ibâdî, Abû Ya'qûb, died 910.

[8] Ibn Râhiweih al-Arjânî, or Arrajânî, according to Steinschneider the same as Isḥâq ibn Ibrâhîm ibn Makhlad al-Mervazî, who died at Nishâpûr in 852/3.

[9] Hilâl ibn Abî Hilâl al-Ḥimṣî, died 883/4. Al-Ḥimṣî means "from Emessa," in Syria.

[10] Sa'îd ibn Ya'qûb al-Dimishqî, Abû 'Oṯmân. He was living in 915.

Abû Kâmil. Between 850 and 930 there lived in Egypt Abû Kâmil,[1] who is known for several works but especially for his treatise on the pentagon and decagon[2] and for his arithmetic and algebra.[3] No writer of his time showed more genius than he in the treatment of equations and in their application to the solution of geometric problems.

About the same time there lived Abû'l-Faradsh Mohammed ibn Ishâq, known as Ibn Abî Ya'qûb al-Nadîm, whose *Kitâb al-Fihrist* (*Book of Lists*), written *c.* 987, is a collection of brief biographies of various prominent mathematicians, both Greek and Mohammedan.[4]

Close of the Golden Age of Bagdad. In a general way it may be said that the Golden Age of Arabian mathematics was confined largely to the 9th and 10th centuries; that the world owes a great debt to Arab scholars for preserving and transmitting to posterity the classics of Greek mathematics; and that their work was chiefly that of transmission, although they developed considerable originality in algebra and showed some genius in their work in trigonometry.

5. The Christian West

The Dark Ages. The period from 500 to 1000 extends from about the time of the fall of Rome (455)[5] to the first reawakening of Europe under Pope Sylvester II (Gerbert). It includes the so-called Dark Ages, the period of the slow civilizing of the northern races, of the development of monastic schools, of the work of Charlemagne, and of the contact with Oriental civilization, chiefly through the Moors in Spain. In mathematics it was the era of the development of the Christian calendar in the West, and of little else. The barbarian had to be civilized, to assimilate slowly the Roman culture which he would have destroyed, and to receive a better religion. The Roman schools

[1] Abû Kâmil Shojâ ibn Aslam ibn Mohammed ibn Shojâ.

[2] H. Suter, *Bibl. Math.*, X (3), 15, 33.

[3] L. C. Karpinski, *Amer. Math. Month.*, XXI, 37, and *Bibl. Math.*, XII (3), 40. See also H. Suter, *Bibl. Math.*, XI (3), 100.

[4] Suter's translation appeared in the *Abhandlungen* (VI, 1) in 1892.

[5] The barbarians entered the city first in 410. The final fall is often given as 476.

had to be supplanted by those of the cathedral and the monastery, and all the mathematics required was limited to the needs of trade, to the keeping of accounts, and to the fixing of dates for Church festivals. In those parts of Europe less subject to Northern influence, such as Marseilles, Arles, and Narbonne, the needs of commerce were still such as to render necessary the arithmetic of exchange in the training of the merchant's apprentice. These cities maintained in this period their trade with Italy, Constantinople, and the Orient, sending dyes, cereals, pottery, and salt to the East, and importing silk from China, pearls from India, and even papyrus rolls from Egypt.[1]

Boethius. Anicius Manlius Severinus Boe'thius,[2] a Roman citizen, a member of the distinguished family of the Anicii, statesman, philosopher, mathematician, man of letters, and founder of the medieval scholasticism, lived at the opening of the period now under discussion. Persecuted for his uprightness, executed for his fearlessness, accepted by the Church as a martyr, his reputation and scholarship gave his books on mathematics high standing in the monastic schools for many centuries.

His greatest work, written while he was in prison, is the *Consolation of Philosophy*.[3] His mathematical works are an arithmetic,[4] a geometry,[5] and a work on music,[6] a subject then

[1] A. Rambaud, *Histoire de la Civilisation Française*, 12th ed., I, 115. Paris, 1911.

[2] Born at Rome *c.* 475; died at Pavia, 524. The more nearly correct Latin form is Boetius.

[3] *De consolatione philosophiae.*

[4] *Boetii de institutione arithmetica libri duo*, ed. Friedlein (Leipzig, 1867); hereafter referred to as *Boethius, ed. Friedlein.* The earliest manuscripts used in this edition are three of the 10th century, a fact worth noting in view of questions as to interpolations discussed later. Readers of Boethius and other Latin writers will find assistance in B. Veratti, "Sopra la Terminologia Matematica degli Scrittori Latini," *Memorie della R. Accad. . . . di Modena*, Vol. V.

[5] *Boetii quae fertur geometria*, in the Friedlein edition cited above. The earliest manuscript used in this edition is one of the 10th century. There is serious doubt as to whether Boethius wrote the *Ars Geometriae* attributed to him. See Tannery, *La Géométrie Grecque*, 128; H. Weissenborn, "Die Boetius-Frage," in the *Abhandlungen* II, 185.

[6] *Boetii di institutione musica libri quinque*, in the Friedlein edition cited above. The earliest manuscript used in this edition was mostly of the 9th century, although Books IV and V were earlier and some parts were missing.

ranked as part of mathematics. The arithmetic was based on the work of Nicomachus, and the geometry on the *Elements* of Euclid. Neither showed any originality in the domain of mathematics, but each was sufficiently successful in its presentation of the subject treated to permit of the general use of

FROM A DRAWING BY RAPHAEL

Fanciful sketches of Ptolemy and Boethius, now in the Accademia in Venice

these books in those monastic schools that had advanced far enough to demand courses in the theory of numbers and in demonstrative geometry.

Minor Writers. It is natural to expect that among the first Christian scholars few would be found with any interest in mathematics or the natural sciences. Their religious faith was too intense, their persecutions too real, and their lives too precarious to permit of speculations in these fields. The names of a few Christians have already been mentioned, but their contributions to mathematics were insignificant. With the close of the 5th century, however, Christianity had become powerful enough to permit of the development of an intellectual class with interests outside of religious faith, and in this class we find the names of several scholars who showed some knowledge of the mathematics of the classical period.

Among these writers was Magnus Aurelius Cassiodo′rus,[1] a descendant of an ancient Roman family.[2] He was a statesman of distinction and was honored both by the last of the Roman rulers and by their Ostrogothic successors. He founded a monastery at Vivarium, and passed his last years within its walls. He insisted upon a high standard of scholarship for the clergy, and his writings show that he himself possessed, within the limits which conditions then imposed, that which he demanded for others. Cassiodorus wrote *De artibus ac disciplinis liberalium literarum*, a trivial sort of compendium of the seven liberal arts,—grammar, rhetoric, and dialectic composing the trivium, and arithmetic, geometry, astronomy, and music composing the quadrivium.[3] This work was widely used in the schools of the Middle Ages,[4] and nothing could better show the low state of learning than this feeble attempt at scholarship. There is also doubtfully assigned to him a *Computus Paschalis sive de indicationibus cyclis solis et lunae*, written in 562, one of the first treatises on the Christian calendar. The plan for the adoption of the Christian era, however, was worked out by Dionysius Exiguus, a Roman abbot, *c.* 525.

[1] Born at Scylaceum (Squillace), *c.* 470; died *c.* 564, a date sometimes given as 585. The name is also spelled Cassiodorius.

[2] For a popular but vivid account of his achievements see M. Crawford, *Rulers of the South*, II, 9.

[3] A common medieval verse reads:

> Gram loquitur, Dia verba docet, Rhet verba colorat,
> Mus canit, Ar numerat, Ge ponderat, As colit astra.

Petrus Pictaviensis, in a verse to Peter of Cluny, writes:

> Musicus, astrologus, arithmeticus, et geometra,
> Grammaticus, rhetor, et dialecticus est.

[4] The first collected edition of his works was published at Paris in 1584 and 1598.

MANUSCRIPT OF THE *ARITHMETICA* OF BOETHIUS

This MS., now in Mr. Plimpton's library, was written *c.* 1294. The scribe has used modern instead of Roman numerals

A little before the time of Cassiodorus there flourished Martianus Mineus Felix Capel'la,[1] author of an encyclopedia known as the *Nuptials of Philology and Mercury*.[2] It is a medley of prose and verse, one part of the work being on geometry and another on arithmetic. In connection with the latter Capella discusses various classes of numbers and the supposed mysteries of the smaller numbers. The book is even more arid than that of Cassiodorus, the only redeeming feature being the statement that Mercury and Venus revolve about the sun instead of the earth.[3]

Before the close of the 5th century there was born in Damascus a Syrian who took his name, Damas'cius,[4] from his birthplace. He was the last of the important Neoplatonists and was a disciple of the Marinus who succeeded Proclus (*c.* 485). In 510 he became director of the school at Athens. When Justinian closed the heathen schools of philosophy in that city (529), Damascius went to Persia, but returned five years later (534). His works were mostly philosophical, but his name has doubtfully been connected with a fifteenth book to be added to Euclid's *Elements*.

Almost the last of the Greeks to show any appreciation of mathematics before the medieval period fairly began was Euto'cius[5] of Ascalon. He wrote commentaries on the first four books of the conics of Apollonius. He also wrote on certain works of Archimedes,—the sphere and cylinder, the quadrature of the circle, and the work on equilibrium; and on the *Almagest* of Ptolemy, this last commentary being lost. These writings of Eutocius are of little value except as they supply certain information relating to Greek mathematics.

[1]Born possibly at Carthage, *c.* 420; died *c.* 490. See E. Narducci, in Boncompagni's *Bullettino*, XV, 505, with biography and bibliography. The name might properly have been given in Chapter IV, but Capella is more closely related to Boethius and Cassiodorus than to the last of the Greeks.

[2]The first edition appeared at Vincenza in 1499, *Opus Martiani Capelle de Nuptijs Philologie & Mercurij libri duo.*

[3]In the *De Astronomia*, the chapter entitled *Tellus quod non sit centrum omnibus planetis.* Fol. 333 of the 1592 edition.

[4] Δαμάσκιος.

[5] Εὐτόκιος. Fl. *c.* 560.

In the 6th century there seems also to have been written the *Codex Arcerianus*,[1] so called from the fact that it belonged at one time (1566–1604) to one Johannes Arcerius in Gröningen. While it relates largely to legal matters of a rural nature, it contains considerable information concerning the Roman surveyors.

There is little else to say for the century. It represents the lowest point on the curve of intellectual progress in Europe. The ecclesiastical element was unable to overcome the general ignorance of the masses, and aside from a faint light in the Irish monasteries, Europe was in darkness.

Isido'rus. The centuries immediately following the death of Boethius saw little interest in the literature and science of the classical period. Even as eminent a man as St. Ouen (*c.* 609–683) spoke of the works of Homer and Vergil as the trifling songs of impious poets[2] and made two distinct personages of Tullius and Cicero; while Gregory of Tours (538–594) uttered the lament: "Unhappy our days, for the study of letters is dead in our midst, and there is to be found no man able to record the history of these times." So debased was civilization that the few who stood for even the remnants of the old Latin cult resorted to doggerel verse, as Capella had done, or diluted their learning in the form of encyclopedias.

Prominent among those who developed the latter plan was Isidorus of Seville,[3] historian, grammarian, orator, theologian, bishop, and general scholar, as well as one of the most remarkable statesmen of the Middle Ages. St. Martin, in his funeral oration, describes him as "generous in his giving, affable in his entertaining, sober in his affections, free in his sentiments, equitable in his judgments, indefatigable in his ministrations," and celebrated for his integrity. A man of fortunate birth, he was helped by his family connections to begin a career of such remarkable success, relative to that of

[1] Mommsen puts it *c.* 450, and Cantor (*Die Römischen Agrimensoren*, p. 95 (Leipzig, 1875); hereafter referred to as Cantor, *Agrimensoren*) thinks it not later than the 7th century. See also Cantor, *Geschichte*, I, chap. 26.

[2] "Sceleratorum neniae poetarum." [3] Born at Seville, *c.* 570; died April 4, 636.

any contemporary, that the Council of Toledo (653), a few years after his death, could truthfully speak of him as "the extraordinary doctor, the latest ornament of the Catholic Church, the most learned man of the latter ages, always to be named with reverence." Since he was the most learned man of his time, it would be expected that his encyclopedia of the trivium and the quadrivium, the seven liberal arts, would contain some mathematics of merit. This work, called by him the *Origines* but often known as the *Etymologies*, consists of twenty books, the third one being on mathematics. The treatment, however, is trivial, the arithmetic being simply a brief condensation of Boethius, and the rest of the work being of as little scientific value.

Bede the Venerable. It was about a century after Isidorus that there was born at Monkton in Northumberland one of the greatest of the Church scholars of the Middle Ages, Baeda (*c.* 673–735), commonly known as Beda Venerabilis, the Venerable Bede,[1] and called by Burke "the father of English learning."

Of him Hallam[2] remarked that he "surpasses every other name of our ancient literary annals; and, though little more than a diligent compiler from older writers, may perhaps be reckoned superior to any man the world (so low had the East sunk like the West) then possessed." Four years before his death he prepared a list of the thirty-seven works which he had written up to that time, and added these words: "I have spent my whole life in the same monastery, and while attentive to the rule of my order and the service of the Church, my

[1] G. F. Browne, *The Venerable Bede* (London, 1880). For a discussion of his scientific works, see J. A. Giles, *Miscellaneous Works of the Venerable Bede*, VI, pp. v, 123 (London, 1843); J. Mabillon, "Ven. Bedae elogium historicum," in the *Opera Omnia* of Bede (Paris, 1862); K. Werner, *Beda der Ehrwürdige und seine Zeit* (Vienna, 2d ed., 1881). Bede was buried at Jarrow, but his remains were moved to Durham *c.* 1050 and his tomb may now be seen in the Galilee Chapel of the Cathedral. A good setting for the study of the education of the period may be found in F. P. Barnard, *Companion to English History* (*Middle Ages*), p. 303 (Oxford, 1902).

[2] *Literature of Europe*, Chapter I, § 7 (London, n. d.).

constant pleasure lay in learning, or teaching, or writing."[1] Taught by Aldhelm and by John of Beverley, two of the heirs to the intellectual and spiritual treasure which Augustine bequeathed to Canterbury,[2] he was also a disciple of Archbishop Theodore of Tarsus and Abbot Adrian, two pioneers in bringing a high grade of scholarship to the monasteries, and thus he was well prepared to render service to the world and to lead a life "consecrated in noiseless activity to God."[3]

In mathematics his interests were in the ancient number theory, the ecclesiastical calendar, and the finger symbolism of number, and his writings include these and other mathematical subjects.[4] To him we are indebted for the best work on the calendar written during the Dark Ages, and for the best work up to his time on digital notation.[5] Certain mathematical recreations have also been attributed to him, but the evidence concerning their authorship is not conclusive.

Alcuin of York. The next great European scholar in mathematics was Al'cuin (735–804). Born in the year of Bede's death, less of a scholar than the latter but more of a man of action, he attained prominence in the State as well as in the Church. He studied in Italy,[6] taught at York,[7] was called (782) by Charlemagne[8] to assist him in his ambitious project for the education of his people, and became abbot of St. Martin

[1] "Semper aut discere aut docere aut scribere dulce habui," words worthy of the one whom Green, the historian, speaks of as "the first great English scholar."

[2] W. F. Hook, *Lives of the Archbishops of Canterbury*, I, 42 (London, 1860); A. Neander, *Church History*, 5th American ed., III, 12 (Boston, 1855); J. E. G. de Montgomery, *State Intervention in English Education*, p. 6 (Cambridge, 1902). See page 187, note 2.

[3] Neander. For a description of Bede's death see *loc. cit.*, p. 153.

[4] *De numeris, De temporum ratione, De numerorum divisione, De circulis sphaerae et polo, De astrolabio.*

[5] His *De temporibus* comes down only to 701/2. His *De temporum ratione* comes down to 726. This second work contains his *De Indigitatione sive de computo per gestum digitorum* and his *De ratione unciarum.*

[6] As shown by one of his letters (XV); see Libri, *Histoire*, I, 89.

[7] On the nature of the schools, see W. W. Capes, *The English Church in the 14th and 15th centuries*, p. 332 (London, 1900).

[8] Who addressed him as "Carissime in Christo praeceptor." Charlemagne reigned as king or emperor from 768 to 814.

of Tours. He wrote on arithmetic, geometry, and astronomy,[1] and his name is connected with a certain collection of puzzle problems[2] which has influenced the writers of textbooks for a thousand years. It is uncertain how much he may have had to do with this set of mathematical recreations, and considerable doubt has been thrown upon his connection with them through recent studies of a certain manuscript at Leyden.[3] This manuscript dates from the first part of the 11th century and is thought to have been written by, or at least inspired by, a monk named Adémar or Aymar, of the ancient house of Chabanais, who was born in 988 and who died on his way to the Holy Land in 1030. He had considerable reputation as a historian and a controversialist[4] and seems to have collected a large amount of material with no scientific care. These problems were very likely part of the medieval versions of Æsop's *Fables*, collections which, although probably begun by Æsop, in Samos, in the 7th century B.C., were modified by Babrius about the 3d century, and were still further corrupted in the Middle Ages. While problems attributed to Alcuin are found here, and probably interested Adémar as they did hundreds of others, there seems to be no good reason to believe that Alcuin may not have collected them from the medieval versions attached to the *Fables*. Certain it is that letters of Alcuin show that he wrote a set of puzzle problems, although there is no direct evidence that this is the one.[5] It would have been in keeping with his ideas to compile a book that should be amusing enough to relieve education of the drudgery of the time.[6]

[1] For his life and works, see G. F. Browne, *Alcuin of York* (London, 1908); C. J. B. Gaskoin, *Alcuin; his life and his work* (London, 1904); R. B. Page, *The Letters of Alcuin* (New York, 1909); A. F. West, *Alcuin and the rise of the Christian Schools* (New York, 1912).

[2] *Propositiones ad acuendos juvenes.*

[3] Cod. Vossianus Lat. oct. 15, edited by G. Thiele and published at Leyden in 1905.

[4] J. Lair, *L'Histoire d'Adémar*, Paris, 1899.

[5] It has been published in the works of Bede as well as in those of Alcuin. The oldest MS. of the work, written early in the 11th century, is now in Karlsruhe. In this are the words: "Dilectissimo fratri siguulfo presbytero alcuinus salutem," but naturally these words are not absolutely conclusive evidence.

[6] When "sub virga degere" meant school life and "pueri subiugales" meant pupils. T. Ziegler, *Geschichte der Pädagogik*, p. 29 (Munich, 1895).

In the collection is to be found, for example, the problem of the hare and hound, already ancient but made the more mysterious by the cipher title,

"De cursu cbnks bc fugb lepprks,"
for "De cursu canis ac fuga leporis."[1]

The continued private wars among petty lords in the 10th and 11th centuries made France a poor field for mathematical or other intellectual progress, and hence these two centuries produced little that was noteworthy.

Decay of British Learning. After the death of Alcuin the brilliant era that started in Great Britain with St. Au′gustine of Canterbury[2] (died *c.* 604 or 613) closed as suddenly as it began. The ravages of the Danes put an end to that feeling of security which makes for intellectual development, and when Alfred (848–900) came to the throne (871) he could only lament, "There was a time when people came to this island for instruction, but now we must obtain it abroad if we desire it." When Aethelstan,[3] the grandson of Alfred, came to the throne (925), however, he showed great interest in the fostering of learning, and in a poem written in the 14th century reference is made to the introduction of Euclid into England in the reign of this powerful ruler:

Thys grete clerkys name wes clept Euclyde,
Hys name hyt spradde ful wondur wide. . . .
The clerk Euclyde on thys wyse hyt fonde,
Thys crafte of gemetry yn Egypte londe;
Yn Egypte he tawȝhte hyt ful wyde,
Yn dyvers londe on every syde . . .
Thys craft com ynto Englond as y ȝow say
Yn tyme of good kynge Adelstonus day.[4]

[1] Cantor, *Agrimensoren*, 139, 142.

[2] Not to be confused with the greater St. Augustine of Hippo (354–430).

[3] Athelston, Ethelstan, Adelstan, Adelston, Edelstan, and other spellings. Born *c.* 895; died 941.

[4] The MS. is in the British Museum (Bib. Reg. 17A, I. p. 32), and was published by J. O. Halliwell, *The Early History of Freemasonry in England* (London, 1840).

Jewish Activity. Probably about the time of Alcuin a Jewish mathematician, Jacob ben Nissim, wrote a work entitled *Sefer Jezira*,[1] which, like various Hebrew writings on mathematics, contains some material on the theory of numbers.

Hrabanus Maurus. Alcuin's most famous pupil was Magnentius Hrabanus Maurus,[2] "Primus praeceptor Germaniae," abbot of the monastery at Fulda (822), and archbishop of Mainz (847). In his younger days he traveled extensively[3] and wrote a worthy treatise on the calendar, based on Bede's work and showing a commendable knowledge of astronomy, a science which included most of the mathematics of his time.[4]

One of his contemporaries, Walafried Strabus[5] (*c.* 806–849), is known to have taught mathematics at Reichenau, near Constance, but he left no works upon the subject.

Remigius of Auxerre. A second great pupil of Alcuin's and a witness to the beneficent influence of the Church in France, was Remi′gius[6] of Auxerre, a Benedictine monk who did much for the schools at Rheims and who founded a school at Paris out of which the university is thought by some to have developed.[7] He wrote a commentary on the arithmetic of Capella,[8] not an important contribution to mathematics, but typical of a period given to useless disputation and empty sophistry.

[1] *Book of Creation.* See M. Steinschneider, "Miscellen zur Gesch. der Math.," in *Bibl. Math.*, III (2), 35; IX (2), 23. An Arabic commentary is known to have been written upon it in the 10th century. The question of the authorship of the *Sefer Jezira* is still unsettled. The work relates chiefly to number mysticism.

[2] Born *c.* 776; died 856. The name appears also as Rabanus Maurus. The date of his birth is also given as *c.* 784.

[3] Ego quidem, cum in locis Sidonis aliquoties demoratus sim." See Neander, *Church History*, III, 457.

[4] He also wrote an encyclopedia, *De universo libri XXII, sive etymologiarum opus.* On his life see J. N. Bach, *Hrabanus Maurus der Schöpfer des deutschen Schulwesens* (Fulda, 1835); D. Türnau, *Rabanus Maurus, der praeceptor Germaniae* (Munich, 1900).

[5] Walafrid Strabo. Cantor, *Geschichte*, I (2), 792.

[6] The name comes from Remy, Remi, *i.e.*, Rheims. Died *c.* 908.

[7] It is also said to have developed from a school of dialectics opened by William of Champeaux, *c.* 1100 to 1110.

[8] The Vatican codex was published in Boncompagni's *Bullettino*, XV, 572.

Hrotsvitha. A certain amount of light is thrown upon the barren field of monastic mathematics of this period by the story of the learned nun Hrotsvitha[1] of the Benedictine abbey of Gandersheim, in Saxony. She wrote several plays and in these she shows a knowledge of the Greek language and of either Greek or Boethian arithmetic. In the *Sapientia* the emperor Hadrian demands the ages of the three daughters of Wisdom (*Sapientia*), namely, of Faith, Hope, and Charity. Wisdom then says that the age of Charity is a defective evenly even number; that of Hope a defective evenly odd one; and that of Faith an oddly even redundant one. Upon Hadrian's remarking, "What a difficult and tangled question has been raised about the mere ages of these girls!" Wisdom replied, "In this is to be praised the great wisdom of the Creator and the marvelous knowledge of the Author of the universe."[2] Hrotsvitha incidentally speaks of three perfect numbers besides 6, namely, 28, 496, and 8128.[3]

Other Writers of the Tenth Century. In the 10th century there may also have been written a treatise on the abacus by

[1]Born *c.* 932; died *c.* 1002. *Hrosvithae Opera*, edited by Winterfeld (Berlin, 1902) (in the *Scriptores rerum Germanicarum*); *Hrotsvithae Opera*, edited by Strecker (Leipzig, 1906); all but one of her works, edited by Conrad Celtes, and with engravings by Dürer, were published at Nürnberg in 1501; there was also an edition by Schurzfleisch (Wittenberg, 1707), and a complete edition by Barack (Nürnberg, 1858). See also Ch. Magnin, *Théatre de Hrotsvitha religieuse Allemande du Xe siècle* (Paris, 1845); E. R. A. Köpke, *Die älteste deutsche Dichterin* (Berlin, 1869), with a refutation of a charge made by Aschbach (1867) that Celtes had forged the works.

The old historian Henricus Bodo referred to her in saying, "Rara avis in Saxonia visa est." The name appears in various other forms such as Roswitha and Hrotsuit. In the Munich MS., apparently contemporary, it appears as Hrotsvith and Hrotsuitha.

[2]A few of the sentences will show the style of the original:

Sapientia. Placetne vobis, O filiae, ut hunc stultum arithmetica fatigem disputatione?

Fides. Placet, mater, . . .

Sapientia. O Imperator, si aetatem inquiris parvularum, Caritas imminutum pariter parem mensurnorum [=annorum] complevit numerum; Spes autem aeque imminutum, sed pariter imparem; Fides vero superfluum impariter parem.

[3]"XXVIII, CCCCXCVI, VIII millia CXXVIII perfecti dicuntur." A perfect number is one that is equal to the sum of its aliquot parts, that is, of its factors and unity; for example, $6 = 1 + 2 + 3$.

Odo of Cluny (879–*c.* 942), although it may be the work of a 12th century writer[1]; but in general the period was a barren one. Only one other writer is worthy of mention, Abbo of Fleury (945–1003), a native of Orléans, who wrote on Easter reckoning,[2] on astronomy, and on the arithmetic of Boethius. His chief title to remembrance, however, is the fact that he was a teacher of Gerbert, the most learned man of his time, whose life and works are considered in the next chapter.

Another example of the ecclesiastical scholar is seen in the case of Bernward, who became Bishop of Hildesheim in 993[3] and who wrote a work on mathematics which was devoted chiefly to the Boethian theory of numbers. A manuscript of this work, possibly the original, is still extant at Hildesheim.

6. The Christian East

Egypt and Constantinople. The eastern countries touching upon the Mediterranean did little for mathematics for a period of five centuries after the fall of Rome. Even the brilliant reign of Justinian (527–565), "the Lawgiver of Civilization," was not able to remove the fears of a barbarian invasion, nor to suppress the disastrous feuds between the Blues and the Greens in Byzantium. Add to this the great fire of 532 and the terrible pestilence of ten years later, and it will be seen that the banks of the Bosporus were not the place for an intellectual revival.

Decay of Alexandria. In Alexandria the chance of progress in the arts and sciences seemed to die out with the fall of Rome, and with the rise of Mohammedanism as a world power the last hope of any revival of the city's ancient glories definitely disappeared. Eighty years after the death of Mohammed his followers had conquered all of northern Africa and had established themselves firmly in Spain. In 642 the great library

[1] S. Günther, *Geschichte der Mathematik*, I, 244 (Leipzig, 1908) (hereafter referred to as Günther, *Geschichte*); Cantor, *Geschichte*, I (3), 843; Th. Martin, "Les Signes Numéraux," *Annali di Mat. pura ed applic.*, V, 50, and reprint, Rome, p. 78 (1864).

[2] *Liber in calculum paschalem.*

[3] H. Düker, *Der liber mathematicalis des Heiligen Bernward.* Hildesheim, 1875.

of Alexandria was destroyed by fire, probably the most serious loss that ever befell any great institution of learning.

Nevertheless a few names appear in the Christian East. Anthemius,[1] an assistant architect in the building of St. Sophia, wrote on conics, and a century later (*c.* 610) Stephen of Alexandria wrote on mathematics and astronomy and taught in Constantinople. In Alexandria, just before the Mohammedan invasion, Asclepias of Tralles (*c.* 635[2]) wrote a commentary on Nicomachus, and Joannes Philop'onus (*c.* 640[3]), known also as Joannes Grammat'icus, did the same and also wrote upon the astrolabe.[4]

Toward the close of the 19th century there was found at Akhmim,[5] in Upper Egypt, a Greek papyrus which seems to have been written about the 7th or 8th century. In this there are tables of unit fractions similar to those found in the Ahmes papyrus, but the work shows no advance over its predecessor of more than two thousand years earlier. Science had long been dead in Egypt except in that part which came under the influence of Alexandria.

School of Cairo. In the early part of the 10th century the Fatimites, a branch of the Mohammedan ruling class, drove their rivals for power out of the city which they thereupon called al-Kâhira, the Victrix,—the modern Cairo. Here they proceeded to establish a school which they ventured to hope would rival that of ancient Alexandria, and which indeed became a center of astronomical activity. With it were connected the names of Ibn Yûnis (p. 175) and al-Ḥaiṯam (p. 175), but it was short-lived, the caliphate of Egypt being destroyed by Saladin in 1171.

[1] Died at Constantinople, 534.

[2] Possibly a century earlier.

[3] The date is very uncertain, being possibly a century too late.

[4] *De vsv astrolabii ejusque constructione libellus.* It was published by H. Hase, *Rheinisches Museum für Philologie*, VI, 127 (Bonn, 1839). His work on Nicomachus was edited by Hoche, Leipzig, 1864, and Wesel, 1867.

[5] Or Ekhmim, the site of the ancient Chemmis or Panopolis. It became a great religious center under the Christians of the early Middle Ages. Nestorius (5th century), the patriarch of Constantinople, was deprived of his honors and banished to Akhmim for heresy. See also Heath, *History*, II, 543.

It is probable that the Jewish scholar Sa'adia ben Joseph[1] studied at Cairo during this period. He wrote on the division of inheritances and on the calendar. He taught in Babylon, where he doubtless met with Isaac ben Salom, who wrote on the Hindu arithmetic and on astronomy.

7. Spain

Oriental Civilization in the West. After the burning of the Alexandrian library (642) the Mohammedans continued their conquests, sweeping along the north coast of Africa and finally entering Spain in 711, defeating the Visigothic king, and establishing themselves for a sojourn of eight hundred years. Bringing with them the Oriental faith in astrology, their primary interest in mathematics was related chiefly to astronomy, trigonometry, and the conics; possessed of esoteric tastes, the mysteries of numbers and of gematria[2] appealed to them; coming into constant relations with the Jews, the cabala doubtless impressed them; inspired by the intellectual brilliancy of Bagdad, the classics of the Greeks found place in their schools. By the time the intellectual supremacy of Bagdad was seriously threatened in the East, Cordova was becoming the intellectual center of Islam in the West. Alhakem II, who reigned from 961 to 976, established a considerable library there, and about the close of the 10th century al-Majrîṭî,[3] a native scholar, wrote on amicable numbers, astronomy, and geometry.

Even in the 10th century the activity in the field of mathematics was not great. The first writer of note was Muslim ibn Ahmed al-Leiṯî, Abû 'Obeida, also called Sâhib al-Qible (died 907/8), a native of Cordova and a writer on astronomy and arithmetic. About the same time Cordova produced Salhab ibn 'Abdessalâm al-Faradî, Abû'l-'Abbâs (died 922/3), an arithmetician of some note.

[1] In Arabic, Sa'îd ibn Yûsuf al-Fayyumî. He died in 941. The Hebrew name Sa'adia Gaon means Sa'adia the Genius (Great).

[2] Largely concerned with the evaluating of names by the numerical value of the letters.

[3] Abû'l-Qâsim Maslama ibn Aḥmed al-Majrîṭî, died 1007/8.

TOPICS FOR DISCUSSION

1. Intercourse between China and other countries, and its possible influence upon mathematics.

2. Progress of Chinese mathematics from 500 to 1000.

3. Nature and sources of early Japanese mathematics.

4. General nature of Hindu mathematics from 500 to 1000.

5. The work of the two Āryabhaṭas.

6. Brahmagupta and the School of Ujjain.

7. The work of Mahāvīra compared with that of Brahmagupta.

8. The Bagdad School, its rise and its relation to the Hindu and Greek learning.

9. The nature of the contributions of the Persian and Arab mathematicians of the ninth and tenth centuries.

10. The life and works of Mohammed ibn Mûsâ al-Khowârizmî.

11. Causes of the decay of eastern Arabic mathematics.

12. Indebtedness of medieval Europe to Oriental mathematics in the Middle Ages.

13. Causes of the low state of mathematics in Europe during the greater part of the Middle Ages.

14. Boethius as a mathematician.

15. The life and mathematical works of Bede.

16. The life, influence, and mathematical works of Alcuin.

17. Evidences of an interest in the Greek theory of numbers in the Middle Ages.

18. The influence of mathematics in the Middle Ages upon the science at present.

19. The mathematics of the quadrivium.

20. The nature of the encyclopedias produced by the Church scholars of the Middle Ages.

21. Mathematical recreations in the Middle Ages.

22. Nature of the mathematics studied in the British Isles in the early part of the Middle Ages.

23. The Church schools as preservers of mathematical knowledge in the early part of the Middle Ages.

24. The mathematical contributions of the Mohammedans of the ninth and tenth centuries in Spain.

25. The relation of medieval astrology to astronomy and also to mathematics in general.

CHAPTER VI

THE OCCIDENT FROM 1000 TO 1500

1. Christian Europe from 1000 to 1200

Religious and Political Influences. Just how much influence the passing of the first Christian millennium had upon the common people it is difficult to say. Historians pay much less attention to the "terreur de l'an Mil" than was formerly the case. It is not probable that many educated persons took literally the biblical remark relating to the period of a thousand years, but it is certain that it was so taken by some. At any rate, the passing of this milestone saw the Christian world aroused to new interests.

Then, too, there were the crusades (1095–*c.* 1270), which have been called "the first Renaissance," and which did for a civilization that had long been dormant one thing which the World War did for the civilization of the 20th century,—it let one part of the race know more of what other parts were doing and thinking and hoping. It was war, but it was in general beyond the boundaries of intellectual Europe.

There was also the potent influence in Europe of a foreign and highly developed civilization in her midst,—the Saracen supremacy in Spain; and it was the Saracen scholars who made known to Latin scholars the best of the Greek and Oriental civilizations.

Moreover, Europe was seeing the folly of her private wars, the "Truce of God" was beginning to make its power felt, and the blessings of peace were once more settling upon France and her neighbors, rendering intellectual pursuits possible.

To these influences there should be added that of the Norman Conquest, which, without prolonged warfare, awakened and united England, and showed her what the Continent had for her in the way of science and art.

As a result of such influences Europe entered upon a new era, one in which cathedral building,[1] church reform, renewed attention to art, political experiment, and scientific achievement played great parts.

Gerbert. Nevertheless, the period was still dominated by the spirit of the earlier centuries of the Middle Ages, "when faith overpowered intelligence" and "authority became the enemy of investigation," when "scholars degenerated into schoolmen" and "science lost itself in the morasses of alchemy or astrology and became anathema to the faithful."[2] This is seen in the attitude of the learned world toward that remarkable churchman and scholar, Gerbert,[3] one of the greatest popes that ever added lustre to the Church and to the city of Rome. Elevated to the papal throne, he reigned under the name of Sylvester II from 999 until 1003. He was born of humble parents,[4] but his natural brilliancy led to his call to study under the monks at Aurillac, and particularly under such a worthy scholar as Abbo of Fleury, and to his being sent to Spain (967) to perfect his education.[5] About 970 he went to Italy, where he was presented to the pope and by him to the emperor, returning to

[1] "It was as though the world had arisen and tossed aside the worn-out garments of ancient time, and wished to apparel itself in a white robe of churches." —Raoul Glaber (985–*c.* 1046).

[2] W. C. Abbott, *The Expansion of Europe*, I, chap. i, New York, 1918.

[3] Born near Aurillac, in Auvergne, *c.* 950; died at Rome, May 12, 1003. The name is pronounced zhĕr-bâr.

[4] "Obscuro loco natum," as an old chronicle states.

[5] For bibliography and for a more elaborate sketch, see Smith-Karpinski, p. 110 seq. See also Cantor, *Geschichte*, II, chap. 39; J. Havet, *Lettres de Gerbert* (983–997), Paris, 1889; N. Bubnov, *Gerberti postea Silvestri II papae opera Mathematica*, Berlin, 1899; A. Olleris, *Œuvres de Gerbert*, Paris, 1867; F. Picavet, *Gerbert, un pape philosophe, d'après l'histoire et d'après la légende*, Paris, 1897; H. Weissenborn, *Gerbert. Beiträge zur Kenntnis der Math. des Mittelalters*, Berlin, 1888; C. F. Hock, *Gerberto o sia Silvestro II Papa ed il suo secolo*, Milan, 1846; A. Nagl, *Gerbert und die Rechenkunst des X. Jahrh.*, Vienna, 1888; G. Friedlein, "Die Entwickelung des Rechnens mit Columnen," *Zeitschrift für Mathematik und Physik*, X, Hl. Abt., 241 (hereafter referred to as *Zeitschrift* (Hl. Abt.)), and *Gerbert, die Geometrie des Boethius*, Erlangen, 1861; K. Werner, *Gerbert von Aurillac*, Vienna, 1878; B. Carrara, *Memorie dell' Accad. d. Nuovi Lincei*, XXVI, 195; K. Schultess, *Papst Silvester II. (Gerbert) als Lehrer und Staatsmann*, s. l. a.

France in 972. He held various offices in the Church, and in 999 was elected to the papacy. He was a man of great learning, was "accused—our learning's fate—of wizardry," combated error, aroused new interest in mathematics, acquired a knowledge of the Hindu-Arabic numerals, gave some attention to the study of astrology (a subject then looked upon as a worthy science), and wrote on arithmetic,[1] geometry,[2] and other mathematical subjects, and probably on the astrolabe.[3]

Minor Church Writers. Contemporary with Gerbert, but living a life as humble as Gerbert's was magnificent, was an English monk of the abbey of Ramsey, Byrhtferth[4] by name. He traveled in France and studied under Abbo of Fleury. Returning to England he found waiting for him at Ramsey a group of students to whom he proceeded to teach astronomy, the calendar, and the principles of mathematics.[5] Times were not propitious for study, however. For three centuries in England (1000–1300) there was an average of a famine every fourteen years, and life was hard. Perhaps the need for the conquest of mind over matter, which such calamities set forth, was one of the influences that made possible the later thinkers of England.

On the Continent, St. Gall was one of the chief centers of monastic learning at this time, and here the well-known scholar Notker Labeo[6] (*c.* 950–1022) translated parts of the encyclopedia of Capella and possibly some of the arithmetic of Boethius, besides writing a computus.[7]

[1] *Regulae de numerorum abaci rationibus; Scholium ad Boethii arithmeticam.*

[2] *Gerberti Isagoge Geometriae.* Some doubt has been expressed as to his authorship, but he probably compiled the work.

[3] *Gerberti Liber de astrolabio*, placed by Bubnov with other works among the *Opera Dubia.*

[4] Or Bridferth. Fl. *c.* 1000.

[5] *De temporum ratione, De natura rerum, De indigitatione, De ratione unciarum, De principiis mathematicis*, the extant MSS. being merely notes of his lectures. The Anglo-Saxon text of his *Handboc* was published by F. Kluge in *Anglia*, VIII, 298. See also the Cologne edition (1612) of Bede's works. There are two other works doubtfully attributed to him.

[6] "Notker the Thick-lipped," so called to distinguish him from earlier scholars of the same family.

[7] A. A. Björnbo in the *Reallexikon der Germanischen Altertumskunde*, IV, 465. Strasburg, 1916.

Of the mathematical pupils of Gerbert the most prominent was Bernelinus of Paris, who wrote an arithmetic[1] in which he explained the use of Gerbert's counters, but concerning his life nothing further is known.[2]

A little later (*c.* 1028) Guido of Arezzo (Aretinus), a Benedictine monk from Pomposa, near Ferrara, wrote on arithmetic,[3] and at about the same time (*c.* 1066) Franco of Liège did the same and, what was not so common at this time, wrote on the quadrature of the circle.[4] Among his contemporaries was Wilhelm, abbot of Hirschau (1026–1091), who taught mathematics and astronomy.

Hermannus Contractus. The most prominent of the successors of Gerbert in the 11th century was Hermannus (1013–1054), son of the Swabian Count Wolverad. His limbs having been painfully contracted from childhood, he is known in history as Hermannus Contractus.[5] Educated in the monastic school at Reichenau, he afterwards joined the Benedictine order, became a lecturer on mathematics, and gathered about him a large number of pupils. He wrote on the astrolabe,[6] the abacus, and the number game of rithmomachia.[7]

Psellus. The period of intellectual activity in the West had very little counterpart in Constantinople. Life was still stagnant there. In the 11th century only a single name stands out as representing any interest whatever in mathematics in the eastern capital,—that of Michael Constantine Psellus[8] (1020–1110), a Greek writer who studied at Athens, became a zealous Neoplatonist, and returned to Constantinople to teach philosophy.

[1] *Liber Abaci.* [2] Gerbert's *Œuvres*, ed. Olleris, p. 357 (Paris, 1867).

[3] B. Baldi, Boncompagni's *Bullettino*, XIX, 590.

[4] *Abhandlungen*, IV, 135.

[5] Treutlein, Boncompagni's *Bullettino*, X, 589, where his *Abacus* is published; Günther, *Math. Unterrichts*, p. 47; Baldi, *Cronica*, p. 70. He is also known as Hermann the Lame.

[6] There is a beautifully written MS. of this work, 12th century, in the British Museum (22,790), first published by Pez in Volume III of his *Thesaurus Anecdotorum* and republished by Migne in Volume CXLIII of *Patrologiae cursus completus.* [7] See page 198.

[8] Ψέλλος, called also Psellus the Younger, there having been another Psellus who taught philosophy *c.* 870. Heath, *History*, II, 545.

He lived during the reigns of several rulers, consulted by the emperors and honored by them with the title of Prince of Philosophers.[1] An introduction to the study of Nicomachus and Euclid is attributed to him, but the authorship is doubtful. Partly because of the fact that he was almost the last of the Greek writers on mathematics, partly because his works were easily read, and partly because of his reputation for learning in general, he is one of the few scholars of his time whose mathematical contributions attracted any attention in the Renaissance period. His leading works on mathematics[2] were published at least thirteen times in the 16th century. The fact that he takes $\sqrt{8}$ as the value of π shows how little he merited his reputation as a scientist.

Rithmomachia. In speaking of the 11th century mention should be made of the number game of rithmomachia.[3] One of the earliest treatises on the subject is due to Fortolfus, a monk, who lived probably at the close of the 11th century,[4] and the indications are that it was not known before that century, although it is occasionally attributed to Boethius and even to Pythagoras. There is a manuscript in the Vatican library on the subject, under the title "Ritmachya," written in 1077 by a monk known as Benedictus Accolytus, and the game is also referred to in a medieval poem *De Vetula*.[5] Among the early

[1] Φιλοσόφων ὕπατος.

[2] *Sapientissimi Pselli opus dilucidum in quattuor Mathematicas disciplinas, Arithmeticam, Musicam, Geometriam, & Astronomiam*, edited by Archbishop Arsenius, Venice, 1532. This was the first edition, the text in Greek. The *Compendium Mathematicum*, containing various works, appeared at Leyden in 1647, but numerous others still remain unedited.

[3] Literally, "combat of numbers." The word is spelled in various ways, rithmimachia, ritmachya, richomachie, and rhythmimachia being among the most common forms.

A work on the subject by Boissière, a French mathematician of the 16th century, is entitled *Nobilissimvs et antiqvissimus ludus Pythagoreus* (*qui Rythmomachia nominatur*), Paris, 1556. See *Rara Arithmetica*, pp. 12, 63, 271, 340.

[4] R. Peiper, "Fortolfi Rythmimachia," *Abhandlungen*, III, 167, 198.

[5] "O utinam ludus sciretur Rythmimachiae!
ludus Arithmeticae folium, flos fructus et eius
gloria laus et honor."

Abhandlungen, III, 222

writers who were interested in the subject were Hermannus Contractus (1013–1054), as already stated, and both Jordanus Nemorarius (died *c.* 1236) and Nicole Oresme (*c.* 1323–1382).

RITHMOMACHIA

From a work published at Paris in 1496. The middle portion of the board is omitted. The part on Rithmomachia may be due to Bishop Shirwood of Durham (died 1494), but is usually ascribed to Faber Stapulensis (1455–1536)

The game is based on the Greek theory of numbers as set forth by Nicomachus. It was played upon a double chess-board, rectangular in form, one side having eight squares and the other sixteen. The pieces were triangles, squares, circles, and pyramids, each possessing a certain value. These pieces were arranged as shown in the illustration (p. 199) from a work of 1496. The numbers were not taken at random, but the plan on which they were arranged is too elaborate for description in this work. Suffice it to say that when we form the triangles we have $81 = 72 + \frac{1}{8}$ of 72, $72 = 64 + \frac{1}{8}$ of 64, $6 = 4 + \frac{1}{2}$ of 4, and $9 = 6 + \frac{1}{2}$ of 6; that in the case of the square pieces, $45 = 25 + 20$ and $15 = 9 + 6$; that the pyramids are superposed squares such that $91 = 6^2 + 5^2 + 4^2 + 3^2 + 2^2 + 1^2$ and $190 = 8^2 + 7^2 + 6^2 + 5^2 + 4^2$. In the case of the squares there is a formula $s = \left(\frac{2n+1}{n+1}\right) s'$, in which the meaning of each letter may be found by looking at the illustration, where $s = 25$, $s' = 15$, $n = 2$; or $s = 81$, $s' = 45$, $n = 4$. The play is very complicated, and for our purposes we may say that the climax of the game was reached in the *Victoria praestantissima*, in which it was necessary to get four numbers in a row, embodying all three of the common progressions,—arithmetic, geometric, and harmonic, the only possible solutions with these pieces being six in number. It will be seen that the game requires such familiarity with the Greek number theory as to make it available only for the élite in mathematics in the Middle Ages. Its popularity is attested by at least three manuscripts of the 11th century and three of the 12th and 13th centuries, besides several printed treatises on the subject, all going to show that there were more scholars in number theory than we should think from the meager list of names that have come down to us.[1]

A Century of Translators. The 12th century was to Christian Europe what the 9th century was to the eastern Mohammedan world, a period of translations. In the case of Bagdad, these translations were from the Greek into Arabic; in the case of

[1] For a description of the game, see D. E. Smith, "Number Games and Number Rhymes," in *Teachers College Record*, XIII (New York), 385, together with a history of "The Great Number Game of Dice." The article on rithmomachia may also be found in the *Amer. Math. Month.*, April, 1911. See also E. Wappler, *Zeitschrift* (Hl. Abt.), XXXVII, 1.

Christian Europe, from the Arabic into Latin. The reasons for this desire to know the science of the East are not difficult to find. The causes already mentioned in connection with the 11th century were even more potent a hundred years later, and the advancement of Moorish Spain in the arts and sciences was already causing intellectual unrest in the higher class of Church schools in France, Italy, and England. The result of this unrest was an influx of students into Spain, an acquiring of some knowledge of Arabic on the part of various scholars, and a strong desire to know and to make known the science of the East. Just as Bagdad never translated the Greek literature, but sought diligently to know Greek science, so Europe gave little attention to Arab letters, but devoted great care to those works on astronomy, arithmetic, trigonometry, optics, astrology, geometry, and medicine that had acquired reputation in the capital of the caliphs. Even the *Elements* of Euclid became known to the scholars of the Latin Church chiefly through its Arabic translation instead of through the original Greek.

Italian and French Translators. In the 12th century Italy and France produced two or three prominent scholars whose knowledge of Arabic and taste for mathematics led them to make known to the Latin world various classics of the Mohammedan and Greek civilizations.

The first of these translators was Plato of Tivoli, or Plato Tiburtinus,[1] who lived *c.* 1120. He translated the astronomy of Albategnius (al-Battânî), the *Spherics* of Theodosius, the *Liber Embadorum* of Abraham bar Chiia (*c.* 1120), and various works on astrology.

About this time Sicily was also active in the translation of Greek and Arabic works.[2] Among the treatises thus brought to the attention of scholars was Ptolemy's *Almagest*, which was turned into Latin by an unknown translator, *c.* 1160,[3]

[1] B. Boncompagni, *Delle versioni fatte da Platone Tiburtino*, Rome, 1851.

[2] C. H. Haskins and D. P. Lockwood, "The Sicilian Translators of the Twelfth Century . . . ," in the *Harvard Studies in Classical Philology*, XXI, 75.

[3] There is in the Vatican a MS. of this translation, written *c.* 1300. It is this that was used by Professors Haskins and Lockwood in the work above cited.

from a Greek manuscript which had formerly been brought from Constantinople to Palermo by a Sicilian scholar.

Some years later Gherardo Cremonense, or Gherardo of Cremona (1114–1187),[1] studied in Italy and then in Spain, learning Arabic in Toledo. With him, as with many other scientists in the Middle Ages and even later, astrology formed a nexus joining medicine and mathematics, his interests therefore lying in all three lines. He translated various mathematical and astronomical works from the Arabic, including Euclid's *Elements* and *Data*, the *Spherics* of Theodosius, a work by Menelaus, and Ptolemy's *Almagest*,[2] "for the love" of which book he journeyed to Toledo.[3] In his translation is found one of the early uses of the word *sinus* for a half chord, this being the first of our modern names for the trigonometric functions.[4] There was a younger Gherardo of Cremona who lived in the 13th century, called da Sabbionetta, who wrote on astronomy.[5]

Among the Italian and French translators there may properly be included Rudolph of Bruges, since most of his work was done under French influence. About his time (*c.* 1143) Hermann of Carinthia translated Ptolemy's *Planisphere*.[6]

[1] Apparently a native of Cremona in Lombardy, although certain Spanish writers have claimed him for Carmona in Andalusia. The name appears in English as Gerard and in Latin as Girardus, with variants. B. Boncompagni, *Della vita e delle opere di Gherardo Cremonense*, Rome, 1851. A considerable amount of information relating to such early Italian mathematicians is given in B. Veratti, *De' Matematici Italiani anteriori all' Invenzione della Stampa*, a pamphlet with bibliographical notes, Modena, 1860.

[2] The translation was finished in 1175, as an old MS. asserts. This was about fifteen years after the Sicilian translation, a work of which Gherardo was apparently ignorant. See also Rose, in *Hermes*, VIII, 332. It was printed in Venice in 1515. On the question of his translations see A. A. Björnbo, *Bibl. Math.*, VI (3), 239.

[3] Amore tamen almagesti, quem apud latinos minime reperiit, Toletum perrexit."

[4] On the question of priority and of the use of the term by Plato of Tivoli, see A. Braunmühl, *Geschichte der Trigonometrie*, I, 49 (Leipzig, 1900, 1903); hereafter referred to as Braunmühl, *Geschichte*. The term was probably first used in Robert of Chester's revision of the tables of al-Khowârizmî. See also *Bibl. Math.*, I (3), 521.

[5] His *Theorica planetarum* was printed at Ferrara in 1472.

[6] This was printed in 1507. See M. Chasles, *Aperçu historique sur l'origine et développement des méthodes en géométrie*, Paris, 1837; 2d ed., 1875, hereafter referred to as Chasles, *Aperçu*; 3d ed., 1889. See also *Bibl. Math.*, IV (3), 130.

English Translators. England produced two or more translators of prominence in the 12th century, and Ireland seems to have produced at least one. Of these the best known is Adelard[1] of Bath (*c.* 1120), a British scholar who studied at Toledo (1130), at Tours, at Laon, and also in the East, and who journeyed through Greece, Asia Minor, Egypt, and possibly Arabia, bringing back numerous mathematical works.[2] He is credited with a knowledge of Greek and was one of the first to translate Euclid into Latin, but he seems to have made this translation from the Arabic.[3] Either he or Campanus seems to have determined the sum of the angles of a stellar polygon, a figure then attracting considerable interest, possibly because of its use in astrology. He probably translated the astronomical tables of al-Khowârizmî, and he is said to have written a commentary on the arithmetic of this author and to have composed a work entitled *Regulae abaci.*[4] Adelard was by no means the first to bring Euclid's name into England, for, as we have seen (p. 187), it was probably known to British scholars in the 10th century.

A few years after Adelard's sojourn in Toledo two other English scholars who were interested in mathematics went to Spain to pursue their studies. The first of these was Robert of Chester (*c.* 1140),[5] who translated al-Khowârizmî's algebra[6] into Latin and prepared several astronomical tables. He was archdeacon of Pampeluna, in northern Spain, and seems also to have traveled in Italy and Greece. He was the first to translate the Koran into Latin (1143).

[1] The older English form was Aethelhard. See C. H. Haskins, "Adelard of Bath," *English Historical Review*, p. 491 (1911).

[2] F. Woepcke, *Journal Asiatique*, I (6), 518.

[3] From certain similarities in the different manuscripts there seems to have been an unknown scholar whose version was consulted by Adelard and various other early translators.

[4] Boncompagni's *Bullettino*, XIV, 1.

[5] Robertus Retinensis, Robertus Ketensis, Robert de Ketene, Robert de Retines, Robertus Cataneus, and other variants. He is known to have been in Spain in 1141 and seems to have been studying at Barcelona with Plato of Tivoli in 1136. As already stated, he translated the astronomy of Albategnius.

[6] For reference to the translation see page 170, note 3.

The second of these English scholars was Daniel Morley,[1] who studied at Oxford in 1180. He went to Paris and thence to Toledo,[2] and wrote on astronomy and mathematics,[3] quoting freely from Arabic authors. That such men were compelled to go abroad for their mathematics at this time is apparent from the records of the work done in the schools of London, this work being chiefly of the nature of grammar and disputation.[4] That they should go to Spain was quite natural, not merely for linguistic reasons but because of the close ties that existed between Castile and England, owing to the marriage of Alfonso VIII (1158–1214) to Lenora, daughter of Henry II.

Other Scholars. One of Adelard's pupils, N. O'Creat,[5] wrote a work on multiplication and division which shows his indebtedness to Arab writers on mathematics. Of O'Creat himself, however, nothing further is known, but the name suggests the country of his birth. The work contains a rule of Nicomachus for squaring a number by using the formula $a^2=(a-b)(a+b)+b^2$, thus: $109^2 = 100 \cdot 118 + 81 = 11{,}881$. He used the Roman numerals, but with both o and a character like the Greek τ for zero.[6]

[1] Daniel of Merlai, Merlac, Marlach. In Latin, Morleius, Merlacus. A MS. in the British Museum (J. O. Halliwell, *Rara Mathematica*, London, 1838–1839, 2d ed., 1841, p. 84; hereafter referred to as Halliwell, *Rara Math.*) begins, "Philosophia magistri Danielis de Merlai ad Johannem Norwicansem episcopum." See also C. Singer in *Isis*, III (1920), 263.

[2] See A. à Wood, *Historia et Antiquitates Vniversitatis Oxoniensis*, I, 56 (Oxford, 1674); hereafter referred to as Wood, *Historia Oxon.*

[3] Probably *De principiis mathematicis.*

[4] See the *Descriptio nobilissimae civitatis Londoniae* written by Fitzstephen (died *c.* 1190), prefixed to his *Life of Becket*, and published in John Leland's *Itinerary* (London, 1770); J. Stow, *Survey of London*, p. 703 (London, 1633).

[5] Probably the same as Joh. Ocreatus. See N. Bubnov, *Gerberti . . . Opera Mathematica*, p. 174, n. 7 (Berlin, 1899); C. Henry, *Abhandlungen*, III, 129. O'Creat begins his work with these words: "N. O. Creati liber de multiplicatione et divisione numerorum ad Adelardum Bathoniensem magistrum suum." There is a 13th century MS. of the work in the Bibliot. nat. in Paris. The prologue begins, "Prologus N. Ocreati in Helceph, ad Adelardum Batensem magistrum suum." On the uncertain meaning of Helceph, see Henry, *loc. cit.*

[6] Possibly from the medieval τζίφρα, from theca (teca), or from τήκειν (to come to naught).

About the year 1125 Radulph of Laon (died 1133) wrote on arithmetic, and a little before this time (*c.* 1090) Gerland, prior of St. Paul, of Besançon,[1] wrote a computus and a brief work on the abacus.[2]

Early in the 12th century there was an astrologer, geometrician, and abacist by the name of Walcherus, a native of Lorraine, who attained considerable prominence in England and wrote a work on astronomy.[3] Such names are of interest simply as they bear witness to the nature of mathematics in the Church schools of the time.

2. Oriental Civilization in the West

Spain. After the year 1000 numerous Moorish scholars appeared in Spain and contributed to the literature of arithmetic and astronomy, and occasionally to that of algebra. A list of a few of the most prominent of these scholars will serve to show their range of interest and achievement in the general field of mathematics.

Ibn al-Ṣaffâr,[4] a native of Cordova, wrote on astronomical tables and instruments. A little later (*c.* 1050) Ibn al-Zarqâla,[5] probably a native of Cordova, wrote on astronomy and astrology, and prepared a set of tables.[6]

[1] An ancient record speaks of "Gerlandus vel Garlandus Prior S. Pauli, annis, 1131, 1132." A document of 1134 records: "Huius praefatae concordiae testes sunt . . . Garlandus magister . . . anno . . . M.C.XXXIIII." He is also mentioned in 1148: "Magistrum quoque Jarlandum Bisuntinum & magistrum Theodericum Carnotensem [i.e., of Chartres], duoa fama & gloria doctores nostri temporis excellentissimos." He is again mentioned in a letter written to him in 1157: "Gerlando scientia trivii, quadriviique onerato & honorato." See Boncompagni's *Bullettino*, X, 654.

[2] See Boncompagni's *Bullettino*, X, 653; Cantor, *Geschichte*, II (2), 843.

[3] In the south aisle of the chancel of the old priory at Great Malvern may still be seen his tomb with this inscription, in part: Philosophvs · dignvs · bonvs · astrologvs · Lotheringvs | vir · pivs · ac · hvmilis · monachvs · prior · hvjvs . . . | . . . geometricvs · ac · abacista :— | Doctor · Walchervs . . . MCXXV · ✠ He should not be confused with the Walcherus of Lorraine who became Bishop of Durham and was murdered in 1075.

[4] Aḥmed ibn 'Abdallâh ibn 'Omar al-Gâfiqî, Abû'l-Qâsim (died 1035).

[5] Ibrâhîm ibn Yaḥyâ al-Naqqâsh, Abû Isḥâq.

[6] Schoner translated one of his works in 1534.

In the latter part of the century Abû'l-Ṣalt,[1] a Spanish physician from Denia, wrote on geometry and astronomy, and Jabir ibn Aflah (died between 1140 and 1150), commonly known as Geber, flourished at Seville and wrote on astronomy, spherical trigonometry, and the transversal theorem of Menelaus.[2] He is often confounded with an alchemist of similar name.

Jewish Scholars of the Eleventh Century. The most learned scholars in Spain at the close of the 11th century, however, were not Mohammedans. The Jewish race, which may conveniently be mentioned in connection with the Oriental civilization in Spain, was generally accorded better treatment under Saracen than under Christian rule, although it had flourished somewhat in Italy before this time. Through the encouragement received from the Moors the Jews contributed in no small degree to the advance of mathematics in Spain, and to them the Christians were indebted for their first knowledge of the Arabic works on the subject.[3] The first of their prominent scholars in this century was Abraham bar Chiia[4] (Abraham Judæus), commonly known as Savasor'da[5] (*c.* 1070–*c.* 1136), a native of Barcelona. He wrote on astronomy, but is chiefly known for an encyclopedia which included arithmetic, geometry, and mathematical geography.[6] Of this only fragments are now extant. He also wrote a work entitled *Liber Embadorum*,[7] treating of geometry but containing numerous definitions used in the theory of numbers. In this he accuses the French

[1] Omeiya ibn ʿAbdelʿazîz ibn Abî'l-Ṣalt, Abû'l Ṣalt (1067/8–1133/4).

[2] His astronomy was translated by Gherardo of Cremona and was printed in 1534.

[3] Libri, *Histoire*, I, 154 n.

[4] Or Chijja, Chiya.

[5] From *Sahib al-Shorta*, "Chief of the Guards." The transliteration to Savasorda is due to Plato of Tivoli (*c.* 1120).

[6] *Iesode ha-Tebuna u-Migdal ha-Emuna.*

[7] This is one of the sources of Fibonacci's geometry. See M. Curtze, *Abhandlungen* XII, where the Latin and German translations are given. The title is medieval Latin from the Greek ἐμβαδόν, an area or surface. For his works and for the contemporary Jewish writers see J. Bensaude, *L'Astronomie Nautique au Portugal*, p. 52 (Bern, 1912); hereafter referred to as Bensaude, *Astron. Portug.* See *Bibl. Math.*, 1896, p. 36.

Jews of being ignorant of geometry and therefore weak in their arithmetic. This work was translated from Hebrew into Latin by Plato of Tivoli.

Rabbi ben Ezra. The second great Hebrew scholar of the period was Abraham ben Ezra.[1] He wrote on the theory of numbers, the calendar, magic squares, astronomy, and the astrolabe, was much interested in the cabala, and is justly ranked as the most learned Jew of his time.[2] He traveled extensively, going at least as far as Egypt to the east and as far as London (1158) to the north. Besides his contributions to astronomy, the calendar, and allied subjects, he wrote three or four works on number: (1) *Sefer ha-Echad*;[3] (2) *Sefer ha-Mispar*,[4] chiefly on arithmetic; (3) *Liber augmenti et diminutionis vocatus numeratio divinationis*, known only in Latin translation and possibly not due to him;[5] (4) *Ta ʿhbula*, containing the Josephus Problem, possibly a separate work, and probably due to him. Of these the *Sefer ha-Mispar* is the only one of importance. It is based on the Hindu arithmetic but uses Hebrew letters for the numerals, with a zero as in algorism. He employed the check of casting out nines, as several of his predecessors had done. The following is an example of his rules: "Whoever would know how great the sum of the numbers is

[1] Born at Toledo, between 1093 and 1096; died at Rome or Rouen, 1167. This is the Rabbi ben Ezra of Browning's poem. He is sometimes confused with Abraham ben Chiia, probably because each was called Abraham Judæus. On his life see M. Steinschneider, *Bibl. Math.*, IX (2), 43; *Abhandlungen*, III, 57; Bensaude, *loc. cit.*, p. 52.

[2] Attention was called to him as a mathematician by O. Terquem, *Journal des mathématiques pures et appl.*, VI, 275. Since then his work has been studied by Luzzato, Rodet, and Steinschneider. See also Smith and Ginsburg in the *Amer. Math. Month.*, XXV, 99.

[3] *Book of Unity*. This has twice been published: Bamberg, 1856; Odessa, 1867.

[4] *Book of Number*. M. Silberberg published a German translation at Frankfort a. M. in 1895. See also *Bibl. Math.*, IX (2), 91.

[5] In favor of his authorship, Cantor, *Geschichte*, I (3), 730; against it, G. Wertheim, *Bibl. Math.*, II (3), 143. See also P. Tannery, *Bibl. Math.*, II (3), 45. The *Liber augmenti . . . divinationis* was published by Libri. His *Liber de nativitatibus* (Venice, 1485) was, however, the first of his works to appear from the press.

which follow one another in a series to a certain number, multiply this by its half increased by $\frac{1}{2}$. The product is the sum."[1]

Although highly esteemed by Jews and Christians alike, his fate was not altogether a happy one, and in his struggle against adversity he voices his lament in words like these:

> Were candles my trade it would always be noon;
> Were I dealing in shrouds Death would leave us alone.

In connection with the Jewish activity of this period there should also be recalled the name of one Hasan, a judge, who may have written in the 10th century, but whose country is unknown, and of Yehuda ben Rakufial, who seems to have been a physician in Spain. Both of these men wrote on the Jewish calendar, and the former is referred to by Rabbi ben Ezra.

Twelfth Century in Spain. The 12th century was even more favorable than its predecessors to the study of mathematics in Spain. The first of the Arab writers was Aver'roës (*c.* 1126–1198/9), as he was commonly called in the Middle Ages,[2] who wrote on astronomy and trigonometry. His most prominent scientific contemporary was Avenpace, as he was called by the Christians,[3] who lived at Seville and Granada *c.* 1140 and wrote on geometry.

Jewish Writers of the Twelfth Century. In this as in the preceding century, however, it was the Hebrew scholar who made the greatest contributions to the advance of mathematics. Aside from Rabbi ben Ezra, two of these scholars are deserving of special mention: Maimonides[4] (1135–1204), a native of Cordova, physician to the sultan, and an astronomer of

[1] That is, $s = n\left(\frac{n}{2} + \frac{1}{2}\right)$. *Sefer ha-Mispar*, ed. Silberberg, p. 24 (Frankfort a. M., 1895).

[2] His name was Mohammed ibn Aḥmed ibn Mohammed ibn Roshd, Abû Velîd.

[3] His name was Mohammed ibn Yaḥyâ ibn al-Ṣâig, Abû Bekr, also known as Ibn Bâjje and as Ibn Ṣâig. The name "Avenpace" (also spelled Avempace) is a Spanish form and, as such, is pronounced ah vân pä'thâ.

[4] Rabbi Moses ben Maimun. He became rabbi of Cairo in 1177.

prominence,[1] and Johannes Hispalensis[2] (fl. *c.* 1140), who professed Christianity and wrote on arithmetic and astrology (1142) and translated various Arabic works on mathematics into Latin.[3]

In the same century there were various Jewish scholars of less prominence, such as Samuel ben Abbas,[4] who wrote on arithmetic,[5] the Hindu numerals and their use,[6] algebra, and geometry. There was also an unknown English Jew who wrote a work called by English historians *Mathematum Rudimenta quaedam.*

Jewish Writers of the Thirteenth Century. The 13th century saw various translations made from the Arabic into Hebrew, and several of the translators are known. Among these was Moses ben Tibbon,[7] whose father and grandfather were celebrated as translators of philosophical and scientific works from the Arabic into Hebrew. He was actively at work about the middle of the century and translated (1259) the astronomy[8] of Alpetra'gius (*c.* 1200) and probably, as stated on page 210, the arithmetic of al-Ḥaṣṣar.

The other Jewish scholars of this period also showed their chief scientific interest in astronomy. Jehuda ben Salomon Kohen of Toledo (died 1247), for example, wrote upon Ptolemy's *Almagest*, although he also prepared a brief extract from Euclid

[1] There is a Jewish calendar of his among the manuscripts in the Bodleian Library. Parts of his works on the calendar were printed at Paris in 1849, at Leipzig in 1859, and at Berlin in 1881.

[2] John of Seville, John of Luna. As in many such cases, the first name is often written Joannes. The full name is also written Johannes Hispanensis or Johannes de Hispania. The date of his death may have been 1153.

[3] At least some were translated into Spanish and were then put into Latin by Domenico Gondisalvi. His *Alghoarismi de Practica Arismetrice* was published by B. Boncompagni, Rome, 1857. It is based on Arab sources, but is not a translation. His translations include works by Alfraganus (al-Fargânî, *c.* 833), Abû 'Alî al-Chaiyât (a prominent astrologer, died 835), and Ṯâbit ibn Qorra (*c.* 875). His works were published at Nürnberg in 1548.

[4] M. Steinschneider, *Bibl. Math.*, X (2), 81. He adopted the Mohammedan faith, his Arabic name being Samû'îl ibn Yaḥyâ ibn 'Abbâs al-Maġrebî al-Andalusî. He died in 1174/5.

[5] *Al-Tab'sira.*

[6] *Al-Qiwâmî*, probably named after a patron, Qiwâm ed-din Yaḥyâ.

[7] *Bibl. Math.*, X (2), 112.

[8] *Kitâb al-hei'a.*

and wrote a commentary upon it, and Isaac ben Sid, of Toledo (died 1256), edited the Alfonsine Tables (see page 228) just before his death.

About the middle of the 13th century there was born in Cordova another descendant of the celebrated Tibbon family,—Jacob ben Machir, known as Prophatius. He lived in Montpellier, wrote on a quadrant which he had invented (the *quadrans Israelis* or *quadrans Judaicus*),[1] translated from the Arabic into Hebrew the *Elements* and *Data* of Euclid and the *Sphere* of Menelaus,[2] and composed a work on the almanac.

Arab Writers of the Twelfth Century. Of the writers on arithmetic among the western Arabs of the 12th century one of the best known was Abû Bekr Mohammed ibn ʿAbdallâh,[3] commonly known as al-Ḥaṣṣâr.[4] His work was so well received that, as already stated, it was translated into Hebrew by Moses ben Tibbon[5] (1259). The work is evidently Western, since it uses the ġobâr numerals.

Arab Writers of the Thirteenth Century and Later. Early in the 13th century Alpetragius,[6] as the Christians called him, lived in Spain, probably in Seville, and wrote on astronomy (*c.* 1200). His theory of planetary motion, which gives him a place in the list of mathematical writers, was translated into Latin by Michael Scott.

Contemporary with Alpetragius there was a certain Ibn al-Kâtib[7] (died 1210/11), who wrote two works which included a little discussion of arithmetic, geometry, and architecture.

[1] There is a good MS. of the work in the Columbia University Library, apparently of the 15th century. The work has been several times translated into Latin.

[2] Boncompagni's *Bullettino*, IX, 595; *Bibl. Math.*, XI (2), 35. The name also appears as Propatius. He died *c.* 1308.

[3] A Gotha MS. gives the name as Abû Zakarîyâ Mohammed ibn ʿAbdallâh ibn ʿAiyâsh. See *Bibl. Math.*, XIII (2), 87.

[4] That is, the Computer; but Suter thinks this a family name. See H. Suter, "Das Rechenbuch des Abû Zakarîjâ el-Ḥaṣṣâr," in *Bibl. Math.*, II, (3), 12, and III (2), 109. See also *ibid.*, XIII (2), 87.

[5] Probably. The Vatican MS. has still to be studied critically.

[6] Nûr ed-dîn al-Beṭrûjî, Abû Isḥâq.

[7] Mohammed ibn ʿAbderraḥmân, Abû ʿAbdallâh.

Of the scholars born in northern Africa in the 13th century, and geographically closely related to the Spanish civilization, the best known is Albanna,, or Ibn al-Banna.[1] From the fact that he is also known as al-Marrâkushî we infer that he was a native of Morocco.[2] He wrote on astronomy, mensuration, algebra, the astrolabe, and proportion. His best-known work is the *Talchîs*, a treatise on arithmetic.[3]

There was also a Mohammedan scholar of Seville, known as Ibn Bedr[4] or Abenbêder, who wrote a compendium of algebra about this time.[5] The date is uncertain, but there is a commentary upon it in verse which was written in 1311/12.

The last of the great Moorish arithmeticians of Spain was al-Qalasâdî,[6] a native of Baza, a town near Granada. He wrote extensively on arithmetic and seems to have had some originality in the treatment of the theory of numbers. He introduced a new radical sign and a sign of equality, and proposed a system of ascending continued fractions.[7]

3. Christian Europe from 1200 to 1300

General Activity of the Thirteenth Century. Whatever may be thought of the mathematics of the 13th century, it is certain that the century itself represents the real awakening of the world after a long period of intellectual torpor. The centuries

[1] That is, Son of the Architect. His full name is Aḥmed ibn Mohammed ibn 'Otmân al-Azdî, Abû'l-'Abbâs. Born *c.* 1258; died in Morocco *c.* 1339.

[2] A. Marre, *Atti dell' Accademia Pontificia dei Nuovi Lincei*, XIX (hereafter referred to as *Atti Pontif.*); M. Steinschneider, Boncompagni's *Bullettino*, X, 313. Suter's list omits the occasional name al-Marrâkushî, and places the date of his birth as *c.* 1258 or later, although it is sometimes given as early as 1252. The father seems to have belonged to a Granada family.

[3] Discussed in Cantor, *Geschichte*, I (3), 806.

[4] Mohammed ibn 'Omar, Abû 'Abdallâh.

[5] José A. Sánchez Pérez, *Compendio de Álgebra de Abenbéder,* Arabic text and Spanish translation, Madrid, 1916.

[6] Ali ibn Mohammed ibn Mohammed ibn 'Alî al-Qoreshî al-Basṭî, Abû'l-Ḥasan. The name "al-Qalasâdî" means the Upright, or Versed in the Law. Suter gives the place and date of his death as Tunis, 1486.

[7] Woepcke in the *Journal Asiatique*, 1854, II, 358, and 1863, I, 58. See also the *Atti Pontif.*, XII, 230, 399.

immediately preceding had produced writers on mathematics in Europe, but they had produced no mathematicians. But now a Spirit of the Times was abroad. The Far East felt its influence, and hence the remarkable revival and development of algebra in China; India felt it, and hence the appreciation of the merit of Bhāskara, now a generation dead; and all of intellectual Europe felt it as never before. It was not a century of great beacon lights, but it was one in which lanterns were hung in all the thoroughfares of the West, promises of the great illumination that was to come with the period of the Renaissance.[1]

Rise of the Universities. The most potent influence in the development of the world's mathematical knowledge has, of course, been the universities, and it is from the 13th century that we trace the rise of these institutions in the modern sense of the term. The earliest medieval universities grew out of the cathedral or Church schools and hence their date of beginning is necessarily obscure. In most cases the years in which they received official privileges from some sovereign, civil or ecclesiastical, are known, however, and are commonly taken as the dates of foundation. In some cases there are two dates, one of the receipt of the privilege from the State and the other that from the Church, the latter giving to the holders of degrees a right to teach. Thus Paris had a charter from the State in 1200 and its degrees were recognized by the pope in 1283. The corresponding dates for Oxford were 1214 and 1296; and for Cambridge, 1231 and 1318. The University of Padua was founded in 1222, and that of Naples in 1224.[2] The 14th and 15th centuries saw a number of other universities established, but we may look upon the 13th century as the one which laid the foundation for this type of higher education, although the mathematics taught was still very meager.[3]

[1] J. J. Walsh, *The Thirteenth, Greatest of Centuries*, New York, 1907.

[2] Some of these dates are uncertain, but they are approximately as stated.

[3] H. Suter, "Die Mathematik auf den Universitäten des Mittelalters," *Festschrift der Kantonschule in Zürich* (Zürich, 1887), p. 39; hereafter referred to as Suter, *Univ. Mittelalt.*

THE TOWER OF KNOWLEDGE

Illustrating the educational system of the Middle Ages. From the *Margarita phylosophica*, 1503

Medieval Curriculum. The student began his study of grammar with Donatus and Priscian, and took his logic from Aristotle and his rhetoric from Cicero. He then entered upon his mathematical studies, such as they were,—arithmetic according to Boethius, music according to Pythagoras, geometry according to Euclid, and astronomy according to Ptolemy. The goal for those who were preparing for church activities was the metaphysics and theology of Peter Lombard (*c.* 1150). This progress was illustrated in a tower of knowledge given by Gregorius Reisch in his *Margarita phylosophica* (1503), as shown on page 213.

LEONARDO FIBONACCI

Modern engraving. The portrait is not based on authentic sources

Leonardo Fibonacci. The first great mathematician of the 13th century, and indeed the greatest and most productive mathematician of all the Middle Ages, was Leonardo Fibonacci, known also as Leonardo Pisano or Leonardo of Pisa.[1]

[1] Born at Pisa, *c.* 1170; died *c.* 1250. On his life and works see B. Boncompagni, *Scritti di Leonardo Pisano*, 2 vols., Rome, 1857–1862 (hereafter referred to as Boncompagni, *Scritti Fibonacci*); *Della vita e delle opere di Leonardo Pisano*, Rome, 1852; *Intorno ad alcune opere di Leonardo Pisano*, Rome, 1854; and *Tre scritti inediti di Leonardo Pisano*, Florence, 1854 (hereafter referred to as Boncompagni, *Tre Scritti*); Cantor, *Geschichte*, II, chaps. xli, xlii; Libri, *Histoire*, I, 156; E. Lucas, "Recherches sur plusieurs ouvrages de Léonard de Pise et sur diverses questions d'arithmétique supérieure," Boncompagni's *Bullettino*, X, 129; G. Loria, "Leonardo Fibonacci," *Gli Scienziati Italiani* (Rome, 1919), p. 4, with excellent bibliography; G. B. Guglielmini, *Elogio di Lionardo Pisano*, Bologna, 1813; F. Bonaini, *Memoria unica sincrona di Leonardo Fibonacci*, Pisa, 1858 (republished in 1867), and also in the *Giornale Arcadico*, Vol. CXCVII (N. S., LII), and in an article by G. Milanesi, *Documento inedito e sconosciuto intorno a Lionardo Fibonacci*, Rome, 1867; V. A. Le Besgue, "Notes sur les opuscules de Léonard de Pise," Boncompagni's *Bullettino*, IX, 583; O. Terquem,

At the time of Leonardo's birth, Pisa ranked with Venice and Genoa as one of the greatest commercial centers of Italy. These towns had large warehouses where goods could be stored and duty paid in all important ports of the Mediterranean, the head of such an establishment being a man of considerable prominence. It was such a position that the father[1] of Leonardo held at Bugia[2] on the northern coast of Africa, and in this town Leonardo received his early education from a Moorish schoolmaster.[3] As a young man he traveled about the Mediterranean, visiting Egypt, Syria, Greece, Sicily, and southern France,[4] meeting with scholars and becoming acquainted with the various arithmetic systems in use among the merchants of different lands. All the systems of computation he counted as poor, however, compared with the one that used our modern numerals.[5] He therefore wrote a work in 1202, *Liber Abaci*,[6]

"Sur Léonard Bonacci de Pise et sur trois écrits . . . ," *Annali di Sci. Mat.*, Vol. VII (reprint, Rome, 1856); M. Lazzarini, "Leonardo Fibonacci," *Bullettino di Bibliogr. di. Sci. Mat.*, VI, 98, and VII, 1; P. Cossali, *Scritti inediti*, ed. Boncompagni, p. 342 (Rome, 1857); Libri, *Histoire*, II, 21.

[1] Guglielmo Bonaccio. But the name "Fibonacci" is thought by Boncompagni and Milanesi to be a family name like Johnson, the form "filius Bonacci" being merely a Latin translation. An ancient document of 1226 has "Leonardo bigollo quondam Guilielmi," in which the Latin form "Bonaccius" does not appear, but in which the grandfather has this name. It seems more reasonable, however, to think that when Leonardo himself wrote "filius Bonacci," "filius Bonaccij," and "filius Bonacii," he knew what the words would mean to Latin readers. Leonardo speaks of his father as being "in duan a bugee,"—in the custom house at Bugia.

[2] Modern Bougie, whence France imported her wax candles (*bougies*). Little of its ancient splendor remains except the Moorish gate (Bab-el-Bahr, "sea-gate") in the old ramparts.

[3] "Vbi ex mirabili magisterio in arte per nouem figuris indorum introductus." *Liber Abaci*, p. 1.

[4] ". . . apud egyptum, syriam, greciam, siciliam et prouinciam."

[5] "Sed hoc totum etiam et algorismum atque arcus pictagore quasi errorem computavi respectu modi indorum," as it appears in the Florentine MS. published by B. Boncompagni, Rome, 1857.

Early writers attributed to him the introduction of these numerals into Italy; thus ". . . e questi fù il primo, che portò nell' Italia i carrateri dei numeri conforme testifica Luigi Colliado." *Aritmetica di Onofrio Pvgliesi Sbernia Palermitano*, p. 12 (Palermo, 1670).

[6] "Incipit liber Abaci Compositus a leonardo filio Bonacij Pisano In Anno M° cc° ij°." This is the title as it appears in the first line of Boncompagni,

in which he gave a satisfactory treatment of arithmetic and elementary algebra. The work is divided into fifteen chapters, and the following brief statement of the contents will serve to show its general scope: 1. Reading and writing of numbers in the Hindu-Arabic system;[1] 2. Multiplication of integers;[2] 3. Addition of integers; 4. Subtraction of integers;[3] 5. Division of integers; 6. Multiplication of integers by fractions; 7. Further work with fractions; 8. Prices of goods; 9. Barter; 10. Partnership; 11. Alligation; 12. Solutions of problems; 13. Rule of False Position;[4] 14. Square and cube roots; 15. Geometry and algebra, the former being devoted to problems in mensuration.

Possibly it was his indulgence in travel that caused him to write his name occasionally as Leonardo Bigollo,[5] since in Tuscany *bigollo* meant a traveler. The word also means blockhead, and it has been thought that he had been so called by the professors of his day because he was not a product of their schools, and that he retaliated by adopting the name simply to show the learned world what a blockhead could do. It would be human to hope that the latter explanation is the correct one, just as it is human to rejoice that the son of a provincial official became the greatest medieval mathematician. Such a remarkable career as Fibonacci's warns us, as Froude so

Scritti Fibonacci, I, 1, from the Codex Magliabechianus. The spelling of *abacus* varies in this and other MSS., often appearing as *abbacus*. B. Boncompagni, *Intorno ad alcune opere di Leonardo Pisano*, p. 1 (Rome, 1854).

[1] "Nouem figure indorum he sunt

9 8 7 6 5 4 3 2 1

Cvm his itaque nouem figuris, et cum hoc signo 0, quod arabice zephirum appellatur, scribitur quilibet numerus." P. 2.

[2] "Incipit capitulum secundum de multiplicatione integrorum numerorum."

[3] "Incipit capitulum quartum de extractione minorum numerorum de maioribus."

[4] "De regulis elchatayn . . . Elchataieym quidem arabice, latine duarum falsarum posicionum regula interpretatur." See Volume II, Chapter VI.

[5] "Incipit flos Leonardi bigolli pisani . . . ." See Boncompagni, *Tre Scritti*, 1. The word *bigolli* also appears as *pigolli*. See also F. Bonaini, *Iscrizione . . . a onore di Leonardo Fibonacci* . . ., Pisa, 1858; 2d ed., 1867.

truly said, "that we should draw no horoscope; that we should expect little, for what we expect will not come to pass."[1]

In the same years and in the same regions in which Leonardo was bringing new light into the science of mathematics, St. Francis, humblest of the followers of Christ, was bringing new light into the souls of men. Each was one of the world's geniuses, and for a genius there is no human explanation.

Fibonacci's Other Works. Leonardo also wrote three other works, the *Practica geometriae*[2] (1220), the *Liber quadratorum*[3] (1225), and the *Flos*,[4] besides which there is extant a letter of his to Theodorus, philosopher to Frederick II, relating to indeterminate analysis and to geometry. These works treat of the theory of numbers in a way that shows that Leonardo was a mathematician of remarkable ability, considering the time in which he lived. His name attaches to the series 0, 1, 1, 2, 3, 5, 8, 13, . . ., in which $u_n = u_{n-1} + u_{n-2}$, where $u_{n-1} : u_n \rightarrow \frac{1}{2}(\sqrt{5} - 1)$.

So far as the schools were concerned, Leonardo's works were like a voice crying in the wilderness. It is probably within the bounds of truth to say that not a professor in the University of Paris, to select what was soon to become the greatest intellectual center of the world, could have made anything whatever out of the fine reasoning of the *Liber Quadratorum* or could have comprehended what the *Flos* was meant to convey to the

[1] "Un brevet d'apothicaire n'empêcha pas Dante d'être le plus grand poète de l'Italie, et ce fut un petit marchand de Pise qui donna l'algèbre aux Chrétiens." Libri, *Histoire*, I, xvi.

[2] On his knowledge of Euclid see G. Eneström, *Bibl. Math.*, V (3), 414.

[3] R. B. McClenon, "Leonardo of Pisa and his Liber Quadratorum," *Amer. Math. Month.*, XXVI, 1. There is a question about the date 1225, although it is given in the MS.

[4] "Incipit pratica geometrie composita a Leonardo pisano de filiis bonaccij anno M°. cc°. xx°."

"Incipit liber quadratorum compositus a leonardo pisano Anni. M. CC. XXV."

"Incipit flos Leonardi bigolli pisani super solutionibus quarumdam questionum ad numerum et ad geometriam uel ad utrumque pertinentium."

These titles are from the Boncompagni editions. *Flos* is a fanciful title,—blossom or flower.

mind. Since the course of study was concerned with little that was scientific,[1] mathematics had no standing there or in the schools of Italy.

Campanus. Roger Bacon speaks highly of a certain Master Nicholas who lived about this time, but concerning whom we know nothing further, and also of Master Campanus de Novaria.[2] The latter is Johannes Campanus[3] (fl. *c.* 1260), sometime chaplain to Urban IV, who reigned as pope from 1261 to 1264. It was he who prepared the translation of Euclid's *Elements* that was used in most of the early printed editions, but which seems to have depended upon at least three earlier translations from the Arabic. Campanus also wrote a *Tractatus de Sphaera*, a *Theoria Planetarum*, a *Calendarium*, a work *De Computo Ecclesiastico*, a work on perspective, and a memoir *De Quadratura Circuli* which seems lost but which was mentioned a century later by Albert of Saxony (*c.* 1370). Contemporary writers have little to say of his life. He held relatively minor positions in the Church, and it is probable that in his later years he was a canon in Paris.[4]

In the appendix to his translation of Euclid he showed how to compute the sum of the angles of a stellar pentagon. It is not improbable that the figures used by the astrologers of this period account for the interest developed by various writers in the study of stellar polygons in general. Campanus also considered the trisection problem, the irrationality of the Golden Section (not yet known by this name), and the angle between a circle and a tangent. He was, therefore, not merely a compiler of translated material but a man genuinely interested in geometry.[5]

[1] The oldest statutes now extant (1215) record: "Non legant in festivis diebus, nisi Philosophos et rhetoricas et quadrivalia et barbarismum et ethicam, si placet." See Suter, *Univ. Mittelalt.*, p. 56.

[2] Of Novara, near Milan.

[3] Giovanni Campano. Cantor, *Geschichte*, II (2), 90; C. S. Peirce, in *Science*, XIII (N.S.) (New York), 809.

[4] Pacioli (1509), in his *De diuina proportione* (I, 4) speaks of him as "el gran philosopho Campan, nostro famosissimo mathematico."

[5] B. Baldi, Boncompagni's *Bullettino*, XIX, 591.

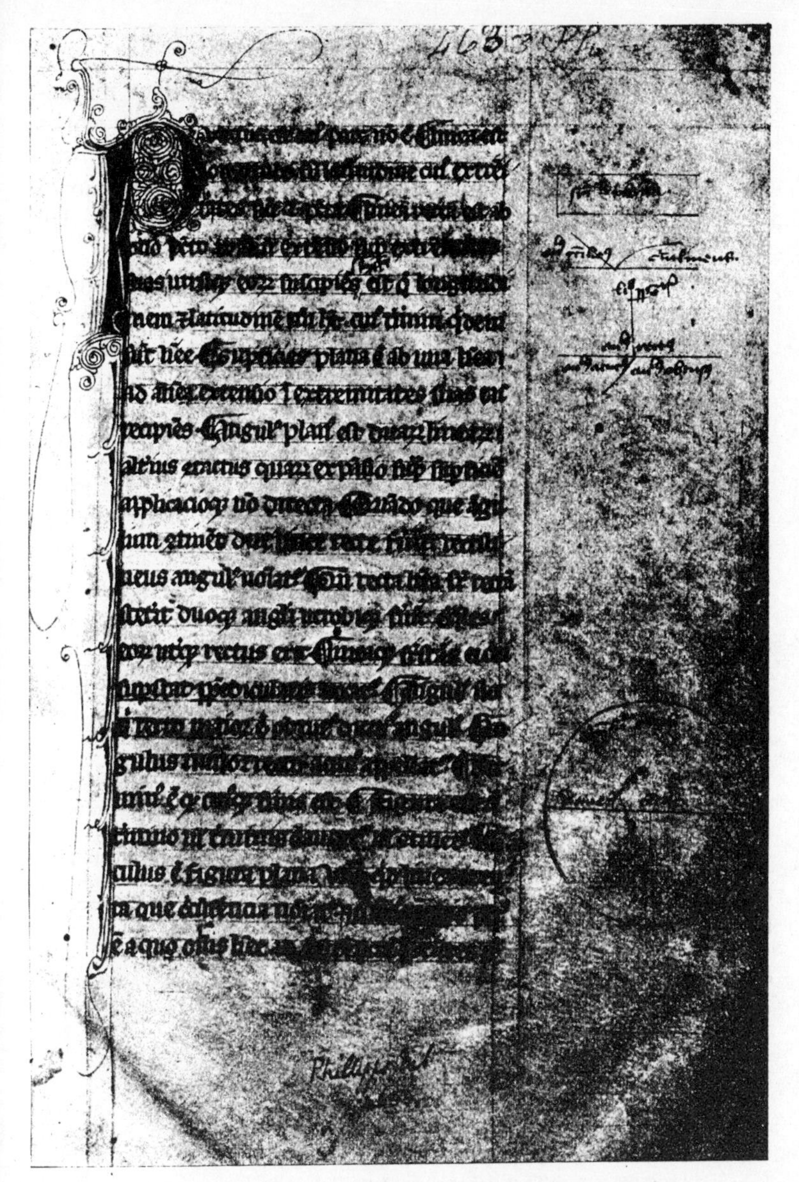

FIRST PAGE OF THE CAMPANUS EUCLID

From the MS. probably given by Campanus to Pope Urban IV, at that time Jacques Pantaléon, Patriarch of Jerusalem. Now in the library of Mr. Plimpton

Other Italian Writers. In the 13th century Italian mathematics consisted almost entirely of astronomy, and of the works bearing upon this subject the most popular one produced was the *Tractatus Sphaerae* of Bartolomeo da Parma, who is known to have been teaching mathematics at Bologna in 1297. He also wrote on geometry and astrology.[1]

Whether Guglielmo de Lunis belonged to the 13th century is uncertain,[2] but he translated an algebra from the Arabic.[3]

Pietro d'Abano[4] (*c.* 1250–*c.* 1316), a professor of medicine[5] at Padua, wrote an *Astrolabium planum*,[6] his interest in the astrolabe being doubtless due to its applications to astrology.

British Scholars.[7] Fibonacci was not a product of the universities, but he speaks of his master[8] as one who had studied in the universities of Oxford and Paris, and no doubt, in his mature years at least, he learned from him. This man was "the wizard" Michael Scott,[9] who had not only studied at the

[1] E. Narducci, Boncompagni's *Bullettino*, XVII, 1, 43, 165.

[2] *Bibl. Math.*, XII (3), 270.

[3] An algebra MS. of the 15th century treats of "la regola de Algebra amucabale . . . secondo ghuglielmo de lunis." See *Rara Arithmetica*, p. 463; *Bibl. Math.*, IV (2), 96, and V (2), 32, 118.

[4] Petrus Aponensis. The dates are sometimes given as 1253–*c.* 1319.

[5] On the relations of mathematics to medicine two or three centuries later, see the author's article on "Medicine and Mathematics in the Sixteenth Century," in *Annals of Medical History*, I, 125.

[6] Published in Venice in 1502.

[7] Cantor, *Geschichte*, II, chap. xlvi.

[8] He dedicates to him the second edition of his *Liber Abaci* in these words: "Scripsistis mihi domine mi magister Michael Scotte, summe philosophe, ut librum de numero, quem dudum composui, uobis transcriberem."

[9] Spelled also Scot. Born possibly at Balwearie, Scotland, *c.* 1175; died *c.* 1234. He was also called Michael Mathematicus.

"In these fair climes it was my lot
To meet the wondrous Michael Scott;
A wizard of such dreaded fame,
That when, in Salamanca's cave,
Him listed his magic wand to wave,
The bells would ring in Notre Dame!"
Scott, *Lay of the Last Minstrel*, II, xiii

"That other, round the loins
So slender of his shape, was Michael Scot,
Practised in every slight of magic wile."
Dante, *Inferno*, XX, Cary translation

universities mentioned, but had learned Arabic and made astronomical observations at Toledo. He was later appointed astrologer to Frederick II and seems to have been employed by this ruler to make known to scholars, through translations from the Arabic, the newly discovered Greek texts.

Upon those who, unlike Michael Scott, studied chiefly in England at this time, some influence may have been exerted through the arrival at Oxford in 1224 of the first of the Franciscans. These men[1] were not learned in the sciences, but they came from the intellectual centers of Southern Europe and could not have been ignorant of what scholars were doing beyond the Alps and the Pyrenees. Their presence in one of the university centers was especially significant.

Sacrobosco. The second of the prominent British scholars of this century was Sacrobosco,[2] who was educated at Oxford and entered the University of Paris *c.* 1230. He afterwards taught mathematics and philosophy in Paris and died there *c.* 1256.[3] He was buried in the Cloister Sodalium Mathurinalium, his astrolabe being placed on his tomb.[4]

Sacrobosco wrote the most popular work on the sphere that had appeared up to that time, and did much, through his

[1] A list is given by Wood, *Historia Oxon.*, I, 67–77, and in A. G. Little, *The Grey Friars in Oxford*, chap. i and p. 176 (Oxford, 1892).

[2] Born at Halifax, Yorkshire, *c.* 1200; died at Paris, *c.* 1256. The name appears in various forms, such as Johannes de Sacrobosco, John of Halifax, John of Holywood, Sacro Bosco, Sacrobusto. Sacrobosco is the Latin for Holywood (Holyfax, Halifax). Widman (1489) writes the name Iohānē vō sacrobusto, and Pacioli (1494) writes it Giouā de sacro busco. J. Aubrey, *Brief Lives*, ed. Clark, I, 408 (Oxford, 1898) (hereafter referred to as Aubrey, *Brief Lives*), says that "Dr. [John] Pell is positive that his name was Holybushe."

[3] The date of his death was formerly given definitely as 1256 on the authority of G. J. Vossius, *De Vniversae Mathesios Natvra & Constitvtione Liber*, p. 179 (Amsterdam, 1650). P. Tannery has shown that the obscure verse from which Vossius obtained this date refers to the completion of his *Compotus*, and moreover that the date should be read 1244 instead of 1256. The verse is

"M Xristi bis C quarto deno quater anno." See *Bibl. Math.*, XIII (2), 32.

[4] J. C. Heilbronner, *Historia Matheseos Universae*, p. 471 (Leipzig, 1742); hereafter referred to as Heilbronner, *Historia*. Wood (*Historia Oxon.*, I, 85) speaks of his teaching there: "Joh. de Sacro bosco. Claruit apud Parisienses in Mathesi & in Philosophia."

Tractatus de Arte Numerandi[1] or *Algorismus*, to make the Hindu-Arabic arithmetic known to European scholars. These books were widely used for three hundred years, and continued in use until the close of the 16th century.[2]

The third of the prominent British scholars of this period was Robert Grosseteste, or Greathead (died October 9, 1253), at one time a student at Paris, later a student and teacher at Oxford, and finally bishop of London.[3] His interest was chiefly in the applications of mathematics to physics and astronomy,[4] but he also wrote a *Praxis geometriae* and a work on Euclid's *Optics*.[5]

Among the Oxford men of this period was John of Basingstoke (died 1252),[6] who learned Greek in Athens (1240) and took back to England some knowledge of the numeral systems and possibly of the mathematics of classical times.

Roger Bacon. The most prominent scholar in England in the 13th century, however, was Roger Bacon (1214–1294), a man of erudition and of prophetic vision. His works show a knowledge of Euclid's *Elements* and *Optics*, of Ptolemy's *Almagest* and *Optics*, of Theodosius on the *Sphere*, of parts of the works of Hipparchus, Apollonius, and Archimedes, and of the works of various Arab writers. He was familiar with the writings of Aristotle and with some of the commentaries upon them. Of

[1] Printed in Halliwell, *Rara Math.*, I. For early editions, see the *Rara Arithmetica*. An edition by M. Curtze appeared in 1897.

[2] Suter, *Univ. Mittelalt.*, p. 67; *Bibl. Math.*, XI (2), 97; P. Riccardi, *Bibl. Math.*, VIII (2), 73.

[3] The variants of his name, as given by Wood (*Historia Oxon.*, I, 81), are interesting. They include such forms as Grossum caput, Groshedius, Grouthede, Grokede, and Groschede. He was also known as Robertus Lincolniensis and Rupartus Lincolniensis.

[4] *Theorica planetarum*, *De astrolabio*, *De cometis*, *De sphaera coelesti*, *De computo*, *Praxis geometriae*, and a *Calendarium*.

[5] See also L. Baur, "Der Einfluss des Robert Grosseteste auf die wissenschaftliche Richtung des Roger Bacon," in A. G. Little, *Roger Bacon Essays*, p. 33 (Oxford, 1914) (hereafter referred to as Little, *Bacon*), and *Die philosophischen Werke des Robert Grosseteste*, Münster, 1912.

[6] Under this date Matthew Paris records: "Obiit magister Johannes de Basingestokes, archidiaconus Legrecestriae, vir in trivio et quadrivio ad plenum eruditus."

FROM A MANUSCRIPT OF SACROBOSCO

This MS. was written in Germany, *c.* 1442. It shows distinctly the numerals as they then appeared. Now in the library of Mr. Plimpton

algebra he knew little except the name.[1] Mathematics as then understood was little more than astronomy, and for the work of most of his contemporaries in this field he had a profound contempt. This contempt was even more pronounced with respect to teaching, in which he asserted that an enormous amount of time was wasted. He stated that he had devoted forty years to study, and that the entire ground could have been covered in from three to six months.[2] The teachers at Paris he could only characterize by their "four defects,—infinite and puerile vanity, ineffable falsity, voluminous superfluity, and the omission of all that is worthy." The charge was untrue, as are most epigrams of the kind, for Bacon was given to dipping his pen in vitriol. It is no wonder that his contemporaries generally hated him. Although spoken of in later times as *doctissimus mathematicus* he contributed nothing to pure mathematics, and his chief work in applied mathematics was a calendar which the world was not ready to appreciate.[3]

John Peckham. Of Bacon's influence upon his pupils, in the direction of mathematics, we have little evidence. There is some reason for thinking that he was possibly the one who inspired John Peckham[4] to take up the study of the science. At any rate Peckham was a scientist of repute and his *Perspectiva communis* was looked upon as a classic for three hundred years.[5] He became archbishop of Canterbury in 1279.

[1] In his *Opus majus*, he says: "Algebra quae est negotiatio, et almochabala quae est census."

[2] "Multum laboravi in scientiis et lingua, et posui jam quadriginta annos postquam didici primo alphabetum . . . et tamen certus sum quod infra quartam anni, aut dimidium anni, ego docerem ore meo hominem sollicitum et confidentem, quicquid scio de potestate scientiarum et linguarum." *Opus Tertium*, cap. xx.

[3] Besides his published works by S. Jebb (1733, 1750), J. S. Brewer (1859), J. H. Bridges (1897), and Robert Steele, consult E. Charles, *Roger Bacon, sa vie, ses ouvrages, ses doctrines* . . . (Paris, 1861), and Little, *Bacon*.

[4] Born in Kent, probably some time before 1240; died at Mortlake, December 8, 1292. The name also appears as Peachamus, Peccamus, and Pithsanus, with various other modifications.

[5] Facio Cardano (1444–1524) edited it under the title *Prospectiua cōis d. Iohānis archiepiscopi Cātauriēsis*, and it was printed s. l. a. (but 1482, at Milan). There are various editions of this work.

Chi commenche algorisme

Ceste signifiance est
appellee algorisme
de le quele nous usons
de teles figures. 9. 8. 7. 6. 5. 4.
3. 2. 1. li premiere fait .1.
la seconde fait .2. la tierce fait
.3. et les autres aussi jusques le
darraine lui est appellee cifre
0. Se tu mes aucune de ces
figures ou premier lieu elle
fera soi simplement. Se tu la
mes ou secont lieu elle fera
.10. fois soi. Et en tierc lieu
ke tu la meteras elle fera
.10. tans plus kel lieu devant
mais ces cifre ne fait rient.
mais elle fait les autres figu
res multeplier. Tu dois savoir
ki sont .3. manieres de nombres
car li .i. sont de git. li autre
article. li autre compost. li di
git sunt dusques .9. li article .10.
et li nombre ki sont multeplie par
.10. li compost sont establi des
articles et des digis .11. 17. 13.
et les autres. Quant tu escri
ras tu dois esgarder le nombre
Se cest digis tu lescriras par
soi. se cest articles tu escri

ras .0. avant et apres larticle
Se cest compost. tu escriras le
digit avant apres larticle
.6. parties sont dalgorisme.
assambler et abatre. doubler
diviser. multeplier. Se tu
assambles ou abas ou dimi
dies: tu commenceras a destre.
Se tu doubles ou multeplies
ou devises. tu commenceras a se
nestre.

Se tu veus assambler .i.
nombre a autre tu es
criras le greignor desseure
et le menour desous. Et en
le maniere ke tu metes
la premiere figure desous soz
la premiere desseure. la secon
de sour la seconde et les autres
ausi en ordene. Se il i a plu
sours figures apres tu assam
bleras les .ii. premiers. Sil en
nient digit tu escriras
ou lieu de le figure desseu
re. Sil en vient articles
tu escriras .0. desseure. et
apres larticle des figures
desseure par ordre. Tu ne dois
pas deviser dautre plus
de .9. fois. De cele division

FIRST PAGE OF THE EARLIEST FRENCH ALGORISM

Written *c.* 1275, and now in the Library of Ste. Geneviève, in Paris

French Scholars. France produced no mathematicians of importance in the 13th century. During a considerable part of the time her great university was a place of rioting rather than a seat of learning. In spite of this fact, however, several respectable scholars appeared, one of the first being Alexandre de Villedieu[1] (*c.* 1225), a Franciscan monk from Bretagne. He wrote *De Sphaera*, *De Computo Ecclesiastico*, and *De Arte Numerandi*, and taught in Paris. He is best known, however, for his *Carmen de algorismo*,[2] a little arithmetic in Latin verse that probably did more to make known the new Hindu-Arabic numerals than any other work of the century.[3] A little later (*c.* 1275) the first algorism in the French language was written.

Among the contemporaries of Alexandre de Villedieu there was Vincent de Beauvais[4] (*c.* 1250), a Dominican, whose encyclopedia, the *Speculum Majus*,[5] written for Louis IX ("Saint Louis"), includes the quadrivium, the subject being very poorly treated.

Roger Bacon mentions as one of the greatest mathematicians of his time (*c.* 1265) a certain Petrus de Maharncuria,[6] but all that is known of him is that he wrote a work on the magnet.

German Writers. Of the German mathematicians of the 13th century only three deserve special mention. Of these the first in order of time and of mathematical ability was Jordanus Nemorarius,[7] who studied at Paris and wrote an *Arithmetica decem libris demonstrata*[8] and possibly an *Algorismus demonstratus*.[9] He also wrote a work on mathematical astronomy,—*Tractatus de sphaera*; one on geometry,—*De triangulis*; and one of the leading books of the Middle Ages on algebra,—*Tractatus*

[1] De Villa Dei, or De Villa Dei Dolensis.

[2] *Song on Algorism*, that is, on al-Khowârizmî's arithmetic methods.

[3] It is printed in full in J. O. Halliwell, *Rara Mathematica* (London, 1838), 2d ed. (1841), p. 73.

[4] Vincentius Bellovacensis.

[5] His *Opera* appeared at Venice in 1494.

[6] Also called Peter de Maharn-Curia and Petrus Peregrinus. Maharncuria seems to have been Maricourt in Picardy. Boncompagni's *Bullettino*, I, 1.

[7] Also known as Jordanus de Saxonia and Jordan of Namur. He was born at Borgentreich in the diocese of Paderborn and died in 1236 or 1237. Cantor, *Geschichte*, II, chap. xliii.

[8] Published at Paris in 1496.

[9] As to the doubt upon this point see G. Eneström, *Bibl. Math.*, V (3), 9.

de numeris datis. The *Arithmetica*[1] is on the theory of numbers as set forth in treatises like that of Boethius, and is the least original of his works. The one noteworthy feature of the book is the use of letters to represent general numbers. This is already found to a certain extent in the works of earlier writers, including Aristotle and Diophantus, but Jordanus uses letters quite as they are used today, letting b, for example, represent any number whatsoever.[2] The *Tractatus de sphaera* was for a long time a classic and several editions were printed. The *De triangulis*[3] is a work in four books containing seventy-two propositions of the usual type, together with propositions on such topics as the center of gravity of a triangle, curved surfaces, and similar arcs.

The *Tractatus de numeris datis* is a system of algebraic rules. The problems[4] generally relate to a *numerus datus*, a given number,[5] which has to be divided in some stated manner, as in many of the problems in our current algebras.[6]

He also wrote a work entitled *De Ponderibus Propòsitiones XIII*, which was printed at Nürnberg in 1533 and contains a brief treatment of statics. He is the Jordanus de Saxonia who, in 1222, became general of the Dominican order.[7]

[1] On the MSS. in the library of the Royal Society and in Oxford, both of which differ from the first printed edition, see J. O. Halliwell, *A Catalogue of Miscellaneous Manuscripts preserved in . . . the Royal Society* (London, 1840), and J. Wallis, *Algebra*, p. 13 (Oxford, 1693).

[2] Cantor, *Geschichte*, II (2), 56; Eneström, *Bibl. Math.*, VII (3), 85.

[3] "Jordani Nemorarii Geometria vel de Triangulis libri IV," in the *Mittheilungen des Coppernicusvereins*, Heft VI (Thorn, 1887).

[4] See P. Treutlein, *Abhandlungen*, II, 135.

[5] "Numerus datus est cuius quantitas nota est."

[6] One of his first problems is practically this: To separate a given number into two parts such that the sum of the squares of the parts shall be another given number. *E.g.*, $x+y=10$, $x^2+y^2=58$, whence $x=7$, $y=3$. *Abhandlungen*, II, 136 (4).

[7] One Oxford MS. distinctly calls him Jordanus de Saxonia. Nicolas Trivet, an English chronicler of the 14th century, under the year 1222, states: "Hoc anno in Capitulo Fratrum Praedicatorum generali tertio, quod Parisiis celebratum est, successor beati Dominici in Magisterio Ordinis Fratrum Praedicatorum factus est frater Iordanus, natione Teutonicus, Dioecesis Moguntinae, qui cum Parisiis in scientiis saecularibus et praecipue in Mathematicis magnus haberetur. . . ."

One of the greatest of the German scholars of this period was Albertus Magnus,[1] Count of Bollstädt, a Dominican priest, and Bishop of Regensburg. He studied at Padua and taught at Bologna, Strasburg, Freiburg, Cologne, and Paris. So versatile was he that he was called "Doctor Universalis." His interests were chiefly in philosophy and physics, but his works include material on astronomy and some reference to Pythagorean arithmetic.[2] Claude Fleury, who wrote an ecclesiastical history in 1691, remarked that he could see nothing great in him but his volumes.

At the close of the century (*c.* 1270) Witelo or Vitello,[3] probably from Thüringen but possibly from Poland,[4] wrote on perspective (optics) and astronomy, a fact which shows the interest in this phase of applied mathematics in Poland in the 13th century.

Other Thirteenth-Century Writers. Of the other scholars of this century whose works touched upon mathematics the most prominent was Alfonso X, King of Castile (1223–1284), known as el Sabio (the Wise). He was an astronomer of merit and his name appears in the Alfonsine Tables,—planetary tables which improved upon the imperfect ones left by Ptolemy. Work upon them began in 1248 and was completed in 1254.[5] Tycho Brahe is said to have deplored the waste of money involved in their compilation, although they unquestionably stimulated the study of mathematical astronomy.

[1] Born at Lauingen, Swabia, 1193 or 1205; died at Cologne, 1280. Albertus Teutonicus, de Colonia, or Ratisbonensis. See Cantor, *Geschichte*, II (2), 86, and Dixon's translation (London, 1876) of his biography by Sighart (Regensburg, 1857).

[2] The first edition of his *Opera Omnia* appeared in Leyden in 1651. The best edition is that of Paris, 1890.

[3] In the oldest MSS. the name appears as Witelo. The forms Vitello and Vitellius are later. There are many variants, such as Witilo, Witulo, Widilo, Wito, and Vitellion.

[4] He speaks of his country, saying: "In nostra terra, scilicet Poloniae habitabili . . . ," and of himself as "Thuringo-polonus" and as "Filius Thuringorum et Polonorum," so that possibly his mother was a Pole.

[5] The date of completion is sometimes placed later than this. The tables were first printed in Venice in 1483.

Another prominent writer of the period was Arnaldo de Villa Nova[1] (*c.* 1235–*c.* 1313), who taught at Paris, Barcelona, and Montpellier. While known principally as a physician and for his twenty works on alchemy, he wrote a *Computus Ecclesiasticus & Astronomicus*,[2] probably being led to a study of the subject through its relation to astrology.[3]

Roger Bacon, in his condemnation of most of his contemporaries, speaks of "the notorious William Fleming who is now in such reputation, whereas it is well known to all the literati at Paris that he is ignorant of the sciences in the original Greek, to which he makes such pretensions." This Flemish writer was William of Moerbecke,[4] chaplain to Clement IV and Gregory X. Among his translations were the catoptrics of Heron and the writings of Archimedes on floating bodies.[5] It is thought that Tartaglia[6] took his translation of Archimedes from this writer.[7] He also wrote on perspective.

Byzantine Writers. In the 13th century the only writer of note in the Near East was Georgios Pachymeres,[8] who may for convenience be classified as European, although born in Asia Minor. He wrote on the *Four Mathematical Sciences*,[9] that is, on arithmetic, music, geometry, and astronomy. The work is important only as showing that interest in learning had not

[1] Arnauld de Villeneuve, Arnald Bachuone, Arnoldus Villanovanus. He was probably born at Villa Nova, Catalonia, but possibly at Villeneuve in Southern France.

[2] Printed at Venice in 1501. With respect to the edition see the *Rara Arithmetica*, p. 73.

[3] On the geometric figures used in astrology the *Ars Magna* of Raymundus Lullus (*c.* 1235–1315), a writer of this period, may be consulted.

[4] Guilielmus Brabantinus or Flemingus. He died *c.* 1281 as archbishop of Corinth.

[5] *De iis quae in humido vehuntur.* See also J. L. Heiberg, *Zeitschrift* (Hl. Abt.), XXXIV, 1–84; XXXV (Hl. Abt.), 41–48, 48–58, 81–100, and later volumes.

[6] See page 297.

[7] Cantor, *Geschichte*, II (2), 514.

[8] Born at Nicæa, in Bithynia, about 60 miles from Constantinople, in 1242; died *c.* 1316.

[9] Περὶ τῶν τεσσάρων μαθημάτων Παχυμεροῦς μεγάλου διδασκάλου. There are various MSS. extant. See E. Narducci, "Di un Codice Archetipo e Sconosciuto dell' opera di Giorgio Pachimere," *Rendiconti della R. Accad. dei Lincei*, VII, 194.

wholly died out in the period between the capture of Nicæa by the Crusaders in 1097 and its downfall before the Turkish invaders in 1330.

Otherwise there is little known of the mathematics of Constantinople in the 13th century. There is evidence, however, to show that her scholars were using Greek numerical characters even as late as the 15th century, augmented by a symbol for zero resembling our inverted *h*. Their problems were trivial, chiefly relating to mensuration.[1] Although they used the Greek forms, they were acquainted with the numerical system of the Arabs and spoke of it as Hindu in origin, but they were not familiar with the numerals which we commonly call Arabic.

4. Christian Europe from 1300 to 1400

General Activity of the Fourteenth Century. After the brilliant beginning of a renaissance of learning in the 13th century, it would naturally be expected that the 14th century would see a notable revival of science and letters. To understand why this expectation was not fully realized, it is necessary to consider the peculiar conditions by which Europe was confronted. As for Italy, this country was at last fully awakened to the beauties of ancient literature, and so Dante (1265–1321), taking Vergil as his master, produced the *Divina Commedia*, the "Epic of Medievalism"; Petrarch (1304–1374) made a notable collection of manuscripts of the ancient classics and started a movement that resulted in a new appreciation of the literature of Greece and Rome; and Boccaccio (1313–1375) showed great zeal in the attempt to collect and study the works of the ancients. In this search Constantinople was drawn upon for Greek manuscripts, with the result that Italian scholars were interested anew in the study of science and letters. Furthermore, the Florentine republic had just become practically a government by the merchant class, owing to modifications in the constitution between 1282 and 1292,—a fact that must have had much to do with the great prominence

[1] J. L. Heiberg, "Byzantinische Analekten," *Abhandlungen*, IX, 161.

of Florentine arithmetic in the schools of the 14th century. A general accumulation of wealth must also have followed, which would naturally tend to foster the arts and sciences. All this was promising, and the result would probably have been the hastening of the period popularly known as the Renaissance, had it not been for two deterring factors.

The first of these factors was the Hundred Years' War (say 1338–1453, although also given as 1328–1491), which overturned the economical and political systems of the two most advanced countries of Europe north of the Alps. The battle of Crécy (1346) struck at something besides feudalism.

The second deterring factor was the terrible ravaging of the Black Death (1347–1349), by which from a third to a half of the population of Europe is thought to have been swept away.

As to the universities in the 14th century, they did little for mathematics. Those of Italy were behind their contemporaries in Paris, England, and Germany, the statutes of 1387 making no mention whatever of the subject.[1] In England, Merton College, Oxford, was the mathematical center[2] and made some pretense at work in this science, while Paris had lectures on algorism, astronomy, and geometry, such as they were. In the newly founded University of Erfurt (1392), which may be taken as a German type, an elementary knowledge of mathematics was offered but apparently was not required.[3]

Italian Writers. No Italian writer of the 14th century stands out as showing any real genius in mathematics, as a brief list will bear witness. Cecco d' Ascoli (1257–1327), also known as Francesco di Simone Stabili and as Francesco degli Stabili, a native of Ascoli in Romagna, was professor of philosophy at Bologna and Rome, wrote a commentary on the *Sphaera* of Sacrobosco,[4] and did much to bring into high repute once more

[1] Suter, *Univ. Mittelalt.*, p. 75. [2] *Ibid.*, p. 83.

[3] W. Hellmann, *Ueber die Anfänge des math. Unterrichts an den Erfurter . . . Schulen*, I, 4. Erfurt, 1896.

[4] On the early Bologna mathematicians in general, consult Silvestro Gherardi, *Di alcuni materiali per la Storia della Facoltà Matematica . . . di Bologna*, p. 17 (Bologna, 1846); hereafter referred to as Gherardi, *Facoltà Mat. Bologna.*

the ancient belief in astrology, a subject which perhaps reached its greatest popularity in this century.

Andalò di Negro (*c.* 1260–*c.* 1340), a native of Genoa, had considerable reputation as a mathematician and astronomer, writing several works on the astrolabe, a book on the planets, and a *Tractatus de sphaera.* There is also ascribed to him a practical arithmetic.[1]

Barlaam (*c.* 1290–*c.* 1348), a native of Seminara in Calabria, Italy, bishop of Geraci, studied in Constantinople and wrote on computing, astronomy,[2] the science of numbers,[3] algebra,[4] and Book II of Euclid.

Paolo Dagomari,[5] a native of Prato, in Tuscany, was prominent in Florence as an arithmetician and astronomer. His *Trattato d'Abbaco, d'Astronomia, e di segreti naturali e medioinali* contained a little commercial arithmetic and gave him a reputation more extended than scientific.[6] That he wrote on algebra is asserted by at least one later writer.[7] He may have been the Paolo Pisano who is said to have lived about this time.

A more worthy writer on mathematics appeared in the person of Rafaele Canacci (*c.* 1380) of Florence, author of an algebra[8] with a number of historical notes.

[1] C. de Simony, Boncompagni's *Bullettino*, VII, 313, 339.

[2] *Libri V logisticae astronomicae.* See also B. Baldi, *Cronica di Matematici*, p. 85 (Urbino, 1707), and Boncompagni's *Bullettino*, XIX, 598. Barlaam's (Barlaamo's) given name may have been Bernardo, but this is uncertain. He is occasionally known, from his birthplace, as Calabro.

[3] *Arithmetica demonstratio eorum quae in secundo libro elementorum (Euclidis) sunt.* It was printed at Strasburg in 1564.

[4] Λογιστικη, *sive arithmeticae, algebraicae libri VI.* It was printed at Strasburg in 1572.

[5] Born at Prato, *c.* 1281; died at Florence, 1365 or 1374. Known also as Paolo dell' Abaco, Paolo Astrologico, Pagolo Astrologo, Paoli il Geometra, Paolo Geometra, and Paolo Arismetra. F. Villani (fl. 1404), in *Le Vite d' Uomini illustri Fiorentini*, 2d ed., Florence, 1826, speaks of him as "geometra grandissimo, e peritissimo aritmetico . . . diligentissimo osservatore delle Stelle e del movimento de' cieli." For a résumé of his work see D. Martines, *Origine e progressi dell' aritmetica*, p. 59 (Messina, 1865); hereafter referred to as Martines, *Origine aritmet.*

[6] For a description of the work, see *Rara Arithmetica*, p. 435.

[7] See *Rara Arithmetica*, p. 463, with reference to "*m*°. paolo fiorj che circha al. 1360. duro."

[8] See *Rara Arithmetica*, p. 459.

There was also a Master Biagio[1] of Parma (died 1416) who wrote an arithmetic and an algebra, but neither has been published. He taught astrology and philosophy at Paris, Pavia, Bologna, Padua, Venice, and Parma, wrote a commentary on Oresme's *De latitudinibus formarum*,[2] and wrote on statics and perspective. The famous educator Vittorino da Feltre (1378–1446, but the dates are doubtful), born in poverty, worked as

FROM DAGOMARI'S TRATTATO D'ABBACO

From an Italian MS. of *c.* 1339. Notice also the early per cent sign, p 100 lano (per 100 the year), and the sign for *lb*. The latter is possibly the origin of the dollar sign

a scullery boy in Biagio's house so as to learn geometry from him, and in turn became one of the best teachers of mathematics of his time.[3]

Toward the close of the century Antonio Biliotti[4] (*c.* 1383) of Florence taught mathematics in Bologna, but he left no works on the subject. Altogether the mathematical output of Italy in this century was not encouraging.

Constantinople. Constantinople was at this time experiencing an intellectual revival similar to the one seen in Italy before the coming of the Black Death. Prominent among her scholars was Maximus Planudes (*c.* 1340), a Greek monk, at one time

[1] Also Biagio da Parma and Pelacani. There was also a "m°. biagio che circha al. 1340. añj morj," as a MS. of *c.* 1440 asserts, although this 1340 may be wrong and the two may be the same. See *Rara Arithmetica*, p. 463.

[2] See page 199. The students of his day in Paris had a phrase "aut diabolus est, aut Blasius Parmensis."

[3] W. H. Woodward, *Vittorino da Feltre*, Cambridge, 1897.

[4] Also called Antonio dall' Abaco.

(1327) ambassador to Venice, who wrote on Diophantus and who also wrote an arithmetic based upon the Hindu-Arabic numerals.[1] He was a man of industry but of no genius, and his arithmetic is of value chiefly as showing the influence of Bagdad upon the mathematical thought of Constantinople. It sets forth the system of notation by the "nine figures received from the Hindus" together with the zero, and is the first of the Greek works to give any attention to modern methods of calculation. Planudes is also deserving of credit as a translator of various Latin classics into Greek.[2]

Among the minor contemporaries of Planudes there was Joannes Pedias′imus (*c.* 1330), also called Galenus, who was keeper of the seal of the patriarch of Constantinople. He wrote a work on geometry in which he attempted to pattern after the style of Heron of Alexandria, and also wrote upon the duplication of the cube, and upon arithmetic.[3] In general, however, his work was literary and philosophical. Among his contemporaries he was known as the "Chief of Philosophers."

There lived in Constantinople a little later than Joannes Pediasimus the celebrated grammarian Manuel Moschopou′-lus,[4] a native of Crete. The dates are uncertain, but he seems to have lived *c.* 1300. Although there were two men of the same name, this one and his nephew, it seems from a manuscript of a work by Nicholas Rhabdas, referred to below, that this is the one who wrote a treatise on magic squares, the earliest contribution to the subject in the Mediterranean countries.[5]

[1] He called his work Ψηφοφορία κατ' Ἰνδούς (*Indian Arithmetic*). There is a Greek edition by C. I. Gerhardt (Eisleben, 1865), and a German translation by H. Wäschke, *Das Rechenbuch des Maximus Planudes* (Halle, 1878) (hereafter referred to as Wäschke, *Planudes*). See also Heath, *History*, II, 546.

[2] Kroll, *Geschichte*, p. 70.

[3] Boncompagni's *Bullettino*, III, 303.

[4] Or Emanuel. In Greek the name appears both as Μανουὴλ and as Ἐμανουὴλ Μοσχόπουλος. Heath, *History*, II, 549.

[5] S. Günther, *Vermischte Untersuchungen zur Geschichte der math. Wissenschaften*, p. 195 (Leipzig, 1876) (hereafter referred to as Günther, *Vermischte Untersuch.*); P. Tannery, "Manuel Moschopoulos et Nicolas Rhabdas," *Bulletin des Sciences math. et astr.*, VIII (2), September 2, 1884.

About this time Nicholas Rhabdas[1] (*c.* 1341), a Greek "arithmetician and geometer"[2] from Smyrna, wrote from Constantinople two letters on arithmetic, and particularly on finger reckoning,[3] a subject first treated of with any completeness by Bede. He also edited a work of Planudes on the Hindu arithmetic,[4] possibly during the lifetime of the latter. With him there flickered out what once had been a great beacon light,—the mathematics of the Greeks,—and at the same time any real appreciation of the language itself almost ceased to exist. Petrarch began to study classical Greek in 1342, with the aid of a monk who had lived in Constantinople, and a learned scholar, Manuel Chrysoloras, lectured upon it in Florence from 1397 to 1400; but it was not until the 16th century that mathematical works of the Greeks began to be known again in the original tongue.

English Writers.[5] England produced several mathematicians of more than ordinary ability in this century, all but one of them doing his real work before the years of the pestilence.

Richard of Wallingford (born c. 1292; died 1336) lectured on the liberal arts at Oxford and wrote on trigonometry[6] and arithmetic.[7] He seems to have been one of the best-known mathematicians of his time.[8] It was no doubt his influence

[1] Nicholas Smyrnaeus, Rhabda, Artabasda, Artabasdes. In one manuscript the name appears as Nicolas Artavasdan. See P. Tannery, "Notice sur les deux lettres arithmétiques de Nicolas Rhabdas," in *Notices et extraits des manuscrits de la Bibliothèque nationale*, Paris, XXXII (1886), 121.

[2] As he describes himself,— "ἀριθμητικοῦ καὶ γεωμέτρου."

[3] Ἔκφρασις τ.ῦ δακτυλικοῦ μέτρου. For a review of this arithmetic see *Bibl. Math.*, I (2), 28. It has been printed several times, as by N. Caussinus, *Eloquentia sacra et humana*, Paris, 1636, and by Morellus, *Nic. Smyrnaei Artabasdae, graeci mathematici*, "Ἔκφρασις *numerorum notationis per gestum digitorum*, Paris, 1614. Heath, *History*, II, 550.

[4] Published by Gerhardt, Eisleben, 1865. [5] Cantor, *Geschichte*, II, chap. xlvi.

[6] *Quadripartitum de sinibus demonstratis*; *De sinibus et arcubus in circulo inveniendo*, *De chorda et arcu*, and *De chorda et versa*. The word *sinibus* often appears in the MSS. as *sinubus*, and *arcibus* commonly as *arcubus*. See Montucla, *Histoire*, I (1), 529; Cantor, *Geschichte*, II (2), 101.

[7] *De rebus arithmeticis* and *De computo*.

[8] One of the medieval writers speaks of him as "in mathesi omnium sui temporis primus."

that led John Manduith[1] (fl. *c.* 1320) to follow in his footsteps and lecture on trigonometry[2] and astronomy at Oxford.

The most prominent of the English mathematicians of the 14th century was Thomas Bradwardine,[3] known as the "Doctor Profundus." He was professor of theology at Oxford, chancellor of St. Paul's cathedral, and an upholder of liberalism, and he died as archbishop of Canterbury. He wrote four works on mathematics. In his *Arithmetica Speculatiua*[4] he followed the Boethian model, the work relating solely to the theory of numbers. His other works were a *Tractatus de proportionibus*, *Geometria speculativa*, and *De quadratura circuli.* His geometry includes some work on stellar polygons,[5] isoperimetric figures, ratio and proportion, irrationals, and loci in space.

About this time there flourished in England a Cistercian monk by the name of Richard Suiceth[6] (*c.* 1345), probably a native of Glastonbury, in Somersetshire. He was educated at Merton College, Oxford, and wrote an obscure work on mathematics.[7] It treats of a subject just beginning to attract attention in England and in France, *De latitudinibus formarum.*[8]

In this period there also lived a well-known writer, Walter Burley,[9] whose work on the lives of the philosophers and

[1] Mandwith, Manduit.

[2] *De chorda et arcu recto et verso, et umbris*, showing that he was acquainted with the use of tangents.

[3] Born at Hertfield (Hartfield), Chichester, *c.* 1290; died at Lambeth Palace, London, August 26, 1349. The name appears in such forms as Bragwardin, Brandnardinus, Bredwardyn, Bradwardyn, de Bradwardina, and de Bredwardina. Pacioli (*Sūma*, 1494, fol. 68, *r.*) calls him Tomas beduardin.

[4] Printed at Paris in 1495.

[5] ". . . figuris angulorum egredientibus."

[6] The first name may possibly have been Roger or Raymund, and the last name appears in such forms as Suisset, Suicetus, Swincetus, Swineshead, and Suineshevedus, a word derived from the Cistercian cloister, Vinshed, on the Holy Island off the coast of Northumberland.

[7] *Opus aureum calculationum . . . Per . . . Iohanē de Cipro . . . emēdat⁹ et explicit*, s. l. a., but Pavia, *c.* 1480; also Pavia, 1498. It may have been this work, which also went by the name "calculator," that led to his being called "calculator acutissimus" by one of the early writers.

[8] See Volume II, Chapter V.

[9] Born at Oxford, *c.* 1275; died *c.* 1357. The dates are very uncertain. Cantor, following Prantl, gives his death as 1337. The Latin spelling, Gualterus Burlaeus, and the late English Burleigh are also used.

poets[1] contains biographical notes on such prominent Greek mathematicians as Pythagoras, Plato, and Ptolemy.

There also flourished in the latter part of the 14th century a celebrated English mathematician and physician,—Simon Bredon.[2] He wrote on astronomy,[3] arithmetic,[4] the calculation of chords,[5] geometry,[6] and other related subjects, and various manuscripts of these works still exist.[7] He was one of the earliest European scholars to pay much attention to trigonometry.

Among the minor writers of the period there were William Reade,[8] of Merton College, who had considerable reputation as a mathematician, and who prepared some astronomical tables, and Walter Bryte,[9] who is said to have written on arithmetic,[10] astronomy, and surgery.

In the 14th century there was written an interesting but anonymous manuscript on the mensuration of heights and distances,[11] beginning: "Nowe sues here a Tretis of Geometri wherby you may knowe the heghte, depnes, and the brede of mostwhat erthely thynges." It is a practical work on shadow reckoning and surveying, using the compass, staff, and quadrant. Like many such works it is divided into three parts, very likely due to the Christian idea of the Trinity.[12]

[1] *De Vita et moribus Philosophorum et Poetarum.* The first printed edition was s. l. a., but Cologne, *c.* 1467. There were at least fourteen editions printed before 1501.

[2] Born at Winchcomb; living in 1386. The name appears as Bridonus and Biridanus.

[3] *In demonstratione Almagesti.*

[4] *Arithmetica theorica.*

[5] *Calculationes chordarum*, and *Tabulae chordarum.*

[6] *Quadratura circuli per Campanum et Simon Bredon.*

[7] See Suter, *Univ. Mittelalt.*, 84.

[8] Reede, Rede. He died in 1385 as bishop of Chichester.

[9] Brithus, Brit, Brytte. Possibly identical with Walter Brute, a lay follower of Wycliffe.

[10] *Tractatus algorismalis, De rebus mathematicis.* There is much doubt as to the authorship of each of the works assigned to him.

[11] Halliwell, *Rara Math.*, 56, where the complete text is given. The MS. is Bib. Sloan. 213. xiv, fol. 120, in the British Museum.

[12] "This tretis es departed in thre. þat es to say. hegh mesure. playne mesure. and depe mesure."

Another anonymous manuscript of considerable interest, *The Crafte of Nombryng*, was written *c.* 1300. It is one of the first works on algorism to appear in the English language.[1]

French Writers.[2] In spite of the calamities of war and plague, France did some noteworthy work in mathematics in the 14th century, not only through those born within her own boundaries but through scholars from other lands who found in Paris a more congenial intellectual atmosphere than they could find elsewhere.

Among those whom France could claim by adoption was Petrus Philomenus de Dacia,[3] a native of Denmark,[4] rector of the University of Paris (1326 or 1327), author of works on algorism[5] and the church calendar,[6] and a compiler of certain tables.[7] Still another of the adopted sons of France was Johannes Saxoniensis, or Johann Danck, who carried on his astronomical work in Paris. He left various writings on astronomy.[8]

Of those born in France, the first in point of time was Johannes de Lineriis or Jean de Lignères (*c.* 1300–1350), professor of mathematics at Paris, who adapted the Alfonsine Tables to the meridian of that city.[9] Joannes de Muris, or Jean de Meurs,[10] was a contemporary of his who studied at the Sorbonne and taught there. He wrote (1321) on arithmetic, astronomy,

[1] This manuscript is described more fully in Volume II.

[2] Cantor, *Geschichte*, II, chap. xlvii.

[3] There is an Easter computation of 1300 attributed to him.

[4] Whence "de Dacia."

[5] *Commentum super algorismum prosaicum Johannis de Sacro Bosco*. M. Curtze, *Petri Philomeni de Dacia in Algorismum Vulgarem Johannis de Sacrobosco Commentarius*, Copenhagen, 1897.

[6] *Computus ecclesiasticus.*

[7] His *Tabula ad inveniendam propositionem cujusvis numeri* contains a multiplication table to 49 × 49. See *Bibl. Math.*, IV (2), 32.

[8] *De astrolabio* and on the Alfonsine Tables. See Boncompagni's *Bullettino*, XII, 352.

[9] His nationality is not certain, nor is it clear whether he is the same as Johannes de Liveriis (or Liverius), whose work on fractions was printed at Paris in 1483, and who may have been a Sicilian. See *Rara Arithmetica*, p. 13; M. Steinschneider, Boncompagni's *Bullettino*, XII, 345, 352, 420.

[10] Born in Normandy *c.* 1290; died after 1360. The name also appears as Johannes de Murs or de Muria. L. C. Karpinski, *Bibl. Math.*, XIII (3), 99.

and music. Of his works on arithmetic,[1]—the *Canones tabula proportionum*, *Arithmetica communis ex diui Seuerini Boetij*, *Tractatus de mensurandi ratione*, *De numeris eorumque divisione*, and *Quadripartitum numerorum*,—the *Quadripartitum* is the most noteworthy. It is partly in verse and contains a certain amount of algebra. Among the algebraic equations solved are $x^2 + 12 = 8x$, with the roots 2 and 6; $3x + 18 = x^2$; and one already given by al-Khowârizmî and by Fibonacci, $2\frac{7}{9}x^2 = 100$. It also contains a close approach to a decimal fraction.[2]

The greatest of the French writers of this period was Nicole Oresme,[3] a native of Normandy, sometime professor and "magister magnus" (1355) in the Collège de Navarre at Paris, protégé of Charles V, dean of Rouen (1361), and finally (1377) bishop of Lisieux, Normandy. He wrote *Tractatus proportionum, Algorismus proportionum, Tractatus de latitudinibus formarum, Tractatus de uniformitate et difformitate intensionum*, and *Traité de la sphère*. He also translated Aristotle's *De coelo et mundo*. In the *Algorismus proportionum* is the first known use of fractional exponents, $2^{\frac{1}{2}}$ being written[4] $\frac{1}{2}\,2^{p}$, and $9^{\frac{1}{3}}$ appearing as $\frac{1}{3}.9^{p}$. He also wrote

$$\boxed{1^{p}\,\tfrac{1}{2}}\;4 \quad\text{and}\quad \boxed{\frac{p.1}{1.2}}\;4$$

for $4^{1\frac{1}{2}}$, stating the value to be 8.[5]

In the *Tractatus de uniformitate* there is set forth a suggestion of coordinate geometry, by the locating of points by

[1] For first printed edition, see *Rara Arithmetica*, p. 117. See also A. Nagl, *Abhandlungen*, V, 135; p. 139 for a list of his works.

[2] L. C. Karpinski, *Science* (N. Y.), XLV, 663.

[3] Born probably at or near Caen, *c.* 1323; died at Lisieux, July 11, 1382. Also known as Nicolaus Oresmus, Horem, Horin, and Oresmius. See M. Curtze, *Die Mathematischen Schriften des Nicole Oresme*, Berlin, 1870.

[4] From the 15th century MS. in the University of Basel, used by Curtze in the work above cited. Other MSS. have slightly different forms.

[5] Cantor, *Geschichte*, II (2), 121.

means of two coordinates.[1] Oresme also stands out prominently as a remarkably clear-thinking economist for his generation.[2]

Of much less importance as a writer on mathematics, but of greater reputation in his lifetime, is Petrus de Alliaco,[3] rector of the University of Paris, bishop of Cambray, and cardinal. His work on astronomy[4] throws considerable light on the early computi.

Other Writers. The other contributors to mathematical literature in this century were in general possessed of less ability than those of France and England.

Early in the century Hauk Erlendssön,[5] a Norwegian official, wrote on algorism. This is the first trace that we have of the Hindu-Arabic arithmetic in Scandinavia. There was also a certain Swedish scholar, Master Sven, or Sunon, who lectured on the sphere in 1340.

The leading Jewish mathematician of the 14th century was Levi ben Gerson[6] (1288–1344), who was also well known as a theologian. His *Work of the Computer*[7] was written in 1321. He also wrote a treatise on trigonometry[8] which was translated into Latin under the title *De numeris harmonicis,* but neither work showed any noteworthy power.[9]

Isaac ben Joseph Israeli was apparently a contemporary of Levi ben Gerson, but we are uncertain as to his dates. He

[1] See Volume II, Chapter V. For an early edition of the *Tractatus de latitudinibus* see *Rara Arithmetica*, p. 117. See also *Zeitschrift* (Hl. Abt.), XIII; *Bibl. Math.*, XIII (3), 115, and XIV (3), 210.

[2] *Tractatus de origine, natura, jure, et mutationibus monetarum*, edited by Wolowski, Paris, 1864.

[3] Born at Compiègne, 1350; died at Avignon, August 8, *c.* 1420. The name also appears as Pierre d'Ailly, Alyaco, and Heliaco.

[4] *Cōcordātia astronomie cū theologia.* First printed at Augsburg in 1490.

[5] Born *c.* 1264; died 1334.

[6] Also called Levi ben Gerschom and Gersonides, Leo Ebraeus, and Ralbag (R L B G, for Rabbi Levi ben Gerson), but more commonly known as Leo de Balneolis or Master Leon de Bagnolo, having been born at Balnaolis or Bagnolas, in Catalonia. See J. Carlebach, *Lewi ben Gerson als Mathematiker*, Berlin, 1910.

[7] *Maassei Choscheb*, edited by G. Lange, Frankfort a. M., 1909.

[8] *De sinibus, chordis, et arcubus.* See *Bibl. Math.*, I (3), 372; IV (2), 73; XII (2), 97.

[9] *Bibl. Math.*, XI (2), 103.

wrote a work on astronomy[1] which contains a chapter on geometry and also serves as a source of information on the activity of Jewish and Arabic scholars in Spain.

Among the lesser Jewish scholars of the period were Joseph ben Wakkar of Seville (died 1396), who worked out certain astronomical tables for Toledo; Jacob Poël of Perpignan (fl. *c.* 1360), who did the same for Perpignan; Imanuel Bonfils of Tarascon (died *c.* 1377), whose astronomical tables were highly appreciated and who wrote on the astrolabe; Jacob Carsono (al-Carsi), who wrote both at Seville and Barcelona (*c.* 1375), and whose tables were known to Tycho Brahe; Isaac Zaddik (al-Shadib), who wrote on the astrolabe and prepared various tables of use to astronomers; and Kalonymos ben Kalonymos, a native of Arles (born 1286), known as Master Calo, whose various translations include a paraphrase of Nicomachus.

Of the German writers, two or three are deserving of special mention. The first of these was Heinrich von Langenstein, Heinrich von Hessen, or Henricus Hessianus,[2] bishop of Halberstadt, who taught mathematics at Vienna and had some reputation as a mathematician and astronomer.[3] The second was Chunrad von Megenberg (*c.* 1309–*c.* 1374), who wrote (*c.* 1350) a work based on Sacrobosco's *De sphaera*.[4] About this time there was another Chunrad (Conrad) who was interested in mathematics,—Conrad von Jungingen (*c.* 1400). According to one of the manuscripts he seems to have been the author of the *Geometria Culmensis*.[5] This work consists of five parts, the first two relating to the mensuration of the triangle, the third to the quadrilateral, the fourth to the polygon, and the fifth to curvilinear figures.

[1] *Liber Jesod Olam sive Fundamentum Mundi*, first published in Berlin in 1777, but with later editions in 1846 and 1848.

[2] Born at Langenstein, near Marburg, 1325; died at Vienna, February 11, 1397 (sometimes given as 1394).

[3] *Quaestio de cometa.* He also wrote on the circle.

[4] F. Müller, *Zeittafeln zur Geschichte der Mathematik*, p. 80, with references (Leipzig, 1892); hereafter referred to as Müller, *Zeittafeln.* See also O. Matthaei, *Konrads von Megenberg Deutsche Sphaera*, Berlin, 1912.

[5] *Liber magnifici principis Conradi de Jungegen, magistri generalis Prusie, geometrie practice usualis manualis.* See Cantor, *Geschichte*, II, chap. xlviii.

Contemporary with the first two of these German writers there was Albert of Saxony,[1] or Albertus de Saxonia, who was educated at Prag and Paris, taught at Paris[2] and Pavia,[3] was the first rector of the University of Vienna (1365), and became bishop of Halberstadt (1366–1390). He wrote several scientific works, among them a theoretical treatment of proportion[4] after the manner of Boethius, *De latitudinibus formarum*,[5] *De maximo et minimo*, and *De quadratura circuli*.[6] Like most medieval writers, he took $3\frac{1}{7}$ for the value of π, apparently without considering it a mere approximation.

5. Christian Europe from 1400 to 1500

Influences leading to the Renaissance. Of the influences leading to that revival of learning known as the Renaissance, the two most potent were the transfer of Eastern scholarship to Italy and the invention of printing. It is customary to speak of the former as dating from the fall of Constantinople, that is, from its capture by the Turks in 1453.[7] When Mohammed II, standing on the banks of the Bosporus, repeated the Persian distich,

> The spider has woven his web in the Imperial palace,
> And the owl has sung her watch-song on the towers of Afrasiab,

he epitomized the situation of Greek culture in ancient Byzantium. There was nothing left for the remnant of the Hellenic civilization to do but to seek refuge in other lands.

Had Stephen Dusan not died when he did (1356), or had he left a worthy successor to the Serbian throne, Constantinople might have fallen to a western instead of an eastern conqueror. It is interesting to speculate as to what would have been the

[1] Born at Riggensdorf, Saxony, *c.* 1325; died at Halberstadt, 1390.

[2] Du Boulay, *Historia Universitatis Parisiensis*, p. 362. Paris, 1668.

[3] F. Jacoli, Boncompagni's *Bullettino*, IV, 495.

[4] *Tractatus proportionum*, first printed *c.* 1478. See *Rara Arithmetica*, p. 9.

[5] Printed at Padua in 1505.

[6] F. Jacoli, Boncompagni's *Bullettino*, IV, 493; H. Suter, *Zeitschrift* (Hl. Abt.), XXIX, 81.

[7] For a vivid description see Gibbon, *Decline and Fall of the Roman Empire*, Vol. VI, chap. lxviii.

result of such an event upon civilization in general and upon mathematics in particular.[1] We have already seen, however, that this transfer of Greek civilization began a century before the fall of the city, and both Rome and Florence had begun to acquire collections of Greek and Latin manuscripts that were to be available when the printing press should be ready to make their contents known.

About the opening of the 15th century Niccolò de' Niccoli (1363–1437) made a noteworthy collection of manuscripts at Florence, and in 1414 Poggio Bracciolini, a secretary of the Roman curia, began the copying and collecting of classical works. During the century there were such munificent patrons of learning as the Medici in Florence and Nicholas V in Rome, and the results of their labors are still seen in the Laurentian and Vatican libraries. There was also Federigo, Count of Montefeltro (1422–1482), who had been filled with enthusiasm by no less a teacher than Vittorino da Feltre himself, and who received from Sixtus IV the title of Duc d'Urbino. His library at Urbino, filled with manuscripts, was a rendezvous for the scholars of Italy. The humanist Giovanni Aurispa (*c.* 1369–1460) also brought some 238 manuscripts of the ancient Greek writers from Constantinople to Venice.

With all this activity, however, Italy showed no native ability in mathematics in the 15th century. What Symonds said of Tuscan culture in general applies in particular to mathematics: "Florence borrowed her light from Athens, as the moon shines with rays reflected from the sun. The revival was the silver age of that old golden age of Greece."

It is evident to one who studies the arithmetics of this century that no national spirit had yet developed. It was the city, not the state as we know it, to which men gave their allegiance. Just as we have Venetian art, so we find Venetian and Florentine and Roman arithmetics, quite as distinct from one another as were the Pisan works from those of Nürnberg.

[1]See J. B. Bury, in the *Cambridge Modern History*, Vol. I, chap. iii (London, 1902). This important work, containing different essays, is hereafter referred to as *Cambridge Mod. Hist.*

Origin of Printing. When we consider the effect of printing upon the development of mathematics, we must recall the fact that the art existed long before Gutenberg's time, about 1450. Not only had the Chinese printed from engraved blocks many centuries earlier than this, but they had also made some use of movable types. In Europe, too, block printing had assumed considerable prominence before movable types were invented.

It is also well to consider how mathematical knowledge was disseminated before it was possible to send it abroad by means of the printed page. There were three general periods in the transmission of such knowledge in the Middle Ages:

1. From the founding of the monastery at Monte Cassino by St. Benedict in 529, and continuing until *c.* 1200, the period in which scholars went to the teacher in the monastery and heard his lectures. This was the period of such men as the Venerable Bede and Alcuin of York.

2. From *c.* 1200 to *c.* 1400, when universities appeared in considerable numbers and claimed control of the scribes and booksellers.[1]

3. From *c.* 1400 to *c.* 1460, when the manuscript trade became more general, when large numbers of copies were made, and when these were sold as we sell books today.

All this was, however, a crude way of disseminating knowledge compared with the circulation of printed books. We should therefore expect the 15th century to begin to make widely known the mathematical classics of the ancient civilization, to meet the mathematical needs of a commerce to which new worlds were opening, and to prepare popular summaries

[1] The scribes had stands (stations) and the booksellers had shops near the university. The bookshops near the Sorbonne in Paris today are relics of the houses of the old *librarii* who, in the 13th and 14th centuries, rented their manuscripts. The *stationarius* was required to employ skilled copyists, who were enjoined to perform their tasks "fideliter et correcte, tractim et distincte, assignando paragraphos, capitales literas, virgulas et puncta, prout sententia requirat." See G. H. Putnam, *Books and their Makers during the Middle Ages*, I, 200 (New York, 1896, 1897); hereafter referred to as Putnam, *Books.* On the whole subject of reproduction and sale see J. A. Symonds, *Renaissance in Italy: The Age of Despots*, II, 129 (New York, 1883).

of the mathematics already accumulated. It would hardly be expected that this century would do more than receive these accumulated treasures and transmit them, postponing any new advance in science until the century following. It should be added that the sack of Mainz by Adolf of Nassau (1462) scattered the printers of that city all over Europe, an event comparable in some respects to the results of the fall of Constantinople.

The Universities. The universities were now beginning to mention mathematics more commonly, but that was about all. In the fragments of the Oxford statutes of 1408, only a little arithmetic was required for the bachelor's examination,[1] and in 1431 there was hardly any improvement.[2] In Paris the work was reorganized by the Papal legate in 1452, but nothing was done to better the condition of this science. The baccalaureate demanded only the reading of a little mathematics,[3] but nothing definite was prescribed. In general, little was required beyond arithmetic and a few pages of Euclid.[4]

Italian Writers. The great commercial activity in Italy in the 15th century gave rise to a large number of mercantile arithmetics, and these set a standard for the treatment of the subject that still influences, in some degree, the textbooks of today. Among the best known of the commercial writers were Matteo da Firenze[5] (*c.* 1400), his son Luca da Firenze (*c.* 1400), Giovanni, the son of this Luca (*c.* 1422), and Andrea di Giovanni Battista Lanfreducci (*c.* 1490), an officer of the

[1] "Algorismus integrorum" and "computus ecclesiasticus."

[2] For the licentiate, "Arithmeticam per terminum anni, videlicet Boëthii," with a little more science and mathematics. Suter, *Univ. Mittelalt.*, p. 90.

[3] "Aliqui libri mathematici."

[4] A MS. of 1515 in the Wolfenbüttler Bibliothek gives the following as representing the arithmetic: "Arithmetica communis ex divi Severini Boetii Arithmetica per M. Ioannem de Muris compendiose excerpta; Tractatus brevis proportionum, abbreviatus ex libro de proportionibus D. Thomae Braguardini Anglici, . . . Algorithmus M. Georgii Peurbachii de integris; Tractatus de Minutiis phisicis compositus Viennae Austriae per M. Ioannem de Gmunden." *Monatsberichte der K. P. Akad. d. Wissensch. zu Berlin*, 1867 (Berlin, 1868), p. 43.

[5] Matthew of Florence. See *Rara Arithmetica*, p. 468.

Republic of Pisa in 1505. The works of these four writers exist only in manuscript and contain little of value except what they tell of the commercial customs of their time.

FROM A MANUSCRIPT OF GIOVANNI DA FIRENZE'S WORK, 1422

This portion relates to the calendar. The MS. contains the statement: "questo d dottobre 1422" and the further date "ad 28 dottobre 1422"

Among the illustrious citizens of Florence one of the 16th century historians[1] mentions a mathematician named Benedetto, commonly known as Benedetto da Firenze. His arithmetic[2] was written *c.* 1460, and relates to the mercantile needs of his native city. It is one of the most complete works of this kind that appeared in the 15th century, but it has never been printed.

Of the Italian writers of this period who had some knowledge of mathematics beyond the field of commercial arithmetic the earliest was Prosdocimo de' Beldamandi,[3] a native of Padua and a student and professor in the university in that city (1422–1428). He wrote on arithmetic, music, and astronomy,

[1] U. Verino, *De illustratione urbis Florentiae libri tres*. Paris, 1583.

[2] "Inchomincia el trattato darismetricha," as in the MS. in the library of Mr. Plimpton. See *Rara Arithmetica*, p. 464.

[3] Born at Padua, *c.* 1370–1380; died at Padua, 1428. The name also appears as Beldomandi, Boldomondo, Beldemandi, and Prosdocimo Padvano. Bernardino Scardeone (*De Antiqvitaie Vrbis Patavii*, Basel, 1560) speaks of him as "è nobili familia Patauina ortus: egregius Musicus, & eximius philosophus, & clarus astrologus."

and also studied medicine.[1] The arithmetic was limited to the ordinary operations and contained no commercial problems.[2]

About 1425 one Leonardo of Cremona, or Leonardo de' Antonii (born *c.* 1380), wrote a brief practical geometry of which three manuscripts are known.[3] A little later (*c.* 1449) a fellow townsman of his, Jacob of Cremona, a teacher at Mantua and Rome, translated some of the works of Archimedes. Towards the close of the century Georgius Valla (1430–1499), a native of Piacenza, lectured on physics and medicine at Pavia and also at Venice, and in his *magnum opus*, which was merely a compendium of knowledge, treated of Boethian arithmetic, Euclidean geometry, optics, and the astrolabe, as well as a variety of other subjects. The work[4] was printed in 1501, but is notable chiefly for its size.

About 1475 an Italian painter, Pietro Franceschi, also called Pietro della Francesca,[5] wrote a work, *De corporibus regularibus*, in which he treated of the mensuration of regular polygons, the sphere, the five regular polyhedrons inscribed in a sphere, and solid figures in general. His problems may be illustrated by the cases of finding the area of a regular octagon circumscribed by a circle of diameter 7,[6] and the finding of the surface of a cube circumscribed about a sphere also of diameter 7.[7]

[1] This on authority of two early MSS.: "Examen medicinae magistri Prosdocimi de padua M° iiij° undecimo," that is, 1411; "Padovani dottori delle arti e Medicina," with the assertion that he "fu esaminato e dottorato nelle arte . . . 1409."

[2] For description, see *Rara Arithmetica*, p. 13; Boncompagni's *Bullettino*, XII, 1.

[3] *Leonardi Cremonensis artis metrice practice compilatio. Primus tractatus.* The date is doubtful. See *Rara Arithmetica*, p. 474; *Bibl. Math.*, IX (3), 280.

[4] *Georgii Vallae Placentini viri clariss. de expetendis, et fvgiendis rebus opus.*

[5] Born *c.* 1410–1420; died 1492. See E. Harzen, "Ueber den Maler Pietro degli Franceschi und seinen vermeintlichen Plagiarius, den Franziskanermönch Luca Pacioli," *Archiv f. d. zeichn. Künste*, II, 231; W. G. Waters, *Piero della Francesca*, London, 1901; and Mancini's monograph mentioned below.

[6] "Diameter circuli qui circumscribit octagonum est 7. Quanta igitur sit octagoni superficies invenire." Tract. I, Problem XL. He uses $\frac{22}{7}$ for π.

[7] Tract. II, Problem XVI. The work is printed in full in G. Mancini, "L'opera 'De corporibus regularibus' di Pietro Franceschi," in the *Atti d. R. Accademia dei Lincei*, XIV (5), 488.

Printing was introduced into Italy in 1464 by Juan Turrecremata,[1] abbot of the monastery of Subiaco, near Rome, the

Inprencipio darte dabacho secondo lostile dinsegnare delmaestro luca di Matteo daffirenze volendo admaestrare molte cose cioe multipricare dividere ragingniere sottrarre zpernumeri rotti ancora diremo alquanti modi perlequali sipoteranno scrivere tutte ragioni dutilita zuna granparte disottilita sichome detto sara innazi sesara piacere diddio zdella vergine maria z pche queste chose sivoghono fare chonfigure tratteremo prima delle figure lequale sono chosi figurate come vedi qui appie cioe.

·1·2·3·4·5·6·7·8·9·10:—

Dobiamo sapere challa prima figura nelprimo luogo amano mancha scrivendo innanzi lariitta pse medesima singnificha uno. La seconda pse medesima singnificha dua. Laterza perse medesima singnifica tre. Laquarta perse medesima singnificha quattro. Laquinta pse medesima singnificha cinque. Lasesta perse medesima singnificha sei. Lasettima pse medesima singnificha sette. Lottava pse medesima singnificha otto. Lanona pse medesima singnificha nove. Ladecima pse medesima singnificha dieci:—

Ora doviamo sapere zintendere ledette figure congiero zsegare zchosi quando adue insieme 10 20 30 40 50 ...

FIRST PAGE OF A MANUSCRIPT OF LUCA DA FIRENZE'S WORK (*c.* 1400)

This MS. dates from *c.* 1475. Now in the library of Mr. Plimpton

first book appearing in the following year.[2] Fourteen years later (1478) the first printed arithmetic appeared at Treviso,

[1] The Latin form of the family name Torquemada. He was a native of Valladolid, in Spain.

[2] Putnam, *Books*, I, 405.

then an important commercial town a day's travel from Venice. Three presses had already been established there, and it was from the one of Manzolo, or Manzolino,[1] that this work appeared. That it was anonymous was merely in keeping with the custom of a land where the glory of the individual was absorbed in the glory of the state.[2] Boncompagni learned of only eight copies of this arithmetic in various degrees of preservation,[3] so that the book is rare. It is quite commercial and never had any noticeable influence on the arithmetics of Italy.[4] It was, however, the first step in a remarkable movement, for up to the close of the 15th century there were printed in Italy at least 214 mathematical works, the number rising to 1527 in the following century.[5]

Three years after the Treviso book appeared, the first purely commercial arithmetic[6] was published. The author seems to have been one Giorgio Chiarino, of whom nothing further is known. The work is a mere compilation of measures and of such customs of exchange as were needed by Florentine merchants,[7] but there was enough in it to lead Pacioli (1494) to borrow freely from it.

Three years later Piero Borghi,[8] a Venetian, published the most noteworthy Italian commercial arithmetic of the century,[9]

[1] Domenico Maria Federici, *Memorie Trevigiane*, p. 73. Venice, 1805.

[2] "When Byron swept with superficial yet brilliant eyes the rolls of Venetian history, what did he find for the uses of his verse? Nothing but two old men, one condemned for his own fault, the other for his son's,—remarkable chiefly for their misfortunes." Mrs. Oliphant, in *The Makers of Venice*.

[3] *Atti Pontif.*, Vol. XVI. The only copy of this arithmetic that ever reached America is in the library of Mr. Plimpton.

[4] For a description, see *Rara Arithmetica*, p. 3.

[5] P. Riccardi, *Biblioteca Matematica Italiana,* Parte seconda, XI, XV, seq. (Modena, 1880), hereafter referred to as Riccardi, *Bibl. Mat. Ital.*

[6] *Qvesto e ellibro che tracta di Mercatantie et vsanze de paesi* (Florence, 1481). See *Rara Arithmetica*, p. 10.

[7] "Finito ellibro de tvcti ichostvmi: cambi: monete: pesi: misvre: & vsanze di lectere di cambi: & termini di decte lectere che nepaesi si costvma et in diverse terre."

[8] Piero Borgi, as the name appears in the first edition. The name is also given as Pietro Borghi and Pietro Borgo. He seems to have died after 1494.

[9] *Qui comenza la nobel opera de arithmethica*, the earliest printed books generally having no title pages. In later editions this was sometimes called the *Libro de Abacho*. The first edition was Venice, 1484.

a work intended for the mercantile class of Venice. One of the features of the book is that, like certain others of the

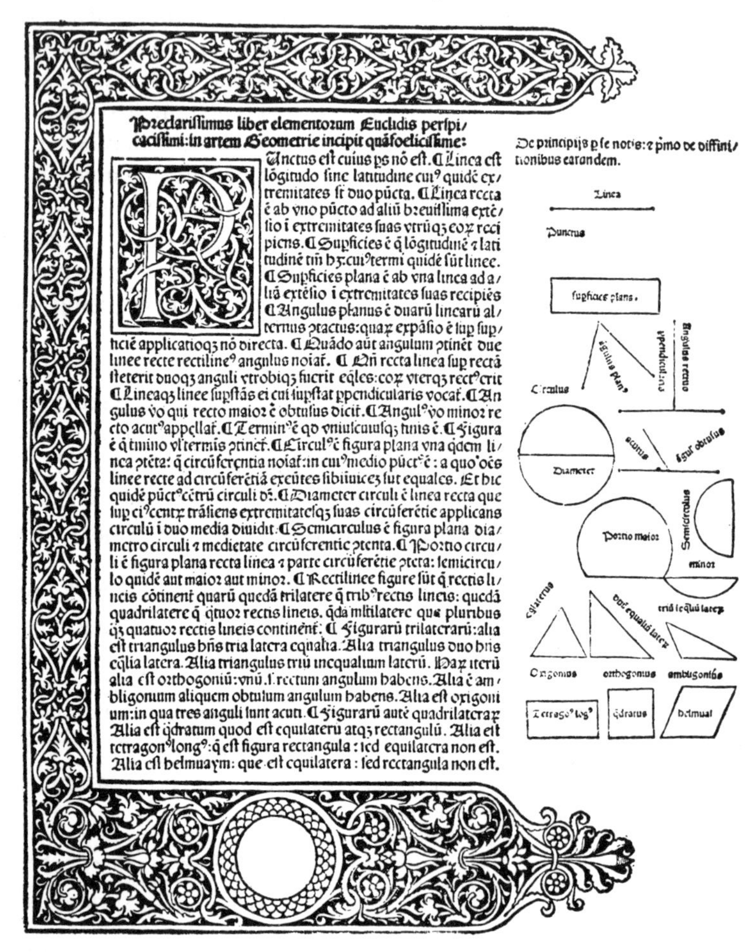

Preclarissimus liber elementorum Euclidis perspi/cacissimi: in artem Geometrie incipit quãfoelicissime:

Punctus est cuius ps nõ est. ¶ Linea est lõgitudo sine latitudine cuiꝰ quidẽ ex/tremitates sũt duo pũcta. ¶ Linea recta ẽ ab vno pũcto ad aliũ breuissima extẽ/sio i extremitates suas vtrũq3 eo℞ reci piens. ¶ Suꝑficies ẽ q̃ lõgitudinẽ ⁊ lati tudinẽ tm̃ hz: cuiꝰ termi quidẽ sũt lineę. ¶ Suꝑficies plana ẽ ab vna linea ad a/liã extẽsio i extremitates suas recipiẽs ¶ Angulus planus ẽ duarũ linearũ al/ternus ꝯtactus: qua℞ expãsio ẽ sup suꝑ/ficiẽ applicatioq3 nõ directa. ¶ Quãdo aũt angulum ꝯtinẽt due lineę rectę rectilineꝰ angulus noiaṫ. ¶ Qñ recta linea sup rectã steterit duoq3 anguli vtrobiq3 fuerit eq̃les: eo℞ vterq3 rectꝰ erit ¶ Lineaq3 lineę supstãs ei cui supstat ꝑpendicularis vocaṫ. ¶ An gulus vo qui recto maior ẽ obtusus dicit. ¶ Angulꝰ vo minor re cto acutꝰ appellaṫ. ¶ Terminꝰ ẽ qd vniuscuiusq3 finis ẽ. ¶ Figura ẽ q̃ tmino vl' termĩs ꝯtineṫ. ¶ Circulꝰ ẽ figura plana vna q̃dem li/nea ꝯtẽta: q̃ circũferẽtia noiaṫ: in cuiꝰ medio pũctꝰ ẽ: a quo oẽs lineę rectę ad circũferẽtiã exeũtes sibiinuicẽ sut equales. Et hic quidẽ pũctꝰ cẽtrũ circuli dr̃. ¶ Diameter circuli ẽ linea recta que sup eiꝰ cent℞ trãsiens extremitatesq3 suas circũferẽtię applicans circulũ i duo media diuidit. ¶ Semicirculus ẽ figura plana dia/metro circuli ⁊ medietate circũferentię ꝯtenta. ¶ Portio circu/li ẽ figura plana recta linea ⁊ parte circũferẽtię ꝯtẽta: semicircu/lo quidẽ aut maior aut minor. ¶ Rectilineę figurę sũt q̃ rectis li/neis cõtineṫ quarũ quedã trilaterę q̃ trib' rectis lineis: quedã quadrilaterę q̃ q̃tuor rectis lineis. q̃dã mltilaterę que pluribus q̃3 quatuor rectis lineis continẽṫ. ¶ Figurarũ trilaterarũ: alia est triangulus hñs tria latera equalia. Alia triangulus duo hñs eq̃lia latera. Alia triangulus triũ inequalium laterũ. Ha℞ iterũ alia est orthogoniũ: vnũ. s. rectum angulum habens. Alia ẽ am/bligonium aliquem obtusum angulum habens. Alia est oxigoni um: in qua tres anguli sunt acuti. ¶ Figurarũ autẽ quadrilatera℞ Alia est q̃dratum quod est equilaterũ atq3 rectangulũ. Alia est tetragonꝰ longꝰ: q̃ est figura rectangula: sed equilatera non est. Alia est helmuaym: que est equilatera: sed rectangula non est.

FIRST EDITION OF EUCLID, 1482

First page, reduced, from the Venice edition of 1482

period, it begins with multiplication, putting addition and subtraction after division and therefore presupposing some ability in computation. The work with integers is followed by fractions, the Rule of Three, and the usual applications of the time,—partnership, profit and loss, barter, and alligation.

These works, however, were mere advance couriers. Arithmetics had been published in Germany and Italy, Euclid's *Elements* had appeared from the famous press of Ratdolt, in Venice,[1] and the psychological moment had arrived for a general treatise summarizing the mathematical knowledge of the time. In such a period of unrest as that centering about the year 1494, when France was at war with most of Italy, when Florence was in arms against her sister towns, when Charles VIII was invading the peninsula and Savonarola was proceeding as an ambassador to Pisa to allay his wrath,—in such a period it would be surmised that a book like this could be prepared only in the peaceful atmosphere of a cloister.

Luca Pacioli. Such was the case, and Luca Pacioli,[2] known from his birthplace as Luca di Borgo, was the author. As a boy he may have come under the influence of his townsman, the artist Pietro Franceschi, already mentioned (p. 247), from whose work he freely took considerable material. At about the age of twenty he went to Venice (1464) and became a tutor to the three sons of a wealthy merchant,[3] and some six years

[1] *Preclarissimus liber elementorum Euclidis*, Venice, 1482. See the facsimile.

[2] Born at Borgo San Sepolcro, Tuscany, *c.* 1445; died probably after 1509. The name is spelled in various other ways, such as Paciolo, Paciolus, and Paciuolo. It does not appear at all in his *Sūma*, and in the *De diuina proportione* (Venice, 1509) it is given only in the Latin genitive,—"Lucae pacioli ex Burgo sancti Sepulchri . . . epistola." In his "Supplica . . . al Doge di Venezia" of December 29, 1508, in which he asks for permission to print the *De diuina proportione*, the name appears as "Luca de pacioli dal borgo sā sepulchro." It is from such contemporary evidence that Boncompagni (*Bullettino*, XII, 420) was led to speak of him as Pacioli, although Bernardino Baldi, writing in 1589, says "Fu de la famiglia de Paciuoli ignobile per quanto mi credo e di poco splendore," which led Cantor to adopt this later spelling. For the contemporary documents, in which he is generally spoken of as Luca or Lucas di Borgo, see Boncompagni, *loc. cit.*

The best sketch of his life is that by H. Staigmüller, "Lucas Paciuolo. Eine biographische Skizze," in the *Zeitschrift* (Hl. Abt.), XXXIV, 81, 121. See also the "Elogio di Fra Luca Pacioli" in B. Boncompagni, *Scritti inediti del P. D. Pietro Cossali* (Rome, 1857), p. 63; B. Boncompagni, in his *Bullettino*, XII, 377.

[3] " . . . nostri releuati discipuli ser Bart°. e francesco e paulo fratelli derōpiasi da la çudeca: degni mercatanti in vinegia: figliuoli gia de ser Antonio." *Sūma*, 1494 ed., fol. 67, *v.*, l. 4.

later he wrote an algebra which he never published.[1] In 1471 he went to Rome and, possibly influenced by his two brothers who had already entered the brotherhood of St. Francis, joined the Minorite order. In 1476 we find him teaching in Perugia and writing a little book for his pupils.[2] Five years later, at Zara, he wrote still another work, "more subtile and rigid,"[3] but neither of these works was printed. He traveled extensively in Italy[4] and possibly in the Orient,[5] and even after he became a Franciscan he was a wanderer. We find him back in Perugia in 1487 and working on his *Sūma*, in Naples in 1494, in Milan in 1496, in Florence and Rome in 1500, and in Venice in 1508.[6]

The great work of Pacioli, summing up not only his previous and unpublished works but also the general mathematical knowledge of the time, appeared in Venice in 1494,[7] having been written seven years earlier, when he was at Perugia.[8] It is a remarkable compilation with almost no originality.[9] He borrowed freely from various sources, often without giving the slightest credit, but in this he merely followed

[1] As to his MSS., now in the Vatican, see G. Eneström, *Bibl. Math.*, XIII (2), 53; B. Boncompagni, in his *Bullettino*, XII, 381, 428.

[2] ". . . alo giouani de peroscia . . . nel. 1476," as he says in the *Sūma* of 1494, fol. 67, *v*.

[3] "E anche in quello che a çara nel. 1481. de casi piu sutili e forti componēmo." *Ibid.*

[4] "Ma da poi che labito indegnamente del seraphyco san francesco ex voto pigliāmo: ꝑ. diuersi paesi ce conuenuto andare peregrinando." *Ibid.*

[5] So Cossali, writing in the 18th century, says that he "per desio di scienza viaggiasse in Oriente, ed in Arabia precipuamente, dove a que' tempi erano le matematiche dottrine in gran fiore"; but there is no contemporary authority for the statement. Certainly the expression "in gran fiore" is without foundation. On the contrary, there is internal evidence from the *Sūma* that he was never in the East.

[6] See also P. Treutlein, *Abhandlungen*, I, 10.

[7] For the title-page, see page 253. The form *Sūma* is a contraction for *Summa*.

[8] "E al presente q̊ ī peroscia . . . correndo glianni del nostro segnore Jesu Christo. 1487." *Sūma*, 1494 ed., fol. 67, *v*.

[9] G. Mancini, "L'opera 'De corporibus regularibus' di Pietro Franceschi," *Atti d. R. Accademia dei Lincei*, XIV (5), 488. See also Chiarino's work, from which he did not hesitate to take what he wished, and Baldi's article in Boncompagni's *Bullettino*, XII, 426.

the custom of the age. His use of material from Euclid, Ptolemy, Boethius, Fibonacci, Jordanus, Sacrobosco, Biagio of Parma, Prosdocimo, and Suiceth shows that he was at any rate rather well-read. He was, however, a careless writer, so much so that Cardan[1] has a chapter devoted to errors in his book.[2] The work includes such algebra as was then known, a large amount of business arithmetic, a poor summary of Euclid, and a treatment of double-entry book-keeping.[3]

In 1497, while at Milan, he wrote his *De diuina proportione*, publishing it at Venice in 1509.[4] This is a work of more merit from the standpoint of

Sũma de Arithmetica Geometria Proportioni ⁊ Proportionalita.

Continentia de tutta lopera.

De numeri e misure in tutti modi occurrenti.
Proportioni e ꝓportiõalita anotitia del.5° de Euclide.e de tutti li altri soi libri.
Chiaui ouero euidentie numero.13.p le q̃tita continue ꝓportiõali del.6°e.7° de Euclide extratte.
Tutte le pti delalgorismo: cioe releuare. ptir. multiplicar. sũmare. e sotrare cõ tutte sue ꝓue i sani e rotti. e radici e progressioni.
De la regola mercantesca ditta del.3. e soi fõdamenti con casi exemplari per cõm° 8. 6. guadagni: perdite: transportationi: e inuestite.
Partir. multiplicar. summar. e sotrar de le proportioni e de tutte sorti radici.
De le.3. regole del catayn ditta positiõe e sua origie.
Euidentie generali ouer conclusioni n° 66. absoluere ogni caso che per regole ordinarie nõ si podesse.
Tutte sorte binomij e recisi e altre linee irratiõali del decimo de Euclide.
Tutte regole de algebra ditte de la cosa e lor fabriche e fondamenti.
Compagnie i tutti modi. e lor partire.
Socide de bestiami. e lor partire
Fitti: pesciõi: cottimi: liuelli: logagioni: egodimenti.
Baratti i tutti modi semplici: compositi: e col tempo.
Cambi reali. secchi. fittitij. e di minuti ouer comuni.

PART OF THE TITLE-PAGE OF PACIOLI'S WORK OF 1494

This shows the general nature of the treatise, forming a kind of table of contents

[1] In his *Arithmetica* of 1539.

[2] "De erroribus F. Lucae quos vel transferendo non diligenter examinavit, vel describendo per incuriam praeteriit, vel inveniendo deceptus est." See also Cossali, "Elogio di Fra Luca Pacioli," in the *Scritti inediti*, p. 63 (published by B. Boncompagni, Rome, 1857). On the variations in each of the two editions (1494 and 1523) see E. Narducci, *Intorno a due edizioni della Summa de Arithmetica*, Rome, 1863.

[3] On this phase of the work see V. Gitti, *Gli Scrittori Classici della Partita Doppia*, a reprint from the *Annali del R. Istituto Indust. e Professionale di Torino*, Vol. V. (Turin, 1877).

[4] German translation by C. Winterberg, Vienna, 1896.

geometry than the *Sūma*, but one that could not, in the nature of the case, be as popular. His figures of the regular solids, as given in this book, were the best that had as yet appeared in print,[1] and have been attributed to Leonardo da Vinci. He also published an edition of Euclid in 1509, but the work is of little merit.

Pictagoras arithmetrice introductor

TITLE-PAGE OF CALANDRI'S ARITHMETIC, 1491

An example of the fanciful portraits of the Greek mathematicians. Such portraits became common during the Renaissance. The anachronisms in this one are evident

Of the Italian astronomers of the 15th century who gave considerable attention to mathematics the most eminent were Giovanni Bianchini (*c.* 1450), professor of astronomy in the University of Padua, Francesco Capuano, Giovanni Batista de Manfredonia (*c.* 1450–1490), and Domenico Maria Novara da Ferrara (1454–1504). Italian science, however, had hardly yet awakened to the need for more advanced mathematics.

[1] Pacioli here takes his material freely from Franceschi's work, already mentioned.

Of the arithmeticians of this time one of the most noteworthy was Filippo Calandri,[1] whose arithmetic appeared in Florence in 1491 and contained the first printed example in long division by our modern method and the first illustrated problems published in Italy. Nothing further is known of his life, but Sfortunati,[2] writing in 1534, speaks of him as a learned man.[3]

Francesco Pellos or Pellizzati, a native of Nice, published a commercial arithmetic at Turin in 1492, in which, as will be shown in Volume II, use is made of a decimal point to denote the division of a number by a power of ten.

98

Eglie una torre che e alta 40 braccia ꝛ dap pie uipassa uno fiume che e largho 30 brac cia. uo sapere quanto sara lungha una fune che sia appicata alla ri ua del fiume ꝛ alla ci ma della torre

40 —— 30
40 30
1600
900
laradice di 2500 sara lunga 50 brac cia

Eglie unalbero insu la riua dun fiume elqua le e alto 50 braccia el fiume e largho 30 bra cia ꝛ per fortuna di uē to siruppe intal luogo che lacima dellalbero toccaua lariua del fiu me. Uo sapere quante braccia sene ruppe ꝛ quanto nerimase ritto

50 —— 30
50 50 30
100 | 900
1600
rimase rito 16 brac cia ꝛ 34 braccia sene ruppe

FIRST ILLUSTRATED PROBLEMS
From Calandri's work, Florence, 1491

Austria and Germany. Before considering the names of those who advanced the cause of mathematics in the German states in the 15th century it is necessary to say a word concerning the influence and work of the Rechenmeisters. In the 13th century there had been a great revival of trade in Germany. The Hanseatic League, a union of commercial towns in the Teutonic countries, had shown that the demands of trade must be recognized. It employed force against the pirates,

[1] Italian writers of the period give this form of the name. Latin writers use Philippus Calender and Philippus Calandrus. Calandri himself gives only the Latin genitive form, Philippi Calandri. See *Rara Arithmetica*, p. 47.

[2] Giovanni Sfortunati (born at Siena, *c.* 1500), *Nvovo Lvme Libro di Arithmetica*, Venice, 1534. See page 306.

[3] "Filippo Calādri Cittadino Fiorentino, huomo certamēte in tale disciplina erudito." From the 1545 edition, fol. 3, *r.*

purchased settlements in foreign cities, and even made war upon Denmark and England in order to protect the interests of its members. The Church schools having failed to prepare boys for business, the League undertook this work, offering not only reading and writing, in which the Church schools gave instruction, but also business arithmetic, in which they gave none. Out of this kind of teaching of apprentices there arose a type of commercial school known as the Rechenschule, presided over by a Rechenmeister. In due time the Rechenmeisters formed guilds, claimed a monopoly in their vocation, and finally came to be looked upon as regular officials of the town. Not infrequently their duties included the sealing of measures, the city gaging of casks, and occasionally the minting of money, in those times a function of the city instead of the state. Occasionally they were also the writing masters, and they finally came to be looked upon as among the dignitaries of the town.[1] In the larger cities the Rechenmeisters' guilds admitted apprentices who served for six years, then becoming Schreibers (writers) with the privilege of becoming assistant teachers. When a *locus* opened, the eldest of these cadets was subjected to an examination[2] and, if successful, was given the position with the rank of Meister. The guild of Schreib- und Rechenmeisters continued in Lübeck, for example, until 1813.

In the 15th century the power of the Rechenmeisters was first shown to any considerable extent, and for the next three hundred years they clung more or less tenaciously to their privileges, demanding that arithmetic should be taught by them instead of by the common schoolmasters. Their influence extended to the commercial towns of Holland, and the Rechenmeister is

[1] Thus, the Bürgermeister and Council of Rostock, in 1627, sent an official call, couched in dignified language, as follows: "Wir Bürgermeister und Rat zu Rostock urkunden hiermit, dass wir den ehrenfesten und wohlgelahrten Jeremias Bernstertz zu unserm und gemeiner Stadt Schreib- und Rechenmeister bestellt und angenommen haben." For the entire document see F. Unger, *Die Methodik der praktischen Arithmetik in historischer Entwickelung*, p. 26 (Leipzig, 1888); hereafter referred to as Unger, *Die Methodik*.

[2] One of the questions from an old paper reads: "Wie wird besagte Chilioheptacosioheptacontatetragonal-Zahl, deren Latus 6 formirt und aus solcher gefundenen Polygonal-Zahl die Wurzel wieder extrahirt?"

frequently mentioned in the Dutch arithmetics of the 17th century.[1] The excellent commercial arithmetics which began to appear in Germany in the 15th century and which continued there and in Holland until the 19th century were the work of the Rechenmeisters or of those whom they influenced.[2]

Owing possibly to the early advance in printing in Germany, possibly to the trend of the classical influence from Constantinople through the Balkan states to Vienna, or possibly to the influence of the leisure which wealth afforded, the Teutonic countries forged ahead in the 15th century, taking rank with Italy in their men of mathematical ability. Four of these men were scholars whose standing was recognized abroad; the rest were mediocrities.

Late in the century Johann Widman (born *c.* 1460), a native of Eger in Bohemia, wrote on arithmetic and algebra. He was a student at Leipzig in 1480, B.A. in 1482, M.B. in 1485, and M.A. in 1486. He may have received a doctor's degree in medicine, for a medical work by a Johann Widman appeared in 1497, but this was probably another person.[3] That he gave lectures on algebra, possibly the first given in Leipzig, is shown by a contemporary manuscript.[4] He was probably the

[1] Thus, Coutereels (1631) addresses a problem to one Ghileyn Pietersz "Schepen [sheriff] deser Stadt, goet Reken-meester"; Cardinael (1659) speaks of himself as "Reecken-meester tot Amsterdam"; the printer of Vander Schuere's arithmetic (1634 ed.) speaks of the book "die ik door de correctie van een goet Reken-Mr. het verbetert"; Eversdyck, who revised Coutereels's arithmetic in 1658, is described as the "Reken-meester ter Reken-Kamer van Zeelandt"; and even as late as 1792 an edition of Bartjens is described as due to "den Wel-ervaren Reekenmeester Klaas Bosch."

[2] The Italian "master of the abacus" was probably suggested by the German title, or vice versa. The expression is found in Italian manuscripts of this period but does not seem to have been used in England. Indeed, even the word "abacus" never had as extensive use there as on the Continent.

[3] *Tractatus clarissimi medicinaꝝ doctoris Johānis widman.* In other works the spelling Widmann is sometimes used. Boncompagni's *Bullettino*, IX, 210.

[4] See *Rara Arithmetica*, p. 36. J. G. Bajerus (J. W. Bayer), *De Mathematvm . . . introdvctione* (1704) (hereafter referred to as Bajerus, *De Math.*), says (p. 9): "Joh. Widemann, natione Noricus, patria Egrensis, disciplina Lipzensis, vir in Mathematicis abunde eruditus. Qui capessis in Philosophia insigniis, cum multa admodum in mathematica, & potissime in speciebus in studio Lipzensi, non sine auditorum summo applausu, aliquot annis volvisset . . ." On his life and works see Cantor, *Geschichte*, II, chap. lv.

author of an *Algorithmus Linealis* (Leipzig, *post* 1489), the first printed treatise on calculation by the aid of counters. He wrote the first important German textbook on commercial arithmetic,[1] and in this appear for the first time in print the signs + and −, not as symbols of operation but to express excess and deficiency in packages of merchandise.

More capable as a mathematician but less known by the populace, Johann von Gmünden[2] was educated at Vienna, taught there, and was the first Austrian to occupy a chair devoted wholly to mathematics. He wrote a treatise on sexagesimal fractions,[3] one on trigonometry,[4] and one on the computus.[5]

A few years later, Nicholas Cusa,[6] son of a fisherman, gave proof of what industry and genius will do even for those born in humble estate. He rose rapidly in the Church and held various positions of honor, including the bishopric of Brescia. He was made a cardinal and became the governor of Rome in 1448. He wrote several tractates on mathematics, including

[1] *Behēde vnd hubsche Rechnung auf allen kauffmanschafft*, Leipzig, 1489.

[2] Born at Gmunden on the Traunsee, Gemünd in Lower Austria, or Gemünd in Swabia, *c.* 1380; died *c.* 1442. The name also appears in such forms as Johannes de Gmunden, Johann von Gemunden, Johannes de Gamundia. He has been identified with the Johann Schindel or Sczindel, called Johannes de Praga, a native of Königgrätz, and possibly the identification is correct. His name has also been given as Wissbier and Nyden, but on doubtful authority. See *Bibl. Math.*, X (2), 4. Bajerus, *De Math.*, has this note (p. 6): "Eo vix A. C. MCCCXCVII. fatis functo, Vindobonam ornare coepit Johannes de Gmunden itidem natione Germanus, (Ricciolus Johannem de Egmunda vocat) Theologus & Astronomus celebris, anno MCCCCXLII. humanis exemtus." See also *Bibl. Math.*, III (3), 140.

[3] *Tractatus de Minucijs phisicis*, first printed at Vienna in 1515.

[4] *De sinibus, chordis, et arcubus.*

[5] It exists in manuscript, closing with the words: "Explicit kalendariū mgrī Joh'is gmünd." See *Rara Arithmetica*, p. 449.

[6] Born at Kues (Cues) on the Mosel, 1401; died at Todi, Umbria, 1464. The name appears in such forms as Nicolaus von Cues, Nicholas Cusanus, Nicholas von Cusa, Nicolaus Chrypffs, Nicolaus Cancer, Nicolaus Krebs. His father's name was Johann Chrypffs or Krebs. "Huomo di mostruoso ingegno impatronissi delle tre lingue megliori, e diede opera all' arti liberali, & alle scienze," Baldi, *Cronica*, 95. Rossi, in his *Niccolo di Cusa* (p. 11), describes him as "originale pensatore in molteplici discipline." There is a good biography by Dr. Schanz, Prog., Rottweil, 1872. The best biographies are those of J. M. Düx, *Der deutsche Cardinal Nicolaus von Cusa*, 2 vols., Regensburg, 1847, and E. Vansteenberghe, *Le Cardinal Nicolas de Cues*, Paris, 1921.

in the subjects treated the quadrature of the circle, the reform of the calendar, the improvement of the Alfonsine Tables, the heliocentric theory of the universe (a theory which was looked upon as a paradox rather than a scientific probability), and the theory of numbers. Wallis[1] asserted that he was the first writer known to have worked on the cycloid, but this is not supported by the evidence.[2] His *Opuscula* appeared *c.* 1490, and his *Opera* appeared in Paris in 1511.[3]

Much better known as a mathematician, Georg von Peurbach[4] studied under Nicholas Cusa and other great teachers, learned Greek from Cardinal Bessarion in order to be able to read Ptolemy, lectured at Ferrara, Bologna, and Padua, and became professor of mathematics at Vienna, making this university the mathematical center of his generation. Although interested primarily in astronomy and trigonometry,[5] he wrote an arithmetic,[6] but this was merely for the use of students in these branches of science. Melanchthon considered the work so excellent that he wrote a preface for the edition of 1534. Peurbach compiled a table of sines, which was extended after his death by his pupil Regiomontanus, and he also wrote various works on astronomy.

The most influential and the best known of the German mathematicians of the 15th century was Johann Müller,[7] generally known, from the Latin name of Königsberg, as Regiomontanus.[8] At the age of twelve he was a student at Leipzig. He

[1] *Philosophical Transactions of the Royal Society of London*, p. 561 (1697); hereafter referred to as *Phil. Trans.*

[2] See *Bibl. Math.*, I (2), 8, 13. [3] *Rara Arithmetica*, p. 42.

[4] Born at Peurbach, Upper Austria, May 30, 1423; died at Vienna, April 8, 1461. The name is also spelled Peuerbach and Purbach.

[5] *Tractatus Georgii Purbachii super Propositiones Ptolemaei de sinubus et chordis*, Nürnberg, 1541; *Theoricae novae planetarvm*, Venice, 1495, with another edition s. l. a. (Nürnberg) and one at Venice, 1499.

[6] *Elementa Arithmetices Algorithmvs de numeris integris*. It went by various other titles. The first printed edition appeared in 1492.

[7] Born at Unfied, near Königsberg, Lower Franconia, June 6, 1436; died July 6, 1476. Cantor, *Geschichte*, II, chap. lv. In the British Museum there is an interesting block-book almanac, not later than 1474, prepared by him, in which his name appears as Magister Johann van Kunsperck.

[8] Also as Joannes de Monteregio.

afterwards studied under Peurbach, lectured at Venice, Rome, Ferrara, and Padua, and lived for a time in Nürnberg.[1] In 1475 he was called to Rome by Pope Sixtus IV on account of one of the frequent attempts to consider a reform of the calendar, and was made titular bishop of Ratisbon. He studied the mathematics of the Greeks in the original, and was "the first who made humanism the handmaid of science." He wrote *De triangulis omnimodis libri V* (*c.* 1464), the first work that may be said to have been devoted solely to trigonometry.[2] He also wrote an *Introductio in Elementa Euclidis* with some supplementary work on stellar polygons, and had certain definite ideas as to the circumnavigation problem.[3]

The earliest known German algorism was compiled in 1445, more than two centuries after Fibonacci prepared a work on the subject, and about seventy years after the first French manuscript relating to it is known to have been written. The earliest example of a German algebra is found in a Munich manuscript of 1461, with text also in Latin. The first printed German arithmetic appeared at Bamberg in 1482.

France. France received the Italian humanism in the spirit of a sister country speaking a kindred language, but for a number of decades after Gregory Tifernas (1458) went to Paris to teach Greek the taste of the learned class was for letters rather than science.[4]

It is perhaps because of this fact that France produced fewer noteworthy mathematicians than Germany or Austria in the 15th century, but it is more probably due to such causes as the constant turmoil in which the University of Paris found herself at this time. With a continual warfare between Church

[1] Ramus speaks of the great glory he brought to Nürnberg: "Noriberga tum Regiomontano fruebatur: mathematici inde & studii & operis gloriam tantam adepta, ut Tarentum Archyta, Syracusae Archimede, Byzantium Proclo, Alexandria Ctesibio non justius quám Noriberga Regiomontano gloriari possit." *Scholarum Mathematicarum libri unus et triginta*, p. 62 (Paris, 1569); hereafter referred to as Ramus, *Schol. Math.*

[2] First printed edition, Nürnberg, 1533.

[3] A. Ziegler, *Regiomontanus . . . Vorläufer des Columbus*, Dresden, 1874.

[4] See also R. C. Jebb, *Cambridge Mod. Hist.*, Vol. I, chap. xvi.

and State over the control of her work, and with continual protests from within her walls, she was in no mood to foster either science or letters.

A type of the best product of the educational system of France at this time is seen in the person of a very mediocre and almost unknown scholar, one Rollandus (*c.* 1424), who, although probably a native of Lisbon, spent his life in Paris. He was a physician and a minor canon of the Royal Chapel[1] and may have been the Rolland who was rector of the university in 1410. He evidently was acquainted with the general field of pure arithmetic and algebra, as is shown by a manuscript of about 1424 still extant.[2] He also wrote on physiognomy and surgery.

The most brilliant of the French mathematicians of the period, however, was Nicolas Chuquet,[3] who, although a native of Paris, lived in Lyons.[4] He wrote (1484) the *Triparty en la Science des Nombres*, a work touching upon the three fields of arithmetic.[5] The first part relates to computation with rational numbers, the second to irrationals, and the third to the theory of equations. Nothing is known about Chuquet except his statement that he was a bachelor of medicine and that he wrote his work at Lyons.[6]

Great Britain. Great Britain produced relatively few writers on mathematics in the 15th century, and none of any special prominence. John Killingworth may be taken as a type of

[1] He speaks of himself as "prebenda capelle palacij regalis parisiensis."

[2] *Scientia de numero ac virtute numeri*, now in the library of Mr. Plimpton. See *Rara Arithmetica*, p. 446.

[3] Born at Paris; died *c.* 1500.

[4] Estienne de la Roche (1520), speaking of the "plusieurs maistres expertz en cest art," mentions "maistre nicolas chuquet parisien."

[5] It was printed in Boncompagni's *Bullettino*, XIII, 555. See Ch. Lambo, "Une algèbre Française de 1484. Nicolas Chuquet," *Revue des Questions Scientifiques* (Brussels, October, 1902).

[6] "Et aussi pour cause quil a este fait par Nicolas chuquet parisien Bachelier en medicine. Je le nomme le triparty de Nicolas en la science des nombres. Lequel fut commance medie et finy a lyon sus les rosne de salut. 1484." From the original copy of his MS. made for Boncompagni by A. Marre, and now in the author's library.

the best scholars of the period. We know little about him,[1] but the records show that he became a fellow of Merton College, Oxford, in 1432 and that he died on May 15, 1445. He seems to have been chiefly interested in astronomy and to have prepared a set of tables for the use of students in this science. There is extant in the Cambridge University Library an algorism written by him in 1444.[2] In this he refers to the use of a slate for purposes of computation, but as to the operations themselves there is no evidence of any originality.

Other Countries. The early mathematics of Russia was largely devoted to questions relating to the calendar and to number puzzles. In some of the medieval manuscripts there are the results of very complicated computations, but we have no knowledge of the methods by which these computations were performed.[3] Several such cases are found in the *Russkaya Pravda*.[4] At the end of the 15th century, work on the calendar is known to have been done by the Metropolitan, Zosima, and by Gennadi, bishop of Novgorod, but the results are not extant. In all this work the numerals seem to have been alphabetic, the system being similar to that of the later Greeks. Even as early as the 14th century, the clergy placed geometry and astronomy under the ban, and not until the 17th century was there any opportunity for the study of mathematics.[5] Even when, in the 18th century, the attempt was made to inaugurate scientific work, all the leaders in mathematics were brought in from abroad.

[1] L. C. Karpinski, "The Algorism of John Killingworth," *English Historical Review* (1914), p. 707.

[2] "Incipit prohemium in Algorismum Magistri Iohannis Kyllyng Worth." The work itself begins: "Obliuioni raro traduntur que certo conuertuntur ordine. Regulas igitur et tabulas ad breua computationem operis calculandi vtiles in formam certam secundum ordinem specierum Algorismi curabo redigere." *Ibid.*, p. 713.

[3] A. N. Peepin, *History of Literature* (in Russian) (Petrograd, 1911 ed.), I, 253.

[4] A work of which there are three versions, one of the 11th century, one of the 12th, and one of the 13th.

[5] V. Bobynin, "De l'étude sur l'Histoire des math. en Russie," in *Bibl. Math.*, II (2), 103.

In the 15th century (*c.* 1450) a Cretan writer commonly known as George of Trebizond[1] (1396–1486) made a new translation of the *Almagest* into Latin and also translated some of Theon's commentaries upon it. He was a quarrelsome man, of little honor to science, to letters, or to manners.

In Hungary there was a certain Georgius de Hungaria who wrote an *Arithmeticae summa tripartita* in 1499,[2] and this perhaps shows the high mark of mathematics in that country in the 15th century.

The Jewish activities in this century were very slight, being chiefly manifest in translations from the Latin. Jacob Caphanton (died by 1439), probably a native of Castile, a physician and a teacher, wrote an arithmetic;[3] and Jehuda Verga (*c.* 1450), known for his compilation on the calamities of the Jews, was also the author of a compendium of the same type.[4] There were various writers who left works on astrology and the calendar, contributions unimportant in themselves but requiring the computation of tables and offering some encouragement to the astronomer.

Sporadic efforts were made here and there in the 15th century to advance the study of mathematics in other countries, but with no effect beyond the establishing of an interest in the science. Thus João (John) II, who was with great difficulty placed (1481) upon the throne of Portugal, sought to elevate scholarship, and particularly astronomy and navigation, by establishing at Lisbon his Junta dos Mathematicos,[5] but the result was not noticeable in pure mathematics. A few scholars, such as the bishops Calsadilha and Don Diogo Ortiz, sought to advance the applications of mathematics in such fields as map drawing, for the purpose of aiding the Portuguese navigators, but this was about all that was done. A little later there were such astrologers as Diogo Mendes Vizinho and

[1] Georgios Trapezuntios. [2] It was reprinted at Budapest in 1894.

[3] *Bar Noten Ta'am le-Chacham* (a Talmudic phrase). See *Bibl. Math.*, XIII (2), 99. [4] *Bibl. Math.*, II (3), 62.

[5] A. Marre, Boncompagni's *Bullettino*, XIII, 560; Bensaude, *Astron. Portug.*, 104; R. Guimarães, *Les Mathématiques en Portugal*, 2d ed., p. 11 (Coimbra, 1909) (hereafter referred to as Guimarães, *Math. Portug.*).

Thomaz Torres, but the former knew little beyond the elements of cartography, and the latter seems to have been interested only in the drawing of horoscopes. It was not until the 16th century that mathematics commanded any noteworthy attention in the Portuguese universities.

Some idea of the new Italian spirit in mathematics and letters succeeded in reaching Spain at about the same time. Men like Barbosa lectured on Greek at Salamanca, and Lebrixa (Nebrissensis) returned from Italy in 1474 and lectured at Seville, Salamanca, and Alcalá, but the interests of the learned were in medieval theology and little attention was given to the advancement of the sciences.

The nature of the Spanish works of the 15th century may be inferred from the *Visiō delectable de la philosophia & artes liberales,* a kind of general encyclopedia, written by Alonso Delatore and published at Tolosa in 1489 and again at Seville in 1538. The fourth chapter of this work is entitled *Dela arismethica y de sus inuētores,* but it has no merit and is little better than the treatment of arithmetic given by Capella and Isidorus several centuries earlier. Spain was at this time occupied in suppressing the bandit nobles, in recodifying her laws, and in banishing the last of the Moors, and it was not until 1492 that the political unification necessary to future peace was effected. Nevertheless the foundations seemed in process of being laid for a new type of commercial mathematics and for the rapid development of the science of navigation. Such voyages as those of Columbus and, in the following century, of Ponce de Leon, led to an expansion of the colonial empire and to a rapid increase in wealth. Unfortunately, however, the Spanish leaders were unable to grasp their opportunity, and they squandered their newly acquired possessions in ambitious schemes that led to profitless wars. The chance which the Medici so happily improved was cast aside by the rulers of Spain, and neither art nor science was fostered. During most of the century, therefore, the situation was such as to dampen any scientific zeal, and not a single mathematical work of any consequence was printed in Spain in the 15th century.

TOPICS FOR DISCUSSION

1. Influences at work in the 11th century to improve the intellectual status of Europe.

2. The life and works of Gerbert, who became Pope Sylvester II.

3. A study of the number game of Rithmomachia.

4. The 12th century as a period of translations.

5. Oriental civilization in Spain in the 11th and 12th centuries.

6. Jewish activity in mathematics in the 11th and 12th centuries.

7. Influences tending to foster mathematics in the 13th century.

8. The life and works of Leonardo Fibonacci.

9. The translations of Euclid and their influence on mathematics in the Middle Ages.

10. The life and works of Johannes Sacrobosco and his influence on mathematics in general and on algorism in particular.

11. Roger Bacon; his contemporaries and his projected reforms.

12. The life of Jordanus Nemorarius and his influence on medieval mathematics in general and on algebra in particular.

13. Influences at work in the 14th century to improve the intellectual status of Europe, particularly in the field of mathematics.

14. The mathematics of England in the 14th century. In what ways did the science meet the needs of the time?

15. A comparison of the mathematics of England, France, and Italy in the 14th century.

16. The various methods of diffusing knowledge among scholars during the Middle Ages.

17. The effect upon mathematics of the influx of Greek manuscripts into Italy in the 15th century.

18. Influences leading to the Renaissance, and their bearing upon the development of mathematics.

19. The general nature of the mathematics of Italy in the 15th century, and its effect upon modern education.

20. Influence of the universities in the 15th century, particularly on the development of mathematics.

21. The leading printed mathematical works of the 15th century.

22. Some of the probable reasons why mathematics made such a slight advance in the 15th century.

23. Reasons for the prominence of commercial arithmetics in the latter half of the 15th century.

CHAPTER VII

THE ORIENT FROM 1000 TO 1500

1. China

The General Period. The most interesting period in the history of Chinese mathematics, and indeed one of the most interesting periods in the history of the world, is that of the five centuries from the year 1000 to the year 1500. It is a period not unlike the contemporary epoch in Europe, a time of awakening from sleep. Europe had seen the glories of the classical period die out; she had passed through a season of darkness; she was awakened by the crusades; she arose with a great feeling of refreshment in the 13th century, and in the 15th century she opened a new world of thought through the invention of printing from movable types, and a new world of commercial activity through the discovery of the Western Hemisphere.

China passed through a similar experience, making great advances in algebra in the 13th century and learning of a new civilization through the work of the Jesuit missionaries three hundred years later.

The Eleventh Century. The 11th century saw little progress made in mathematics in China. Indeed, the only name of any importance to be found in her annals is that of Ch'ön Huo (1011–1075), president of the Bureau of Astronomy, a minister of state, and the author of a work in which there is found, perhaps for the first time in China, the summation of a series of any difficulty. This summation appears in the solution of a problem on the number of wine kegs in a pile whose form is a truncated pyramid, the upper row containing 2^2 kegs, the lower one 12^2, and there being 11 rows. The author considers also the question of finding the length of a circular arc in terms of the radius, and speaks of the difficulty of the problem.

In 1083 the government showed its interest in mathematics by printing Liu Hui's (*c.* 250) *Sea Island Classic*,[1] and a year later (1084) it printed the arithmetic of Ch'ang K'iu-kien[2] (*c.* 575).

The Twelfth Century. The 12th century was about as barren as its predecessor. The *Huang-ti K'iu-ch'ang*[3] seems to have been printed (*c.* 1115), and Ts'ai Yüan-ting[4] wrote on the *I-king*, but these events simply emphasize the fact that the century was barren of achievement. What is much more significant is the fact that the East was again coming in contact with the West; commerce was exchanging the problems of business, and the astrologer was doing the same for the higher and more mysterious strata of mathematical knowledge. We know that in this century metals were transported from Arabia to China, and in 1178 a Chinese work[5] was written in which a great export trade is described and a list of the merchandise is given. In 1128 an immense army from China invaded Turkestan and found there many Chinese residents.[6] The stimulus of world intercourse was becoming stronger, and the effect of this stimulus was to appear in the century following.

Exchange of Thought in the Thirteenth Century. The increase of opportunity for the exchange of thought in the 12th century was carried over into the 13th century, quite as the peripatetic scholars in Europe increased in number and in activity in the same period. Metal was imported from Persia (Po-sz') and from the country of the Arabs (Ta-shi).[7] In 1219

[1] The *Hai-tau Suan-king* (see page 142). This was, of course, from blocks, not from movable type.

[2] The *Ch'ang K'iu-kien Suan-king* (see page 150).

[3] "Yellow Emperor's Nine Sections." One of the versions of the *K'iu-ch'ang Suan-shu* (see page 31).

[4] Born 1135; died 1198; student and historian (see Giles, *loc. cit.*, No. 1985).

[5] The *Ling-wai-tai-ta* of Chóu K'ü-feï. Upon this work Chau Ju-kua's great treatise was based. See the edition by Hirth and Rockhill, Petrograd, 1911.

[6] E. Bretschneider, *Mediaeval Researches*, 2 vols., I, 232 (London, 1910); hereafter referred to as Bretschneider, *Mediaeval Res.*

[7] This is related in the *Si Shi ki* (travels) of Ch'ang-ti. Mention is made of *pin t'ie*, probably steel. See Bretschneider, *loc. cit.*, I, 146.

the conquering Chinghiz Khan[1] carried out his great expedition to Western Asia, established roads connecting Eastern Mongolia with Persia and Russia, sent armies even to Eastern Europe, and opened the way to trade between the Orient and all parts of the Occident. He captured Bokhara, Samarkand, Herat, Merv, Nîshâpûr, Kiev, and probably Moscow (Moscoss). He invaded Poland, Galicia, Silesia, and Hungary, and his astrologers must have mingled freely with their guild in the capitals of Eastern Europe. In 1221 the Chinese traveler K'iu Ch'ang ch'un reached Samarkand[2] and, what is especially significant, records the fact that "Chinese workmen are living everywhere" in the city, and that he saw "peacocks and great elephants which had come from Yin-du" (India). He also speaks of meeting and conversing with an astronomer of Samarkand, as we should naturally expect of a man of his learning. In 1236 the Mongols invaded Bulgaria; in 1238 France received a Mussulman ambassador asking aid against the Oriental forces; in 1241 the Mongols invaded Galicia, and in 1259 they burned Cracow for the second time.

The fact is also significant that in 1266 the king of Ceylon had Chinese soldiers in his service, while about this time the Chinese traveler Chau Ju-kua speaks of the Hindus from his personal knowledge as being "good astronomers and calculators of the calendar."[3]

Europeans in the East. Moreover, Europeans were frequently seen in the East, and the Grand Duke Yaroslav, in 1246, met one Fra Plano Carpini, a Franciscan, at the Mongol court. Only a few years later Rubrouck, another of the friars minor, visited the same court, and both he and Carpini have left accounts of their travels. There were also Eastern travelers who went to the West, such as the Taoist monk Ch'ang ch'un, from 1220 to 1224; Ch'ang-ti, who was sent out in 1259 by Mangu Khan, and who visited Bagdad; and Ye-lü Hi-liang, who traveled in Central Asia from 1260 to 1262.

[1] Jenghiz Khan, Genghes Khan, Jinghis Khan; originally Temuchin.

[2] The Semiscant of medieval writers, known to the Chinese as Sie-mi-sz'-kan and Sün-sz'-kan.

[3] Hirth and Rockhill, *loc. cit.*, p. 111.

The second half of the century saw other travelers from the West in the courts of Mongolia and China. Haithon (Hethum), king of Little Armenia, visited Mongolia in 1254, and the account of his journey is well known. Marco Polo, who set out from Venice in 1271, spent seventeen years in China and traveled in a leisurely way through Arabia, Persia, and India. Rashid ed-dîn, vizier of the Persian Empire in 1298, wrote a history of the Mongols showing thorough familiarity with their country. All these events, unimportant in detail, are very significant taken as a whole, because they afford evidence of the free interchange of ideas between the East and the West. Thus they serve to clear up many questions as to whether the algebra of China, for example, could have found its way to Italy in the 12th century. We repeat that it would be a cause for wonder if it had failed to do so. Moreover, the 13th century was a period of wealth, of luxury, and also of opportunity. Chinghiz Khan, in a letter to Ch'ang ch'un the Taoist traveler, asserted that "Heaven had abandoned China owing to its haughtiness and extravagant luxury."[1] This may have been true; a conqueror would be apt to say so; but it is certain that it was a century in which mathematics ought to have flourished in both continents. We shall see that this was the case.

Mathematical Activity in the Thirteenth Century. The 13th century, in China as in Europe, was a period of awakening. Indeed, it may be said that it was the period of the highest development of native mathematics in the East. Whether as a result of an interchange of thought with the West, or of the leisure which wealth brought to the country before the invasion of Chinghiz Khan, or of the growth of idealism which war is said by its advocates to foster, or of various other causes, China made a noteworthy advance in algebra at this time.

Perhaps the foremost scholar in this movement was Ch'in Kiu-shao, a man whose intimate history is quite unknown, but who was a soldier in his early days, was in government service

[1] Bretschneider, *Mediaeval Res.*, I, 37.

in 1244, was governor of two provinces, and wrote the *Nine Sections of Mathematics* in 1247.[1] The work relates chiefly to numerical higher equations, in which the author anticipates to some extent Horner's Method (1819); but it also considers indeterminate equations and the application of algebra to trigonometry. The "nine sections" of his work are not the same as those of the one mentioned on page 31. For the value of π the author gives 3, $\frac{22}{7}$, and $\sqrt{10}$. In his work, too, the symbol ○ is used for zero and the place value is used in the writing of all numbers. The author shows little interest in applying his knowledge of algebra to the solution of practical problems, preferring to look upon it as a pure science.

Li Yeh's Work. The next noteworthy step in this direction was taken by Li Yeh (1178–1265). He wrote the *Sea Mirror of the Circle Measurement*[2] in 1249, the *I-ku Yen-tuan* in 1259, and various other works. In his early life he was engaged in public service, and in 1232 was governor of Chün Chou. He was later held in high esteem by Kublai Khan, whose reign began in 1260.[3] Li Yeh directed his son to burn all his works except the *Sea Mirror*. The *I-ku Yen-tuan* was also preserved, however, and each has since been looked upon as among the great works of China. While Ch'in Kiu-shao had given his attention chiefly to the solution of abstract equations, Li Yeh devoted himself to the forming of equations representing various complicated problems, the solution being neglected.

[1] Chang Ch'i-mei (1616) says that the author called the work *Su-shu* or *Su-hsiao*, but in the Chinese bibliographical works the title is *Su-shu Kiu-ch'ang*. Biernatzki (pp. 13, 28) transliterates the title as *Su schu kiu tschang* and the author as Tsin Kiu tschaou. Vanhée gives the date as 1257 and the author as Ts'in K'ieou-Chao. For details as to his methods see Mikami, *loc. cit.*, p. 63.

[2] *Ts'ö-yüan Hai-king*. Vanhée (*T'oung-pao*, XIV, 537) gives the date as 1248 and the title as *Ts'é yuen hai king*. Biernatzki gives the name of the author as Le Yay and Le yay Jin king, and the title as *Tsih yuen hä king*. Li Yeh also used the *nom de plume* Ching-chai.

[3] The Yuen Dynasty, founded by the Mongols, began in 1280, however. In 1260 the Mongols issued paper money, an event bearing somewhat on the history of arithmetic problems.

About this time another scholar, Liu Ju-hsieh, wrote a treatise on algebra, but the work is not extant.[1]

Chinese Astronomers. That astronomy was the mathematical subject of chief interest at this time, as indeed at all times in early Chinese history, is shown by the considerable list of names of scholars devoted to the science. Among the leaders at this particular period was Ye-lü Ch'u-ts'ai,[2] who lived *c.* 1230. He established a great school at Peking (then Yen-king) and accompanied Chinghiz Khan to Persia, occupying himself with the calculation of eclipses and coming into contact with the Persian astronomers.

Yang Hui's Work. In 1261 Yang Hui wrote *The Analysis of the Arithmetic Rules in Nine Sections*,[3] a work in which he explained some parts of the original *Nine Sections* (see page 31). In this work he gives a graphic representation of the summation of an arithmetic series. In another work[4] he gives rules for summing the series

$$1 + (1 + 2) + (1 + 2 + 3) + \cdots + (1 + 2 + 3 + \cdots + n)$$

and

$$1^2 + 2^2 + 3^2 + \cdots + n^2,$$

but offers no explanations. He wrote several other works,[5] and among his problems are such as that of the hare and hound and several involving simple and compound proportion.

Yang Hui's teacher, Liu I, a native of Chung-shan, wrote (*c.* 1250) a work which is known only by name,[6] but it doubtless related to numerical higher equations.

[1] He is mentioned in a work of 1303 as a prominent contemporary. His work is said to have been written before 1300. Since a commentary upon it was written by Yuen Hao-wen, who was a friend of Li Yeh (1178–1265), the date may be taken as *c.* 1260.

[2] Rémusat, in his *Nouveaux Mélanges Asiatiques*, II, 64, does not speak of his mathematical attainments, but in the *Yuen-Shih* (*Historical Records of the Yuen Dynasty*) these are discussed.

[3] *Hsiang-kieh K'iu-ch'ang Suan-fa.* Yang Hui is also known in European works as Yang Hwuy, Yan Hui, Yang Houei, Yang Hwang, and Kien-kouang.

[4] *Suan-fa T'ung-pien Pen-mo.*

[5] His six works on arithmetic, thought to be lost, were discovered in Shanghai in 1842.

[6] *I-ku Kon-yüan.*

Kóu Shóu-king. In 1267 the Mongols are known to have employed various Arab artillery officers,[1] so that contact with Arabia existed through the army. But this contact is the more apparent in the scientific achievements of Kóu Shóu-king[2] (1231–1316), a man well versed in the astronomy of the Arabs. He was a native of Hsing-t'ai[3] and was remarkable for his attainments even in early childhood, seeming to have inherited the scholarly qualities of his grandfather, Kóu Yung, a mathematician of repute. In early manhood Kóu Shóu-king developed into one of the greatest engineers of China. He was appointed by Kublai Khan to reform the calendar,[4] and for this purpose he replaced the armillary sphere which had been made about 1050, and which was calculated for Peenking, the former capital, differing about 4° in latitude from Peking, by the earliest of the great bronze instruments now on the wall of the latter city. The instruments constructed by him included several that were adapted to observations made in the daytime as well as at night,[5] and show that he had considerable knowledge of spherical trigonometry. Only two of his instruments seem to be extant, and these were found and described by Matteo Ricci[6] when he visited Peking early in the 17th century. Ricci also speaks of having seen similar instruments at Nanking.

With Kóu Shóu-king may be said to have begun the study of spherical trigonometry in China, a subject already far advanced in the Arab schools.

Further contact with the West was made at this time through the sojourn of Friar Odoric in Canton from 1286 to 1331.

[1]Two are called by the names I-se-ma-yin (probably Ismael) and La-pu-tan in the Chinese records.

[2]Also transliterated as Ko-cheou-king, Kouo Cheou-kin, and Kou Shou-ching. Since his cognomen is given as Jŏ-sze, it is possible that he is identical with the Liu Ju-hsieh already mentioned.

[3]Hing-tae, with other transliterations, a district in the prefecture of Shun-tĭh.

[4]He devised this calendar, the *Shóu-shï-li*, in 1280 and it was adopted in 1281.

[5]For a list and for the general subject of Kóu Shóu-king see A. Wylie, "The Mongol Astronomical Instruments in Peking," *Travaux de la 3e session du Congrès internat. des Orientalistes*, Vol. II.

[6]See page 303.

Chu Shï-kié. The 13th century in China closed with the remarkable work of Chu Shï-kié,[1] a native of Yen-shan. As to his private life, we know only that for more than twenty years he was a wandering teacher. He wrote two works, the *Introduction to Mathematical Studies*[2] in 1299 and *The Precious Mirror of the Four Elements*[3] in 1303. With him the old abacus algebra, in which the coefficients were represented by sticks placed on a checkered board,[4] reached its highest mark. In the first of his works there appears the algebraic rule of signs and an introduction to algebraic processes in general. In the second treatise, however, he considers a variety of new questions in higher algebra. He begins with what is called at present Pascal's Triangle, giving the values of the binomial coefficients and referring to the scheme as an old one. He considers higher equations with more than one unknown quantity, his treatment showing some knowledge of elimination by a determinant notation. He shows much ingenuity in his solution of numerical higher equations by the method already used by Ch'in Kiu-shao and which resembles Horner's Method.

2. Japan

Close of the Dark Ages. For Japan, as well as for Europe, there was a period which may properly be spoken of as the Dark Ages. For a thousand years after Buddhism was introduced into Japan there were few other events of intellectual significance to record. Japan was awaiting the world's rebirth, and this period of rebirth came to the East at about the same time as to the West.

Only two men stand out as worthy of mention in the history of Japanese mathematics between the year 1000 and the year 1500. The first of these is Fujiwara Michinori, a daimyō or

[1] Tchou Che-kié, Tchou Che-Kié, Tschu Schi kih, Choo Ché-kié, Chu-Shih-Chieh, Tchou Che-kie, and various other transliterations. He is also called Chu Sung ting.

[2] *Suan-hio-ki-möng*, or *Suan-hsiao Chi-mêng*.

[3] *Szu-yuen Yü-kien*, or *Szu-yuen Yu-chien*. The introduction was written by Tsu Yi Chi Hsien Fa.

[4] See Volume II, Chapter VI.

feudal lord in the province Hyūga, who wrote a work on permutations[1] between the years 1156 and 1159. The work is now lost, but it was thought important enough to be considered by the leaders of mathematics in Japan in the 17th century.

The second name worthy of note is that of the Buddhist priest Genshō, who lived in the first part of the 13th century. No trace remains of any of his writings, but tradition says that he was possessed of remarkable arithmetical ability.

3. India

Śrīdhara. The first of the Hindu writers of this period seems to have been Śrīdhara, commonly known as Śrīdharācārya, Śrīdhara the Learned, who was probably born in 991.[2] His work is known as the *Gaṇita-Sara* (*Compendium of calculation*) but is more commonly designated by the subtitle *Triśatika*, a name referring to its three hundred couplets.[3] The subjects considered are numeration, measures, rules, and problems, and the order bears very close resemblance to the one followed about a century later by Bhāskara in his *Lilāvati*. The latter writer was acquainted with Śrīdhara's work, as he himself testifies in his *Bija-Gaṇita*.

Under the general topics above mentioned are included series of natural numbers, multiplication, division, zero, squares, cubes, roots, fractions, Rule of Three, interest, alloys, partnership, mensuration, and shadow reckoning. The statement relating to zero is noteworthy as the clearest one to be found among the Hindus: "If zero is added to a number, the sum is that number itself; if zero is subtracted, the number remains unchanged; if zero is multiplied, the result is zero; and if a number is multiplied by zero, the product is zero only." The question of division by zero is not considered.

[1] The theory known as *Keishizan*.

[2] The date is uncertain, being placed by one writer three centuries earlier. The question is discussed by N. Ramanujacharia and G. R. Kaye in the *Bibl. Math.*, XIII (3), 203.

[3] The Sanskrit text was published in 1899. The English text is given in the article mentioned above. The name may also have come from the fact that it originally had 103 couplets.

For dividing by a fraction Śrīdhara gives the rule of multiplication by the inverted divisor, a rule already known to Mahāvīra (*c.* 850). Like the latter, too, he uses $\sqrt{10}$ for π.

Bhāskara (1114–*c.* 1185). There is only one other writer who stands out prominently in the history of Hindu mathematics from 1000 to 1500, and that is Bhāskara, commonly known

ŚRĪDHARA'S TRIŚATIKA, *c.* 1025

Two pages from the copy used by Colebrooke in his works on Hindu mathematics. From Kaye's *Indian Mathematics*

as Bhāskara the Learned (Bhāskarācārya),[1] a native of Biddur[2] in the Deccan, but working at Ujjain. An ancient temple inscription refers to him in the following terms: "Triumphant is the illustrious Bhāskarācārya whose feet are revered by the wise, eminently learned, . . ., a poet, . . . endowed with good fame and religious merit. . . ."[3]

[1] The name is variously transliterated. Thus, we have Bhascara Acharya, Bhaskaracharya, and other forms. See J. Garrett, *loc. cit.*, p. 92; Taylor, *Lilawati*, 1; T. W. Beale, *The Oriental Biographical Dictionary*, Calcutta, 1881; Colebrooke, *Bhāskara*. [2] Probably the modern Bidar.

[3] Kaye, *Indian Math.*, 37.

Bhāskara's Lilavati. Bhāskara wrote chiefly on astronomy, arithmetic, mensuration, and algebra. His most celebrated work is the *Lilāvati*, a treatise based upon Srīdhara's *Triśatika* and relating to arithmetic and mensuration.[1] This work was translated into Persian by Fyzi[2] in 1587 by direction of the emperor Akbar, a great patron of letters. Fyzi states, though it does not appear upon what authority, that Lilāvati was the name of Bhāskara's daughter and that the astrologers predicted that she should never wed. Bhāskara, however, divined

PALM-LEAF MANUSCRIPT OF THE LILĀVATI

Showing the form in which the Hindu manuscripts appeared before paper became a common medium. This manuscript was copied *c.* 1400. From the author's collection

a lucky moment for her marriage and left an hour cup floating on the vessel of water. This cup had a small hole in the bottom and was so arranged that the water would trickle in and sink it at the end of the hour. Lilāvati, however, with a natural curiosity, looked to see the water rising in the cup, when a pearl dropping from her garments chanced to stop the influx.

[1] The first translation into English is that of Taylor (1816), already mentioned.

[2] Also spelled Faizi and Feizi. He was a brother of Akbar's secretary, Abu Fazil. The work was printed at Calcutta in 1827.

So the hour passed without the sinking of the cup, and Lilāvati was thus fated never to marry. To console her, Bhāskara wrote a book in her honor, saying: "I will write a book of your name which shall remain to the latest times; for a good name is a second life and the groundwork of eternal existence."

The work begins, as is the custom in the East, with an address to the Deity: "Salutation to the elephant-headed Being

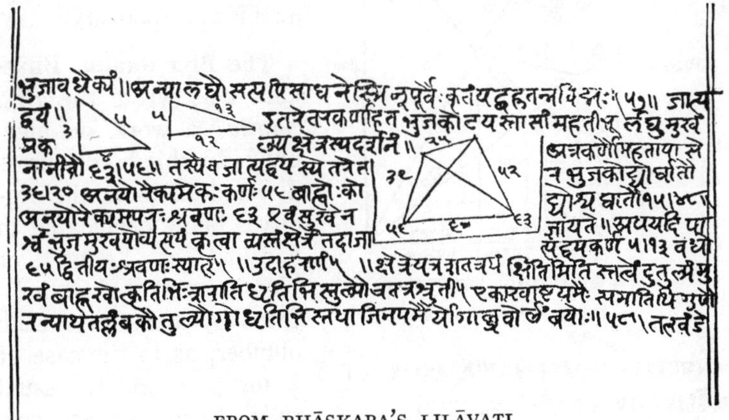

FROM BHĀSKARA'S LILĀVATI

From a manuscript of *c.* 1600. The original work was written *c.* 1150. The illustration shows the form of Hindu manuscripts just following the use of palm-leaf sheets. This page has the following statement: "Assuming two right triangles [as shown], multiply the upright and side of one by the hypotenuse of the other: the greatest of the products is taken for the base; the least for the summit; and the other two for the flanks. See" [the trapezoid]. Colebrooke's translation, page 82

who infuses joy into the minds of his worshipers, who delivers from every difficulty those who call upon him, and whose feet are reverenced by the gods." The book includes notation, the operations with integers and fractions, the Rule of Three, the most common commercial rules, interest, series, alligation, permutations, mensuration, and a little algebra. The rules relating to zero are also given, to the effect that $a + 0 = 0$, powers of 0 are 0, and $a \cdot 0 = 0$. The statement that $a \div 0 = 0$ (corrected by his commentators) was evidently not clear to him, for his statement is "A definite quantity divided by cipher is the

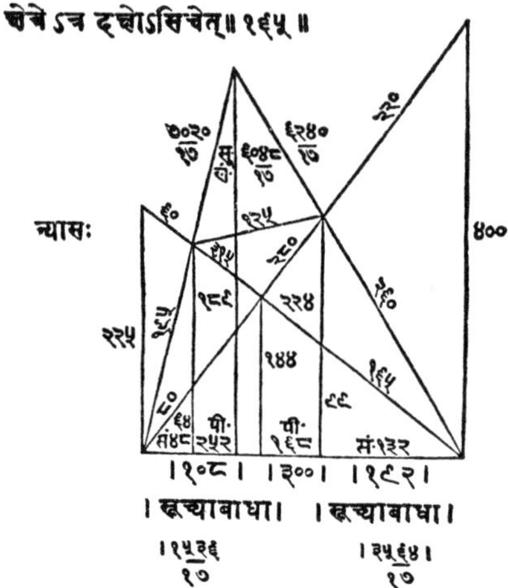

PAGE FROM THE FIRST PRINTED SANSKRIT EDITION OF BHĀSKARA'S LILĀVATI

Printed at Calcutta, 1832. This is a continuation of the portion shown in manuscript on page 277. The statement, as translated by Colebrooke, is as follows: "Length of the base, 300. Summit, 125. Flanks, 260 and 195. Perpendiculars, 189 and 224." From these the other parts are found

sub-multiple of nought," while his illustrations are that $10 \div 0 = \frac{10}{0}$ and $3 \div 0 = \frac{3}{0}$,[1] the latter being accompanied by the statement that "this fraction, of which the denominator is cipher, is termed an infinite quantity."

The Bija Gaṇita. Bhāskara also wrote the *Bija Gaṇita*,[2] a work on algebra.[3] In this he discusses directed numbers, the negatives being designated in Sanskrit as "debt" or "loss"[4] and being indicated by a dot over each number, as in the case of $\dot{3}$ for -3, and the usual rules being stated correctly. The imaginary is dismissed with the statement, "There is no square root of a negative quantity: for it is not a square." Where several unknown quantities are used, they are mentioned as colors: "'so much as' and the

[1] Colebrooke, *loc. cit.*, pp. 19, 20, 137.

[2] Variously transliterated as Vījaganita, Vija-Gañita, and the like. The term literally means "seed counting" or "seed arithmetic." The work was translated by Colebrooke, *Bhāskara*, p. 129, with the subtitle *Avyacta-Gañita*.

[3] Translated into Persian by Ata Allah Rusheedee in 1634. The English version of this translation was published by Edward Strachey, *Bija Ganita*, London, n. d. (*c.* 1812), apparently with the aid of a translation made, with the help of a pundit, by S. Davis, *c.* 1790.

[4] *Rīna* or *cshaya*.

colors 'black, blue, yellow, and red,' and others besides these, have been selected by venerable teachers for names of values

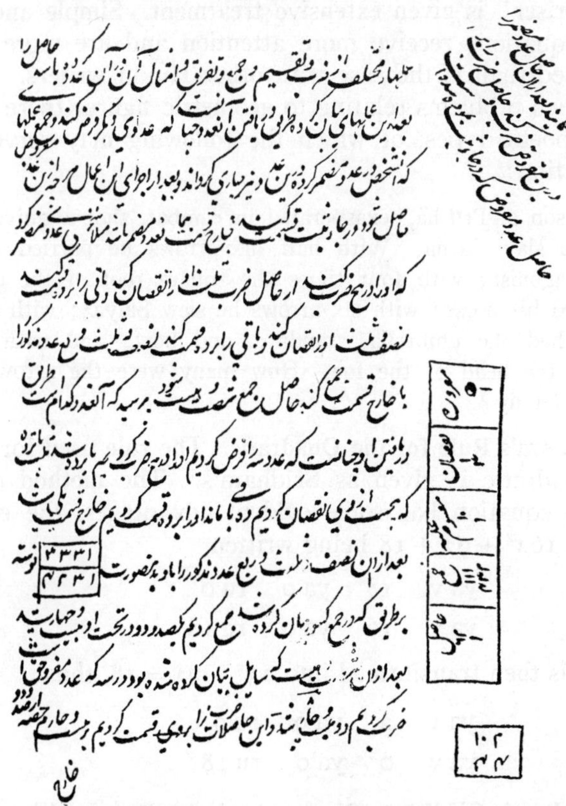

FROM A MANUSCRIPT OF FYZI'S TRANSLATION OF THE LILĀVATI

Fyzi, counselor of Akbar, made the translation into Persian in 1587. This manuscript is dated 1143 A.H., or 1731 A.D. The Persian manuscripts were written in our ordinary book form. The above is page 32 of Taylor's translation of 1816. The problem is: "A number is multiplied by 5; from the product is subtracted one-third of itself, and the remainder is divided by 10; to the quotient is added $\frac{1}{2}$ of $\frac{1}{3}$ of $\frac{1}{4}$ of the assumed number, and the result is 68. What is the number?"

of unknown quantities."[1] Surds are treated extensively, as in many medieval works on algebra, the difficulty of handling

[1] The initial Sanskrit syllables of the names of the colors are used.

irrationals of all kinds being particularly great in the era of poor symbolism. As with Aryabhaṭa and Brahmagupta, the "pulveriser" is given extensive treatment. Simple and quadratic equations receive more attention and are more clearly discussed than is the case with other Hindu writers. Besides numerous problems relating to geometric figures there are the usual poetic types, of which the following may serve as an illustration:

The son of Prĭt'há,[1] exasperated in combat, shot a quiver of arrows to slay Carńa. With half his arrows he parried those of his antagonist; with four times the square root of the quiverful he kílled his horse; with six arrows he slew Salya[2]; with three he demolished the umbrella, standard, and bow; and with one he cut off the head of the foe. How many were the arrows which Arjuna let fly?

Śrīdhara's Rule for the Quadratic. The rule used for solving the quadratic is given as Śrīdhara's.[3] The method of writing an equation has some evident advantages, the equation $18x^2 = 16x^2 + 9x + 18$ being written

ya v	18	ya o	ru o
ya v	16	ya 9	ru 18

which is then transformed into $2x^2 - 9x = 18$, thus:

ya v	2	ya $\dot{9}$	ru o
ya v	o	ya o	ru 18[4]

Bhāskara's Siddhānta Śiromani. A third work of importance written by Bhāskara is the *Siddhānta Śiromani* (*Head jewel of accuracy*)[5] in which, in the book *Goladhia* (*Theory of the*

[1] The son's name was Arjuna, mentioned later in the problem.

[2] The charioteer of Carńa. [3] See page 274.

[4] *Ya v* is for *yāvat-tāvat*, "as many of" (the unknown), and is used for the highest power. *Ya* is a color, green, and is used here for the first power. *Ru* is for *rūpa*, the known number. See Volume II and Colebrooke, *Bhāskara*, p. 139.

[5] Various notes on this work will be found in H. T. Colebrooke, *Miscellaneous Essays*, 2 vols., 2d ed., Vol. II (Madras, 1872). It should be stated that the *Siddhānta Śiromani* is thought by various scholars to include the other works mentioned, but this is merely a question of division of material.

Sphere), he treats of astronomy and asserts, as various ancient Greek philosophers have done, the sphericity of the earth.

The ancient inscription referred to on page 275 relates that Bhāskara's grandson, Chaṅgadwa, was chief astrologer to King Siṁghana, and that in his time a college was founded to expound the doctrines of Bhāskara.[1]

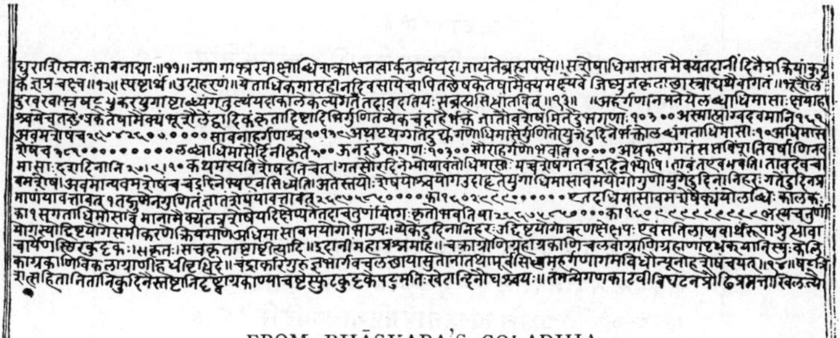

FROM BHĀSKARA'S GOLADHIA

From a manuscript of Bhāskara's work on astronomy, being the fourth and last chapter of the *Siddhānta Śiromani*. This work was written *c.* 1150. The reproduction is greatly reduced

In the forming of Pythagorean triangles Bhāskara follows Brahmagupta in stating the relations

$$\sqrt{m},\quad \frac{1}{2}\left(\frac{m}{n}-n\right),\quad \frac{1}{2}\left(\frac{m}{n}+n\right),$$

and adds the two further relations

$$m,\quad \frac{2mn}{n^2-1},\quad \frac{m(n^2+1)}{n^2-1}$$

and

$$\frac{m(n^2-1)}{n^2+1},\quad \frac{2mn}{n^2+1},\quad m.$$

Other Parts of South Asia. In Eastern Sumatra (the Chinese San-fo-ts'i) it is related by Chinese travelers of the 13th century that the people were able mathematicians and could calculate future eclipses of the sun and moon.[2] No doubt the

[1] G. R. Kaye, *Indian Mathematics*, p. 37.

[2] Hirth and Rockhill, *loc. cit.*, p. 64.

same could have been said at that time of other parts of Asia whose records have not come down to us, such calculations

॥ सिद्धान्तशिरोमणिः ॥ १४३

उदाहरणं ।

येयाताधिकमासहीनदिवसा येचापि तच्छेषके तेषा मैक्यमवेक्ष्य जिष्णुजकृताच्छास्त्राद्यथैवागतं । भूशैलें दुखखाभ्रपट्करयुगाष्टाब्ध्यङ्ग ।६४८४२६०००१७१ । तुल्यं यदा । काले कल्प गतं तदा वदति यः सब्रह्मसिद्धा न्तवित् ॥ १२ ॥

अहर्गणानयने ये लब्धा अधिमासाः क्षयाहाश्च येच तच्छेषके तेषामैक्यं भूशैलेंदुखखाभ्रपट्करयुगाष्टाब्ध्यङ्ग ।६४८४२६०००१७१ । तुल्यं कृताष्टाष्टिगोब्ध्यादिभि । ८६३ ३७४४९१६८४ । गुणितं जातं । ५५९८३४६८२९२३२६४२ २०७७९६४ । व्येकचन्द्राहै । १६०२९९८९९९९९९ । भक्तं लब्धं । ३४९२४१९३२३३६ । जातोवशेषमितोगतें दुद्युगणाः ।१०३००० । अस्मात्प्राग्वदवमानि । १६१ । अव मशेषञ्च । २६७४२६०००००० । सावनाहर्गणश्च । १०१३९ । अथ पृथग्गतेंदुद्युगणोधिमासै गुणितो युगेंदुदिनै भक्तो लब्धं गताधिमासाः । १० । अधिमासशेषञ्च । ३८१०००००००००० । लब्धाधिमासैः दिनीकृतै । ३०० । रूनइं दुद्युगणः । १०३०० । सौराहर्गणो भवति । १०००० । अतः कल्पगतं सप्तविंशति । २७ । वर्षाणि नव । ९ । मासाः दश । १० । दिनानि अस्योपपत्ति बीजगणितेन एको हरश्चेद्गुणकौ विभिन्नावित्यादिना कथमस्य विषय इति चेत् उच्यते गतसौरदिनेभ्यो यावन्तोधिमासाः यच्च शेषं गतचन्द्रदिनेभ्योपि तावन्त एव भवन्ति तावदेव चाव

FROM THE FIRST PRINTED EDITION OF BHĀSKARA'S GOLADHIA

This shows in print the same page shown in manuscript on page 281. The work was printed at Calcutta in 1842

having been a part of the general stock in trade of the astrologers in various countries for many centuries preceding this particular period.

4. Persia and Arabia

Decay of Bagdad. For two centuries after the golden age of the first three caliphs Bagdad continued to be a center of scientific activity, in spite of the fact that it began to lose political prestige after the death of al-Mâmûn (833). By the year 1000, however, the spiritual supremacy of the city had passed and the seats of learning of the Western Arabs had begun to take the place of the capital of the caliphs in Mesopotamia. The Seljuk Turks, an intolerant Tartar tribe, overran much of the territory formerly so well governed by the caliphs, captured the holy cities of Palestine, and by their ruthless behavior gave excuse for the crusades. In 1258 the Mongols took Bagdad, and thenceforth it was little more than a name. Brute force had put an end to the idealism that had been so noticeable in the eastern Mohammedan empire.

Al-Karkhî. Among the last of the real contributors to mathematics in the city of the caliphs was al-Karkhî,[1] who died *c.* 1029. His first work of note was an arithmetic, the *Kâfî fîl Hisâb*,[2] probably written between 1010 and 1016, and drawn largely if not exclusively from Hindu sources.[3] It not only contains the elements of arithmetic as set forth by many writers of the time, but gives the rule of quarter squares,

$$\left(\frac{a+b}{2}\right)^2 - \left(\frac{a-b}{2}\right)^2 = ab,$$

a rule probably due to the Hindus. It also gives such methods of multiplication as are expressed by the formulas

$$(10\,a + a)(10\,b + b) = [(10\,a + a)b + ab]\,10 + ab$$

and $$(10\,a + b)(10\,a + c) = (10\,a + b + c)a \cdot 10 + bc.$$

[1] Mohammed Abû Bekr ibn al-Ḥasan (or al-Ḥosein), al-Karkhî.

[2] *Book of Satisfactions.* A. Hochheim, *Kâfî fîl Hisâb des Abu Bekr Muhammed Ben Alhusein Alkarkhî*, 3 parts, Halle a. S., 1878–1880; hereafter referred to as Hochheim, *Kâfî fîl Hisâb.*

[3] H. Weissenborn, *Gerbert*, p. 196 seq. (Berlin, 1888). But see Cantor's earlier opinion that it came from Greek sources; *Geschichte*, I (1), 655.

In the approximations for roots al-Karkhî gives, among others,

$$\text{for} \quad m = a^2 + r, \quad \sqrt{m} = a + r/(2a + 1);$$
$$\text{for} \quad r \gtreqless a, \quad \sqrt{m} = a + r/2a.$$

He also considers the mensuration of plane figures, particularly as it involves surd numbers, and includes the Heron formula $\sqrt{s(s-a)(s-b)(s-c)}$.

The work closes with a treatment of algebra, including quadratic equations and the usual Arabic explanation of the terms *al-jabr* and *al-muqâbala*, discussed in Volume II.

Al-Karkhî's Fakhrî. Al-Karkhî is best known, however, for his algebra, the *Fakhrî*,[1] which includes the usual operations on algebraic quantities, roots, equations of the first and second degrees, indeterminate analysis, and the solution of problems. The quadratic equations include such forms as $x^4 + 5x^2 = 126$, and the solution of quadratics in general depends upon rules such as that represented by the equation

$$ax^2 + bx = c$$

and the formula
$$x = \left[\sqrt{\left(\frac{b}{2}\right)^2 + ac} - \frac{b}{2}\right] : a.$$

The rules are explained geometrically, as in the works of the earlier Arab writers. Various problems given by him are apparently suggested by al-Khowârizmî and Diophantus, and these include such cases as the finding of integral solutions for

$$x^3 + y^3 = z^2,$$
$$x^2 - y^2 = z^3,$$
$$x^2y^3 = z^2,$$

and
$$x^3 + 10x^2 = y^2,$$

and the finding of fractional solutions for

$$x^2 - y^3 = z^2$$

and
$$x^3 + y^2 = z^3.$$

The work ranks as the most scholarly algebra of the Arabs.

[1] F. Woepcke, *Extrait du Fakhrî . . . par . . . Alkarchî* (Paris, 1853). For an explanation of the term *Fakhrî* see Volume II, Chapter VI.

Minor Writers of the Eleventh Century. Of the minor writers of the 11th century the following deserve brief mention: Mohammed ibn al-Leiṯ[1] lived about 1000, was interested in the trisection problem, and wrote on the construction of regular polygons of seven and nine sides.

Ḥâmid ibn al-Khiḍr[2] wrote on the astrolabe and asserted that the equation $x^3 + y^3 = z^3$ cannot be solved.

Manṣûr ibn 'Alî[3] wrote on astronomical instruments, trigonometry, spherical sines, and Ptolemy's *Almagest*.

Al-Nasavî wrote on Hindu arithmetic and on the works of Archimedes.[4]

One of the most brilliant writers on the contemporary history of mathematics at the opening of the 11th century was Alberûnî.[5] He was one of the *munajjimin*, or astrologer-astronomers of the Arabs. He visited India and made a careful study of that country and of its work in mathematics and the other sciences. He summarized the debased state of knowledge of his day in the words, "What we have of sciences is nothing but the scanty remains of bygone better times." In later life he wrote his work on India, and to this we are indebted for the best summary of Hindu mathematics that the Middle Ages produced.

Avicenna. Among the contemporaries of Alberûnî there was the famous physician and philosopher known in Christian Europe as Avicen'na[6] (980–1037). He was born in Safar, near

[1] Mohammed ibn al-Leiṯ, Abû'l Jûd.

[2] Ḥamid ibn al-Khiḍr, Abû Maḥmûd, al-Khojendî, died *c.* 1000.

[3] Manṣûr ibn 'Alî ibn 'Irâq, Abû Naṣr, *c.* 1000.

[4] 'Alî ibn Aḥmed, Abû'l-Ḥasan, al-Nasavî, *c.* 1025. Woepcke in the *Journal Asiatique*, 1863, p. 496. He was born at Nasa, in Khorâsân. On his arithmetic see Suter in *Bibl. Math.*, VII (3), 113.

[5] Mohammed ibn Aḥmed, Abû'l Rîḥân (or Raiḥân), al-Bêrûnî. Born probably in Khwarezm, 973; died 1048. He may have been born at Byrun in the valley of the Indus, his name also appearing as al-Bîrûnî. On his life and works, see his *Athâr-el-Bâkiya, Chronology of Ancient Nations* (London, 1879), and his *India*; Boncompagni's *Bullettino*, II, 153; S. Günther, *Zeitschrift* (Hl. Abt.), XXI, 57; E. C. Sachau, *Zeitschrift d. deutsch. Morgenl. Gesellschaft*, XXIX.

[6] Al-Ḥosein ibn 'Abdallâh ibn al-Ḥosein (or Ḥasan) ibn 'Alî, Abû 'Alî, al-Sheich al-Ra'îs, ibn Sînâ. See K. Lokotsch, *Avicenna als Mathematiker* (Erfurt, 1912).

Kharmîtan, not far from Bokhârâ. He wrote on Aristotle, Euclid, astronomy, music, medicine, and arithmetic, his treatment of numbers being based upon Greek models.

Ibn al-Ṣalâḥ,[1] who died in 1153/54, was one of the later generation of the Persian scholars who made Bagdad so famous. Like so many mathematicians of the East he was also learned in philosophy and medicine,—the latter partly on account of the supposed connection between the healing art and astrology. He was born in Bagdad and finally went to Damascus and died there. He wrote on geometry, and manuscripts of his works, apparently fragmentary, are still extant.

Omar Khayyam. The 12th century saw less attention to mathematics in the ancient Arab seats of learning, and more attention to the science in Persia. Of those whose names added luster to Persian mathematics and letters the most prominent was the poet who is generally known to English writers as Omar Khayyam[2] (*c.* 1100). While he is known to the Western world chiefly as the author of the *Rubaiyat*,[3] he wrote on Euclid and on astronomy,[4] and contributed a noteworthy treatise on algebra.[5]

Minor Writers. A little later than Omar Khayyam another Persian, a native of Khorâsân, made for himself a great name. This writer was al-Râzî,[6] known as one of the leading philosophers, physicians, and mathematicians of the Persians. His contributions to mathematics were chiefly in the domain of geometry.

[1] Aḥmed ibn Mohammed ibn al-Surậ Nejm ed-dîn, Abû'l-Futûḥ.

[2] 'Omar ibn Ibrâhîm al-Khayyâmî, Giyâṯ ed-dîn, Abû'l-Fatḥ. He was born at Nîshâpûr (Nishapur, Nishabur) *c.* 1044 and died there in 1123/24.

[3] Largely through the remarkable but not very exact translation of Edward Fitzgerald, London, 1859.

[4] "Ah but my computations people say
Reduced the year to better reckoning."

[5] *L'Algèbre d'Omar Alkhayyâmî*, Arabic and French texts, by F. Woepcke, Paris, 1851. On his life, see various editions of the Rubaiyat, and also J. K. M. Shirazi, *Life of Omar al-Khayyám*, Edinburgh, 1905.

[6] Mohammed ibn 'Omar ibn al-Ḥosein, Abû 'Abdallâh, Faḥr ed-dîn al-Râzî, ibn al-Khatîb, 1149/50–1210.

Of the Arab scholars of the 12th century one of the best known was Kemâl ed-dîn ibn Yûnis, or ibn Manʿa,[1] who was born at Moṣul on the Tigris river. His works on the theory of numbers and conic sections were highly esteemed by his Arab contemporaries.

Contemporary with the last-named writer was Taʿâsîf,[2] a native of upper Egypt, a jurist, an engineer, and a mathematician. He showed his interest in the foundations of mathematics by writing upon Euclid's postulates.

The Mongol Scourge. If the Seljuk Turks had been intolerant in the 11th and 12th centuries, rendering difficult the leading of an intellectual life, the great Mongol scourge, led by Chinghiz Khan between 1206 and 1227, rendered such a life well-nigh impossible. His conquests and his son's included a considerable part of the civilized world from Northern China through Turkestan and Persia, and down to the banks of the Indus. The son, Oktai (died 1241), quite as brutal as his father, ravaged nearly half of Europe, and the result of the total conquest was the impoverishment of all the intellectual centers that were once the glory of central and western Asia. To be sure, two of the successors of these tyrants, Kublai Khan (1216–1294) and Timur, or Tamerlane (1336–1405), contributed to the better things of life, but in general the record of the Mongol invasions for two centuries is one of the blackest in all history.

Decay of Learning. In the 13th century only one Persian writer deserves special mention, and even he spent his closing years in Bagdad. This writer was Naṣîr ed-dîn,[3] a native of Ṭûs, in Khorâsân. He was an all-round scholar, writing upon trigonometry, astronomy, computation, geometry, and the construction and use of the astrolabe.

[1] Mûsâ ibn Yûnis ibn Mohammed ibn Manʿa, Abû'l-Fatḥ, Kemâl ed-dîn. Born at Moṣul, 1156; died at Moṣul, 1242.

[2] Qaiṣar ibn Abî'l-Qâsim ibn ʿAbdelġanî ibn Musâfir, ʿAlam ed-dîn, known under the name of Taʿâsîf. Born 1179 (possibly 1169); died 1251.

[3] Mohammed ibn Mohammed ibn al-Ḥasan, Abû Jaʿfar, Naṣîr ed-dîn al-Ṭûsî, 1201–1274.

Of the Arab writers, Ibn al-Yâsimîn,[1] who lived in Morocco, is known chiefly for the influence of a poem which he wrote on algebra, the *Arjûza*. Several manuscripts still exist, and it seems to have had some such influence in popularizing algebra as the *Carmen de Algorismo* (p. 226) had with respect to algorism. Ibn al-Lubûdî,[2] a native of Ḥaleb (Aleppo), was known in his century for works on arithmetic, algebra, and Euclid. The Arab interest in learning was rapidly waning, however, and only one other name deserves mention in the record of the 13th century,—that of al-Ṭûsî,[3] another prominent native of Tûs. He wrote on geometry and algebra and invented one form of astrolabe known as "Ṭûsî's staff." Islam had lost its hold upon mathematics; the mathematical world was becoming a hyperbola having one focus in China and the other in Christian Europe, with nothing between the branches.

The most notable of the Christian writers in the Near East at this time was Bar Hebræus,[4] whose father, a Jew named Aaron, had entered the Christian church. When the son was twenty years old (1246) he was made Jacobean bishop of Gubos, near Malatia, and later he occupied other positions of ecclesiastical importance. He wrote on astronomy and lectured on Euclid and Ptolemy.[5]

In the 14th century only three Mohammedans of any considerable prominence appear among the world's mathematicians, and no one of these was a genius. Two lived at least part of the time in Egypt, Ibn al-Hâ'im,[6] a writer on

[1] 'Abdallâh ibn Mohammed ibn Ḥajjâj, Abû Mohammed. He died *c.* 1203–1205.

[2] Yaḥyâ ibn Mohammed ibn 'Abdân ibn 'Abdelvâḥid, Abû Zakarîyâ Nejm ed-dîn, 1210/11–1267/68.

[3] Al-Moẓaffar ibn Mohammed ibn al-Mozaffar Sharaf ed-dîn al-Ṭûsî. He died *c.* 1213.

[4] That is, Son of the Jew. His Arabic name was Jûḥannâ Abû'l-Faraj Bar-Hebræus. Born at Malatia, in Eastern Asia Minor, 1226; died at Moṣul, July, 1286.

[5] F. Nau, in No. 121 of the *Bibliothèque de l'École des Hautes Études*. Paris, 1899.

[6] Aḥmed ibn Mohammed ibn 'Imâd, Abû'l-'Abbâs Shihâb ed-dîn. Born at Cairo, 1352 or 1355; died at Jerusalem, 1412.

arithmetic, and Ibn al-Mejdî,[1] who wrote on astronomy, trigonometry, arithmetic, the calendar, and mathematical tables. A third Arab writer of the period, commonly known as Ibn al-Shâṭir,[2] left works on trigonometry, the astrolabe, and astronomy, and prepared a few mathematical tables.

Ulugh Beg, the Royal Astronomer. Of the representatives of the Arab-Persian interest in mathematics who lived in the 15th century the only one who seems to have possessed any genius was Ulugh Beg[3] (1393–1449), and even this genius was rather perseverance than any unusual endowment of intellect. He was a Persian prince, born at Sultanieh, and his interest in astronomy and astronomical tables was shown in the observatory which he founded at Samarkand. The tables which were worked out under his direction were highly esteemed in Europe as well as in the East.[4] His assistant, al-Kashî,[5] wrote a short treatise in Persian on arithmetic and geometry.[6]

Summary of Arab Achievements. With these names the achievements of the Mohammedan writers practically close. As we sum up these achievements we are struck by the interest of the Arabs in science but by their lack of originality. They received their astronomy first from the Hindus and then from the Greeks, their geometry solely and their algebra chiefly from the Greeks, and their trigonometry largely from the Hindus in connection with astronomy. As already stated (p. 177), they originated nothing of importance either in arithmetic or in geometry, they systematized algebra to some extent,

[1] Aḥmed ibn Rajeb ibn Tîboġâ, Shihâb ed-dîn Abû'l-'Abbâs. Born 1359; died in Egypt, 1447.

[2] 'Alî ibn Ibrâhîm ibn Mohammed al-Moṭ'im al-Anṣârî, Abû'l-Haṣan. Born 1304; died 1375/76 or 1379/80.

[3] Ulûġ Beg.

[4] See L. P. E. A. Sédillot, *Prolégomènes des Tables Astronomiques d'Ouloug Beg*, Paris, 1847; T. Hyde, *Tabulae Longitudinis et Latitudinis Stellarum Fixarum ex Observatione Ulugbeighi*, Oxford, 1665; E. B. Knobel, *Ulugh Beg's Catalogue of Stars*, Washington, 1917.

[5] Jemshîd ibn Mes'ûd ibn Maḥmud, Ġiyât̲ ed-dîn al-Kâshî. He is also known as Kazi Zadeh al Rumi and Ali Kushi. Died *c.* 1436.

[6] H. Hankel, *Geschichte der Mathematik*, 289 (Leipzig, 1874) (hereafter referred to as Hankel, *Geschichte*); Taylor, *Lilawati*, Introd., p. 14. The introduction to his *Miftâh al-ḥisâb* (*Key of arithmetic*) was translated by Woepcke.

they improved upon the astronomy of their predecessors, and they made some real contributions to trigonometry. All these matters will be discussed in the appropriate chapters. But on the whole the Arabs of this period were still transmitters of learning rather than creators, and to them Europe is chiefly indebted for preserving in their translations many of the important works of the Greeks. To this rather sweeping assertion, however, one noteworthy exception may justly be taken, for it seems quite certain that it is to an Arab, or rather to a Turkish, scholar that we owe the first actual use of a decimal fraction. This step was taken independently at a later period by European arithmeticians, but the decimal fraction seems certainly to have been in use in Samarkand early in the 15th century. In a work by al-Kashî, or Jemshîd (p. 289), the ratio of the circumference to the radius of a circle is given, in part, as follows:

Integer
6 28318 . . . ,

the full result being correct to sixteen decimal places.

Justice also requires that the Arabs of the four centuries beginning with the year 800 should be judged not with respect to the great achievements of the golden age of Greece but rather in comparison with the very meager results secured by their contemporaries in Europe and the Far East. If we consider Europe during the same period, we shall find the names of few original scholars in the domain of mathematics. The number of Arab, Persian, and Turkish scholars in this period exceeds, so far as we yet know, that of their European contemporaries, and their achievements were more significant. It was only during the period from 1200 to 1400 that European mathematics forged ahead, and even then the Arab influence was one of the prominent moving causes.

Justice further requires the admission that for lucidity of statement the scholars of Bagdad surpassed their contemporaries both in the East and in the West during a period of about six centuries, and were quite their equals in originality.

TOPICS FOR DISCUSSION

1. A comparison of the general nature of Chinese mathematics in the Middle Ages with that of the mathematics of Europe.

2. Evidences of the interchange of thought between the East and the West in the Middle Ages, and the possible effect of this interchange upon the mathematics of Europe and Asia.

3. Influence of astrology upon astronomy and upon pure mathematics, both in the East and in the West.

4. Nature of astronomy in the Middle Ages in the East and the border line between this science and astrology.

5. Nature of the mathematics of China in the period of its greatest development, the 13th century.

6. The reliability of Chinese texts in the Middle Ages, based upon a general study of Chinese literature.

7. The justice of the claims of China in the field of numerical higher equations.

8. The nature of the problems that interested Chinese scholars in the Middle Ages.

9. Early history of mathematics in Japan. Reasons for the failure of the science to advance.

10. The nature of the mathematics of India in the Middle Ages. A comparison of this mathematics with that of China.

11. The nature of the problems that interested Hindu scholars in the Middle Ages.

12. The works of Bhāskara; their nature and influence.

13. Causes of the decay of mathematics in the Mohammedan countries, beginning with the 11th century.

14. Nature of algebra as developed by Mohammedan writers after the Golden Age of Bagdad.

15. The contributions to mathematics made by the Persian poet Omar Khayyam.

16. The life and works of the prince-astronomer Ulugh Beg and the reasons why his influence was not more powerful in Mohammedan lands.

17. A summary of the contributions of the Arabs to the science of mathematics.

18. General position of Hebrew mathematics in the Near East in the Middle Ages.

CHAPTER VIII

THE SIXTEENTH CENTURY

1. General Conditions

The Sixteenth Century in General. Until about the year 1500 the mathematics of the world was, so far as any records tell us, limited to a small number of individuals in each century. Printing having only just been invented, there was no simple way for comparatively obscure workers, even in the 15th century, to make their contributions or their interest known, and so they left no record of their achievements. In the 16th century, however, the printed page began to perpetuate names, and it now becomes impossible to do more than select a few out of the many for such comment as space may allow.[1]

Science and Letters. Furthermore, historical events now began to be recorded more freely, the world moved more rapidly, and the influences that bear upon the development of mathematics become more difficult to trace. That the opening of a new world would greatly increase the interest in commercial mathematics is evident, and the fact is abundantly proved by the printed books of the 16th century; but the influence of the great literary movement illustrated by such writers as Shakespeare, Cervantes, and Camoens, or of such world events as the defeat of the Spanish Armada, is not so apparent in the scientific field. Indeed, we may say that it was the influence of such scientists as Leonardo da Vinci, Copernicus, Palissy, and Tycho Brahe that stimulated the literary renaissance of the period, rather than the reverse.

[1] The reader who has access to the seventh edition of the *Encyclopædia Britannica* will find that the "Dissertation Third," by John Playfair, on "The Progress of Mathematical and Physical Science since the Revival of Letters in Europe" forms a very good background for his studies of this period.

Mathematical Conditions. The conditions in the field of mathematics were such as to mark out the course of progress. Euclid's *Elements* had appeared in print in 1482 and the *Conics* of Apollonius was known in manuscript,[1] and hence the most promising field in pure mathematics was in the domain of analysis. The quadratic equation had been fully solved, and so the next step was to attempt the solution of the cubic equation, with which the Greeks and Arabs had been successful only in special cases and by having recourse to the intersection of conics. It was here, then, that mathematics would naturally be expected to advance. Along with this there would be expected to develop a better symbolism, and one that was also suited to the needs of typography. After the cubic was solved, the mathematical world would be expected to try the equations of the fourth and higher degrees. We should anticipate, for geographic reasons, that the lead would be taken in Italy, and we should also expect that the demands of astronomy and navigation would require a more rapid development of trigonometry. All these suggested expectations, as we shall see, were fulfilled. Indeed, in the space of a single century mathematics made more advance than had been achieved since the days when the Alexandrian School dominated the scholastic world.

MATHEMATICS IN THE 16TH CENTURY

Concept of the range of the science. From the title-page of Coignet's *Arithmetica*, Antwerp, 1580

[1] It was first printed in Venice in 1537. The Commandinus edition appeared at Bologna in 1566, and the Halley edition at Oxford in 1710.

2. Italy

Leonardo da Vinci. No list of the Italian mathematicians of the 16th century would be complete without some mention of Leonardo da Vinci[1] (1452–1519), and such were the remarkable attainments of this gifted man that his name may properly be given in its chronological order, thus standing first on the record. He was born at Vinci, near Florence, resided in Florence, Milan, and Rome, went to France at the invitation of the king in the year 1516, and died near Amboise in 1519. Famous as a painter, sculptor, goldsmith, investigator of the circulation of the blood, general scientist, architect, and writer on mechanics, optics, and perspective, he would have ranked as a worthy mathematician had not his talents in this direction been obscured by his unusual gifts in these other lines. In applied mathematics he may be looked upon as one of the founders of the modern theory of optics. In geometry he distinguished between curves of single and double curvature, gave much attention to the subject of stellar polygons, was interested in constructions with a single opening of the compasses, and gave various correct or approximate constructions of regular polygons. In physics he knew the theory of the inclined plane, found the center of gravity of a pyramid, worked in the field of capillarity and diffraction, knew the camera obscura without a lens, and studied the resistance of the air and the effect of friction. The world has rarely produced such an all-round genius.

Early Workers in the Field of Equations. Since the solution of the equations of the third and fourth degrees was the chief mathematical achievement in Italian mathematics of the 16th century, it is proper to group together those scholars who were most prominent in this work. Although the details of their chief contributions will be reserved for Volume II, a brief statement of their achievements will now be given.

[1] P. Duhem, *Études sur Léonard de Vinci*, 2 vols., Paris, 1906–1913. For a popular sketch see D. Merejkowski, *The Romance of Leonardo da Vinci*, New York [1902].

Scipione del Ferro,[1] a native of Bologna, whom Cardan calls by his Latin name of Scipio Ferreus, was professor of mathematics in the city of his birth. In geometry he was interested in constructions depending on a single opening of the compasses. In algebra he found a method of solving the cubic equation for the special case of $x^3 + ax = b$.

In 1506 he revealed this method to his pupil Antonio Maria Fior,[2] a Venetian, who proceeded to turn the information to account in the popular mathematical contests that were then in vogue. Of the life of Fior little is known, but he is said by Tartaglia[3] to have been living in 1536.

Zuanne de Tonini da Coi[4] (*c.* 1530) was a teacher in Brescia and was interested in mathematics from the standpoint of problem solving.[5] In 1530 he sent as a kind of challenge to Tartaglia the two equations

$$x^3 + 3x^2 = 5$$

and

$$x^3 + 6x^2 + 8x = 1000.$$

For some time Tartaglia was unable to solve them, but, as we shall see in Volume II, he finally succeeded in doing so, this being an important step in the general problem of the cubic equation.

Cardan. The first of the two prime movers in the solution of the cubic was Giro'lamo Carda'no.[6] He was the illegitimate

[1] Born *c.* 1465; died at Bologna between October 29 and November 16, 1526. L. Frati, *Bollett. di bibliogr. d. sci. matem.*, XII, 1. The name also appears as Ferri and as Ferreo.

[2] This on the statement of Tartaglia (1546). Cardan (1545) says it was about 30 years earlier than the time of his writing, which would make it *c.* 1514 or 1515. The name appears with various spellings, particularly in the Latin form of Antonius Maria Floridus and in the form of Antoniomaria Fior.

[3] Pronounced tar tä'lya. See page 297.

[4] Also Zuane, Giovanni, Giovanno, John, with his last name sometimes given as Colle or in the Latin form of Colla.

[5] In a letter written in 1540, Tartaglia speaks of "that devil" having returned: "Eglie ritornato qui quel diauolo de Messer Zuanne Colle."

[6] Born at Pavia, 1501; died at Rome, September 21, 1576. The name appears as Hieronymus Cardanus and Jerome Cardan. He is commonly called Cardan by writers of English. H. Morley, *The Life of Girolamo Cardano*, 2 vols., London, 1854 (hereafter referred to as Morley, *Cardan*); Cantor, *Geschichte*, II, chaps. 64, 65, 66; V. Mantovani, *Vita di Girolamo Cardano*, Milan, 1821; Gherardi, *Facoltà Mat. di Bologna*, 47 seq.

son of a jurist, Facio Cardano (1444–1524), who was professor of jurisprudence and medicine in Milan and who edited Peckham's *Perspectiva communis.* Girolamo was a man of remarkable contrasts. He was an astrologer and yet a serious student of philosophy, a gambler and yet a first-class algebraist, a physicist of accurate habits of observation and yet a man whose statements were extremely unreliable, a physician and yet the father and defender of a murderer, at one time a professor in the University of Bologna and at another time an inmate of an almshouse, a victim of blind superstition and yet the rector of the College of Physicians at Milan, a heretic who ventured to publish the horoscope of Christ and yet a recipient of a pension from the Pope, always a man of extremes, always a man of genius, always a man devoid of principle. A certain bitter rival said of Voltaire, "Ce coquin-là has one vice worse than all the rest; he sometimes has virtues." So it was with Cardan.

PORTRAIT OF CARDAN
From the title-page of the first edition (1539) of his arithmetic

His *Ars Magna*,[1] the first great Latin treatise devoted solely to algebra, appeared at Nürnberg in 1545 and set forth the theory of algebraic equations so far as it was then known, including the solution of the cubic, which he seems to have secured from Tartaglia under pledge of secrecy and then dishonorably to have published, and the solution of the biquadratic which had been discovered by his pupil Ferrari. He also wrote on arithmetic,[2] astronomy,[3] physics,[4] and various other branches of knowledge,[5] proving himself a man of remarkable versatility and learning.

Tartaglia. Nicolo Tartaglia,[6] one of the greatest mathematicians of Italy in the 16th century, was born at Brescia. Although known as Tartaglia, we learn from his will[7] that his brother's name was Fontana. It is said that he was present as a child at the taking of Brescia by Gaston de Foix (1512) and at that time received a saber cut in the face which caused an imperfection in his speech. This gave him the nickname of Tartaglia ("the stammerer"), which name he formally used in his published works. He was self-educated but acquired such proficiency in mathematics that he earned a livelihood by teaching the science in Verona, Vicenza, Brescia, and Venice (1535).

Tartaglia seems to have substantially completed the solution of the cubic equation and, as already stated, to have imparted the secret to Cardan, who, in violation of his oath, published it in 1545.[8]

1 *Artis Magnae, sive de regvlis algebraicis, liber vnvs*, Nürnberg, 1545; Basel, 1570.

2 *Practica arithmetice*, Milan, 1539.

3 *De revolutione annorum, mensium et dierum . . . liber*, Nürnberg, 1547; *De temporum et motuum erraticarum restitutione*, Nürnberg, 1547; *Aphorismorum astronomicorum segmenta septem*, Nürnberg, 1547.

4 *De subtilitate*, Nürnberg, 1550, with a Paris reprint in 1551.

5 His *Opera* appeared in ten volumes, Lyons, 1663.

6 Born *c.* 1506; died at Venice, December 13/14, 1557. The name is also spelled Tartalea. It is spelled in the text as it appears on the title-page of his work of 1556.

7 Published by Boncompagni in 1881.

8 On his possible indebtedness to Ferro see G. Eneström, in *Bibl. Math.*, VII (3), 38.

Tartaglia was the first to apply mathematics to artillery science,[1] a subject just being perfected by the great French masters Galiot de Genouillac and Jean d'Estrées. He also wrote the best treatise on arithmetic[2] that appeared in Italy in his century, containing a very full discussion of the numerical operations and the commercial rules of the Italian arithmeticians. The life of the people, the customs of merchants, and the efforts at improving arithmetic in the 16th century are all set forth in this remarkable work. Tartaglia also published (1543) editions of Euclid and Archimedes.

NICOLO TARTAGLIA

From the title-page of *La Prima Parte del General Trattato*, Venice, 1556

On the question of the publication by Cardan of the Tartaglia solution of the cubic after a pledge of secrecy, a biographer of the former, but one who could not feel the influences of the 16th century, has this to say:

The attempt to assert exclusive right to the secret possession of a piece of information, which was the next step in the advancement of a liberal science, the refusal to add it, inscribed with his own name,

[1] *Nuova scienza, cioè Invenzione nuovamente trovata, utile per ciascuno speculativo matematico bombardiero . . .*, Venice, 1537; *Qvesiti ed invenzioni diverse*, Venice, 1546.

[2] *General Trattato di nvmeri, et misvre*, 2 parts (volumes), Venice, 1556–1560; *Tutte l'opere d'arithmetica del famosissimo Nicolò Tartaglia*, Venice, 1592, being substantially Volume I of the *General Trattato*.

to the common heap, until he had hoarded it, in hope of some day, when he was at leisure, of turning it more largely to his own advantage, could be excused in him only by the fact that he was rudely bred and self-taught, and that he was not likely to know better.

TARTAGLIA APPLIES MATHEMATICS TO ARTILLERY SCIENCE
From *Il Primo Libro delli Qvesiti, et Inventioni diverse*, Venice, 1546; 1562 ed.

Any member of a liberal profession who is miserly of knowledge, forfeits the respect of his fraternity. The promise of secrecy which Cardan had no right to make, Tartalea had no right to demand.[1]

As already remarked, however, it is difficult to imagine conditions in the year 1545, and it is hardly just to apply modern ethics to a situation so different from our own.

[1] Morley, *Cardan*, I, 270; Rixner and Siber, *Leben und Lehrmeinungen berühmter Physiker am Ende des XVI. und am Anfange des XVII. Jahrhunderts*, Sulbach, 1820; Firmiani, *Girolamo Cardano*, Naples, 1904. There is a brief but good biography of Tartaglia in D. Martines, *Origine aritmet.*, p. 61 n. See also A. Favaro, "Per la biografia di Niccolò Tartaglia," *Archivio storico Italiano*, 1913, and "Di Niccolò Tartaglia," in *Isis*, I, 329.

Ferrari. Lodovico Ferrari,[1] born in humble circumstances, was taken into Cardan's household in Milan at the age of fifteen. Cardan soon recognized his remarkable ability and made him his secretary. In spite of his ungovernable temper and his blasphemous habits he was later accepted by Cardan as his pupil and friend. Mathematical Italy would have given much to be in his place in Cardan's household; but, such were their quarrels, it would have given more to be out again. At the age of eighteen even Ferrari was glad to sever all relations with his patron and to begin teaching by himself in Milan. He was so successful there and in the mathematical contests of the day as to attract the attention of the court and of the Cardinal of Mantua. Through the favor of the latter he secured a position that brought him abundant means. He then became professor of mathematics at Bologna, but died there in the first year of his service, at the age of thirty-eight, probably poisoned by his only sister.

Zuanne de Tonini da Coi had proposed a problem which involved the equation

$$x^4 + 6x^2 + 36 = 60x.$$

This problem Cardan attempted to solve, and having failed he gave it to Ferrari. The latter succeeded in finding a method, and thus the solution of the equation of the fourth degree was discovered. Ferrari left no written works on mathematics, but Cardan published in the *Ars Magna* (1545) this noteworthy contribution to the theory of equations.

Bombelli. Rafael Bombelli (born *c.* 1530), a native of Bologna, was the last of those Italian mathematicians of the 16th century who contributed in any noteworthy way to the solution of the cubic and biquadratic equations. He wrote *L'Algebra parte maggiore dell' arimetica divisa in tre libri* and

[1] Born at Bologna, 1522; died *c.* 1560. The date of his death is also given as 1562 and 1565. Morley gives 1560. The Christian name is also written Luigi. Cardan left an unpublished *Vita Ludovici Ferrarii Bononiensis.* See also G. de' Sallusti, *Storia dell' Origine e de' Progressî delle Matematiche,* I, 38 (Rome, 1846).

published the work in Bologna in 1572.[1] In this work Bombelli set forth the reality of the three roots of a cubic equation in the case in which the cube root of an imaginary expression is involved in the result secured by the Tartaglia-Cardan rule. The book contained the most teachable and the most systematic treatment of algebra that had appeared in Italy up to that time.

Of Bombelli's life almost nothing is known. He is thought, from the introduction to his algebra, to have been an engineer in the service of the patron to whom he dedicates his book, Alessandro Rufini, bishop of Melfi.

FRANCESCO MAUROLICO
Engraved after a portrait from life

Francesco Maurolico. Of the mathematicians of this period who were interested in the Greek writers the most prominent was Frances'co Mauroli'co,[2] a native of Messina, Sicily, but of Greek parentage. He was a priest, at one time an abbot, and for some years professor of mathematics at Messina. He

[1] A second edition appeared at Bologna in 1579 with another title, *L'Algebra Opera*, with the dedicatory letter reset, but otherwise using the same sheets as the 1572 edition. The name is spelled as above in both editions of this work.

[2] Born at Messina, Sicily, September 16, 1494; died at Messina, July 21, 1575. Latin, Franciscus Maurolycus or Maurolykus; also known as Marullo. D. Scina, *Elogio di Francesco Maurolico* (Palermo, 1808); Martines, *Origine aritmet.*, 65. A life of Maurolico, written by his nephew, was published at Messina in 1613.

translated into Latin the works of Theodosius and Menelaus, the treatise of Autolycus on the sphere,[1] and the *Phaenomena* of Euclid, and published works on Apollonius[2] and Archimedes.[3] He also wrote various general works on mathematics and arithmetic,[4] wrote on mathematical induction,[5] and was a man of some creative power. A few of his more prominent contemporaries are listed below, with a brief statement of their contributions.

Italian Geometers. Federigo Commandino of Urbino (1509–1575) is known as one of the leading translators and editors of the Greek classics in mathematics. His editions of Euclid, Archimedes, Apollonius, Aristarchus, Heron, Ptolemy, and Pappus are highly esteemed.

Frances'co Baroz'zi[6] (*c.* 1538–*post* 1587), a Venetian nobleman, edited the commentary of Proclus on the first book of Euclid.[7] He also wrote on cosmography and geometry[8] and translated Heron's works.

Giambattista Benedetti[9] (1530–1590), a Venetian by birth, wrote on the geometry of a single opening of the compasses,[10] on the gnomon, on optics, and on the theory of numbers, and gave excellent graphic treatments of various problems.[11]

[1] Published at Messina, 1558. [2] Messina, 1654.

[3] Palermo, 1670 (the edition being lost in a shipwreck) and 1685.

[4] *Opuscula Mathematica*, 2 vols., Venice, 1575, although written in 1553. The *Arithmeticorum libri duo* (Venice, 1575, but written in 1557) was the second volume of the *Opuscula* and was republished in 1580. See also the *Elogia* above mentioned, p. 114.

[5] G. Vacca, "Maurolycus, the first discoverer of the principle of mathematical induction," *Bulletin of the Am. Math. Soc.*, XVI, 70; W. H. Bussey, "Origin of Mathematical Induction," *Amer. Math. Month.*, XXIV, 199.

[6] Franciscus Barocius, Francesco Barocci.

[7] *Procli Diadochi . . . in primum Euclidis . . . librum comment.*, Padua, 1560.

[8] *Geometricum problema tredecim modis demonstratum*, Venice, 1586.

[9] Giovanni Battista Benedetti, Joannes Baptista Benedictus.

[10] *De resolutione omnium Euclidis problematum aliorumque . . .*, Venice, 1553. On the general history of this important subject see W. M. Kutta, "Zur Geschichte der Geometrie mit constanter Zirkelöffnung," *Nova Acta . . . Abh. der K. Leop.-Carol. Deutschen Akad. der Naturforscher*, Halle, LXXI, 71, and *Bibl. Math.*, X (2), 16.

[11] *Diversarvm Specvlationvm Mathematicarum, & Physicarum Liber*, Turin, 1580. See *Rara Arithmetica*, p. 364.

Cosimo Bar'toli (1503–1572), a Florentine geometer, translated into Italian the works of the French mathematician Oronce Fine[1] and wrote a popular work on mensuration.[2] Among the features of the book is a table of squares to 662^2.

Pietro Antonio Catal'di[3] (1548–1626) was a native of Bologna and spent the closing years of his life there. He was professor of mathematics and astronomy at Florence (1563), Perugia (1572), and Bologna (1584). He wrote several mathematical works and to him are due the first steps in the theory of continued fractions, although not the first idea of these forms. His *Prima Parte della Pratica Aritmetica* (Bologna, 1602) and *Trattato dei numeri perfetti* (Bologna, 1603) were printed under the pseudonym of Perito Annotio, formed by transposing the letters in his given names. The second part of the *Pratica* appeared under his own name in 1606, and similarly for the third and fourth parts (1617, 1616). He also edited the first six books of Euclid's *Elements*,[4] wrote a brief treatise on algebra,[5] and contributed to the theory of roots (1613), the quadrature of the circle (1612), and various other subjects.

Matteo Ricci. Among the contemporaries of Cataldi the one who did most for the spread of mathematics in remote lands was Matteo Ricci, a man of remarkable energy and of great influence through his work in China.[6] He entered the Jesuit order in 1571, left Rome for China in 1577, and reached

[1] *Opere di Orontio Fineo del Delfinato; Diuise in cinque Parti: Arimetica, Geometria, Cosmografia, & Oriuoli, Tradotte Da Cosimo Bartoli, Gentilhuome, & Academico Fiorentino*, Venice, 1587 (posthumous). See *infra*, page 308.

[2] *Del Modo di Misvrare le distantie, le superficie, i corpi, le piante, . . .* Venice, 1564, with possibly an earlier edition.

[3] Although the spelling Cattaldi is given on the title-page of his *Dve Lettioni di Pietr' Antonio Cattaldi* (Bologna, 1577), the name is generally given in his other works as Cataldi. The above-named book is curious because the printer, not having fraction forms, was obliged to insert all fractions by hand. The name also appears as Cataldo.

[4] Bologna, 1620.

[5] *Regola della Quantità, o Cosa di Cosa*, Bologna, 1618.

[6] Born at Macerata, Ancona, October 6, 1552; died at Peking, May 8 (or 11), 1610. His Chinese name was Li-ma-to, derived from Ri (Chinese Li, for Ricci) and Matteo. H. Bosmans, *Revue des Quest. scient.*, January, 1921.

Canton in 1578.[1] Here he did more than any of his predecessors to make known in that country the mathematics and astronomy of the West. With the help of native scholars, the most prominent of whom were two learned mandarins, Hsü Kuang-ching (1562–1634) and Li Chi Ts'ao (died 1631), he translated (1603–1607) into Chinese the first six books of Euclid's *Elements*.[2] He also wrote an arithmetic,[3] which he dedicated to his assistant, Li Chi Ts'ao, and compiled various astronomical works.[4]

Minor Writers. In the second half of the century Silvio Belli (died 1575) wrote on practical geometry[5] and on the theory of proportion.[6] His geometry was very popular, six editions appearing in the 16th century.

Another writer whose work attracted considerable attention in the 16th and 17th centuries was Petrus Bongus,[7] to take the Latin form of his name as it appears in the first edition of his work. He was a native of Bergamo and became canon of the cathedral in that city. His work of nearly 500 pages on the mystery of numbers[8] went through several editions. It contains a mass of information upon such subjects as the religious significance of three, seven, and other numbers.

The Italian Arithmeticians. In all the leading countries many arithmetics were printed in the 16th century,[9] but for our present purposes it suffices to mention only a few of the more prominent Italian writers.

Girolamo and Giannantonio Tagliente,[10] Venetian arithmeticians of *c.* 1500, wrote a work on commercial arithmetic[11] which

[1] Pietro Tacchi Venturi, S. J., *L'Apostolato del P. Matteo Ricci*, 2d ed., Rome, 1910. [2] *Ki-ho-yüan-pen*. [3] *Tung-wen-suan-ki*.

[4] P. F. S. Vella, "Del P. Matteo Ricci," *Memorie, Pontificia Accad. dei Nuovi Lincei*, Rome, XXVIII, 51.

[5] *Libro del misurar con la vista*, Venice, 1565, with later editions in 1566, 1569, 1570, 1573, and 1595.

[6] *Della proportione, et proportionalita*, Venice, 1573.

[7] Died at Bergamo, September 24, 1601.

[8] *Mysticae Nvmerorvm significationis liber in dvas divisvs partes*, Bergamo, 1583–1584.

[9] For the list, consult *Rara Arithmetica*. [10] Pronounced tä lyen'tā.

[11] *Opera che insegna A fare ogni Ragione de Mercãtia*.

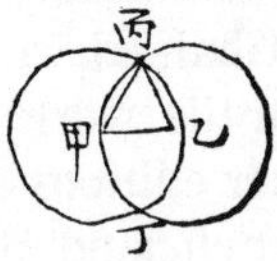

以乙為心。則乙甲線。與乙丙乙丁線亦等。何者。凡為圜

自心至界各線俱等。故。界說十五既乙丙等于乙甲。

而甲丙亦等于甲乙。即甲丙亦等于乙丙。一公論

三邊等。如所求。凡論有二種。此以是為論者。正論也。下倣此。

其用法不必作兩圜。但以甲為心。乙為界作

近丙一短界線。乙為心。甲為界亦如之。兩短

界線交處即得丙。

諸三角形。俱推前用法作之。詳本篇廿二

第二題

一直線線或內或外有一點。求以點為界作直線與元線

FROM RICCI'S TRANSLATION OF EUCLID, 1603–1607

From a manuscript copy of this translation, made in the 17th century. This page shows the first proposition of Book I

appeared in Venice in 1515 and was so popular that it went through more than thirty editions in the 16th century.

Francesco Ghaligai,[1] a Florentine arithmetician, published in 1521 a mercantile work entitled *Summa De Arithmetica*,[2] of which two other editions appeared later.

Francesco Feliciano da Lazesio[3] published an elementary work on arithmetic, algebra, and practical geometry at Venice in 1517/18.[4] This work went through at least fourteen editions (including the revision of 1526) in the 16th century and several in the century following.

Giovanni Sfortunati[5] published at Venice in 1534 a work on commercial arithmetic[6] which was well received and went through several editions.

The first mercantile tables that had great popularity were published in 1535 by a Venetian arithmetician, Giovanni Mariani,[7] under the title *Tariffa perpetva*. The work was often reprinted.

3. France

Centers of Activity. The centers of mathematical activity in France in the 16th century were Paris and Lyons, the former because of its ancient importance in all matters intellectual and the latter because of its commercial supremacy and the desire to cultivate some of the idealism of the northern capital. The theoretical books appeared more frequently from the Paris presses, while the output of what is known as practical mathematics was fairly large among those whom a recent writer has called "the morose and inhospitable Lyonese, . . . in whose

[1] Pronounced gä lē gä'ē. He died February 10, 1536.

[2] The edition of 1548 had the title *Practica d'Arithmetica*.

[3] Born at Lazisa, near Verona, *c.* 1490; was living in 1536. The family name is pronounced fa lē'chē ä no. Lazesio also appears as Lazisio.

[4] *Libro de Abaco*. There was a second work (Venice, 1526), called the *Libro di Arithmetica & Geometria . . . Intitulato Scala grimaldelli*, but this was only a revision of the work above mentioned.

[5] Pronounced sfor tōō nä'tē. He is also known by the Latin form Johannes Infortunatus. Born at Siena, *c.* 1500.

[6] *Nvovo Lvme Libro di Arithmetica*.

[7] Born at Venice, *c.* 1500. The name also appears as Zuane Mariani.

esteem the pick of humanity is the prosperous silk merchant." Paris was not yet a metropolitan city in the ecclesiastical sense, the ancient Roman political divisions having been retained by the Church, and the capital city being then and until 1622 subordinate to the metropolitan city of Sens. All this had no effect, however, upon her intellectual and political supremacy over all France.

Theoretical Works. The first of the French writers who sought to maintain the standing of the Greek mathematics in the intellectual atmosphere of France was Jacques le Fèvre d'Estaples,[1] known in his Latin works as Jacobus Faber Stapulensis. He was a "Doctor Sorbonnicus," a priest, vicar of the bishop of Meaux, lecturer on philosophy at the Collège Lemoine in Paris, and tutor to the son of François I. He wrote an introduction to the arithmetic of Boethius[2] and a work on geometry, edited (1499) Sacrobosco's *Sphere* and a description of the number game of Rithmomachia, and published various other works. His own writings were heavy and theoretical, and expressed the dying body of medieval mathematics.[3]

Charles de Bouelles,[4] canon and professor of theology at Noyon, wrote on geometry[5] and the theory of numbers.[6] The latter work includes a book on perfect numbers. He is particularly worthy of attention, however, because of his work

[1] Born at Estaples, near Amiens, *c.* 1455; died at Nérac, *c.* 1536. The French forms also appear as Febvre and Étaples.

[2] *Introductio Jacobi fabri Stapulēsis, in Arithmecam Diui Seuerini Boetij*, published in a volume with other works, Paris, 1503, the above title being from the edition of *c.* 1507. In these two editions his geometry and perspective also appear. The first edition of his compendium appeared in 1488. For details, see *Rara Arithmetica*, pp. 27, 30, 80, 82.

[3] Baldi's estimate of his merits is exaggerated: "D'ingegno felicissimo attese con gran frutto ad ogni sorte di dottrina, e giunse all' eccellenza di maniera che fu giudicato meraviglia del suo secolo." *Cronica*, 107.

[4] Born at Saucourt, Picardy, *c.* 1470; died at Noyon, *c.* 1553. The name appears also as Charles Bouvelles, Boüelles, Bouilles, and in the Latin form of Carolus Bovillus.

[5] *Geometricae introductionis libri VI*, Paris, 1503; *Livre singulier et utile, touchant l'art et pratique de Géométrie*, Paris, 1511, with several later editions.

[6] *Liber de duodecim numeris*, part of his general work published at Paris in 1509/10.

on the cycloid, he being one of the first to consider this figure from the scientific standpoint. He also wrote on regular convex and stellar polygons.

Minor Writers. One of the most pretentious of the mathematicians of his time, and one of the least worthy, was Oronce Fine,[1] more commonly known by the Latin form of his name, Orontius Fineus.[2] In his young manhood (1518–1524) he was imprisoned on account of his opposition to the Concordat, an agreement between France and the Pope. Upon being released he devoted himself to teaching, and about 1532 became professor of mathematics in the newly founded institution which was later known as the Collège de France.[3] He wrote extensively on astronomy and produced several works on arithmetic and geometry,[4] including one on the quadrature of the circle.[5] Some of his works were translated into Italian by Cosimo Bartoli. While he enjoyed some reputation, he died in poverty and his works were soon forgotten.

Among several French physicians of the 16th century who devoted much attention to mathematics the only one of great distinction is Jean Fernel[6] (1497–1558). He received his degree in medicine at Paris in 1530, and four years later he had so risen in his profession as to be called to a chair in the faculty. His admirers spoke of him as the modern Galen, and his *Universa Medicina* went through more than thirty editions. In the field of mathematics he published (1528) a work of the

[1] Born at Briançon, 1494; died in Paris, October 6, 1555.

[2] As spelled in the first edition of his *Protomathesis*, Paris, 1530–1532. The name is also spelled Finaeus. There is no warrant for the spelling Finé.

[3] Ramus speaks of him as the one "qui primus regia professione in Galliam mathematicas artes retulit." Introduction to his *Libri dvo*, p. 3 (Basel, 1569).

[4] Both of these subjects and some parts of astronomy are considered in his *Protomathesis*. Among his other works are *In sex priores libros geometricorum elementorum Euclidis demonstrationes*, Paris, 1536; *De re et praxi Geometrica libri III*, Paris, 1555; and *De rebus mathematicis, hactenus desideratis, Libri IIII*, with a biography, Paris, 1556.

[5] *De quadratura circuli*, Paris, 1544. This was severely attacked by Buteo. See *Bibl. Math.*, XII (3), 250.

[6] Joannes Fernelius, as the Latin form appears in his *De Proportionibus Libri duo*, Paris, 1558.

Boethian type on proportion, and his computation of the length of a degree of the meridian was so satisfactory[1] as to entitle him to a worthy place in the history of geodesy.

Among those who may properly be called the dilettanti mathematicians of the time was Claude de Boissière,[2] who wrote on poetry and music as well as on astronomy and arithmetic. His arithmetic,[3] a combination of the medieval theory and the contemporary practice in calculation, is one of the many books of the time that, as a result of the Hundred Years' War, related mathematics to the science of warfare.

At about the same time François de Foix, Comte de Candale (*c.* 1502–1594), another of the dilettanti and a bishop in southern France, was interested in a better translation of Euclid's *Elements*,[4] but he contributed nothing to the general theory of geometry.

Ramus. Pierre de la Ramée,[5] better known by the Latin form of his name, Petrus or Peter Ramus, descended from a noble but impoverished family. His grandfather had been driven from his estates in Burgundy and had been forced to become a charcoal burner, and his father was a humble peasant. Pierre early showed unusual intellectual powers, and after many struggles obtained employment as a servant to a rich student in the Collège de Navarre at Paris. By working in this capacity during the day and studying at night he made his way to the master's degree. It was on his examination in 1536, when he was only twenty-one years old, that he attracted

[1] He made it 56,746 French toises instead of 57,024.

[2] Claudius Buxerius. Born in the province of Grenoble, *c.* 1500.

[3] *L'art d'Arythmetiqve contenant tovte dimention, tres-singvlier et commode, tant pour l'art militaire que autres calculations*, Paris, 1554.

[4] *Euclidis . . . Elementa geometrica, Lib. XV . . . restituta. His accessit decimus sextus liber . . .*, Paris, 1566. There was another edition, "Novissime collati sunt XVII^{us} et XVIII^{us} priori editione . . .," Paris, 1578. His name appears in the Latin phrase "Auctore Francisco Flussate Candalla." For various other editors and translators of Euclid in this period see P. Riccardi, *Saggio di una Bibliografia Euclidea*, Bologna, 1887; hereafter referred to as Riccardi, *Saggio Euclid.*

[5] Born at Cust (Cultia, Cusia, Cus, Cuz, Cuth, Cut), Picardy, 1515; died August 26, 1572.

the attention of intellectual Europe by his audacious attack upon one of its idols, his thesis being "All that Aristotle has said is false."[1] Soon after this he began his career of teaching and was not long in attaining a high position in his profession. For many years (from 1546) he was principal of the Collège de Presles and held a professorship (from 1551) in the Collège de France. He was an orator of great power and a skillful debater, but his brilliant career was closed at the massacre of St. Bartholomew's Day, August 26, 1572. Although his work was chiefly in philosophy and the humanities, he devoted much attention to mathematics,[2] editing the *Elements* of Euclid[3] and writing on theoretical arithmetic,[4] geometry,[5] and optics.

Vieta. The greatest of all the French mathematicians of the 16th century was François Viète,[6] Seigneur de la Bigotière, better known by the semi-Latin name of Vieta. As a young man he practiced law in his native town, afterward taking up a political career and becoming a member of the Bretagne parliament. His first work on mathematics appeared in Paris in 1579. In 1580 he became master of requests at Paris, and later was a member of the king's privy council. Under these circumstances he was able to devote much leisure time to the

[1] *Quaequmque ab Aristotele dicta essent, commentia esse.*

[2] There is a good summary of his life in a work by F. P. Graves, *Peter Ramus*, New York, 1912, with a bibliography (p. 219) of the publications of Ramus and of secondary sources of information.

[3] Paris, 1545 and 1549. See Boncompagni's *Bullettino*, II, 389.

[4] *P. Rami, eloquentiae et philosophiae professoris regii, arithmeticae libri tres*, Paris, 1555; Paris, 1557; Basel, 1567; Paris, 1584. Also a *Libri Duo*, Paris, 1569, 1577, and 1581; Basel, 1580; Frankfort, 1586, 1591, 1592, 1596, and 1599; Lemgo, 1599; English translation, London, 1593. See *Rara Arithmetica*, pp. 263, 330, 335.

[5] Paris, 1577; English translation, London, 1636. See also his *Scholarvm Mathematicarvm, Libri vnvs et triginta*, Basel, 1569, in which he criticizes Euclid's arrangement of the *Elements* from the point of view of logic.

[6] Born at Fontenay-le-Comte (Fontenay-Vendée), 1540; died in Paris, December 13, 1603. The French form also appears as Viet, Viette, and de Viette, and the Latin as Vietaeus. There is a good sketch of his life in the *Penny Cyclopaedia*, London, 1843, by De Morgan, with a summary of all his works. See also J. L. F. Bertrand, "La vie d'un savant au XVI. siècle, François Viète," in his *Éloges académiques*, p. 143 (Paris, 1902); F. Ritter, *François Viète*, Paris, 1905; G. Gambier, *Le mathématicien François Viète*, La Rochelle, 1911.

study of mathematics, and the results were such that he ranks as one of several notable instances of a man attaining high standing in this science although not devoting himself chiefly to it until rather late in life. Vieta, indeed, remarked in a letter to Adriaen van Roomen that he did not profess to be a mathematician, but was merely one to whom mathematical studies were delightful in his hours of leisure.

FRANÇOIS VIÈTE (VIETA)
From an old lithograph of a portrait from life

Vieta wrote chiefly on algebra,[1] but he was also interested in geometry,[2] the calendar, and mathematics in general. In connection with the Gregorian reform of the calendar he acquired much unfortunate notoriety through his bitter antagonism to Clavius and through his wholly unscientific attitude. He was an expert in deciphering, for the government, the cryptic writing of diplomatic correspondence.

[1] *Isagoge in artem analyticam, De aequationum recognitione et emendatione libri duo, De numerosa potestatum purarum atque adfectarum ad exegesin resolutione tractatus,* all published privately by Vieta, and republished by Frans van Schooten, Leyden, 1646, in the *Opera mathematica* of Vieta. See also Boncompagni's *Bullettino*, I, 223, 245. There was a French translation of his treatment of equations made by one J. L. de Vaulezard, *Les Cinq Livres des Zetetiques*, published at Paris in 1630. There was an *Algebre de Viete d'vne methode novvelle, claire, et Facile*, by one James Hume (Iac Hvmivs), a Scotchman, Paris, 1636, but it is only after the style of Vieta.

[2] *Effectionum geometricarum canonica recensio*, and *Supplementum Geometriae*, Paris, 1593.

Vieta's Work in Algebra. It will be shown in Volume II that Vieta contributed extensively to the development of algebra and trigonometry, but a brief reference to his work is appropriate in this connection. He was among the first to employ letters to represent numbers in algebra, often using vowels for the unknowns and consonants for the knowns. He found the formula for sin $n\phi$ in terms of sin ϕ; made an advance towards proving that an equation of the nth degree is made up of n linear factors; showed how to increase, decrease, multiply, or divide the roots of the equation $f(x)=0$ by k; gave one of the earliest methods of evaluating π by infinite products; applied algebra to geometry in such way as to lay a foundation for analytic trigonometry; indicated powers more simply than his predecessors had done, using Aq for the square of the unknown, Ac for its cube, Aqq for its fourth power, and so on; and showed clearly the relation between the problems of the trisection of an angle and the solution of a cubic equation. His interesting combination of infinite products and series is seen in his statement that

$$\frac{2}{\pi}=\sqrt{\tfrac{1}{2}}\cdot\sqrt{\tfrac{1}{2}+\tfrac{1}{2}\sqrt{\tfrac{1}{2}}}\cdot\sqrt{\tfrac{1}{2}+\tfrac{1}{2}\sqrt{\tfrac{1}{2}+\tfrac{1}{2}\sqrt{\tfrac{1}{2}}}}\cdots.$$

Minor Writers. Of the writers on the theory of mathematics at this period the only one of any note who published his works at Lyons was Joannes Buteo,[1] a brother and afterward general of the order of St. Anthony. He wrote chiefly on geometry[2] and arithmetic.[3] His geometry refuted various pretensions of Oronce Fine as to the quadrature of the circle.

[1] Born at Charpey, *c.* 1485–1492; died at Caam, *c.* 1560–1572. These two places are in Dauphiné, France. The dates given by various writers differ greatly. See also Boncompagni's *Bullettino*, XIII, 258, 265 n. This is the form in which the family name appears in his *Logistica*, Lyons, 1559, that is, *Ioan. Bvteonis Logistica*. It is also given as Boteo, Jean Butéon, Bateon, Borrel, and Borell.

[2] *Opera Geometrica*, Lyons, 1554; *De Quadratura Circuli, Libri II*, Lyons, 1559.

[3] *Logistica qvae & Arithmetica vulgò dicitur in libros quinque digesta*, Lyons, 1559; the most original of his works. See *Rara Arithmetica*, p. 292; G. Wertheim, "Die Logistik des Johannes Buteo," *Bibl. Math.*, II (3), 213.

Another instance of a French mathematician of some ability publishing outside of Paris, and in this case outside of France, is that of Francesco dal Sole (born *c.* 1490). He wrote in Ferrara, Italy, and published an arithmetic in Venice.[1] The only feature of the book worth mentioning is the combination which it makes of the concepts of number and space.[2]

Perhaps the most elaborate arithmetics published in France in the 16th century, and among the least practical, were those of Pierre Forcadel.[3] Little is known of this writer except that he lived for a long time in Italy and finally, through the efforts of Ramus, was called to Paris as professor of mathematics in the Collège Royal. He translated Euclid I–VI and parts of the works of Proclus, Archimedes, and other writers.

Practical Mathematics. The earliest of the Lyons school of arithmeticians, and one of the most brilliant as well as most unscrupulous, was Estienne de la Roche,[4] known as Villefranche, although a native of Lyons. He was a pupil of Chuquet, and in his arithmetic[5] he appropriated a large amount of material from a manuscript of the latter which has since been published. Perhaps no other French arithmetic of the 16th century gives a better view of the methods of computation and of the commercial applications of the subject.[6]

The second noteworthy writer of the southern school was Jacques Peletier,[7] a native of Le Mans, who is also known by the Latin name of Peletarius. He was head of a college at Bayeux (1547), secretary to the bishop of Le Mans, a physician at Bordeaux (1550), Poitiers, Lyons, and Paris, and finally head of a college at Le Mans. He contributed to general

[1] *Libretti nvovi con le regole Di Francesco Dal Sole Gallo*, Ferrara, 1546. The first edition was Venice, 1526. There was a third edition in 1564.

[2] Thus he has "Regola delle additione in generalita, tanto geometrica, quanto arithmetica."

[3] Born at Béziers; died in Paris, 1574. For description of the various works see *Rara Arithmetica*, p. 284.

[4] Born at Lyons, *c.* 1480.

[5] *Larismethique nouellement composee par maistre Estienne de la roche dict Villefrãche natif de Lyō*, Lyons, 1520. *Rara Arithmetica*, 128; Cantor, *Geschichte*, II, chap. 59.

[6] A. Marre, Boncompagni's *Bullettino*, XIII, 573.

[7] Born at Le Mans, 1517; died in Paris, July, 1582.

literature and to elementary mathematics. His arithmetic[1] was published both at Poitiers (1549) and at Lyons (1554). He also wrote on algebra,[2] Euclid's *Elements*, the geometry of lines and angles, and the circle. He equated the terms of an equation to zero and stated that, when all roots are integral, any root is a factor of the last term.

Another Lyons arithmetician of considerable note appeared in the person of Ian Trenchant (born *c.* 1525). His arithmetic[3] includes the usual commercial applications and the operations with counters as well as with common numerals.

Among the practical tables published at Lyons are a set prepared by Monte Regal Piedmontois, professor of mathematics in the University of Paris. These tables[4] are beautifully printed on vellum and copies are very rare. The work contains the products of numbers to 100 × 1000, and the author speaks of having published part of the tables in Venice in 1575.

4. England

English Writers. England was later than Italy or France in her appreciation of mathematics, or at least in her publication of works on this subject.[5] Although there is some mention of arithmetic in the early works,[6] it was not until 1522 that a book devoted wholly to mathematics was printed in Great Britain,—the erudite but dull arithmetic[7] of Tonstall.

[1] The 1607 edition has the title *L'Arithmetiqve de Iacqves Peletier dv Mans, Departie en quatre liures.* There were other editions. *Rara Arithmetica,* 245.

[2] *L'algèbre départie en deux livres*, Lyons, 1554.

[3] *L'Arithmetiqve de Ian Trenchant, Departie en trois liures*, Lyons, 1566. This title is from the 1578 edition. There were several editions. On an edition of 1558 see *Bibl. Math.*, II (3), 356.

[4] *Invention novvelle et admirable, pour faire toute sorte de cōpte*, Lyons, 1585.

[5] See also R. C. Jebb, in the *Cambridge Mod. Hist.*, Vol. I, chap. xvi, and J. Gairdner, *ibid.*, chap. xiv.

[6] For example, in Caxton's *Mirrour of the World or Thymage of the same*, London, 1480, translated from the French, there is a chapter (10) beginning, "And after of Arsmetrike and whereof it proceedeth," but this cannot be called a treatise on the subject. There is a recent edition by O. H. Prior, London, 1913 (for 1912).

[7] *De Arte Svppvtandi libri qvattvor Cvtheberti Tonstalli*, London, 1522.

TITLE-PAGE OF THE TONSTALL WORK OF 1522

Tonstall. Cuthbert Tonstall[1] was recognized as a man of great learning and influence. He was born in Hackforth, Yorkshire, went to Oxford in 1491, left on account of the plague and entered Cambridge, and afterward went to Padua, where he took the degree of doctor of laws. In 1511 he became vicar-general to the Archbishop of Canterbury, was presented at court, and soon thereafter received various ecclesiastical and diplomatic appointments. Erasmus speaks in the highest terms of his remarkable attainments at this period of his career. In 1522 he was promoted to the bishopric of London, but continued for some time in his diplomatic career. In 1530 he was made bishop of Durham. In the troublous times of Edward VI, however, no man of prominence was safe, and Tonstall was accused of conspiracy in 1552, was deprived of his bishopric, and was imprisoned in the Tower. On the accession of Mary he was restored to his bishopric of Durham, but under Elizabeth he was again deprived of the honor (1559) and died a few months later.

Tonstall relates in his work that in his dealing with certain goldsmiths he suspected that their accounts were incorrect and therefore renewed his study of arithmetic so as to check their figures. Having been educated in part at Padua, in a country that was still the leader in commercial arithmetic, he was familiar with business methods of computing and was quite prepared to write a book on the subject. On his appointment to the See of London he bade farewell to the sciences by publishing this work. He dedicated the book to one of the greatest scholars and one of the noblest men of his generation, Sir Thomas More, who, in the *Utopia* (1516), "the only work of genius that she [England] can boast in this age," as Hallam characterized it, had spoken of him as his "colleague and companion . . . that incomparable man Cuthbert Tonstal," whose "learning and virtues are too great for me to do them justice."

[1]Born 1474; died at Lambeth Palace, November 18, 1559. The name is also spelled Tonstal and Tunstall. The first name is commonly spelled Cuthbert, but the spelling in the first edition of his work is Cuthebert. The border of the title-page of this edition, shown on page 315, was engraved by Holbein, whose initials appear at the left. The border had been used in an earlier book.

Some further idea of the intellectual group in which Tonstall moved may be obtained from the fact that it was his friend Margaret Roper, More's daughter, whom Erasmus addressed as the "ornament of thine England."[1]

The arithmetic of Tonstall was not original, the material being confessedly drawn from such Italian writers as Pacioli, but the arrangement was good and the presentation was clear even if unnecessarily extended. As for an arithmetic written in Latin, however, the time had passed for such a book to be popular in England.[2]

ROBERT RECORDE

From a recently discovered oil portrait on wood, apparently made from life, and now in the possession of W. F. Bushell, Fleetwood, Lancashire. The painting bears the inscription "Robt Record. M.D. 1556," but this is now so darkened by age as not to show in the photograph

Recorde. The most influential English mathematician whose works were published in the 16th century was Robert Recorde.[3] He entered Oxford *c.* 1525 and became a fellow of All Souls College in 1531. In 1545 he received the degree of M. D. at Cambridge. He taught mathematics in private classes both at Oxford and at Cambridge, but after receiving his degree in medicine he went to London and became physician to Edward VI and Queen Mary, perhaps absorbing

[1] "Margareta Ropera Britanniae tuae decus."

[2] John Gill, *Systems of Education*, p. 1. London, 1876.

[3] Born at Tenby, Pembrokeshire, *c.* 1510; died in London, 1558. In some editions of his works the name is spelled Record.

some of his educational ideas from Roger Ascham (1515–1568), who was then the Latin secretary to each of these rulers. He was also versed in the law, but this did not keep him from dying in prison. In his will he describes himself as "Robert Recorde, doctor of physicke, though sicke in body yet whole of mynde." The cause of his imprisonment is not known. Although a doubtful tradition says that it was for debt, it is more probable that it was for some misdemeanor in connection with the mines in Ireland, where he was for a time "Comptroller of Mints and Monies." Recorde may be said to have been the founder of the English school of mathematics, inasmuch as he wrote in the English language and showed originality in the treatment of his subjects and in his method of presentation.

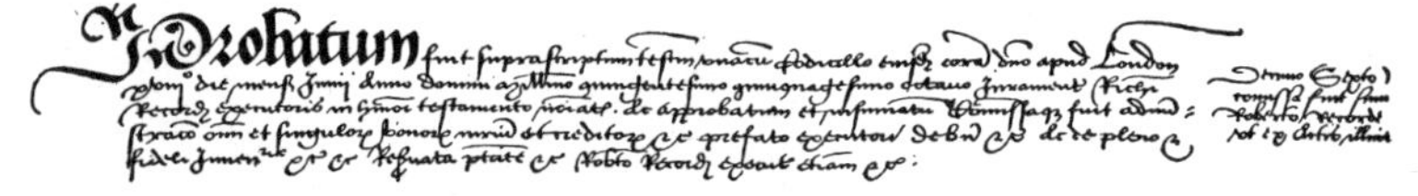

FROM THE PROBATE OF RECORDE'S WILL

It is often stated that the will bears the date June 28, 1558. The will is not in existence, so far as known, but the official copy is preserved, and the record shows that this date of probate should be read as xviii die mensis Junii

Recorde's Mathematical Works. Recorde published four books on mathematics and one on medicine, all of which are now extant, and he seems to have written others which have been lost and which were probably never printed. The four mathematical works were written in dialogue, a custom not uncommon at that time,[1] and bore various fanciful names, as shown in the following list:

1. *The Grovnd of Artes*, printed in London between 1540 and 1542, of which no copy of the first edition is known to be extant. This was one of the most popular arithmetics printed in the 16th century. It went through at least eighteen editions

[1] The custom appears in the Middle Ages as well, for example, in a 10th century MS. at Munich (S. Günther, *Geschichte des mathematischen Unterrichts im deutschen Mittelalter*, p. 26 n. (Berlin, 1887), hereafter referred to as Günther, *Math. Unterrichts*), but of course goes back to Plato. It appears in the various editions of the anonymous *Rithmimachia* (*Rara Arithmetica*, p. 63), in the arithmetic of Thierfelder (Nürnberg, 1587), and in numerous other works.

before 1601 and at least eleven more in the next century. Well did a writer of 1662 remark that this book was "entail'd upon the People, ratified and sign'd by the approbation of Time." The work includes computation by counters as well as by written figures, and contains the usual commercial topics which European countries north of the Alps had derived from Italy. It is commonly but incorrectly referred to as the first arithmetic printed in the English language, and it is in fact the earliest one of much lasting influence; but Recorde mentions the existence of other works of this kind, saying:

> I doubt not but some will like this my booke aboue any other English Arithmetike hitherto written, & namely such as shal lacke instructers, for whose sake I haue plain-ly set forth the exāples, as no book (that I haue seene) hath hitherto.

The whetstone of witte,

whiche is the seconde parte of Arithmetike: containyng thextraction of Rootes: The *Cosßike* practise, with the rule of *Equation*: and the woorkes of *Surde Nombers.*

Though many stones doe beare greate price,
The whetstone is for exersice
As neadefull, and in woorke as straunge:
Dulle thinges and harde it will so chaunge,
And make them sharpe, to right good vse:
All artesmen knowe, thei can not chuse,
But vse his helpe: yet as men see,
Noe sharpenesse semeth in it to bee.
The grounde of artes did brede this stone:
His vse is greate, and moare then one.
Here if you list your wittes to whette,
Moche sharpenesse therby shall you gette.
Dulle wittes hereby doe greately mende,
Sharpe wittes are fined to their fulle ende.
Now proue, and praise, as you doe finde,
And to your self be not vnkinde.

These Bookes are to bee solde, at the Weste doore of Poules, by Jhon Kyngstone.

TITLE-PAGE OF RECORDE'S ALGEBRA, 1557

The first work devoted chiefly to algebra and the theory of numbers to be printed in England

2. *The Castle of Knowledge*, printed in London in 1551, a work on astronomy, and one of the first to bring the Copernican system to the attention of English readers.

3. *The pathewaie to knowledge*, printed in London in 1551, containing an abridgment of Euclid's *Elements*.

4. *The whetstone of witte*, printed in London in 1557, "containyng the extraction of Rootes: The *Cossike* practise, with the rule of *Equation*: and the woorkes of *Surde Nombers*." The "cossic art" was another name for algebra, or, as Recorde says, the subject that begins with "The rule of equation, commonly called Algebers Rule." In this work the modern sign of equality first appears in print.

Such fanciful names as he uses for his titles were then the fashion, as is seen in numerous other works of the time.[1]

As to his arithmetic, England was just beginning to feel the need for such a book. In Elizabeth's reign, extending from 1558 to 1603, mercantile England came to the front, the native powers of the country were developed, new manufactures were introduced, and artisans from all over Europe were encouraged to enter the employ of her nobles and her merchant princes.[2] Never was there a better opportunity for a commercial arithmetic, and never was the opportunity more successfully met.

As already stated, the *Grovnd of Artes* was not the first popular arithmetic in the English language. In 1537 there appeared at St. Albans an anonymous work entitled *An Introduction for to lerne to reken with the Pen and with the Counters, after the true cast of arismetyke or awgrym in hole numbers, and also in broken*, and this was reprinted in 1539, 1546, 1574, 1581, and 1595, although it never ranked with Recorde's work either in scholarship or in popularity.

Recorde's other works were, naturally enough, not so successful, but they filled the needs for which they were written. It was no slight honor to have it said that "he was the first that ever writ of astronomy in the English tongue,"[3] even though the statement is exaggerated.

[1] *E.g.*, *Mirror for Magistrates*, *A Gorgeous Gallery of Gallant Inventions*, *Groat's Worth of Wit*, *Pap with a Hatchet*. See G. Saintsbury, *History of Elizabethan Literature*, p. 11 (London, 1887).

[2] W. Cunningham, in the *Cambridge Mod. Hist.*, Vol. I, chap. xv. London, 1902.

[3] Aubrey, *Brief Lives*, II, 200.

The first rival to Recorde's *Grovnd of Artes*, and the only serious one that appeared in Great Britain for a hundred years, was *The Well spring of Sciences*,[1] written by one Humphrey Baker[2] in 1562 and published in 1568. It went through five editions before 1601 and several after that date. The work is a commercial arithmetic, is evidently under many obligations to Recorde, and was written to meet the criticism of continental scholars on the backward state of the subject in England.[3]

Leonard and Thomas Digges. For about a hundred years, beginning in the second half of the 16th century, the effect of

MILITARY MATHEMATICS IN 1572

From the *Arithmeticall Militare Treatise, named Stratioticos*, by Leonard and Thomas Digges, London, 1572. This cut is from the 1579 edition

the continental wars showed itself in the textbooks on arithmetic and practical geometry, particularly in England and France. Problems relating to military affairs became more numerous, and even the titles of mathematical works bore evidence of this tendency. The most notable example in England is seen in a work by Leonard Digges (died *c.* 1571) and

[1] The form in which the title appears in the 1580 edition.

[2] He was a native of London and died after 1587. In some editions the name is spelled Humfrey.

[3] See the quotations in *Rara Arithmetica*, p. 327.

his son Thomas Digges (died 1595), entitled *An Arithmeticall Militare Treatise, named Stratioticos* (London, 1572). Leonard Digges came of an ancient Kentish family. He studied at University College, Oxford, but took no degree. He was a mathematician of ability, his chief interest lying in the application of the science to surveying, military engineering, and architecture. He published in 1556 a work on mensuration, the *Tectonicon*. He also wrote another work on geometry which his son Thomas published in 1571 and again in 1591 under the title *A Geometricall Practise, named Pantometria.*[1] The *Stratioticos* was begun by him and completed by his son. Thomas matriculated at Cambridge as pensioner of Queens' College in 1546, taking his B. A. in 1551 and his M. A. in 1557, and attained first rank among the English mathematicians of his time. His own works related chiefly to astronomy and navigation, and, like Recorde, he was probably a believer in the Copernican theory, although he did not openly advocate the doctrine.

JOHN DEE

After a portrait from life

Of the 16th century writers on applied mathematics one of the earliest was Richard de Benese, whose *Boke of Measuring of Lande* was published between 1562 and 1575, probably the earliest book on surveying printed in England.

[1] "At the end he discourses of regular solids, and I have heard the learned Dr. John Pell say it is donne admirably well." Aubrey, *Brief Lives*, I, 233.

John Dee. Toward the close of the century there lived one of the most curious characters to be found among the scientific men of his time. This man was John Dee (1527–1608), a native of London and a student of St. John's College, Cambridge. He took his B.A. in 1546 and was selected as one of the original fellows of Trinity College. His studious nature may be inferred from his own record of his rules,—"only to sleep four houres every night; to allow to meate and drink (and some refreshing after) two hours every day; and the other eighteen hours (except the tyme of going to and being at divine service) was spent in my studies and learning." He afterwards traveled on the continent, forming friendships with Gerard Mercator, Gemma Frisius, Jean Fernel, and other scholars, taking courses for two years at Louvain, and publicly lecturing on Euclid before large audiences in Paris. He became interested in alchemy and astrology and his relations to the occult made his life a romantic one.[1] He wrote the preface to the first English translation of Euclid's *Elements*,[2]

The first Booke

E passe vnder the line B D: namely, that the whole ſhalbe equall to the part: which is alſo impoſſible. Wherefore C D is one right line: which was required to be proued.

The 8. Theoreme. The 15. Propoſition.

If two right lines cut the one the other: the hed angles ſhalbe equal the one to the other.

Demonſtration.

SVppoſe that theſe two right lines A B and C D, do cut the one the other in the point E. Then I ſay, that the angle A E C, is equall to the angle D E B. For foraſmuch as the right line A E, ſtandeth vpon the right line D C, making theſe angles C E A, and A E D: therefore (by the 13. propoſitiō) the angles C E A, and A E D, are equall to two right angles Agayne foraſmuch as the right line D E, ſtandeth vpon the right line A B, making theſe angles A E D, and D E B: therfore (by the ſame propoſition) the angles A E D, and D E B, are equall to two right angles. and it is proued, that the angles C E A, and A E D, are alſo equall to two right angles. VVherfore the angles C E A, and A E D, are equall to the angles A E D, and D E B. Take away the angle A E D, which is common to them both. VVherefore the angle remayning C E A, is equall to the angle remayning D E B. And in like ſort may it be proued, that the angles C E B, and D E A, are equall the one to the other. If therefore two right lines cut the one the other, the hed angles ſhalbe equall the one to the other: which was required to be demonſtrated.

A C E D B

Thales Mileſius the firſt inuenter of this propoſition.

Thales Mileſius the Philoſopher was the firſt inuenter of this Propoſition as witneſſeth *Eudemius*, but yet it was firſt demonſtrated by *Euclide*. And in it there is no conſtruction at all. For the expoſition of the thing geuē, is ſufficient inough for the demonſtration.

No conſtruction in this propoſition.

What hed angles are.

Hed Angles, are appoſite angles, cauſed of the interſection of two right lines and are ſo called, becauſe the heddes of the two angles are ioyned together in one pointe.

The conuerſe of this propoſition after Pelitarius.

The conuerſe of this propoſitiō after Pelitarius.

If fower right lines being drawen from one point, do make fower angles, of which the two oppoſite angles are equall: the two oppoſite lines ſhalbe drawen directly, and make one right line.

Suppoſe that there be fower right lines A B, A C, A D, and A E, drawen from the poynt A, making fower angles at the point A: of which let the angle B A C be equall to the angle D A E, and the angle B A D to the angle C A E. Then I ſay, that B E and C D are onely two right lines: that is, the two right lines B A and A E are drawen directly and

FROM THE FIRST ENGLISH EDITION OF EUCLID

The Billingsley or Dee translation, London, 1570

[1] Charlotte Fell Smith, *John Dee*, London, 1909.

[2] *The Elements of Geometrie of the most auncient Philosopher Euclide of Megara. Faithfully (now first) translated into the Englishe toung, by H. Billingsley, Citizen of London. . . . With a very fruitfull Praeface made by M. I. Dee*, London, 1570.

and may, indeed, have made the translation in whole or in part himself, although it is attributed to Sir Henry Billingsley, who was later (1596) Sheriff and Lord Mayor of London.[1] Dee was a man of erudition and of remarkable powers of exposition, and his influence on mathematics in his day must have been considerable.

Minor Writers. Among the books containing algebra and printed in England in the 16th century there was an arithmetic written by Thomas Masterson and published at London in three books in 1592–1595. Concerning Masterson himself nothing is known. He planned a treatise in six books but completed only three, the third consisting of that part of algebra that has to do with the powers and roots of numbers.

The century closes with the name of Thomas Blundeville, who published (1594) a work entitled *Exercises, containing sixe Treatises*,[2] one of these parts being on arithmetic and the rest being on cosmography. The arithmetic contains "a briefe description of the tables of the three speciall right lines belonging to a circle, called sines, lines tangent, and lines secant," the first fairly complete treatment of trigonometry in England.[3] He also published several other books, partly on cosmography.

5. Germany

Nature of the German Mathematics. When we compare the mathematics of Germany with that of France in the 16th century, we are struck by the same difference that existed in the art of the two countries. The mathematics of Germany was Gothic, unpolished, but virile; the mathematics of France was Renaissance, polished, but generally weak. Germany produced a notable group of arithmeticians; France produced hardly more than one. Germany produced two strong algebraists; France produced one,—a dilettante but a brilliant one. Germany made a definite advance in geometry, in the study of

[1] On this question see A. De Morgan, *British Almanac and Companion* for 1837, p. 38 of the *Companion*.
[2] The seventh edition appeared in 1636.
[3] A. De Morgan, *British Almanac and Companion* for 1837, p. 42 of the *Companion*; *Arithmetical Books*, p. 30.

higher plane curves; France was content to contemplate the past and possibly to dream of the century just ahead.

Of course the greatest influence for advance in the 16th century was printing; but there was also Erasmus (1467–1536),

FROM THE MARGARITA PHYLOSOPHICA (1503)
Showing Arithmetica between the ancient counter reckoning and the modern algorism. This is from the edition of 1504

who, although born in Rotterdam, lived in Germany, England, France, and Italy, and was the world's scholar of the first third of the 16th century; and there was also Martin Luther (1483–1546), who set Germany thinking, not always for the best. It was a century of intellectual awakening and of breaking away from traditions, and all this showed itself in the mathematical activity of the time. As in other countries, the number of names connected with this activity now becomes so great as to allow for only a limited selection.

The Margarita Phylosophica. The first modern encyclopedia of any note, based upon the late Latin models, was the *Margarita Phylosophica*,[1] first published at Freiburg in 1503. The author was Gregorius Reisch (died 1523), who studied at Freiburg in 1487 and took his bachelor's and master's degrees there. He became a Carthusian and was made prior at Freiburg and confessor to Maximilian I. The work consists of twelve books, and includes considerable material upon arithmetic, geometry, and astronomy. Its popularity is shown by the fact that there were sixteen editions in the course of a century.

Albrecht Dürer. It is not often that the artist of today is confessedly a mathematician; the mathematician is more frequently an artist. But the 16th century has upon its roster the names of several great artists who did something in mathematics, generally in architecture or in perspective. One man, however, stands out with special prominence as a great artist and at the same time as a mathematician with distinctly new interests, and this is Albrecht Dürer.[2] His work as a painter and engraver is well known, but it was in his treatises on geometry,[3] fortification,[4] and human proportion[5] that he showed his

[1] The title of the second (1504) edition, printed at Strasburg, reads, *Aepitoma omnis phylosophiae. alias Margarita phylosophica tractans de omni genere scibili: Cum additionibus: Quę in alijs non habentur.*

[2] Born at Nürnberg, May 21, 1471; died at Nürnberg, April 6, 1528. Cantor, *Geschichte*, II, chap. 63; S. Günther, *Die geometrischen Näherungskonstruktionen Albrecht Dürers*, Ansbach, 1886; S. Günther, "Albrecht Dürer, einer der Begründer der neueren Kurvenlehre," *Bibl. Math.* (1886), p. 137; H. Staigmüller, *Dürer als Mathematiker*, Prog., Stuttgart, 1891; F. Amodeo, "Albrecht Dürer precursore di Monge," *Atti della R. Accademia delle Soc. Fis. e Mat.*, XIII (2), No. 16; W. B. Scott, *Albert Dürer, his Life and Works*, London, 1869.

[3] *Underweysung der messung mit dem zirckel und richtscheyt in Linien, ebnen, vnnd gantzen corporen*, Nürnberg, 1525, with Latin editions, Paris, 1532 and 1555, and Arnheim, 1605; *Institutiones Geometricae*, Paris, 1532.

[4] *Etliche vnderricht zu befestigung der Stett, Schloss und Flecken*, Nürnberg, 1527, 1530, and 1538, with a Latin edition, Paris, 1535.

[5] *Hierin sind begriffen vier Bücher von menschlicher Proportion*, Nürnberg, 1528, with a Latin edition, Nürnberg, 1532.

mathematical powers. The geometry was the first printed work to consider the subject of higher plane curves, and the first to discuss scientifically the question of such approximate constructions as that of the regular heptagon.

Johann Stöffler,[1] although professor of mathematics at the University of Tübingen, can be ranked as a mathematician only because of his computation of astronomical tables. He was one of the first to show how the Julian calendar could be brought into harmony with astronomical events. His calculations led him to the absurd prediction, however, that the Deluge would be repeated in the year 1524. The announcement stirred all Europe, and the number of schemes to protect the race was legion. The people of Toulouse even went so far as to build an ark. Stöffler, however, seems to have survived the storm of protest that ensued upon the failure of his prediction, for he published in the year of his death a new ephemeris, and left a commentary on the *Sphere* of Proclus, which was published posthumously in 1534.[2]

Stifel. The first German writer of the century to devote his life to mathematics and to acquire an enviable reputation in this field was Michael Stifel.[3] A lover of mathematics from his childhood; brought up in Esslingen, a veritable bulwark of the ancient faith; trained in the local Augustine convent and taking holy orders; giving promise of success in the Church; he was finally captured by the eloquence of Luther, and thought himself a reformer when he was really a fanatic. Starting for heaven with a group of peasants on the day which he had prophesied would see the blotting out of this world, he ended ignominiously behind the bars of a jail. Since Luther had launched him on a career that landed him in prison, it was proper that he should get him out, which he did. The state of

[1] Born at Justingen, Swabia, December 10, 1452; died at Blaubeuern, February 16, 1531.

[2] The curious case is discussed in Bayle's *Dictionaire*, under *Stofler*.

[3] Stiefel, Styfel, Stiffelius. Born at Esslingen, April 19, 1487 (some say 1486); died at Jena, April 19, 1567. See Th. Müller, *Der Esslinger Mathematiker Michael Stifel*, Prog., Esslingen, 1897; Cantor, *Geschichte*, II, chap. 62.

Stifel's mind may further be seen by the following line of reasoning in which he indulged:

1. The Latin for Leo Tenth is Leo Decimus.
2. This may be written Leo DeCIMVs.
3. These capitals may be arranged thus: MDCLVI.
4. We may take away M for Mystery, and add X because it is Leo X, and we then have DCLXVI.
5. But this is 666, the "Number of the Beast" in the Book of Revelations; and hence Leo X is the Beast.

Yet this is the man who, in the next few years, produced some of the most original and vigorous mathematical works to be found in the 16th century.

Stifel wrote five works on mathematics, these works treating chiefly of the mysticism of numbers,[1] arithmetic,[2] and algebra.[3] His arithmetic, which is largely on algebra, is a more scholarly work of the kind than any that had yet appeared in Germany, doing for that country what Cardan's and Tartaglia's treatises were doing for Italy.

Christoff Rudolff. Stifel's chief work on algebra was his edition (1553–1554) of a book known as the *Coss,* which appeared in 1525 and was the first algebra of any moment to be published in Germany. It was written by Christoff Rudolff,[4] concerning whose life very little is known. Rudolff published three books, the *Coss* (1525), the *Kunstliche rechnung*[5] (1526),

[1] *Ein Rechen Büchlein*, Wittenberg, 1532.

[2] *Arithmetica Integra*, Nürnberg, 1544, to which Melanchthon wrote the preface; *Deutsche Arithmetica*, Nürnberg, 1545; and *Rechenbuch von der Welschen vnd Deutschen Practick*, Nürnberg, 1546. Before the World War of 1914–1918 there was preserved in the University of Louvain a copy of the *Arithmetica Integra* with marginal notes by Gemma Frisius. Fortunately these notes were copied by H. Bosmans, S. J., and several of the most important ones were published before the destruction of the library.

[3] *Die Coss Christoffs Rūdolffs*, Königsberg, 1553–1554, 1571, and 1615. While, as stated above, this is Rudolff's work, it contains Stifel's commentary. The *Arithmetica Integra* also contains the treatment of radicals that is now considered part of algebra. The word *Coss* is from *cosa* (*causa*), which, as explained in Volume II, refers to the first power of the unknown quantity.

[4] Born at Jauer, *c.* 1500.

[5] The title of the third edition (Nürnberg, 1534) is *Kunstliche rechnung mit der ziffer vnnd mit den zal pfenningē.* There were eleven editions in the 16th century.

and a collection of problems[1] (1530). Of these the *Coss* was the most important, doing for algebra in Germany what Pacioli's *Sūma* had done for the subject in Italy. All such works, however, were for the scientific élite of the universities, not for the elementary Latin schools. Of forty-six Schulordnungen of the 16th century only twenty-four gave mathematics any place, and most of these were issued only in the second half of the century.

Johann Scheubel. In marked contrast to Stifel was his contemporary, but slightly his junior, Johann Scheubel.[2] Stifel was brilliant, Scheubel was scholarly; Stifel was popular, Scheubel was heavy; Stifel was eccentric, Scheubel was balanced; and Stifel was effusive, while Scheubel was a man of dignity and poise. The University of Tübingen called Scheubel to a professorship of mathematics at about the same time that Stifel was sent to prison, and at Tübingen he wrote his arithmetics,[3] Latin works that were too heavy for commercial purposes and too light for his own students. Here, too, he wrote his algebras, one of which he published,[4] leaving the other in manuscript,[5] and here he edited the seventh, eighth, and ninth books of Euclid's *Elements* (1558). He also left in manuscript a copy of Robert of Chester's translation of al-Khowârizmî's algebra.[6] He gave the so-called Pascal Triangle a century before Pascal wrote upon it, and extracted roots as high as the 24th by a process similar to the one which employs the Binomial Theorem.

Grammateus. Of about the age of Scheubel and with similar interests, but not so learned, Heinrich Schreyber[7] made for

[1] *Exempel-Büchlin*, Augsburg, 1530, a commercial arithmetic.

[2] Scheybl, Scheubelius, Scheybel. Born at Kirchheim, Württemberg, August 18, 1494; died February 20, 1570.

[3] *De Nvmeris et Diversis Rationibvs seu regulis computationum opusculum, a Ioanne Scheubelio compositum*, Leipzig, 1545; *Compendivm Arithmeticae Artis*, Basel, 1549 (this title from the 1560 edition).

[4] *Algebrae compendiosa facilisque descriptio*, Paris, 1551.

[5] This is now in the library of Columbia University, New York City. His other manuscripts were left to the University of Tübingen.

[6] This is bound with the Columbia MS. above mentioned, and the translation by Professor Karpinski was published in 1915.

[7] Born at Erfurt as early as 1496. The name appears as Schreyber in one of his works of 1523, but is given by modern writers as Schreiber.

himself a worthy place in German history. He was better known in maturity by his Latinized Greek name of Henricus Grammateus, and in his young manhood by the Latin name of Henricus Scriptor. He studied at Cracow and was later (1507)[1] enrolled in the University of Vienna, where he was afterwards an instructor. He took his bachelor's degree in 1511 and his master's degree in 1518, thereafter teaching for a time in the University and privately, and having Rudolff for one of his pupils.[2] Driven from Vienna by the plague (1521), he went to Nürnberg and Erfurt, but returned a little later (1525) and devoted himself to writing. His best-known work was an arithmetic in the German language.[3] It includes arithmetic computations with counters and figures, a little work in the theory of numbers, a chapter on bookkeeping, a few of the simplest rules of algebra, and a brief treatment of the gaging of casks. He is the first German writer to make free use of the signs + and − in the treatment of algebraic expressions, although the symbols had long been used for other purposes. He also published other works on arithmetic,[4] and wrote on the theory of proportion[5] and on mensuration.[6]

Ludolf van Ceulen. One German writer, Ludolf van Ceulen,[7] may quite as well be classified among the Dutch mathematicians, since he spent most of his life in Holland. He seems to

[1] For a record of the time reads: "Anno domini millesimo quingentesimo septimo . . . Henricus Scriptoris de Erfordia."

[2] Rudolff writes: "Ich hab von meister Heinrichen so Grammateus genennt der Coss anfengklichen bericht emphangen. Sag im darumb danck."

[3] *EYn new künstlich behend vnd gewiss Rechenbůchlin vff alle Kauffmanschafft* . . . ¶ *Bůchhalten durch das Zornal* . . . ¶ *Visier růten*. . . . *M. Henricus Grammateus*, Vienna, 1518, with later editions. This title is from the 1535 edition. C. F. Müller, *Henricus Grammateus*, pamphlet, Zwickau, 1896.

[4] *Behend unnd khunstlich Rechnung nach der Regel und welhisch practic*, Nürnberg, 1521; *Algorismus de integris Regula de tri cum exemplis*, Erfurt, 1523; *Eynn kurtz newe Rechenn unnd Visyrbuechleynn*, Erfurt, 1523, a Visierbuch being a work on the gaging of casks.

[5] *Algorithmus proportionum*, Cracow, 1514.

[6] *Libellus de compositione regularum pro vasorum mensuratione*, Vienna, 1518.

[7] Born at Hildesheim, January 18, 1540; died at Leyden, December 31, 1610. The name also appears as Ludolph van Collen, Cuelen, and Keulen. See D. Bierens de Haan in Boncompagni's *Bullettino*, XIV, 571.

have left Germany in his childhood and to have been educated under Dutch influences. He taught mathematics in Breda, Amsterdam, and Leyden, and became professor of military engineering in the University of Leyden in 1600. He is known chiefly for his value of π, at first given[1] to 20 and then to 35 decimal places.[2] He also published (1615) a work on arithmetic and geometry.

Mention should also be made of Johann Werner,[3] a priest, who was interested chiefly in astronomy but who wrote the first original work on conics to appear in the 16th century.[4]

Pitiscus. The last of the German writers on mathematics whose work falls chiefly in the 16th century and whose contributions entitle him to special mention is Bartholomäus Pitiscus.[5] He was a clergyman by profession but a mathematician by preference. His trigonometry[6] was the first satisfactory textbook published on the subject and the first book to bear this title. He also edited and perfected the table of sines of Rhæticus.[7]

The Classical Group. The name of Philip Melanchthon,[8] the friend and colleague of Luther and professor of Greek at Wittenberg, does not ordinarily suggest the science of mathematics. His wide range of human interest, however, led him not only

[1] *Van den Circkel,* Delft, 1596, with editions in 1615 and (Latin) 1619.

[2] *De arithmetische en geometrische fondamenten*, Leyden, 1615, with a Latin edition by Snell, *Fundamemta* (*sic*) *Arithmetica et Geometrica*, Leyden, the same year.

[3] Born at Nürnberg, February 14, 1468; died at Nürnberg, 1528.

[4] *Libellus super viginti duobus elementis conicis*, Nürnberg, 1522. He also wrote *De Triangulis Libri IV*; see *Abhandlungen*, XXIV.

[5] Born at Schlaun, near Grünberg, Silesia, August 24, 1561; died at Heidelberg, July 3, 1613. The name also appears as Petiscus.

[6] *Trigonometriae sive de dimensione triangulorum libri quinque*, Frankfort, 1595, as an appendix to the astronomy of Abraham Scultetus, or Abraham Schultz (1566–1625). Complete editions by Pitiscus were published at Frankfort in 1599, 1608, and 1612, and at Augsburg in 1600. An English translation appeared in London in 1630. In the edition of 1612 Pitiscus makes use of a decimal point, sin 10″ being given as 4.85 for $r = 100{,}000$.

[7] *Thesaurus mathematicus sive canon sinuum ad radium 1,0000 0000 0000* . . . Frankfort, 1593. On Rhæticus, see page 333.

[8] Greek form for Schwartzerd, his family name. He was born at Bretten, Baden, February 16, 1497; died at Wittenberg, April 19, 1560.

into the fields of philosophy and religion, but also to the study of astronomy and mathematics. He edited (1521–1560) several works on astronomy by Aratus, Peurbach, Schoner, al-Farġânî, Sacrobosco, and Ptolemy. His activity in pure mathematics is shown in a preface to Stifel's *Arithmetica Integra* (1544) and in a work on the mathematical disciplines.[1] He is to be valued, however, for his influence rather than for his contributions in the field of the exact science.

With the name of Melanchthon is naturally connected that of his friend and associate, Joachim Camerarius,[2] a distinguished classicist and a professor at Tübingen and Leipzig. His edition of Nicomachus[3] appeared at Augsburg in 1554, and a work on computation, *De logistica*, was published there at the same time. He also wrote some astronomical verses, which were published with Melanchthon's work of 1540.

Another one of the classical group who did a certain amount of work in mathematics was Jacobus Micyllus.[4] His arithmetic[5] was written for the Latin schools, and it contains a considerable amount of ancient material. His treatment of sexagesimal fractions is unusually extensive.

A little later than Micyllus there lived the well-known scholar Michael Neander,[6] who became professor of mathematics and Greek (1551) in the University of Jena, and later (1560) professor of medicine in the same institution.[7] He wrote an

[1] *Mathematicarvm disciplinarvm, tvm etiam astrologiae encomia*, Leyden, 1540.

[2] Born at Bamberg, April 12, 1500; died at Leipzig, April 17, 1574. The family name was Liebhard, but the office of chamberlain to the Prince-Bishop of Bamberg being hereditary in this family, he took the Latin name of Camerarius (chamberlain).

[3] The title of the Daventer (1667) edition is *Explicatio Ioachimi Camerarii Papebergensis in dvos libros Nicomachi Geraseni Pythagorei Deductionis Ad Scientiam Numerorum.*

[4] Born at Strasburg, April 6, 1503; died at Heidelberg (?), January 28, 1558. The name also appears as Moltzer, Molshem, Molsehm, and Molshehm.

[5] *Arithmeticae logisticae libri duo*, Basel, 1555.

[6] Born in the Joachimsthal, April 3, 1529; died at Jena, October 23, 1581. The family name was Neumann, whence the classical form Neander.

[7] On the relation of medicine to mathematics at this time, see D. E. Smith, "Medicine and Mathematics in the Sixteenth Century," *Annals of Medical History*, New York, 1917, p. 125.

excellent work on metrology (Basel, 1555), one of the first to treat of the subject historically and in a scholarly manner. He also wrote on spherics.[1]

Still another of the classical group was Guilielmus Xylander,[2] professor of Greek at Heidelberg. He translated the first six books of Euclid's *Elements* into German (Basel, 1562), and various works from Greek into Latin, including the *Arithmetica* of Diophantus (Basel, 1575) and the work of Psellus (Basel, 1556). His *Opuscula Mathematica* appeared at Heidelberg in 1577 and contains a certain amount of work on astronomy, arithmetic, algebra,[3] and geometry.

Mathematical Astronomers. Although mathematics and astronomy were no longer synonymous terms, the 16th century produced one or two scholars whose interests in the two sciences were apparently about equal. The first of these was Petrus Apianus.[4] He wrote chiefly on astronomy, but his arithmetic[5] is interesting because it contains the first triangular arrangement of the binomial coefficients (the Pascal Triangle) to appear in print. This appeared some years before Stifel mentioned the subject. Apianus was professor of astronomy at Ingolstadt, was interested in the teaching of trigonometry, and was one of the few university professors of his time to give instruction in the German language.

The leading mathematical astronomer in the Teutonic countries in the middle of the 16th century was Georg Joachim Rhæticus.[6] He studied at Zürich and Wittenberg and was

[1] There was another Michael Neander of the same period (1525–1595), who wrote on physics, theology, and philology.

[2] Wilhelm Holzmann. Born at Augsburg, December 26, 1532; died at Heidelberg, February 10, 1576.

[3] "De svrdis, qvos vocant, nvmeris iis, qvi a qvadratis primò nascuntur, Institutio docendo explicanda."

[4] Born at Leisnig, 1495; died at Ingolstadt, April 21, 1552. Known also by his German name, Peter Bienewitz or Bennewitz.

[5] *Eyn Newe Vnnd wolgegründte vnderweysunġ aller Kauffmanss Rechnung,* Ingolstadt, 1527. His *Cosmographia* appeared in 1524.

[6] Georg Joachim von Lauchen, his last name being derived from his home region, the ancient Rhætia. Born at Feldkirch, in Vorarlberg, February 16, 1514; died at Kaschau, Hungary, December 4, 1576.

professor of mathematics in the latter university from 1537 to 1542. He published an arithmetic at Strasburg in 1541, but most of his work was on astronomy and trigonometry.[1] He visited Copernicus in 1539, studied with him, and did much to make his theories known.

CHRISTOPHER CLAVIUS

Engraved after a portrait from life

Clavius. Probably the man who did the most of all the German scholars of the 16th century to extend the knowledge of mathematics, although doing little to extend its boundaries, was Christopher Clavius,[2] a Jesuit, who passed the later years of his life in Rome. He was an excellent teacher of mathematics, and his textbooks were highly esteemed because of their arrangement, particularly in the Latin schools. Sixtus V testified to his standing, saying: "Had the Jesuit order produced nothing else than this Clavius, on this account alone should it be praised." His arithmetic[3] was published in Rome in 1583, was translated into Italian in 1586, and went through several editions. His

[1] *Opus Palatinum de Triangulis*, Neustadt a. Hardt, 1596; *Thesaurus Mathematicus*, Frankfort a. M., 1613, both posthumous.

[2] Christoph Klau. Born at Bamberg, 1537; died at Rome, February 6, 1612.

[3] *Epitome Arithmeticae Practicae*, Rome, 1583.

algebra[1] appeared in 1608 and was one of the best textbooks on the subject that had been written up to that time. He published an edition of Euclid in 1574.[2] While not precisely a translation, the book proved very valuable because of the great erudition shown in the extensive scholia which it contains.

Clavius was one of the mathematicians engaged in the reform of the calendar (1582) under the direction of Pope Gregory XIII. His collected works[3] contain, in addition to his arithmetic and algebra, his commentaries on Euclid, Theodosius, and Sacrobosco, his contributions to trigonometry and astronomy, and his work on the calendar.

Johann Schoner. Among the minor German writers on astronomy and mathematics in this period Johann Schoner[4] and his son Andreas Schoner (1528–1590) are perhaps the best known. Johann was for a time a preacher at Bamberg, and was later a teacher in the Gymnasium at Nürnberg. He edited a well-known medieval arithmetic[5] and wrote on geometry, astrology, and astronomy. Andreas wrote on astronomy and dialing and edited his father's works.[6]

German Arithmetic. The German arithmetics in the 16th century were very practical.[7] Commerce was active, and Frankfort was one of the great trading centers of Europe, her agents pushing out through the Hansa towns to England, France, and the northern countries, as well as through Austria and Italy to the Orient. The great house of Fugger multiplied its capital tenfold between 1511 and 1527, possessing five times the wealth which made the Medici so powerful in the preceding

[1] *Algebra Christophori Clavii Bambergensis*, Rome, 1608.

[2] *Euclidis Elementorum Libri XV*, as the title appears in the Frankfort edition of 1654. The earlier editions were 1574, 1589, 1591, 1603, 1607, 1612, which show the popularity of the work.

[3] *Opera Mathematica*, 5 vols., Mainz, 1611 and 1612.

[4] Born at Karlstadt, near Würzburg, January 16, 1477; died at Nürnberg, January 16, 1547. The name also appears as Johannes Schonerus or Schöner.

[5] *Algorithmvs demonstratvs*, Nürnberg, 1534.

[6] *Opera Mathematica*, Nürnberg, 1561.

[7] Hugo Grosse, *Historische Rechenbücher des 16. und 17. Jahrhunderts*, Leipzig, 1901.

century. But whereas the Medici had been the patrons of art and of letters, the Fuggers were little beyond accumulators of wealth. As a result of this national spirit, the German mathematics of this period was more largely commercial than it might have been under more favorable circumstances.

While the classical group was carrying mathematics to the intellectual aristocracy a group containing a few skillful writers was carrying it to the intellectual democracy. One of them, Johann Böschensteyn,[1] like Stifel a native of Esslingen, taught both Luther and Melanchthon,—a sufficient honor for any man. He was a professor of Hebrew at Ingolstadt, Heidelberg, Nürnberg, and Antwerp, but his mathematical contribution was of a humble nature, consisting merely of an arithmetic of a commercial kind, printed at Augsburg in 1514.

Contemporary with Böschensteyn, and indeed but two years his senior, was one whose influence on the people was far greater,—Jakob Köbel,[2] a native of Heidelberg. He and Copernicus were fellow students at Cracow, and each had varied lines of interest. Köbel began as a teacher of arithmetic, a printer, a woodcarver, a poet, and a student of law, and ended as a petty officeholder; while Copernicus began as a priest, physician, and astronomer, and ended as a giant among the thinkers of the world. Köbel wrote three arithmetics, although the editions varied so much as to give an impression of a larger number. These arithmetics were the *Rechenbiechlin*[3] (1514), *Mit der Kryden*[4] (1520), and the *Vysierbuch*[5] (s.a. but 1515). Of these the first was the most important, passing through no less than twenty-two editions in the 16th century. It is purely commercial but shows more vigor and originality than any arithmetic that had yet appeared in Germany. Among other

[1] Among the variants of the name are Beschenstein, Boeschenstain, Bossenstein, Boechsenstein, Buchsenstein, Poschenstein, and Besentinus. Born at Esslingen, Swabia, 1472; died 1540.

[2] Kobel, Kobelius, Kobelinus. Born at Heidelberg, 1470; died at Oppenheim, January 31, 1533.

[3] As spelled on the title-page; *Rechenbüchlein* as spelled in the colophon.

[4] *Mit der Krydē od' Schreibfedern durch die zeiferzal zu rechē Ein neüw Rechēpüchlein*, Oppenheim, 1520.

[5] On gaging. In the 1537 edition, *Eyn new Visir Bůchlin.*

features are the crude illustrations, although similar ones had already been used by Widman as early as 1489.

Adam Riese. The greatest of all the Rechenmeisters of this century, however, was Adam Riese.[1] He was the most influential of the German writers in the movement to replace the old computation by means of counters ("auff der Linien") by the more modern written computation ("auff Federn"). He wrote four arithmetics,[2] all of a commercial nature and the second ranking as one of the most popular schoolbooks of the century. It was to Germany what Borghi's book was to Italy, Recorde's to England, and Gemma's to the Latin schools of the Continent. So famous

ADAM RIESE

Germany's best-known Rechenmeister. From the title page of his *Rechenung nach der lenge/auff den Linihen vnd Feder*, Leipzig, 1550

[1] Born possibly at Staffelstein, near Bamberg, *c.* 1489; died at Annaberg, March 30, 1559. The name also appears as Ryse, Ris, and Reis. See B. Berlet, *Adam Riese, sein Leben, sein Rechenbücher und seine Art zu rechnen*, new ed., Leipzig, 1892.

[2] *Rechnung auff der linihen*, printed in 1518 and again in 1525 and 1527. Title from the 1525 edition.

Rechnung auff der Lynihen vn Federn in zal/mass/vnd gewicht auff allerley handierung gemacht, Erfurt, 1522, with at least thirty-seven editions before 1600.

Ein Gerechent Büchlein, Leipzig, 1536.

Rechenung nach der lenge/auff den Linihen vnd Feder, Leipzig, 1550.

was the author that the phrase "nach Adam Riese" is still used in Germany to signify arithmetical accuracy or skill.

It was due in no small degree to the encouragement given by Luther that books like Riese's met with their great success. At the very time that the great Rechenmeister was preparing the second edition of his most successful work[1] Luther was laying down his famous doctrine that all children should study mathematics,[2] a thing unheard of before.

The only other Rechenmeister of the century to deserve special mention is Simon Jacob,[3] who wrote two commercial arithmetics. Bürgi mentions Jacob's treatment of series, and apparently the former's table of antilogarithms, the *Progress Tabulen,*[4] was suggested by the nature of exponents as laid down in these and similar books of the 16th century.

6. The Netherlands

Geographical Limits. When we speak of the Low Countries, the Netherlands, it should be borne in mind that boundaries and governments were constantly shifting at this time, and that when Charles V (died 1558) inherited this territory from his grandmother, Mary of Burgundy, it included seventeen provinces, each with its own government. When Philip II (died 1598) inherited it from Charles, the new ruler was accepted with a pronounced expression of discontent, and finally (1568) the provinces rebelled under William of Orange. The period was therefore one of uncertain geographical limits, and in speaking of the Netherlands it must be understood that we speak of territory of which a part was under Spanish rule during the 16th century. In general, therefore, we shall include for our present purposes what is now Holland and Belgium.

[1] That is, in 1524, the edition appearing in 1525.

[2] This is in his *Schrift an die Ratsherren aller Städte Deutschlands,* wherein he says: "Wenn ich Kinder hätte, und vermöchts, sie müssten mir nicht allein die Sprachen und Historien hören, sondern auch singen und die Musika mit der ganzen Mathematika lernen."

[3] Born at Coburg; died at Frankfort a. M., June 24, 1564. His arithmetics appeared at Frankfort a. M. in 1557 and 1565.

[4] Discussed in Volume II, Chapter VI.

General Conditions. The general conditions in the country, aside from those imposed by dynastic ambitions, were favorable to commercial and intellectual advance. The Netherlands had shown considerable commercial activity in the 15th century. The economic decline in the latter part of that period, owing to British rivalry, had been somewhat overcome by their success in navigation and in the fishing industry.[1] Their arithmetics reflect this commercial activity,[2] showing an extensive trade with Nürnberg, Frankfort, Augsburg, Danzig, and other German towns; with Cracow, Venice, and Lyons; and with Spain.

In matters of scholarship the work of that great humanist of the North, Erasmus of Rotterdam (1467–1536), had told in Holland as it had told in every other intellectual center of Europe. Leyden was becoming one of the great forces of the world. The Netherlands were ready to do their part.

Mathematics in the Netherlands. The first of the writers of consequence in this territory in the 16th century was Joachim Fortius Ringelbergius,[3] a student at Louvain and a teacher of philosophy and mathematics in various places in France and Germany. He wrote on astronomy, optics, and arithmetic, and his collected works,[4] encyclopedic in character, were published at Leyden in 1531. The book was filled with the usual erudition of the time but contained nothing that advanced the bounds of science.

Late in the 16th century Adriaen van Roomen,[5] a student of both medicine and mathematics in Italy and Germany, became

[1] A. W. Ward, in the *Cambridge Mod. Hist.*, I, chap. xiii; W. Cunningham, *ibid.*, chap. xv.

[2] For a list of these books, not complete, see D. Bierens de Haan, "Bibliographie néerlandaise historico-scientifique," in Boncompagni's *Bullettino*, XIV, 519; XV, 355; hereafter referred to as Bierens de Haan, *Bibliog.* See also the *Rara Arithmetica* for a list of books on arithmetic. There is also a list of Dutch arithmetics in the preface to the 1690 edition of Coutereel's *Cyffer-Boeck.*

[3] Born at Antwerp, *c.* 1499; died *c.* 1536. In the vernacular the name appears as Joachim Sterck van Ringelbergh.

[4] *Ioachimi Fortii Ringelbergij Andouerpiani opera*, Leyden, 1531.

[5] Born at Louvain, September 29, 1561; died at Mainz, May 4, 1615. The Latin form Adrianus Romanus and the French form Adrien Romain are often

professor of these two sciences in Louvain. He then became professor of mathematics at Würzburg, and finally was appointed royal mathematician (astrologer) in Poland. While at Louvain he published the first part of a general work on mathematics,[1] and in this he gave the value of π to seventeen decimal places, an unusual achievement at the time. His other works include one on the treatment of the circle by Archimedes (1597) and one on spherical triangles (1609).

There lived in Holland in the middle of the century an engineer who also had the name of Adriaen. His father's name was Anthonis, and so he was called Adriaen Anthoniszoon. From some connection with Metz,[2] he was known as Metius, a name also used by his sons. He suggested $\frac{355}{113}$ as a convenient value of π, probably oblivious of the fact that it had already been given (1573) by a minor writer in Germany,[3] and certainly ignorant of its use in China several centuries earlier. This Adriaen had a son, also named Adriaen, who was called, after the custom of the time, Adriaen Adriaenszoon, but was also called by his father's geographical name of Metius. This younger Adriaen Metius[4] studied both law and medicine and became (1598) professor of mathematics and medicine in the University of Franeker. Although he wrote on mathematics,[5] his chief contributions were to astronomy. He published his father's value of π, and hence it is commonly attributed to the son, although not due in the final analysis to either.

used. The first name is also spelled Adriaan. See H. Bosmans, *Annales* of the *Société Scientifique*, XXVIII and XXIX (Brussels), and *Bibl. Math.*, V (3), 342.

[1] *Ideae mathematicae pars prima, seu Methodus polygonorum*, Antwerp, 1593.

[2] He himself was probably born at Alkmaar, *c.* 1543. He died at Alkmaar, November 20, 1620. This reason, commonly given for the name Metius, is doubtful. The family name was Van Schelvan (haycock; Latin, *meta*).

[3] Valentinus Otto, or Valentin Otho, also called Parthenopolitanus, a native of Magdeburg. See *Bibl. Math.*, XIII (3), 264.

[4] Born at Alkmaar, December 9, 1571; died at Franeker, September 6 (16, or 18), 1635. The Latin form, Adrianus, is also used.

[5] *Praxis Nova Geometrica per usum circini*, Franeker, 1623; *Arithmetica et Geometrica nova*, Leyden, 1625; *Maet-Constigh Liniael*, Franeker, 1626, describing a kind of slide rule; *Doctrinae Sphericae Libri V*, Franeker, 1591; and other works.

A more humble writer but a more progressive teacher appeared early in the century in the person of Giel Vander Hoecke. Although of no marked scholarship, his arithmetic[1] is worthy of study because of its early use of the plus and minus signs as symbols of operation. Widman had already used them as signs of excess and deficiency, and Grammateus had used them with their modern significance, but they now appear for the first time in the Low Countries.

GEMMA FRISIUS

From a contemporary engraving

Gemma Frisius. The most influential of the various Dutch mathematicians of this century was Gemma Regnier.[2] Having been born in Friesland (Frisia), he was called the Frisian, and was known as Gemma Frisius. He was thirty-two years old when his arithmetic[3] was published, and so favorably did this work strike the popular taste, combining as it did the commercial with the theoretical, that it went through at least fifty-nine editions in the 16th century, besides several thereafter. He also wrote on geography and astronomy, suggesting the present method of obtaining longitude by means of the

[1] *Een sonderlinghe boeck in dye edel conste Arithmetica*, Antwerp, 1514; 2d ed., 1537.

[2] Born at Dockum, East Friesland, December 8, 1508; died at Louvain, May 25, 1555. The name is variously spelled, as Rainer, Renier, Reinerus.

[3] *Arithmeticae Practicae Methodvs Facilis*, Antwerp, 1540. From certain internal evidence (1575 ed., fol. B, 5, *v.*) it is probable that the book was written *c.* 1536.

difference in time, and taking one of the first steps toward the modern methods of triangulation.[1] He became professor of medicine at Louvain in 1541, and his son, Cornelius Gemma Frisius (1535–1577), carried on his work, becoming professor of medicine and astronomy in the same university. Cornelius edited one of his father's works and wrote on astronomy and medicine.

While Gemma Frisius wrote for the Latin schools, a man of the people was needed to write for the common schools, and this man was found in the person of Valentin Menher, a native of Kempten. He wrote in the French language three or four arithmetics[2] that occupied the same position in the Netherlands that Borghi's did in Italy and Recorde's in England. His arithmetic of 1573 includes a certain amount of work in geometry and trigonometry.

Among the Belgians who brought a high degree of scholarship into their work in the editing of the early classics of mathematics the one who stood highest at the beginning of the 16th century was Jodocus Clichtoveus, a native of Nieuport, in Flanders. He spent most of his time in France, assisted in editing Boethius on arithmetic, and wrote a *Praxis Numerandi* which was merely an edition of the algorism of Sacrobosco. He died at Chartres in 1543.

The last of the Dutch writers of the century was Jacob Van der Schuere of Meenen (*c.* 1550–1620), a teacher in a French school in Haarlem. His arithmetic,[3] commercial in character, was one of the first of a long series of popular textbooks of this type that appeared in Holland from 1600 to about 1750.

Stevin. The most influential of all the mathematicians produced by the Low Countries in the 16th century was Simon Stevin.[4] In his younger days he was connected with the government service in Bruges. He traveled in Prussia, Poland,

[1] *Libellus de locorum describendorum ratione*, Antwerp, 1533.

[2] See *Rara Arithmetica*, pp. 250, 346.

[3] *Arithmetica, Oft Reken-const, . . . Door Iacqves Van Der Schvere van Meenen, Nu ter tijdt Francoysche School-meester tot Haerlem*, Haarlem, 1600.

[4] The name also appears in such forms as Stevinus, Steven, Stephan, Stevens. Born at Bruges, *c.* 1548; died at Leyden or The Hague, *c.* 1620.

and Norway and later became a quartermaster general in the Dutch army and director of certain of the public works.[1] His most influential but not his most popular work was an arithmetic,[2] first published in Flemish at Leyden in 1585, and republished the same year in a French translation. The importance of this work lies in the fact that it was the first one to set forth definitely the theory of decimal fractions, a subject that had been slowly developing for a century. Stevin also made the first translation into a modern language of the work of Diophantus, apparently from the Latin text of Xylander.[3]

What made Stevin best known among his contemporaries, however, was his contribution to the science of statics and hydrostatics,[4] a subject naturally occupying much attention in a country like Holland.

7. Spain

Spanish Writers. Spain furnished several native mathematicians of considerable merit in the 16th century.[5] The intellectual atmosphere was not favorable to the development of mathematics, however, and many Spanish scholars settled in France and Italy or at least published their works abroad. It is no reflection upon the honesty of purpose of the Church to say

[1] F. V. Goethals, *Notice historique sur la vie de S. Stevin*, Brussels, 1842; M. Steichen, *Mémoire sur la Vie et les travaux de Simon Stevin*, Brussels, 1841.

[2] The title of the 1585 French edition, published at Leyden, is *L'Arithmetiqve de Simon Stevin de Brvges: Contenant les computations des nombres Arithmetiques ou vulgaires: Aussi l'Algebre, auec les equations de cinc quantitez. Ensemble les quatre premiers liures d'Algebre de Diophante d'Alexandrie, maintenant premierement traduicts en François.*

[3] Xylander's edition, Basel, 1575; Stevin's translation, Leyden, 1585.

[4] *De Beghinselen der Weeghconst*, Leyden, 1586; *De Beghinselen des Waterwichts*, Leyden, 1586; *Weeghdaet*, Leyden, 1586. See also his *Wisconstighe Gedachtenissen Inhoudende 't ghene daer hem in gheoeffent heeft* . . . , Leyden, 1605–1608, with a Latin edition by W. Snell (Leyden, 1608) and a French edition by A. Girard (Leyden, 1634). He also wrote *Problematum Geometricorum Libri V* (Antwerp, 1583) and other works.

[5] J. Rey Pastor, *Los matemáticos españoles del siglo XVI*, Oviedo, 1913; F. Picatoste y Rodríguez, *Apuntes para una Biblioteca Científica Española del Siglo XVI*, Madrid, 1891 (hereafter referred to as Picatoste, *Apuntes*); Acisclo Fernández Vallín, *Cultura Científica de España en el Siglo XVI*, Madrid, 1893. The first and second of these works are particularly valuable. See also G. Loria, "Le Matematiche in Ispagna," *Scientia*, XXV (May, June, 1919).

that the religious fervor of Spain from the 15th to the 18th century turned the thoughts of the intellectual class from mathematics, although such work as was done was due to the clergy. This was especially the case after the compact made at Bologna in 1530 between Charles V and Clement VII.[1] To this influence there should be added that which came from the expulsion of the Jews, a race which had done so much in the Middle Ages to foster the science of mathematics, at least with respect to astrology and the theory of the calendar.

Ciruelo. The earliest Spanish mathematician of the century was Pedro Sánchez Ciruelo,[2] who was professor of theology and philosophy at Alcalá, and was later canon of the cathedral of Salamanca. He published an arithmetic[3] at Paris in 1495, and a general work on mathematics[4] in 1516. He also edited the theoretical work on arithmetic of Bradwardine in 1495 and the *Sphaera* of Sacrobosco in 1498, and in general was a learned exponent of the old school of mathematicians that was then in favor in Paris.

Ortega. The second of the early Spanish writers was Juan de Ortega, a Dominican from Aragon, concerning whose career we know little except that he was living in 1512 and in 1567. He wrote an arithmetic[5] which was published in 1512, both in Barcelona and in Lyons, being the first book on commercial computation known to have been printed in France. It was a popular work, being reprinted in Rome, Messina, Seville, Paris, and Granada. It is purely commercial and includes the usual treatment of computation and the common applications of the time.

[1] R. C. Jebb, in the *Cambridge Mod. Hist.*, Vol. I, chap. xvi.

[2] Born at Daroca, Aragon, *c.* 1470; died 1560.

[3] The 1505 edition has the title *Tractatus Arithmetice Pratice qui dicitur Algorismus.*

[4] *Cursus quattuor mathematicarvm artiū liberaliū*, Paris, 1516. First Spanish edition, Alcalá, 1516, but wanting the geometry, at least in the copy examined by the author.

[5] The title of the Rome edition of 1515 is *Svma de Arithmetica: Geometria Pratica vtilissima.* The geometry consists simply of a little mensuration. In the privilege of Leo X (1515), in this edition, the author is addressed as "DIlecto filio Iohāni de Ortega Hispano Clerico Palētino."

Joannes Martinus Blasius,[1] a Spanish astrologer and arithmetician, published in Paris in 1513 a work on computation. It was popular enough to warrant four editions. The author was one of the earliest writers whose works appeared in print with the spelling *substractio* for "subtraction," a custom followed quite generally by the Dutch and English arithmeticians for several generations. He showed a good knowledge of the classical writers in the domain of mathematics.

Another Spanish scholar who found the scientific work more stimulating in Paris than in his native country was Gaspar Lax.[2] He took a course in theology at Saragossa, taught in the University of Paris, and finally returned to Saragossa as a teacher. His principal work was a prolix treatment of theoretical arithmetic[3] based on Boethius. He also wrote on the Greek and medieval theory of proportion.[4]

One of the most noteworthy treatises on mathematics produced in Spain in the 16th century is the work[5] of Juan Perez de Moya (1562). This writer was born in San Stefano, in the Sierra Morena, studied at Alcalá and Salamanca, and became a canon at Granada. His *Arithmetica* includes calculation, applied arithmetic, algebra, and practical geometry, and contains a considerable amount of interesting historical material.

The last of the Spanish writers of any note, in the 16th century, Jerónimo Muñoz,[6] received his bachelor's degree at Valencia in 1537. He traveled in Italy, taught at Ancona, returned to Spain, and taught for ten years in the University of Valencia. He wrote on arithmetic (1566) and Euclid, but his chief work was on astrology.

[1] This is the name as it appears in the 1513 edition of his arithmetic. In the 1519 edition the name appears as "Ioannes Martinus, Scilicevs." The usual Spanish form is Juan Martínez Silíceo. The original name was Juan Martínez Guijeno, the word *guijeno* meaning *silex*, whence Sileceus, Sileceo, or Silíceo. See V. Reyes y Prósper, "Juan Martinez Silíceo," *Revista* of the *Soc. matem. española*, I, 153.

[2] Born at Sariñena, *c.* 1487; died at Saragossa, 1560.

[3] *Arithmetica speculatiua magistri Gasparis Lax Aragonensis de sarinyena duodecim libris demonstrata*, Paris, 1515.

[4] *Proportiones magistri Gasparis lax*, Paris, 1515.

[5] *Arithmetica practica, y specvlatiua del Bachiller Iuan Perez de Moya*, Salamanca, 1562.

[6] Hieronymus Munyos.

Loss of the Jews. As for the Jews who had once added to the brilliancy of Spain, hardly one of their descendants remained in that country at the opening of the 16th century. The edict of banishment of 1492 had driven out hundreds of thousands of this race, some to slavery, some to death at the hands of pirates, some to the plague-stricken towns of Italy, and some to starvation. Persecution after persecution had accomplished the purpose of those in power, but the result had sapped the strength of Spain, and some of the best thought that would have made for the advance of mathematics was turned to the solving of the problem of self-preservation.[1]

8. Other European Countries

Poland. In the 16th century Poland was one of the most progressive countries of Europe in the field of arithmetic, producing several works by native writers and reprinting a number by foreign scholars.[2] The first of her own arithmetics was the *Algoritmus* of Tomas Kl̄os, which appeared at Cracow in 1538.[3] Later in the century (1561) Benedictus Herbestus (1531–1593), a Jesuit priest, published an arithmetic in Latin.[4] One in the Polish language, chiefly commercial in nature, was written by Giřjka Gőrla z Gőrlssteyna and published at Czerny in 1577. It was not through works like these, however, that Poland contributed to human knowledge, but through those of her greatest astronomer, or perhaps we should say the world's greatest astronomer.

Copernicus. Not all of those who have aided in the progress of mathematics have been primarily mathematicians. As we

[1] For a description of one phase of this movement see J. A. Symonds, *Renaissance in Italy: The Age of Despots*, p. 399 (New York, 1883). On the unpublished material relating to the Jewish contributions to mathematics and astronomy, beginning in the 11th century and closing in the 16th century, see B. Cohen, "Ueber unveröffentlichte Schriften jüdischer Astronomen des Mittelalters," *Jahrbuch der Jüdisch-Literarischen Gesellschaft*, XII (1918), 1.

[2] *Rara Arithmetica*, pp. 32, 97, 123, 190, 260, 303, and 353.

[3] A reprint, edited by M. A. Baraniecki, was published at Cracow in 1889.

[4] *Arithmetica Linearis, eiqꝫ adiuncta Figvrata, cum quibusdam ex compvto necessarijs*, Cracow, 1561. This title is from the edition of 1577.

have already seen, the science of astronomy has always contributed to her sister science, not merely in those centuries in which the name "mathematician" meant an astronomer or an astrologer, but in more recent times, when mathematics and astronomy each outgrew the possibility of mastery by the disciples of the other. Among those whose interest was primarily in astronomy but who stimulated the mathematician to seek for new applications of his science none stands higher than Nicholas Copernicus.[1] He was educated at the University of Cracow (1491–1495), spent some time in the study of law, medicine, and astronomy in the universities of Padua and Bologna, and went to Rome for the purpose of continuing his work in astronomy under the patronage of the pope, Alexander VI. He returned to Poland in 1505, took holy orders, and obtained a canonry at Frauenburg. By 1530 he had completed his theory of the universe, the most significant step ever taken in the science of astronomy, but it was not until 1543 that he published his doctrines.[2] Gutenberg made the free spread of

COPERNICUS
From an early engraving

[1] Born at Thorn, on the Vistula, February 19, 1473; died at Frauenburg, May 24, 1543. He was named after his father, Niklas Koppernigk (died 1483), a native of Cracow. The English form of the name, Copernicus, is so familiar that it is used throughout this work, although the spelling Coppernicus is nearer the original. On his life consult L. Prowe, *Nicolaus Coppernicus*, 3 vols., Berlin, 1883–1884. The literature relating to him is extensive.

[2] *De revolutionibus orbium coelestium*, Nürnberg, 1543.

thought possible, but Copernicus gave the thought; Columbus opened a new world, but Copernicus opened millions of worlds.

The work of Copernicus necessitated the improvement of trigonometry, and for this reason he wrote a treatise on the subject,[1] his single contribution to the literature of pure mathematics.

A generation after Copernicus a Danish mathematician, Thomas Finke (1561–1646), whose name also appears as Finck, Fink, and Finchius, published a work called *Geometria Rotundi* (Basel, 1583), in which he made a number of contributions to trigonometry.

Switzerland. The best known of the Swiss mathematicians of this century was Henricus Loritus Glareanus,[2] whose last name probably comes from the name of his native canton of Glarus. He was a professor in Basel (1515–1521), at the Collège de France (1521–1524), and later in both Freiburg and Basel. He wrote on arithmetic, metrology, and music.

The only other Swiss writer of the period who need be mentioned is Cunradus Dasypodius.[3] He was professor of mathematics at Strasburg and canon of St. Thomas's Church in that city. He had in mind the editing of all the Greek mathematical works, and made a beginning in that direction. His edition of Euclid's *Elements* appeared at Strasburg in 1564, and he wrote a mathematical dictionary.[4]

Portugal. Only a single Portuguese mathematician acquired any considerable reputation in the 16th century. This man was Pedro Nunes (better known by his Latin name of Nonius but

[1] *De lateribus et angulis triangulorum libellus*, Wittenberg, 1542.

[2] Heinrich Loriti Glarean. Born at Mollis, in Glarus, June, 1488; died at Freiburg, Breisgau, May 28, 1563. H. Schreiber, *Heinrich Loriti Glareanus*, Freiburg, 1837. The date of his death may have been March 27, 1563.

[3] Name as given in his *Lexicon*. This is the Latin-Greek form for his family name of Rauchfuss (rough-foot), Greek δασυπόδειος (of a hare), a δασύπους (rough foot) being a hare. Born at Frauenfeld, *c.* 1530; died at Strasburg, April 26, 1600.

[4] *Λεξικον seu Dictionarium Mathematicum, M. Cunrado Dasypodio,* Strasburg, 1573. He also wrote *Institutionum Mathematicarum voluminis primi Erotemata Logisticae Geometriae Spherae Geographiae*, 2 vols., Strasburg, 1593, 1596; *Volvmen primum: mathematicum disciplinarum principia*, 2 vols., Strasburg, 1567, 1570; *Brevis Doctrina de Cometis* (on astrology), Strasburg, 1578.

often called by the Spanish name of Núñez,[1] a scholar of Jewish origin. He studied at the University of Lisbon and later (1530) became professor of moral philosophy in that institution. He was (1544–1562) also professor of mathematics in the University of Coimbra and held the posts of cosmographer to the king, Don João III, and tutor to the royal princes.[2] His only mathematical work was devoted to algebra, arithmetic, and geometry,[3] but he is best known for his works on navigation and astronomy,[4] and for the instrument for the reading of small angles, often called the "nonius," which was the forerunner of the vernier that is seen on transits and calipers.[5] He left several manuscripts on geometry and navigation.

In the early part of the 16th century the interest in the great voyages of the Portuguese led to the publication of various treatises on the sphere and the use of the astrolabe for nautical purposes. Among the earliest of these works was an anonymous one[6] on the astrolabe and the quadrant, the instruments then chiefly used in navigation.

[1] Born at Alcacer do Sal, 1502; died at Coimbra, August 11, 1578. The year of his birth is often given as 1492 and that of his death as 1577. Nunes himself records, however, ". . . sit anno Domini 1502 quo ego natus . . . ," and a 16th century MS. note in a book in the National Library at Lisbon reads: "Natus est hic Doctor año Dñi 1502. Obiit vèro tertio idus Augusti año Dñi 1578." Nothing more authoritative is known. See Guimarães, *Les Math. Portug.*, p. 16, with authorities given. The name also appears as Pedro Núñez Salaciense, and in French as Pierre Nugne. See also Bensaude, *Astron. Portug.*, p. 59; Picatoste, *Apuntes*, p. 218.

[2] His dedicatory epistle of 1564 speaks of himself as "Cosmographo Mayor del Rey de Portugal, y Cathedratico Jubilado en la Cathedra de Mathematicas en la Vniversidad de Coymbra."

[3] *Livro de algebra em arithmetica y geometria*, written *c.* 1532, published at Antwerp in 1564, and reprinted there in 1567 with variations in the spelling of words in the title. The above title is substantially as in Guimarães, *Les Math. Portug.*, pp. 21, 105, 396; but see H. Bosmans, "Sur le 'Libro de algebra' de Pedro Nuñez," *Bibl. Math.*, VIII (3), 154; Cantor, *Geschichte*, II, chap. 59; Picatoste, *Apuntes*, p. 221.

[4] *Tratado da Esphera com a Theorica do Sol e da Lua*, Lisbon, 1537; *Tratado sobre certas duvidas da Navegação*, an appendix to the preceding work.

[5] This is described in his *De Crepusculis Liber unus*, Lisbon, 1542.

[6] *Regimento do estrolabio & do quadrante*, published *c.* 1509. This was reproduced in facsimile in Volume I of J. Bensaude, *Histoire de la Science Nautique Portugaise*, Munich, 1914. See this publication for other works of similar nature in that period.

Jewish Writers in the East. With the expulsion of the Greek Christians at the fall of Constantinople (1453) there returned to the city many Jews who had been subject to persecution under their régime. This movement continued as occasion offered, and notably when the Jews were driven from Spain in 1492 and from Portugal in 1496. The sultan Mohammed II made Moses Kapsali chief rabbi over all the Turkish Jews, and on the latter's death he was succeeded in office by Elia Misrachi.[1] This learned rabbi wrote an arithmetic[2] based to a considerable degree on a work by Rabbi ben Ezra. Knowing both Greek and Arabic, he also drew from each of these sources, particularly from the latter. He was interested in making arithmetic practical and in having the processes thoroughly understood.[3] He seems to have been the first Hebrew writer to treat of finding the sum of the cubes of the first n natural numbers. He also wrote commentaries on the works of Euclid and Ptolemy.

Russia. The only mathematical works known to have been written in Russia in the 16th century were a geometry and an arithmetic, both of which date from about 1587 to 1594.[4] Translations of monographs on western mathematics also appeared in this century, but they only show the low state of science at that time in Russia. This is seen, for example, in the fact that the *Origines* of Isidorus of Seville (570–636) was thought worthy of translation.

9. The Orient

Close of the Dark Ages. We have seen that the East had its dark periods in history just as the West had them, and with each the darkness was greatest just before the dawn. In Asia the gloom was particularly oppressive in the 16th century.

[1] Born probably at Constantinople, *c.* 1455; died at Constantinople, 1526.

[2] *Sefer ha-Mispar* (*Book of Numbers*).

[3] G. Wertheim, *Die Arithmetik des Elia Misrachi,* p. 6. Frankfort, 1893.

[4] N. M. Karamzin, *History of the Russian Empire* (in Russian), X, 259, 436 (Petrograd, 1824); Lavrovsky, *Ancient Russian Schools* (in Russian), p. 180 (Kharkov, 1854); Russian *Encyclopedia* (in Russian).

India was intellectually dead. China was just becoming aware of the extent of Western learning, and seemed discouraged in the effort to advance independently. Japan was not yet awake. For these reasons the 16th century was a dark one for Oriental mathematics.

Islam. Of the heirs to the glory of the scholars of Islam in the field of mathematics the name of only one of any note is found in the records of the century. Behâ Eddîn,[1] as he is generally called, was probably a Persian. He wrote on a variety of subjects and among his works was an elementary textbook on arithmetic[2] and the first part of an exhaustive treatise on the subject.[3]

A single other writer may properly be mentioned with Behâ Eddîn, namely, Mohammed ibn Ma'rûf ibn Aḥmed, Taqî eddîn (1525/1526–1585), who seems to have lived in Constantinople. He wrote on algebra, arithmetic, and astronomy.

India. After the death of Bhāskara (*c.* 1175) there was no great interest shown in the advance of mathematics in India except so far as it related to astronomy. From the 16th century only two names of any note have come down to us. Suryadaśa, who flourished *c.* 1535,[4] and Ganeśa,[5] who lived about the same time, were both commentators on the works of Bhāskara. Suryadaśa refers to Śrīdhara's method of finding the area of a cyclic quadrilateral, and Ganeśa quotes one of Śrīdhara's rules for the area of a segment of a circle.

China. China at this time was experiencing a kind of calm before that influx of European mathematics which was heralded by the great Jesuit leader, Matteo Ricci. K'u Ying-hsiang (*c.* 1550), governor of Yünnan, wrote on algebra and geometry; T'ang Shun-ki (1507–1560) wrote on the mensuration of the

[1] Behâ ed-dîn al-'Âmilî, Mohammed ibn Ḥosein. Born probably at Amul (Amol), near the Caspian Sea, 1547; died at Ispahan, 1622. Nazâm ed-dîn Aḥmed, in his biography, says that he was born at Baalbek. See A. Maare, *Beha Eddin*, 2d ed., Rome, 1864.

[2] The *Kholâsat al-Ḥisâb* (*Essence of Arithmetic*). The work has no particular merit. There was an Arabic-Persian edition, Calcutta, 1812; an Arabic-German edition, Berlin, 1843; and a French translation by Marre, Rome, 1864.

[3] The *Bâhr al-Ḥisâb* (*Ocean of Arithmetic*), left unfinished.

[4] *Bibl. Math.*, XIII (3), 205.

[5] *Ibid.*, 205. Also written Ganeça.

circle; Hsin Yun-lu (*c.* 1590) contributed to the subject of the calendar, more in quantity than in quality; and Ch'êng Tai-wei (1593) wrote a work on arithmetic[1] in which we find the earliest description of the *suan-pan* computation. These writers lacked the genius of some of their immediate predecessors; and they contributed nothing that was commensurate in importance with the productions of the century following.

Japan. The 16th century did not see the awakening of the intellectual Japan as it saw the awakening of Europe. The East was at this time about a century behind the West in this great world movement. Perhaps it is more nearly accurate to say that the 16th century in Japan corresponds more closely to the 13th century in the West; it was a century of preparation.

Probably the chief cause which contributed to this preparation in the field of mathematics was the journey to China made by one Mōri Kambei Shigeyoshi, a scholar in the service of two of the powerful lords of Japan. The story goes that the great hero Toyotomi Hideyoshi, better known as Taikō, having subdued all of the country, decided that he would make his court a great intellectual center. In pursuance of this purpose he sent Mōri to China to acquire and bring back that mathematical knowledge which was so lacking in Japan. Mōri, being a man of humble birth, was not well received in China, and for this reason Taikō made him Lord of Dewa,[2] hoping thus to give him high standing among Chinese scholars. Owing to political and military difficulties, chiefly Taikō's invasion of Korea (1592), Mōri's mission was not successful, but he brought back with him a considerable amount of material and is said by some to have made the Chinese abacus[3] known in Japan. His last years were spent in Kyōto in teaching the use of this instrument.

Although this is the story as often told, there is a question as to whether Mōri really visited China or went only to Korea.

[1] *Suan-fa Tong-tsung* (*A systematized treatise on arithmetic*). His name also appears as Tch'eng Ta-wei. [2] *Dewa no Kami.*

[3] The *suan-pan*. It developed later into the Japanese *soroban*. Both of these instruments are discussed in Volume II.

It seems certain, however, that he knew something of Chinese mathematics, that he was an expert with the abacus, that he advertised himself as "the leading instructor in division in the world," and that he was a very successful teacher. Among his pupils were three men,[1] known to their contemporaries as "The Three Arithmeticians," who will be mentioned later. That Mōri was the first who took the abacus from China to Japan is very doubtful, but he seems to have made it popular.

10. The New World

General Conditions. One would not expect to find a treatise on mathematics printed in the New World within sixty-four years of its discovery by Columbus, and still less would he think that only forty-five years after this great discovery there was set up a press for the disseminating of knowledge among the inhabitants of the western hemisphere. Each of these events stands out, however, as a historic fact, a testimony to the zeal and foresight of the early conquerors.

Among the adventurous band organized by Cortés for his first expedition to Yucatan in 1518 was a young chaplain whose name appears in his work of 1556 as Juan Diez. He was of a literary turn of mind, as is shown by three or four books which he published. One of these works was on mathematics, and this appeared in Mexico under the following title:[2]

¶ Sumario cõpẽdioſo delas quẽtas de plata y oro q̃ en los reynos del Piru ſon neceſſarias a los mercaderes: y todo genero de tratantes. Cõ algunas reglas tocantes al Arithmetica.
Fecho por Juan Diez freyle.

It should be stated, however, that there were several writers at this time by the name of Juan Diez (Diaz), two of them

[1] Yoshida Shichibei Kōyū, Imamura Chishō, and Takahara Kisshu.
[2] D. E. Smith, *The Sumario Compendioso of Brother Juan Diez*, Boston, 1921.

apparently being in Mexico, and there is much uncertainty among the best Spanish biographers as to which was the author of the *Sumario*.

Printing Established in Mexico. The first viceroy of New Spain, which included the present Mexico, was a man of remarkable genius and of prophetic vision,—Don Antonio de Mendoza. He assumed his office in 1535, and for fifteen years administered the affairs of the colony with such success as to win for himself the name of "the good viceroy." He founded schools, established a mint, ameliorated the condition of the natives, and encouraged the development of the arts. In his efforts at improving the condition of the people he was ably assisted by Juan de Zumárraga, the first bishop of Mexico. Among the various activities of these leaders was the arrangement made with the printing establishment of Juan Cromberger of Seville whereby a branch should be set up in the capital of New Spain.

The idea of setting up a press in Mexico seems to have been considered as early as 1534, even before Mendoza became viceroy, doubtless at the suggestion of Juan de Zumárraga; but it was not until 1536 that the plan was carried out. Juan Cromberger then sent over as his representative one Juan Pablos, a Lombard printer, and so the "casa de Juan Cromberger" was established, prepared to spread the doctrines of the Church to the salvation of the souls of the unbelievers. Cromberger himself never went to Mexico, but his name appears either on the *portadas* or in the colophons of all the early books. From and after 1545, however, the name is no longer seen, Cromberger having died shortly before this time.

Nature of the Sumario Compendioso. The *Sumario Compendioso* consists of one hundred and three folios, generally numbered. After the dedication there is an elaborate set of tables, including those relating to the purchase price of various grades of silver, to per cents, to the purchase price of gold, to assays, and to monetary affairs of various kinds.

The mathematical text consists of twenty-four pages besides the colophon. Of these pages, eighteen relate chiefly to arithmetic and six to algebra. The arithmetic includes problems in the reduction of maravedis to pesos, of ducats to crowns, and the like; but it also considers questions relating to the preceding tables and to simple commercial transactions in general. There are also problems relating to the theory of numbers, some of them involving rules similar to those found in the works of Fibonacci and Diophantus. Of the latter type the following is an example:

Una q̃rta	p̃s	t.xlvj.mr̃s.	
media on	p̃s	j.t.xxxvij.	ij
j.on	p̃s	iiij.t.xviij.	iij
ij.on	p̃s	vj.t.xxxvij.	ij
iij.on	j.p̃s	ij t.	
iiij.on	j.p̃s	v.t.xviij,	iij
v.on	ij p̃s	t.xxxvij.	ij
vj.on	ij.p̃s	iiij.t.	
vij.on	ij.p̃s	vij.t.xviij.	iij.
j.mr̃os	iij p̃s	ij.t xxxvij	ij
ij.mr̃os	vj p̃s	v.t.xviij.	iij.
iij mr̃os	x. p̃s	t.	
iiij.mr̃os	xiij.p̃s	ij.t.xxxvij.	ij
v.mr̃os	xvj.p̃s	v.t.xviij.	iij.
vj.mr̃os	xx.j p̃s	t.	
vij.mr̃os	xxiij.p̃s	ij.t.xxxvij.	ij
viij mr̃os	xxvj.p̃s	v.t xviij.	iij.
ix.mr̃os	xxx. p̃s	t.	
x mr̃os	xxxiij. p̃s	ij t.xxxvij.	ij
xx.mr̃os	lx vj.p̃s	v.t.xviij.	iij.
xxx.mr̃os	c. p̃s	t.	
xl. mr̃os	cxxxiij p̃s	ij.t.xxxvij	.ij.
xl j.mr̃os	cxxxvj p̃s	v.t.xviij.	iij
xlij.mr̃os	cxl. p̃s		
xliij mr̃os	cxliij.p̃s	ij.t.xxxvij.	ij
xliiij mr̃os	cxlvj.p̃s	v.t.xviij.	iij.
xlv.mr̃os	cl. p̃s		

PART OF THE TABLE OF THE *SUMARIO*

The entire table covers nearly one hundred and eighty pages, and the above facsimile fills half of one page

Give me a number which, increased by 15, is a square number; and decreased by 4 is also a square number. *Rule*: Add 15 and 4, making 19; then add 1 to this result, making 20. Now take the half of this number 20, which is 10; square this result thus: 10 times 10 is 100. From this subtract 15, and we have 85, and this is the number required, that is, the one from which if you subtract 4 you have 81, the root of

25. Re. y da.		24.
100. Re y da.		96.
169. Re. y da.		120.
225. Re. y da.		216.
289. Re. y da.		240.
400. Re. y da.		384.
625. Re. y da.	336\y,	600.
676. Re. y da.		480.
841. Re. y da.		840.
900. Re. y da.		864.
1156. Re. y da.		960.
1225. Re. y da.		1176.
1212. Re. y da.		1080.
1681. Re. y da.		720.
2025. Re. y da.		1944.
2500. Re. y da.	1344	2400.
2602. Re. y da.		2160.
2704. Re. y da.		1920.
2809. Re. y da.		2520.
3025. Re. y da.		2905.
3364. Re. y da.		3360.
3600. Re. y da.		3456.
3221. Re. y da.		1320.
4225. Re. y da.		2026.

CONGRUENT AND CONGRUOUS NUMBERS FROM THE *SUMARIO* OF JUAN DIEZ

He defines the terms thus: "A congruous number is such a square number that, subtracting from or adding to it another number, called a congruent number, it will still be a square." There are five errors in the table here given

which is 9. The same thing happens if you add the 15, the result being a hundred, the root of which is 10; for 10 times 10 is 100, which checks.

Some idea of the work in the theory of numbers may be obtained from the table of congruous and congruent numbers here shown in facsimile, and of the nature of the algebraic problems from the following example:

> A man has mares and cows in quintuple proportion, in such a way that if you square the number of mares and square the number of cows, the products added will be 1664. Required the number of mares and the number of cows.

Apparently the author had some taste for mathematics and was fairly well versed in algebra as it was then known in the best schools of Spain and Italy. When we reflect that only two important treatises on algebra had at that time been issued from the European presses, and that the *Sumario* was the first mathematical work to be published on another continent, the credit due to Diez is the more apparent.

TOPICS FOR DISCUSSION

1. Reasons for the great advance made in mathematics in the 16th century.

2. Causes of the prominence of Italy in mathematical research in the 16th century.

3. Influences leading to the predominance of algebra in the field of mathematics in the 16th century.

4. The leading features in the progress of algebra in the 16th century, with a consideration of the countries and individuals most closely connected with this progress.

5. A comparison of the mathematics of England, France, Italy, and the Teutonic countries in the 16th century.

6. Influences leading to the advance in mathematics in the Netherlands at this time.

7. The leading treatises on algebra in this century, with a consideration of the important features of each.

8. The life and works of Robert Recorde. His influence upon British mathematics.

9. The way in which the arithmetics of the various countries met the commercial needs of the time.

10. The life and works of Michael Stifel.

11. Influences leading to the development of trigonometry in Europe at this time.

12. Effect of the Renaissance upon the nature of mathematics in the 16th century.

13. The nature of Oriental mathematics in the 16th century.

14. A consideration of the causes of the backward state of mathematics in Spain and Portugal in the 16th century.

15. The life and works of Copernicus and his influence upon mathematics in general.

16. The field of mathematical activity developed by the Jewish scholars in the 16th century.

17. The general nature of the literary activity in Mexico in the 16th century, with special reference to the need which produced the work of Juan Diez.

18. The revival of the study of the works of the classical writers in the 16th century, with special reference to its influence upon mathematics.

CHAPTER IX

THE SEVENTEENTH CENTURY

1. General Conditions

Political Situation. In order to comprehend the causes of the remarkable advance of mathematics in the 17th century it is necessary to consider briefly the political and social influences that made this century conspicuous in science, in letters, and in the development of human rights. This was the century that saw broken forever in the Anglo-Saxon civilization the doctrine of the divine right of kings; that saw the beginning of the end of this same doctrine in France in the brilliant reign of Louis XIV; and that saw Russia amalgamated into a powerful nation by a powerful leader, Peter the Great. In this century, too, we may possibly find one of the early steps toward the World War of 1914–1918, namely, the founding of the military machine of Prussia at the hands of the Great Elector of Brandenburg. This was also the period in which Europe saw the turning back of the Turks by the Hapsburgs in Austria; in which the New World was definitely opened to colonization and trade; and in which the Thirty Years' War (1618–1648) disturbed the political and religious life of a considerable part of Europe. Great world activities like these could not but affect natural science as well as political, and abstract science as well as natural.

The Trend to the North. There was also the general reason why mathematics, like all intellectual pursuits, should trend to the north,—the reason that had slowly led it away from the warmer countries from time immemorial,—the ability to conquer the cold and the darkness of the long winter nights. Heat and light have always joined with the soil and the distribution of moisture to make conditions favorable for intellectual work. While coal was known in England as early as the 9th century, the 16th and 17th centuries saw a great advance in the comforts of living north of the Alps, and hence in the ability to utilize the long winter nights in intellectual pursuits. Other influences were evidently more powerful, but this one must be recognized as being somewhat significant.

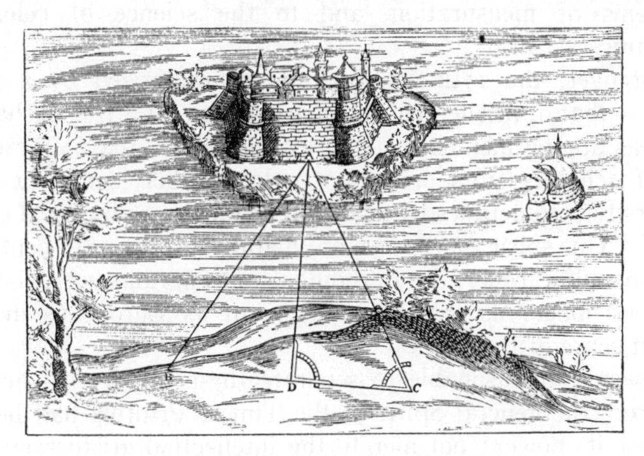

MATHEMATICS IN MILITARY AFFAIRS

Measuring the distance to a castle by the aid of a form of quadrant. From Mario Bettino's *Apiaria*, Vol. I, Bologna, 1645

Mathematical Situation. Mathematics does not develop by centuries any more than by political units or by religious faiths. Vieta and Harriot were quite as much of the 16th as of the 17th century, just as Erasmus was quite as much of England and Germany and France as of the Low Countries, and as Stifel the Lutheran was quite as good an algebraist as Stifel the monk. Nevertheless it is convenient to mark off blocks of time, and centuries serve the purpose fairly well in mathematics as in art and in politics.

It is impossible to say with truth that this century or that is the greatest in the development of any human interest, but it is entirely within the range of truth to assert that few if any centuries did so much for mathematics as that one which saw Fermat begin the modern theory of numbers and, with Descartes and Harriot, invent the analytic geometry; Cavalieri pave the way for Newton and Leibniz, who, in their turn, established the calculus; Pascal and Desargues open new fields for pure geometry; Napier reveal to the world a new method of computation; and a large number of brilliant scholars apply the theories thus developed to the study of curves, to difficult problems of mensuration, and to the science of celestial mechanics.

Moreover, the 17th century was characterized by a new spirit,—that of intellectual internationalism, of a free exchange of ideas among members of the learned class, and of the calling of scholars from one country to another, sometimes by universities, sometimes by academies, sometimes by royal command. This spirit was even more manifest in the 18th century, when men like Euler, the Bernoullis, and Lagrange were looked upon no longer as national assets, but as Europeans in the largest sense of the term.

It seems unreasonable to separate these intellectual activities from the general Spirit of the Times. Printing had begun to show its power; not merely the intellectual aristocracy but the people in general were beginning to think; the scholar could now make his discoveries known, and his audience was no longer composed literally of those who heard his voice; and with the breaking of religious and political canons came also the breaking of the canons of science.

As an incident in the military activity of this period the mathematics of warfare became even more prominent than it was in the 16th century,—a symptom of the terrible part that it was to play some three hundred years later in the great World War.

Effect of Skepticism in General. Skepticism (not religious skepticism in particular, but skepticism with respect to tradition in general,—skepticism which Buckle has called "hardness of belief") has been the *sine qua non* of progress. This is as true in mathematics as it was in the combat with the physics of Aristotle, with the music of Boethius, with the canons of

Giotto, or with the divine right of kings. Just so long as Euclid was sacrosanct in elementary geometry, or Apollonius in conics, or Ptolemy in astronomy, or Boethius in arithmetic, the world could not progress in mathematics. It was only when men began to doubt the infallibility of these ancient leaders that they developed a new conception of geometry, a new way of handling conics, a new system of the universe, and a new

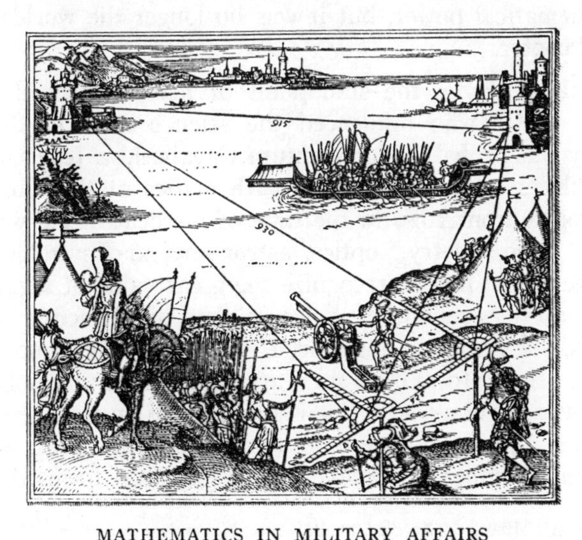

MATHEMATICS IN MILITARY AFFAIRS

From Leonhard Zubler's work on geometric instruments, Zürich, 1607

view of arithmetic. The world began to get skeptical of authority in the 16th century, and the leaven of new ideas worked so rapidly that the opening of the 17th century saw the time ripe for an entire recasting of mathematical theories.

2. Italy

Shifting the Center. The 17th century saw a shifting of the center of mathematical activity from Italy northward. One cause has already been mentioned, and other causes are not difficult to see, being largely political. In general, mathematics flourishes where the environment is favorable, and in the 17th century the political environment was more favorable in France and England than in Italy. The glory of Venice was rapidly fading out; Pisa was no longer a seaport; Florence had ceased to be the source of business customs; Rome had lost her hold upon the most vigorous parts of Europe; and the rivalry between the Italian states was not of that healthy nature which makes for intellectual progress. Italy was still a mathematical power, but it was no longer the world's intellectual center.

Cavalieri. From the standpoint of mathematics alone the Italian writer who influenced the science most in the 17th century was probably Bonaventura Cavalieri,[1] a Jesuit, a pupil of Galileo's, and professor of mathematics at the University of Bologna from 1629 to the time of his death. He wrote on conics,[2] trigonometry,[3] optics, astronomy, and astrology, and was one of the first to recognize the great value of logarithms. His greatest contribution, however, was his principle of indivisibles,—a principle announced by him in 1629 but not set forth in printed form until six years later.[4] The theory is based upon the assertion that a line is made up of an infinite number of points, a plane of an infinite number of lines, and a solid of an infinite number of planes. The theory thus forms the basis

[1] Born at Milan, 1598; died at Bologna, November 30, 1647. The year of his birth may, however, be 1591. He may have died December 1, 1647, but it was the night of November 30–December 1. See P. Frisi, *Elogio del Cavalieri*, Milan, 1778; G. Piola, *Elogio di Bonaventura Cavalieri*, Milan, 1844; F. Predari, *Della Vita e delle Opere di Bonaventura Cavalieri*, Milan, 1843; A. Favaro, *Bonaventura Cavalieri nello studio di Bologna*, Bologna, 1888. The name appears also as Cavallièri, Cavaglieri, Cavalerius, and de Cavalleriis. His place of birth is given by Piola as Milan, although others give it as Bologna. The tomb records the date of his death and the fact that he was a native of Milan.

[2] *Lo specchio ustorio, overo Trattato delle settioni coniche*, Bologna, 1632.

[3] *Directorium generale uranometricum in quo Trigonometriae logarithmicae fundamenta ac regulae demonstrantur, astronomicaeque supputationes ad solam fere vulgarem eruditionem reducuntur*, Bologna, 1632; *Compendio delle regole dei triangoli colle loro dimostrazioni*, Bologna, 1638; *Trigonometria plana et spherica*, Bologna, 1643.

[4] *Geometria Indivisibilibus continuorum nova quadam ratione promota*, Bologna, 1635; 2d ed., Bologna, 1653; *Exercitationes Geometricae sex*, Bologna, 1647.

of a crude kind of calculus, and by its aid Cavalieri found it possible to solve many problems in mensuration that would now be solved by the more scientific methods of integration. The term "indivisible" is ancient, and Cavalieri made no claim to originality in its use.

BONAVENTURA CAVALIERI

From an engraving by G. A. Labus after a drawing by A. Alfieri

Galileo. Much more widely known than Cavalieri,—more widely known, indeed, than most men of his time,—Galileo Galilei[1] was destined to bring great glory to Italy in general and to Florence in particular. Born on the day of Michelangelo's death, and dying in the year of Newton's birth, he seemed to fill the gap between the lives of these two great leaders, a stirring period in the history of art, letters, politics, science, and religious thought; and in the numerous controversies relating to all these lines he played a major part.

He was the son of a certain Florentine nobleman, a dilettante in music and mathematics, whose estate had become so greatly

[1] Born at Pisa, February 18, 1564; died at Florence, January 8, 1642. The literature relating to Galileo is extensive. A good bibliography, particularly as to his life and the controversy over the inquisition question, is given by Karl von Gebler, *Galileo Galilei and the Roman Curia,* English ed., London,

reduced as to indicate for Galileo a life devoted to the restoration of the family fortune,—the life of a cloth merchant. By

GALILEO GALILEI, 1564–1642

From the painting in the Pitti Gallery in Florence, school of Sustermans (Suttermans, 1597–1681). Robinson's *Medieval and Modern Times*

some good chance, however, he was sent to the convent of Vallombrosa, and here he displayed such unusual powers that

1879, p. xxix. See also P. Frisi, *Elogio del Galileo*, Milan, 1774; G. P. C. de' Nelli, *Vita e commercio letterario di Galileo Galilei,* 2 vols., Lausanne, 1793; A. Favaro, *Galileo Galilei e Suor Maria Celeste,* Florence, 1891, with valuable material relating to his contemporaries. Viviani's sketch of the life of his great teacher may be found in the Milan edition of the *Opere di Galileo Galilei* of 1808–1811, Vol. I. The best edition is the one published under the editorship of A. Favaro, Florence, 20 vols., 1890–1909, with a biographical index in Volume XX.

his father changed his mind and decided that he should study medicine. He entered the University of Pisa in 1581, in a period of great intellectual upheaval, but his medical studies were soon put aside by an incident that was to help change the scientific thought of the world. The beautiful lamp that still hangs in the cathedral at Pisa had been moved from its vertical position in order the more readily to light it, and Galileo noticed that the oscillations were at first considerable but gradually became less and less. They seemed, however, to be made in equal periods of time, and this inference he confirmed by comparing them with his pulse. Thus he was able to establish the approximate isochronism of the vibrations (a fact which had been asserted by the Arabs) and to make a beginning in medical diagnosis by accurately timing the arterial beats. Galileo was also led at this time, it seemed by chance, and certainly against his father's wishes, to the study of geometry. His success was such that at last he secured parental consent to give up medicine and devote himself to science. He soon became well known throughout Italy, and in 1589 was made professor of mathematics at Pisa. It is indicative of the low esteem in which mathematics was then held that, whereas the professor of medicine received the equivalent of about $2150 a year, Galileo was rewarded by a salary equivalent to only $65. It was here that he began his experimental work in physics, but owing to local controversies he resigned his chair in 1591. The next year he was offered the professorship of mathematics at Padua, and here he was enabled to carry on some of his most important scientific work.

Of his controversies in astronomy, of his construction of the first satisfactory telescope,[1] of his invention of the modern type of microscope, and of his work in physics this is not the place to speak. It should be said, however, that his interest in mathematics was maintained throughout his stormy life. While at Padua he invented the proportional

[1] On the invention of the telescope the first and one of the best of the standard works is P. Borel, *De Vero Telescopii Inventore*, The Hague, 1655–1656.

compasses,[1] an instrument which was in great favor for a century or more but which of late has been generally discarded.

Torricelli. Of those physicists who sat at the feet of Galileo and who also showed an interest in pure mathematics no one was more celebrated than Evangelista Torricelli.[2] Although more than forty years the junior of Galileo, he survived him by only five years, dying at the age of thirty-nine. He had studied under a pupil of this great master, but was also privileged to receive instruction from the latter himself, then blind and enfeebled by age, in the last year of his life. Not only did Torricelli write on physical questions and give to the world the barometer, but he contributed to geometry as well,[3] anticipating the work of Roberval on the method of tangents. Père Mersenne had announced to Galileo in 1638 that Roberval had squared the cycloid, a curve to which Galileo had first called attention. Galileo thereupon sent the letter to various friends, and Torricelli responded by squaring the cycloid, and Viviani by determining the tangent.

EVANGELISTA TORRICELLI

After a contemporary engraving

[1] *Le Operazioni del Compasso Geometrico et Militare*, Padua, 1606. See also A. Favaro, "Per la storia del compasso di proporzione," in the *Atti* of the *Istituto Veneto*, LXVII, 2, 723.

[2] Born in or near Faenza, possibly in the village of Modigliana, October 15, 1608; died at Florence, October 25, 1647. See the *Opere* of Torricelli, ed. Loria and Vassura, *Introduzione*, Faenza, 1919; G. Loria, *Atti d. R. Accad. dei Lincei*, XXVIII (5), 409.

[3] *Opera geometrica*, Florence, 1644; F. Jacoli, "Evangelista Torricelli ed il metodo delle tangenti detto Metodo del Roberval," Boncompagni's *Bullettino*, VIII, 265.

Vincenzo Viviani. This Vincenzo Viviani[1] was also a disciple of Galileo's. He was interested in physics and in the applications of mathematics, but his tastes led him even more strongly to geometry.[2] His first work (1659) established his reputation, and he was honored by the Medici, made mathematician to Ferdinand II, the grand duke of Tuscany, elected to membership in various learned societies, and invited to France by Louis XIV and to Poland by King Casimir,—invitations which he felt compelled to decline. In 1692 he proposed to scholars a problem which attracted wide attention, and which may be stated briefly as follows: There is among the ancient monuments of Greece a temple dedicated to Geometry. The plan is circular and the temple is surmounted by a hemispherical dome which has four equal windows of such size that the rest of the surface can be exactly squared. Required to find how this is possible.

The problem appeared in the *Acta Eruditorum* under a designation which is an anagram of the words "A postremo Galilei Discipulo," a title which Viviani was always proud to bear.[3] Of this problem there were submitted correct solutions by Leibniz, Jacques Bernoulli, l'Hospital, Wallis, and David Gregory, but Viviani himself gave the simplest one of all. He also solved the trisection problem by the aid of the equilateral hyperbola.

Minor Writers. Among the minor Italian writers of the 17th century Giovanni Antonio Magi'ni,[4] a friend of Kepler, was widely known as the maker of the most perfect maps of Italy up to that time, and as professor of astrology, astronomy, and mathematics in the University of Bologna for nearly thirty

[1] Born at Florence, April 5, 1622; died at Florence, September 22, 1703.

[2] *De Maximis, et Minimis Geometrica Divinatio in Qvintvm Conicorum Apollonii Pergaei adhvc desideratvm*, Florence, 1659; *Quinto libro degli Elementi d' Evclide*, Florence, 1674; *De Locis Solidis Secunda Divinatio Geometrica*, Florence, 1673 and 1701; and other works.

[3] Viviani's solution appeared in his *Formazione e misure di tutti i cieli, con la struttura e quadratura esatta dell' intero, e . . . uno degli antichi delle volte regolari degli architetti*, Florence, 1692.

[4] Born at Padua, June 13, 1555; died at Bologna, February 11, 1617 or 1615.

years (from 1588). While his interest was chiefly in astronomy, on which he wrote extensively, he also contributed to the theory of numbers[1] and to trigonometry.[2]

Marino Ghetaldi,[3] another of the minor writers of the period, was descended from a patrician family and divided his time between scientific and diplomatic pursuits, working in the field of mathematics and acting as ambassador for Venice to Rome and to Constantinople. In his chosen field of science he still further divided his interests, this time between pure mathematics and its physical applications. He wrote upon geometry[4] and algebra[5] but contributed little that was original.

Giovanni Alfonso Borelli,[6] a third in the list of lesser mathematicians, studied at Pisa and then taught philosophy and mathematics at Messina (1649). In 1656 he was recalled to Pisa to take the chair of mathematics. He was also a physician, and his posthumous work *De motu animalium* (1680–1685) was highly esteemed. He edited some of the Greek classics on mathematics.[7]

The Cassini Family. Although classified among the minor writers from the standpoint of their contributions to pure mathematics, several of the members of the remarkable Cassini family would stand among the leaders if we considered

[1] *Tabula tetragonica seu quadratorum numerorum, cum suis radicibus,* Venice, 1592.

[2] *De Planis Triangvlis Liber Vnicus*, and *De Dimetiendi ratione per Quadrantem, & Geometricum Quadratum, Libri Qvinqve*, Venice, 1592; *Tabulae et canones primi mobilis; item calculus triangulorum sphaericorum,* Venice, 1604.

[3] Born at Ragusa, 1566; died probably at Constantinople, but possibly at Ragusa, 1626 or 1627.

[4] *Nonnullae propositiones de parabola*, Rome, 1603; *Apollonius redivivus seu restituta Apollonii pergaei de inclinationibus Geometria*, Venice, 1607; and two other works. See also H. Wieleitner, *Bibl. Math.*, XIII (3), 242; E. Gelcich, *Abhandlungen*, IV, 191.

[5] *De resolutione et compositione mathematica, libri quinque, Opus posthumum,* Rome, 1630,—a work on algebra applied to geometry.

[6] Born at Castelnuovo, near Naples, January 28, 1608; died at Rome, December 31, 1679.

[7] *Euclides restitutus*, Pisa, 1658; *Apollonii Pergaei conicorum libri V, VI, et VII*, Florence, 1661; *Elementa conica Apollonii et Archimedis opera*, Rome, 1679.

their work in the field of astronomy. At any rate they deserve to be mentioned in this connection because of their skillful application of mathematics in their chosen field of science. The founder of the astronomical line was Giovanni Domenico Cassini,[1] professor of astronomy at Bologna (1650). Louis XIV asked that he be sent to Paris (1669), and soon after

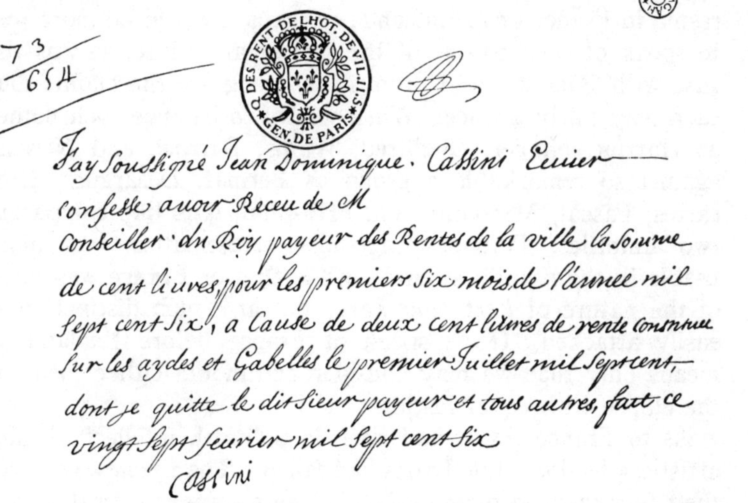

73/654

J'ay Soussigné Jean Dominique Cassini Escuier
Confesse auoir Receu de M
Conseiller du Roy payeur des Rentes de la ville la Somme
de cent liures pour les premiers six mois de l'annee mil
Sept cent six, a Cause de deux cent liures de rente Constituee
sur les aydes et Gabelles le premier Juillet mil Sept cent,
dont je quitte le dit Sieur payeur et tous autres, fait ce
vingt Sept feurier mil Sept cent six
Cassini

AUTOGRAPH OF GIOVANNI DOMENICO CASSINI

From a receipt written on parchment, February 27, 1706. The French form of the name was adopted after he went to France in 1669

he became (1671) the first astronomer royal of France. Since he became naturalized in France and his son Jacques Cassini[2] was born there, the line now ceases to be Italian. Jacques, the second son, succeeded his father (1712) as astronomer royal, and, like him, wrote numerous monographs on astronomy. César-François Cassini de Thury,[3] the son of Jacques, succeeded his father (1756) as astronomer royal, and was in turn

[1] Born at Perinaldo, June 8, 1625; died in Paris, September 14, 1712.
[2] Born in Paris, February 18, 1677; died at Thury, April 16, 1756.
[3] Born in Paris, June 17, 1714; died in Paris, September 4, 1784.

succeeded by a son, Jacques Dominique Cassini de Thury.[1] Each of these members kept up the traditions of the remarkable family with respect to its contributions to science.

3. France

France and England. It is of little moment whether we say that the center of mathematical activity in the 17th century rested in France or in England; perhaps it would be more just to speak of each as one of the foci of an ellipse, as was the case with Bagdad and Cordova, so strong are the claims that each may fairly adduce. When we try to balance such names as Harriot, Napier, Oughtred, Wallis, Barrow, and Newton against so remarkable a group as Fermat, Desargues, Descartes, Pascal, Mersenne, and l'Hospital, it is like comparing two infinities. Perhaps England's mathematics was more usable in the natural sciences, while that of France was more of the nature of *l'art pour l'art*; but any such distinction is easily attacked. If we speak of France before England, it means only that we may consider at random either focus of the ellipse with equal justice.

As to France, Paris had risen in political, intellectual, and artistic splendor, while Lyons had fallen. The Lyonese still had their four great fairs annually, and were content with their lot. The two cities still represented, however, the mathematics of the country.

Early Writers. Among the early French writers of this century, Denis Henrion[2] published the first logarithmic table to appear in France.[3] Contemporary with him was the learned scholar Claude Richard,[4] who entered the Jesuit order in 1606, taught mathematics in Lyons for a number of years, was called to Spain in 1624, and became professor of mathematics at

[1] Born in Paris, June 30, 1748; died at Thury, October 18, 1845.

[2] Born *c.* 1590; died *c.* 1640.

[3] *Traicté des logarithmes*, Paris, 1626. He also published a *Collection . . . de divers Traictez Mathematiques*, Paris, 1621; the *Logocanon, ou Regle Proportionelle*, Paris, 1626; and other works.

[4] Born at Ornans, Burgundy, 1589; died at Madrid, October 20, 1664.

Madrid. He wrote commentaries on Euclid's *Elements* (1645), the *Conics* of Apollonius (1655), and other Greek works.

Another writer of this period, Pierre Hérigone, whose life was one of comparative obscurity, published a work on general mathematics[1] which stands out as a good summary for the time, and which displayed considerable originality in the field of algebra.

Descartes. If one were asked to name the man who was most influential in the revolutionizing of mathematics in the 17th century, he would naturally find it difficult to answer. Probably the name of Newton would lead in any ballot among scholars. Newton's modest assertion that he had seen farther only by standing on the shoulders of giants[2] is capable of easy proof, as is the case with most men of eminence. It is the genius, however, who can pick out his giants. Certainly among those selected by Newton or by Fate was René Descartes,[3] a man whose varied genius led philosophers to rank him primarily as one of themselves, physicists to claim him for their guild, and mathematicians to look upon him as one of the greatest geniuses in their domain, each group being fully justified in its own opinion.

Descartes was fortunate in his birth, his father, Joachim (died 1640), being a counselor in the parliament[4] of Bretagne (from 1586) and possessed of sufficient means to give the son

[1] *Cursus mathematicus*, 5 vols., Paris, 1634–1637, with a *Supplementum*, 1642–1644.

[2] D. Brewster, *Life of Newton*, I, 142. Edinburgh, 1855.

[3] Born at La Haye, Touraine, March 31, 1596; died at Stockholm, February 11, 1650. Latin, Renatus Cartesius, whence we speak of the Cartesian geometry and the Cartesian philosophy. The earlier English and French writers frequently used the form Des Cartes. Other branches of the family used the older spelling Des Quartes and Des Quartis. René was also called, against his wish, M. du Perron, from a small *seigneurie* belonging to the family and situated near Poitou. This name was used by the family in his younger days to distinguish him from his brother.

The most recent biographies are those of G. Milhaud, *Descartes, Savant*, Paris, 1921, containing various essays theretofore published; Elizabeth S. Haldane, *Descartes, His Life and Times*, London, 1905; and C. Adam, *Descartes*, Paris, 1910. See also J. Millet, *Descartes, sa vie, ses Travaux, ses découvertes, avant 1637*, Paris, 1867.

[4] *Parlement*, little more than a local court.

an opportunity for early advance. When only a child René was placed in a recently founded Jesuit school at La Flèche,

RENÉ DESCARTES

After the painting by Franz Hals, now in the Louvre

in Maine, where he remained for eight years (1604–1612) and where he came to know Mersenne (p. 380), who was seven or eight years his senior. The two established there a friendship

Monsieur

iay receu le contenu de la lettre de change quil
vous a pleu menuoyer et vous en remercie.
ie lauroys gardé vn peu plus long tems pour
tascher de vous le remettre auec quelque
proffit, mais ie ne doute point quil ne profite
dauantage estant entre vos mains quil ne
pourroit faire entre les miennes, et ie suis sur le
point de partir dicy. Ie ne scay que respondre
a la courtoisie de Monsieur Huguens sinon que ie
cheris lhonneur de sa connoissance comme lune de
mes meilleures fortunes, et que ie ne seray iamais
en lieu ou ie puisse auoir le bien de le voir que
ie nen recherche les occasions, ainsy que ie feray
tousiours celles de vous tesmoigner que ie suis

Monsieur

dAmsterdam ce 23 May 1632

Vostre treshumble et
tresaffectionné seruiteur
DesCartes

AUTOGRAPH LETTER OF DESCARTES

Written about the time he was completing the manuscript of *La Géométrie*. The M. Huguens mentioned was the father of Christiaan Huygens, then a child of three years

that endured through life. He was sixteen years old when he finished his course at La Flèche and returned to his home "overwhelmed by the blessings and praises of his teachers." There he remained for a year, his father then deciding that he should complete his education in Paris. Here he renewed his acquaintance with Mersenne, now become a Minimite, but the acquaintance was soon interrupted by his friend's departure for Nevers. He now set about to cultivate the study of geometry, made the acquaintance of Mydorge (p. 378), and came in contact with the best exponents of the mathematical thought of the day. It was not long, however, before he felt the need for a broader knowledge of the world and so decided to enlist in army service. Not sympathizing with the aims of the French nobility, he joined the army of Maurice, Prince of Orange, a man who had already begun to attract to himself a group of scholars, and in this way he became acquainted with Stevin (p. 342). He soon became weary of army life, however, and in 1621 withdrew from the service. He now devoted four years to travel, visiting the German states, Denmark, Holland, Switzerland, and Italy, returning to Paris in 1625. His old friend Mersenne had also returned, Mydorge was still there, Desargues (p. 383) had joined the group, and besides these men of the mathematical coterie he met Balzac and the literary world generally, and was presented to Richelieu. Paris was not the place for his meditations, however, and in 1628 he decided to take up his residence in Holland, staying there until 1649, when, at the invitation of Christina of Sweden, he went to Stockholm, where he remained until his death (1650) a few months later.

His first years in Holland (1629–1633) were given to the preparation of a work on philosophy.[1] On finishing this treatise he learned of the condemnation of Galileo for daring to tell the truth, and so he, quite unlike his Italian contemporary, decided that discretion was the better part of valor. As a result, the work was not published during his lifetime. He then devoted himself to the preparation of his great treatise on

[1] *Le Monde*, Paris, posthumously published in 1664.

method in science,[1] a work which included three appendixes, one of which bore the modest title *La Géométrie.*

Descartes's Geometry. It was in this appendix, a small handbook of only about a hundred pages, that analytic geometry first appeared in print.[2] The fundamental idea in Descartes's mind was not the revolutionizing of geometry so much as it was the elucidating of algebra by means of geometric intuition and concepts; in a word, the graphic treatment of the equation. His imagination extended far beyond this, however, to the establishing of a universal mathematics in which algebra, geometry, and arithmetic should be closely related members. He began by extending the ancient idea of latitude and longitude, showing that any point in a plane is uniquely determined by two coordinates, x and y, the equation $F(x, y) = 0$ expressing a property which is true for every point of the curve. By studying the equation, therefore, he could, through the principle of a one-to-one correspondence, transfer his results at any time to the curve itself. It constituted what John Stuart Mill[3] says was "the greatest single step ever made in the progress of the exact sciences." The *Géométrie* is divided into three books. In the first book he relates the fundamental operations of arithmetic to geometry, his use of a certain unit of length being the only novelty in this feature; in the second book[4] he classifies curves and considers the methods of finding tangents and normals; and in the third book[5] he deals with the nature

[1] *Discours de la méthode pour bien conduire sa raison et chercher la vérité dans les sciences*, Leyden, 1637. The appendixes were *La Dioptrique*, *Les Météores*, and *La Géométrie.* A Latin translation of the geometry, by Frans van Schooten, appeared at Amsterdam in 1649.

[2] The *Opera Omnia* of Descartes appeared at Amsterdam in 9 volumes, 1690–1701, with a second impression in 1713. It also appeared at Paris in 13 volumes, 1724–1726. The V. Cousin edition of *Les Œuvres* appeared at Paris in 11 volumes, 1824–1826, with a new edition by J. Simon in 1844. The latest edition, edited by Charles Adam and Paul Tannery, was printed at Paris, Volume I appearing in 1897. For a bibliography of the most important works relating to Descartes see Haldane, *loc. cit.*, p. 387.

[3] *An Examination of Sir W. Hamilton's Philosophy*, p. 617. London, 1878.

[4] In the Latin edition, "De natura linearum curvarum."

[5] In the Latin edition, "De constructione Problematum Solidorum, & Solida excedentium."

of the roots of equations, considering among other things the rule of signs that has since been known by his name. The idea of analytic geometry had already been worked out by Fermat, or at least was conceived by him and Descartes at about the same time; but Fermat only thought, while Descartes not only thought but wrote. Some conception of the plan was probably also in the mind of Harriot, and Descartes was familiar with Harriot's work;[1] but, after all, the real idea of functionality as shown by the use of coordinates was first clearly and publicly expressed by Descartes.

Some idea of his range of knowledge and interest may also be obtained from his work on anatomy, begun in 1634, and of which a Latin edition appeared at Leyden in 1664.[2]

Descartes the Man. Descartes inherited, apparently from his mother, a feeble constitution. He speaks of his "dry cough and pale complexion," which remained with him until he entered the army, and which led his physician to predict that his life would be a short one. He overcame his early ailments, however, and although he died at fifty-four he was able to say in his later life that for thirty years he had been free from any illness that deserved the name. John Stuart Mill remarked of him: "Descartes is the completest type which history presents of the purely mathematical type of mind—that in which the tendencies produced by mathematical cultivation reign unbalanced and supreme." Although the statement may well be questioned, it is interesting as a striking assertion if for no other reason.

Early Commentators on Descartes. Of the two leading commentators on Descartes in the 17th century the first (1649) was his warm personal friend, Florimond de Beaune,[3] an officeholder at Blois. He also wrote on algebra,[4] being one of the

[1] *Artis Analyticae Praxis*, London, published posthumously in 1631.

[2] *De Homine figuris, et Latinitate Donatus a Florentino Schuyl*, Leyden, 1664. This edition has some engravings of great merit.

[3] Born at Blois, 1601; died at Blois, 1652.

[4] *De aequationum constructione et limitibus*, published posthumously at Amsterdam in 1659. There was also an edition, Amsterdam, 1683.

first to treat scientifically of the superior and inferior limits of the roots of a numerical equation. He endeavored to deduce the nature of curves from the properties of their tangents and was also interested in the improvement of astronomical instruments. The second and more prominent of the commentators was Frans van Schooten[1] (*c.* 1615–1661), who will be considered later.

FERMAT
After an old lithograph

Fermat. The life and works of Pierre de Fermat[2] illustrate the fact that no one can account for a genius. Why should the greatest writer on the theory of numbers, at least after the time of Diophantus, suddenly appear in the person of a modest, retiring, punctilious counselor of the parliament of Toulouse, and in the 17th century? Why should this obscure officeholder succeed in making such a name for himself while apparently giving no serious attention to mathematics until he was over thirty? And why, aware of his powers as he must have been, was he content to make his results known chiefly through letters to men like Mersenne, Roberval, Pascal, and Descartes instead of publishing them for the benefit of scholars in general? One answer to each of these questions is that genius is eccentric.

It may have been Bachet's translation (1621) of Diophantus that directed Fermat's attention to the theory of numbers, for

[1] Known also as Franciscus van Schooten.

[2] Born at Beaumont de Lomagne, near Toulouse, *c.* 1608; died at Castres or Toulouse, January 12, 1665. The date of his birth as given by different writers varies from 1590 to 1608. His tombstone in the church of the Augustines in Toulouse (later in the museum) gives the date of death as above and the age as fifty-seven years.

E. Brassinne, *Précis des œuvres mathématiques de Pierre Fermat,* Paris, 1853; *Œuvres de Fermat*, edited by P. Tannery and C. Henry, Paris, 1891–1912. See also C. Henry in Boncompagni's *Bullettino*, XII, 477 and XIII, 437; P. Tannery in Darboux's *Bulletin*, VII (2); G. Libri, "Fermat," in *Revue des Deux Mondes*, May 15, 1845.

he left a series of notes and letters on this work which were published in the form of a commentary (Toulouse, 1670) after his death. At any rate he showed remarkable ability in this field of mathematics. He asserted that no integral values of x, y, and z can be found to satisfy the equation $x^n+y^n=z^n$ if n is an integer greater than 2, and this is commonly known as Fermat's Theorem. No satisfactory demonstration has ever been published, and it is not known whether Fermat himself demonstrated it, few of his proofs having been preserved. What knowledge we have of these proofs is due in part to marginal notes made by him in his reading, and in part to a certain manuscript of Huygens found at Leyden in 1879. He seems to have claimed that he had proved this particular theorem. Fermat's letters show that he had developed the idea of analytic geometry before Descartes published (1637) his work upon the subject. Descartes proposed to represent a curve by an equation, to study this equation, and in this way to discover the properties of the curve itself; while Fermat did substantially the same thing, designating the equation as the "specific property" of the curve and deriving all other properties from it.

In connection with his study of curves Fermat proceeded to apply the idea of infinitesimals to the questions of quadrature and of maxima and minima as well as to the drawing of tangents. In this he seems to have anticipated the work of Cavalieri, but the date of his discovery is unknown.[1]

Mydorge. Of the influential members of the brilliant group of mathematicians that brought the science prominently to the attention of Paris in the first half of the 17th century Claude Mydorge,[2] a friend of Descartes, must be named among the first. He was a man of means, an official of the government, a

[1] The esteem in which he was held by Descartes is shown in one of the letters which the latter wrote to him: "Je n'ai jamais connu personne, qui m'ait fait paraître qu'il sût tant que vous en géométrie." Cantor, *Geschichte*, II, chap. 79. Fermat's *Varia Opera Mathematica*, 2 vols., edited by his son, Samuel Fermat (Toulouse, 1679), was reprinted in facsimile in Berlin in 1861.

[2] Born in Paris, 1585; died in Paris, July, 1647.

physicist of recognized standing, and a mathematician of fair attainments, writing upon optics, conics,[1] and the recreations of mathematics.[2]

Other Writers on the Theory of Numbers. Fermat was not alone in his interest in the theory of numbers at this time. Indeed, he was not the earliest French scholar of the century to consider the subject, for Diophantus had been made known in France before Fermat showed any interest in the subject. The man responsible for this initial step in the theory of numbers was Claude-Gaspar Bachet, Sieur de Méziriac,[3] mathematician, philosopher, theologian, poet, and one of the ablest writers of his day. He came from an ancient and noble family[4] and passed some part of his youth in Italy. He took the first steps toward entering the Jesuit order, but abandoned the idea and went to Paris, where he became a member of the Académie des Sciences. His work on mathematical

MERSENNE
Engraving by Duflos

[1] *De Sectionibus Conicis Libri IV*, Paris, 1631. See also the "Problèmes de Géométrie Pratique" in Boncompagni's *Bullettino*, XVI, 514.

[2] He edited the popular *Recreations Mathematiqves* (title as in the 1628 ed.) of Leurechon (1st ed., 1624). Mydorge's editions were Paris, 1630, 1634, 1639, and later. See Boncompagni's *Bullettino*, XIV, 271.

[3] Born at Bourg-en-Bresse, October 9, 1581; died at Bourg-en-Bresse, February 25, 1638.

[4] For particulars and for biographical information in general see P. Bayle, *Dictionaire historique et critique*, Paris, 1734, with numerous bibliographical references; hereafter referred to as Bayle, *Dictionaire*.

I

recreations[1] was the best of all that appeared in the 17th century and is still looked upon as a classic in that field, both in style and in content. His well-known translation of Diophantus from the Greek into Latin was published in 1621.[2]

AUTOGRAPH LETTER OF MERSENNE

Written about 1640

Another contributor to the theory of numbers at this time appeared in the person of Marin Mersenne,[3] a Minimite friar, who taught philosophy and theology at Nevers and Paris and was in constant correspondence with the greatest mathematicians

[1] *Problemes plaisans et delectables, qui se font par les nombres*, Lyons, 1612. There were later editions,—Lyons, 1624; Paris, 1874; Paris, 1879; Paris, 1884.

[2] *Diophanti Alexandrini Arithmeticorum libri sex*, Paris, 1621. The text appears in Greek and Latin. Fermat's edition appeared at Toulouse in 1670. Xylander's translation had already appeared at Basel in 1575.

[3] Born at Oizé, Maine, December 8, 1588; died in Paris, September 1, 1648. F. H. D. C., *La Vie dv R. P. Mersenne*, pp. 2, 6 (Paris, 1649).

of his day. He was a voluminous writer, editing some of the works of Euclid, Apollonius, Archimedes, Theodosius, Menelaus, and various other Greek mathematicians. He also wrote on a variety of other subjects, including physics, mechanics, navigation, geometry, and mathematical and philosophical recreations. It is in the theory of numbers, however, particularly with respect to prime numbers and perfect numbers,[1] that he made contributions of real value.

BLAISE PASCAL

After a contemporary drawing

Pascal. Blaise Pascal,[2] whom Bayle[3] appreciatively calls "one of the most sublime spirits in the world," was blessed in having a father who could and did start him in the right direction. Étienne Pascal was an able mathematician and was so desirous of giving the best advantages to his only son that he relinquished his post of Président à la Cour des Aides of his province and went to Paris in 1631. Educated solely under his care, Blaise showed phenomenal ability in mathematics at an early age, and although his father wished him first to have

[1] His *Cogitata Physico-Mathematica*, Paris, 1644, appeared four years before his death. On the nature of Mersenne's Numbers see W. W. R. Ball, *Messenger of Mathematics*, XXI, 34, 121; but consult also L. E. Dickson, *History of the Theory of Numbers*, I, 12 and 31 (Washington, 1919), hereafter referred to as Dickson, *Hist. Th. Numb.*

[2] Born at Clermont-Ferrand in Auvergne, June 19, 1623; died in Paris, August 19, 1662.

[3] Bayle, *Dictionaire*, IV, 500, with an unusually good biography for Bayle, who is commonly more erudite than helpful. See also A. Maire, *L'œuvre scientifique de Blaise Pascal*, Paris, 1912; A. Desboves, *Étude sur Pascal*, Paris, 1878.

a thorough grounding in the ancient languages, and therefore took from him all books on mathematics, he succeeded in beginning geometry by himself and in making considerable progress before his efforts were discovered. Various anecdotes of his youthful activities in mathematics are told by his sister, Madame Périer, who wrote his biography. She relates that he discovered independently most of the first book of Euclid, that his intuition in mathematics seemed miraculous, and that geometry was simply his recreation. He played with conics as other children play with toys, but with the divine enjoyment of discovering eternal truths. When Descartes was shown a manuscript which Pascal wrote on conics at the age of sixteen, he could hardly be convinced that it was not the work of the father instead of the son. At the age of nineteen he invented a computing machine that served as a starting point in the development of the mechanical calculation that has become so important in our time. That he should have been permitted to present one of these machines to the king and one to the royal chancellor shows the esteem in which he must have been held. At the age of twenty-three he became interested in the work of Torricelli in atmospheric pressure, and soon established for himself a reputation as a physicist. Among his discoveries was the well-known theorem which bears his name, that the three points determined by producing the opposite sides of a hexagon inscribed in a conic are collinear,—a theorem from which he deduced over four hundred corollaries. He also wrote (1653) so extensively on the triangular arrangement of the coefficients of the powers of a binomial, which had already attracted the attention of various writers, that this arrangement has since been known as Pascal's Triangle. In connection with Fermat he laid the foundation for the theory of probability.[1] He also perfected the theory of the cycloid and solved the problem of its general quadrature.

[1] I. Todhunter, *A History of the Mathematical Theory of Probability*, Chap. II (Cambridge, 1865), hereafter referred to as Todhunter, *Hist. Probability*. On Pascal's use of induction in this theory see W. H. Bussey, *Amer. Math. Month.*, XXIV, 203. Pascal's work on the triangular arrangement of coefficients was published posthumously in 1665.

His contributions to science and letters often appeared under the nom de plume of Louis (Lovis) de Montalte (as in his *Lettres provinciales*) and the anagram on the same name, Amos Dettonville (as in various problems which he proposed). It is on this account that Leibniz occasionally speaks of him as Dettonville.

Having already, at the age of twenty-five, made for himself an imperishable reputation in mathematics and physics, he suddenly determined to abandon these fields entirely and to devote his life to a study of philosophy and religion. The life of penance which he lived thereafter seems strange, since all must feel that he had little of which to repent, but at any rate it speaks well for the faith of men that he was sincere in his belief and irreproachable in his conduct.

Desargues. In the line of pure geometry the most original contributor of the 17th century was Gérard Desargues,[1] of whose life not much is known except that he was for a time an officer in the army, that he then lived in Paris (1626), where he gave some public lectures, that he was an engineer, and that his later years were spent upon his estate near Condrieux. He published several works, but is known chiefly for his treatise on conics.[2]

Perhaps because it appeared at about the same time as the great work of Descartes, perhaps because the chief interest in mathematics shown by Desargues had been in its applications to the study of perspective, this masterpiece seems to have attracted little general attention, although appreciated by both Pascal and Descartes. At any rate the work was soon forgotten and remained almost unknown until Chasles happened to find a copy in 1845, since which time it has been looked upon as one of the classics in the early development of modern pure geometry. In this work he introduced the notions of the point at infinity, the line at infinity, the straight

[1] Born at Lyons, 1593; died at Lyons, 1662. N. G. Poudra, *Desargues . . . Œuvres*, 2 vols. Paris, 1864.

[2] *Brouillon projet d'une atteinte aux éuénemens des rencontres d'un cone avec un plan*, Paris, 1639; M. Chasles, *Aperçu*, 74; Poudra, *loc. cit.*, pp. 97, 303.

line as a circle of infinite radius, geometric involution, the tangent as a limiting case of a secant, the asymptote as a tangent at infinity, poles and polars, homology, and perspective, thus laying a substantial basis for the modern theory of projective geometry.

MARQUIS DE L'HOSPITAL

He had much to do with the introduction of Newton's mathematical ideas into France

L'Hospital. Guillaume François Antoine de l'Hospital,[1] Marquis de St.-Mesme, a man of ancient and honorable family, was one of the world's infant prodigies in mathematics. When only fifteen he was one day at the Duc de Roanne's and heard some mathematicians speaking of a difficult problem of Pascal's. To their surprise he said that he thought he could solve it, and in a few days succeeded. A career which he sought in the army proved impossible owing to his defective sight, and the latter part of his life was given to his favorite study. He was a pupil of Jean Bernoulli's and introduced the ideas of the new analysis into France.[2] He also wrote on geometry, algebra, and mechanics, most of his works being published after his death.

[1] Born in Paris, 1661; died in Paris, February 2, 1704. He is also known as the Marquis de l'Hospital. The family also spelled the name Lhospital and, somewhat later, l'Hôpital.

[2] *Analyse des infiniment petits*, Paris, 1696.

Frénicle de Bessy and De la Loubère. Among the correspondents of Fermat, Bernard Frénicle de Bessy,[1] an office-holder and a member of the Académie des Sciences at Paris, was known for his work on Pythagorean numbers, that is, numbers which form the sides of a right-angled triangle,[2] and for his interest in magic squares. At about the same time Antoine de la Loubère,[3] a Jesuit and a lecturer on mathematics, rhetoric, theology, and the humanistic subjects, was showing much interest in the study of curves. This interest is seen in his quadrature problem[4] and in his study of the cycloid.[5] His method of tangents, in which the tangent is taken as the direction of a moving point, was quite forgotten until its value was recognized in its applications in kinematics.[6]

Roberval. Among the contemporaries of De la Loubère, Gilles Persone de Roberval[7] became well known for his discoveries in the field of higher plane curves and for his method of drawing a tangent to a curve (already suggested in substance by Torricelli), which was a definite step in the invention of the calculus. He was professor of philosophy in the Collège Gervais at Paris, and later professor of mathematics in the Collège Royal. His chief interest was in physics, but he also wrote on the cycloid (his "trochoid") and other curves, on algebra and indivisibles,[8] and (1644) on the astronomy of Aristarchus.

[1] Born in Paris, *c.* 1602; died 1675.

[2] *Traité des triangles rectangles en nombres*, Paris, 1676 (posthumous).

[3] Born at Rieux, Languedoc, 1600; died at Toulouse, 1664. The name also appears as Laloubère, Lalouère, Lovera, Lalovera, Lalouvère.

[4] *Elementa tetragonismica seu demonstratio quadraturae circuli et hyperbolae ex datis ipsorum centris gravitatis*, Toulouse, 1651. See also Montucla, *Histoire*, II (2), 77.

[5] *Propositio 36ª excerpta ex quarto libro de cycloide nondum edito*, Toulouse, 1659; *Veterum geometria promota in septem de cycloide libris*, Toulouse, 1660.

[6] See Chasles, *Aperçu*, 58, 96; F. Jacoli, Boncompagni's *Bullettino*, VIII, 265.

[7] Born at Roberval, near Beauvais, August 8, 1602; died in Paris, October 27, 1675. The name Roberval, by which he is commonly known, was merely that of his birthplace. The family name, Persone, appears also in the Latin form of Personerius, whence a derived French form is Personier.

[8] *De geometrica planarum et cubicarum aequationum resolutione*, *De recognitione aequationum*, and *Traité des indivisibles*. These and other of his memoirs were collected, published in 1693, and republished in the *Mémoires de l'ancienne académie*, *Vol. VI*.

Other Writers of the Period. At about the same time Claude François Milliet Dechales,[1] for some time a Jesuit missionary in Turkey, taught in the schools of his order in Marseilles, Lyons, and Chambéry. He is chiefly known for his editions of Euclid's *Elements*[2] and for a general work on mathematics,[3] but his original contributions to the subject were slight.

Noted as a voluminous writer on theology, Antoine Arnauld (1612–1694)—"the great Arnauld" as the Jansenists called him, hated by the Jesuits and the Calvinists alike because of his bitter attacks upon their beliefs—deserves at least some slight mention for his encouragement of mathematics. He was interested in the works of his great contemporaries, such as Descartes and Pascal, wrote on geometry (1667) and magic squares, and showed interest in the theory of numbers.[4]

Among those who in the latter part of the century did much to make geometry popular was Philippe de Lahire,[5] a pupil of Desargues's and a man of scattered genius. He was at first a painter and architect, then a *pensionnaire astronome* of the Académie des Sciences at Paris, then professor of mathematics in the Collège Royal and the Académie de l'Architecture, and in his later years (from 1679) was connected with the geodetic survey of France. He wrote several works on conics, algebra,[6] and astronomy, besides contributing a large number of memoirs on mathematics, astronomy, and physics to the Académie des Sciences. He also wrote on epicycloids (1694) and roulettes (1694, 1706), and summarized what was then known on magic squares (1705).

[1] Born at Chambéry, Savoy, 1621; died at Turin, March 28, 1678. The name also appears as Deschales and De Challes.

[2] Latin ed., Lyons, 1660; French ed., Paris, 1677.

[3] *Cursus seu Mundus Mathematicus,* Lyons, 1674, with a later edition in 1690.

[4] K. Bopp, "Antoine Arnauld . . . als Mathematiker," *Abhandlungen,* XIV, 187.

[5] Born in Paris, March 18, 1640; died in Paris, April 21, 1718. The *Biographie Universelle* gives the date of his death as 1719, but most authorities give it as 1718. The name is often written La Hire. See Chasles, *Aperçu,* 118, 550, 553; Curtze, in *Bibl. Math.,* II (2), 65.

[6] *Théorie des coniques,* Paris, 1672; *Nouvelle Méthode de Géométrie,* Paris, 1673; *Nouveaux élémens des sections coniques*; *Les Lieux Géométriques*; *La Construction ou effection des équations,* Paris, 1679, with an English translation, London, 1704; *Sectiones conicae in novem libros distributae,* Paris, 1685.

Among the contemporaries of de Lahire, but a few years his junior and like him a *pensionnaire* of the Académie des Sciences, Michel Rolle[1] made for himself a worthy name. He was connected with the war department and apparently was not concerned with teaching. He wrote on both geometry[2] and algebra, his publications on geometry appearing in the form of numerous memoirs, and those on algebra in memoirs and in two books.[3] To him is due the theorem that $f'(x) = 0$ has at least one real root lying between two successive roots of $f(x) = 0$.[4]

As a representative of the other French mathematicians of this period there may be named Pierre Nicolas,[5] a pupil of De la Loubère's and rector of the Jesuit college at Béziers, who wrote on the logarithmic spiral and on conchoids.[6]

In the field of textbook making the most popular French writer of this time was François Barrême, a native of Lyons, who died in Paris in 1703. His *Arithmétique* (Paris, 1677) went through many editions, and his name is still a synonym for a ready reckoner (*barème*).

4. Great Britain

Great Britain in the Seventeenth Century. As already remarked, the two foci of the ellipse that bounded mathematical Europe in the 17th century were located in France and Great Britain. The British center was Cambridge, although the contributions of Edinburgh in the notable discoveries of Napier

[1] Born at Ambert, Auvergne, April 21, 1652; died in Paris, November 8, 1719. The date of his death is also given as July 5.

[2] A certain curve, $xy^2 = a(y - mx)^2$, has recently come to bear his name, but apparently with no justification.

[3] *Traité d'algèbre,* Paris, 1690; *Méthode pour résoudre les questions indéterminées de l'algèbre*, Paris, 1699. F. Cajori, "What is the origin of the name 'Rolle's Curve'?" *Amer. Math. Month.*, XXV, 291, and *Bibl. Math.*, XI (3), 300. See also *Bibl. Math.*, IV (3), 399, and *L'Intermédiaire des Mathématiciens*, V, 76, and XVI, 244 (Paris), hereafter referred to as *L'Intermédiaire.*

[4] *Démonstration d'une Methode pour résoudre les Egalitez de tous les degrez*, Paris, 1691.

[5] Born at Toulouse, *c.* 1663; died *c.* 1720.

[6] *De novis spiralibus exercitationes*, Toulouse, 1693; *De Lineis logarithmicis spiralibus hyperbolicis*, Toulouse, 1696; *De conchoidibus et cissoidibus*, Toulouse, 1697.

and Gregory, and of Oxford in the works of such men as Harriot, Briggs, Halley, Wren, and Wallis were such as to challenge the supremacy of the Cambridge school.

Harriot. It is rather surprising to think that the man who surveyed and mapped Virginia[1] was one of the founders of algebra as we know the science today. Such, however, is the case, for Thomas Harriot[2] was sent by Sir Walter Raleigh to accompany Sir Richard Grenville (1585) to the New World, where he made the survey of that portion of American territory. He returned to England (1587) and published (1588) a report upon the colony, and some years later wrote a work that helped to establish the English school of algebraists. Harriot took his B.A. at Oxford in 1579. After his return from America he was introduced to the Earl of Northumberland and was received with other scholars into his household. The earl allowed him a pension of £300 a year for the rest of his life. He was prominent as an astronomer and corresponded with Kepler. He discovered the solar spots, and his observations of the satellites of Jupiter were independent of those made by Galileo at the same time. His great work on algebra[3] was published ten years after his death. In this work he assisted in setting the standard for a textbook in algebra which has been generally recognized since that time. The work includes the formation of equations with given roots, the law as to the number of roots, the relation of the roots to coefficients, the transforming of equations into equations having roots differing from the original roots according to certain laws, and the solution of numerical equations. He used small consonants for the known quantities and small vowels for the unknown quantities.

[1] Or rather North Carolina, since the present boundaries had not yet been fixed. On his map of Virginia see P. L. Phillips, *Virginia Cartography*, Washington, 1896. His report was published in London in 1588. A second edition appeared in Hakluyt's *The Principall Navigations*, London, 1589. F. V. Morley, *The Scientific Monthly*, XIV, 60.

[2] Born at Oxford, 1560; died near Isleworth, July 2, 1621. The name appears as Hariot in several early works.

[3] *Artis Analyticae Praxis, ad Æquationes Algebraicas nouâ . . . Methodo resoluendas*, London, 1631. It was probably written *c.* 1610.

He also took some steps in the direction of analytic geometry. In the further matter of symbols, he used $\sqrt{3}$ for the cube root,[1] and the characters $>$ and $<$ for "is greater than" and "is less than" respectively.

Napier. When Hume, the historian, wrote his appreciation of Napier as "the person to whom the title of 'great man' is more justly due than to any other whom his country has ever produced," he spoke without exaggeration. Burns can be appreciated only in the vernacular, Scott appeals chiefly to a single period in the development of the individual, the fame of those whose effigies justly have place along Princes Street in Edinburgh is generally only national; but Napier's remarkable invention affects the whole world with constantly increasing power. The artisan who carries in his pocket the slide rule is relatively as much indebted to the genius of the Laird of Merchiston as the astronomer, the engineer, the physicist, and the mathematician.

JOHN NAPIER
Engraved by Stewart after the original painting in Edinburgh

John Napier[2] was descended from a strong ancestry which included men who had held positions of prominence because they deserved to do so. Merchiston Castle, built in the 15th

[1] As in $\underbrace{\sqrt{3.)26 + \sqrt{675}}}$ for $\sqrt[3]{26 + \sqrt{675}}$.

[2] Born at Merchiston Castle, now in the city of Edinburgh, 1550; died there, April 4, 1617. The name also appears as Naper, Naperus, Neper, and Neperius.

century, was one of the two strongholds on the outskirts of Edinburgh, and was enlarged from time to time until it became an imposing structure, symbolic of the Napier house.

Napier was born in the period of greatest strife between Protestantism and Catholicism in Scotland. John Knox began his mission only three years before Napier's birth, and the seeds of the bitter antagonism which the latter felt towards Rome were early planted in his soul. In 1563, when only thirteen years old, Napier was sent to the University of St. Andrews, but left without taking a degree. He probably studied abroad, but in 1571 was back again in Scotland, this time at Gartness, in Stirlingshire, where his father had some property. It was probably here that he wrote his popular theological work, *A Plaine Discouery of the whole Reuelation of Saint Iohn* (Edinburgh, 1593),[1] a bitter attack upon the Church of Rome. The common people accused him of dealing in the black art, while the intellectuals recognized him as a man of remarkable ingenuity; but the end was the same. He planned to use burning mirrors, like those of Archimedes but so potent that they should destroy an enemy's ships "at whatever appointed distance"; a piece of artillery that should, as an early writer described it, "clear a field of four miles circumference of all the living creatures exceeding a foot of height"; a chariot which should be like "a living mouth of mettle and scatter destruction on all sides"; and "devises of sayling under water," all of which is of interest in view of the engines of destruction used first in the World War of 1914–1918. It is of interest to compare the mind of Napier, as seen in his vision of future achievements, with that of Roger Bacon. Each seemed to many of his contemporaries, and perhaps to most of those who knew him, as mentally unbalanced and as a mere visionary in his contemplation of future warfare, and yet each prophesied with remarkable success respecting many inventions of the present time.

[1] Its great popularity is shown by the fact that it was reprinted in London in 1594, 1611, and 1641; in French translation at La Rochelle in 1602, with a fourth edition in 1607; in Dutch translation at Middelburgh in 1600; and in German translation at Frankfort a. M. in 1611.

Napier on Logarithms. Napier wrote two works on logarithms,[1] besides one on computing rods[2] and one on algebra.[3] Of all his works he probably thought the *Plaine Discouery of the whole Reuelation of Saint Iohn* the most important, but the world has long since forgotten it. The popular verdict of his day was that the *Rabdologia* was his greatest work, but it is now looked upon only as one of the curiosities of history. The scientific world looked upon his *Descriptio* as epoch-making, and the scientific world was right.[4]

Briggs. The man who did most to start the invention of Napier on its road to success was Henry Briggs,[5] whom Oughtred rather absurdly called the English Archimedes.[6] He entered St. John's College, Cambridge, in 1577, took the degrees of B.A. in 1581 and M.A. in 1585, and was made a fellow in 1588. He was the first professor of geometry at Gresham College, London (1596–1619), after which he became professor of astronomy at Oxford. He saw at once the great importance of Napier's invention, and in a letter to James Ussher, archbishop of Armagh, dated March 10, 1615, he speaks of himself as being "wholly employed about the noble invention

[1] *Mirifici Logarithmorum Canonis descriptio . . . Authore ac Inventore Ioanne Nepero, Barone Merchistonii, &c. Scoto*, Edinburgh, 1614, with other editions in 1616, 1619, and 1889; Leyden, 1620; London, 1616 and 1618; *Mirifici ipsius canonis constructio*, which appeared posthumously and was added to the edition of 1619 mentioned above. On all these editions and on the subject in general see C. G. Knott, *Napier Tercentenary Memorial Volume*, London, 1915.

[2] *Rabdologiae, sev nvmerationis per virgulas libri dvo*, Edinburgh, 1617, published the year of his death. For description, see Volume II. There was an edition at Leyden, 1626; an Italian translation, Verona, 1623, in which Napier is quoted as ascribing all glory and honor "alla Beatissima Vergine Maria," which is the last thing he would have dreamed of doing; and a German translation, Berlin, 1623. There was a free English translation of part of the work by John Dansie (*A Mathematicall Manuel*, London, 1627), the first English version to appear.

[3] Preserved in the Napier family and published in Edinburgh in 1839 under the title *De Arte Logistica Joannis Naperi Merchistonii Baronis Libri Qui Supersunt*.

[4] The subject is considered at length in Volume II.

[5] Born at Warley Wood, Yorkshire, February, 1560/61 (1561 N. S.); died at Oxford, January 26, 1630/31. The date of birth is often given as 1556, but the parish register shows that it was 1560/61.

[6] Aubrey, *Brief Lives*, I, 124.

of logarithms." In the following year (1616) he made a visit to Edinburgh for the purpose of meeting with Napier, and repeated his visit in 1617. It was on the first of these visits that Briggs suggested the base 10, of which Napier had already thought, this being the base of the common system of logarithms that has been in use ever since that time, and on his return to Oxford he prepared a table accordingly.

Briggs published ten works and left six others unpublished. The published works include treatises on navigation, Euclid's *Elements*,[1] logarithms,[2] and trigonometry.[3]

Gellibrand. Briggs's friend Henry Gellibrand,[4] who edited his trigonometry (1633), entered Trinity College, Oxford, in 1615, was granted the degrees of B. A. in 1619 and M. A. in 1623, took holy orders, and entered upon church work. Having heard one of Sir Henry Savile's lectures, he was so impressed that he gave up his curacy and devoted himself entirely to mathematics. Besides editing the trigonometry left unpublished by Briggs he also wrote on navigation and the variation of the magnetic needle, and composed a trigonometry of his own.[5]

Oughtred. One of the greatest of the writers of the early part of the 17th century in his influence upon English mathematics was William Oughtred.[6] As in the case of Harriot,

[1] *Elementorum Euclidis libri VI priores*, London, 1620.

[2] *Arithmetica logarithmica*, London, 1624. The final (French) edition appeared at Gouda in 1628.

[3] *Trigonometria Britannica, sive de doctrina triangulorum libri duo*, Gouda, 1633, published posthumously by his friend Henry Gellibrand.

[4] Born in the parish of St. Botolph, Aldersgate, London, November 17, 1597; died in London, February 16, 1636/37.

[5] *An Institution Trigonometricall wherein . . . is exhibited the doctrine of the dimension of plain and spherical triangles . . . by tables . . . of sines, tangents, secants, and logarithms*, London, 1638. This title is from the second edition (1652).

[6] Born at Eton, March 5, 1574; died at Albury, June 30, 1660. He also wrote his name Owtred, and John Locke preferred the form Outred, from which spellings we infer the pronunciation. On his life and works see F. Cajori, *William Oughtred*, Chicago, 1916, hereafter referred to as Cajori, *Oughtred*. The date of his birth is from Aubrey, *Brief Lives*,—"Gulielmus Oughtred natus 5 Martii 1574, 5 h. P. M."

he was not a professor of mathematics; but, like Harriot also, he knew more mathematics than most professors of his day. He entered King's College, Cambridge, in 1592, became a fellow in 1595, and received the degrees of B. A. in 1596 and M.A. in 1600. Speaking of his college work he says:

The time which over and above those usuall studies I employed upon the Mathematicall sciences, I redeemed night by night from my naturall sleep, defrauding my body, and inuring it to watching, cold, and labour, while most others tooke their rest.

He vacated his fellowship in 1603 and in the following year began his ministry. He gave much of his time, however, to mathematics and to correspondence with mathematicians. Aubrey,[1] whose gossiping biographies are always entertaining, gives this description of him:

He was a little man, had black haire, and blacke eies (with a great deal of spirit). His head was always working. He would drawe lines and diagrams on the dust . . . did use to lye a bed till eleaven or twelve a clock . . . Studyed late at night; went not to bed till 11 a clock; had his tinder box by him; and on the top of his bed-staffe, he had his inke-horne fix't. He slept but little. Sometimes he went not to bed in two or three nights.

He thus seems to have violated many of the usual canons of health, and probably continued to do so until his death at the ripe old age of eighty-six.

The Clavis Mathematicæ. Oughtred's best-known work is his *Clavis mathematicæ*,[2] a brief treatise on arithmetic and algebra, composed (*c.* 1628) for the purpose of instructing the son of the Earl of Arundel. It was published in London in 1631. He had already written (*c.* 1597) a treatise on dialing, but this was not published until 1647, when it appeared in an edition

[1] Aubrey, *Brief Lives*, II, 106.

[2] For the full title and for a study of the various editions see Cajori, *Oughtred*, p. 17; H. Bosmans, "La première édition de la 'Clavis Mathematica' d'Oughtred," *Annales de la Société scientifique de Bruxelles*, XXXV, 2me partie, p. 24.

of the *Clavis*.[1] The influence of the latter work was very great. In it appear contracted multiplication and division, the distinction between the two uses of the signs + and −, the symbol (: :) for proportion, and the symbols × for multiplication (already known) and ∽ for the absolute value of a difference.

The Slide Rule. The invention of the slide rule seems unquestionably due to Oughtred,[2] and there also seems to be good reason for believing that he is the author of the *Appendix to the Logarithmes* printed with the English translation of Napier's work (London, 1618) and containing the first natural logarithms. He also wrote a trigonometry[3] to which reference will be made later, and a work on gaging, and he translated and edited Leurechon's French work on mathematical recreations. Among his many pupils were John Wallis and Sir Christopher Wren, the former of whom wrote of the *Clavis* that it "doth in as little room deliver as much of the fundamental and useful part of geometry (as well as of arithmetic and algebra) as any book I know."

Gunter. Connected with the general movement to simplify calculation, initiated by Napier and Briggs, there stands out prominently the name of Edmund Gunter,[4] who left the ministry to become professor of astronomy (1619) at Gresham College, London. He published the first table of logarithmic sines and tangents to the common base,[5] suggested to Briggs the use of the arithmetic complement, invented the surveyors' table and the chain which until recently was commonly known as

[1] It was translated into Latin by Sir Christopher Wren. Another of his works on dialing (1600) was published in 1632.

[2] *The Circles of Proportion and The Horizontall Instrument*, London, 1632. See F. Cajori, *History of the Logarithmic Slide Rule*, New York, 1909, and "On the History of Gunter's Scale and the Slide Rule," *University of California Publications in Mathematics*, I, p. 187.

[3] *Trigonometrie, or, The manner of calculating the Sides and Angles of Triangles, by the Mathematical Canon, demonstrated*, London, 1657.

[4] Born in Hertfordshire, 1581; died in London, December 10, 1626.

[5] *Canon triangulorum or Table of Artificial Sines and Tangents*, London, 1620.

Gunter's chain, and devised a kind of slide rule known as Gunter's scale. Aubrey relates the following incident:

When he was a student at Christ Church, it fell to his lott to preach the Passion sermon, which some old divines that I knew did heare, but they sayd that 'twas sayd of him then in the University that our Saviour never suffered so much since his passion as in that sermon, it was such a lamentable one—*Non omnia possumus omnes.*[1]

The Savilian and Lucasian Professorships. Mention is so often made of the Savilian professorships of geometry and astronomy at Oxford and of the Lucasian professorship at Cambridge that a brief statement concerning them is desirable.

Sir Henry Savile,[2] the founder of two chairs at Oxford, was warden (1585) of Merton College, Oxford, and in 1596 was appointed provost at Eton. He lectured on Euclid,[3] and although he contributed little to mathematics, he contributed much to the extending of the knowledge of the science through the professorships which he founded in 1619, and which have been held by a distinguished line of scholars for three centuries.

Henry Lucas[4] was for a time a student at St. John's College, Cambridge, but seems not to have matriculated. He was, however, admitted M.A. in 1635–1636 and was elected to represent the university in parliament in 1639–1640. By his will he directed that lands yielding £100[5] a year be purchased to found the professorship that bears his name. It was in the year after the founding of the professorship that Barrow was elected (1664) as the first occupant of the chair, and six years later he was succeeded by Newton (1670). In his inaugural address Barrow speaks of Lucas as "a new and benignant star, shining with a ray both true and propitious, such as has not for many years risen above the academical horizon."[6]

[1] *Brief Lives*, I, 276.

[2] Born at Over-Bradley, near Halifax, Yorkshire, November 30, 1549; died at Eton, February 19, 1622.

[3] *Praelectiones XIII in principium elementorum Euclidis, Oxoniae habitae 1620*, Oxford, 1621.

[4] Died in London, June 22, 1663. [5] In 1860 it had risen only to £155.

[6] Whewell's translation. For Kirby's translation see the English edition of the *Mathematical Lectures*, p. vii (London, 1734).

Barrow. To Isaac Barrow[1] it was given to rank as one of the best Greek scholars of his day, to attain the highest honors as professor of mathematics, to be looked upon as one of the leading theologians of England, to be recognized as a profound student of physics and astronomy, to acquire fame as a preacher and controversialist, and to be perhaps the first to recognize the genius of Newton and to develop his great talents, but his reputation might have been higher had he worked exclusively in only one of his fields of interest, particularly in optics or geometry. He was prepared for the university at Charterhouse, London, and at Felstead School, in Essex. He entered Trinity College, Cambridge, in 1644, took his degree of B. A. in 1648 (when only eighteen), was elected fellow the year following, and left in 1655. His ready wit is illustrated by an incident that occurred in his examination for holy orders, the dialogue being said to have run as follows:

Chaplain. *Quid est fides?*
Barrow. *Quod non vides.*
Chaplain. *Quid est spes?*
Barrow. *Magna res.*
Chaplain. *Quid est caritas?*
Barrow. *Magna raritas.*

The chaplain is then said to have given up in despair and to have reported the candidate's lack of reverence to the bishop. Fortunately the latter had a sense of humor and Barrow was duly admitted.[2]

Barrow traveled for some time after taking orders and in 1662 was elected professor of geometry at Gresham College. Soon after, as already stated, he was elected (1664) the first Lucasian professor. Six years later (1670)[3] he resigned in favor of Newton and devoted his attention to theology. He

[1] Born in London, October, 1630; died in London, May 4, 1677.
[2] W. W. R. Ball, *Cambridge Papers*, p. 109. London, 1918.
[3] These two dates are often given as 1663 and 1669, on account of the style of calendar. His inaugural lecture was given March 14, 1664. See W. Whewell, *The Mathematical Works of Isaac Barrow*, p. 5 (Cambridge, 1860).

ISAACUS BARROW
VIR SUO TEMPORI
PIETATIS PROBITATIS FIDEI ERUDITIONIS
MODESTIÆ SUAVITATIS EXEMPLUM:
PROFESSOR MATHESEOS IN HAC ACADEMIÂ
ET EO QUIDEM NOMINE NEWTONI ANTECESSOR
SED MELIORE TITULO OB PRÆCLARA SUA
INVENTA MATHEMATICA:
THEOLOGUS ARGUMENTORUM GRAVITATE
ET SERMONIS COPIÂ PRÆCELLENS:
COLLEGIUM HOC PRÆFECTUS ILLUSTRAVIT
JACTIS BIBLIOTHECÆ FUNDAMENTIS AUXIT
OBIIT IV. DIE MAII ANNO DOM. MDCLXXVII
ÆTATIS SUÆ XLVII.
AMORIS ERGÔ IN COLLEGIUM SUUM.

BARROW

Statue in Trinity College Chapel, Cambridge. It stands just north of the statue of Newton

became chaplain to Charles II in 1670, master of Trinity College in 1673,[1] and vice-chancellor of the university in 1675. He edited Euclid's *Elements*[2] and the *Data*,[3] the works of Archimedes,[4] the conics of Apollonius, and the spherics of Theodosius. His own contributions to science appear chiefly in his general lectures[5] and in his works on optics[6] and geometry.[7] He also gave a new method for determining tangents,—one that approached the methods of the calculus.[8] In this work he made use of a "differential triangle," which is still essentially the basis of the initial work in differentiation and which we shall consider at some length in Volume II.

Newton. Isaac Newton[9] was born in the stirring times of the Cromwell rebellion and was the posthumous son of a farmer in Lincolnshire. He was a small and feeble child, and as a boy he gave little promise of success in the battle of life. At the age of twelve he was taken from the local day school and placed in a school at Grantham. According to his own statement he was at first extremely inattentive to his studies and ranked among the lowest in the school. His chief interests seemed to be in carpentering, mechanics, the writing of verses, and drawing. Later, however, he began to show considerable

[1] W. W. R. Ball, *Cambridge Papers*, p. 171.

[2] *Euclidis Elementorum Libri XV*, Cambridge, 1655, with many later editions.

[3] *Euclidis Data*, Cambridge, 1657.

[4] The Archimedes, Apollonius, and Theodosius were published together at London in 1675.

[5] *Lectiones Mathematicae XXIII*, London, 1683 (posthumous).

[6] *Lectiones XVIII, . . . in qvibvs Opticorum Phaenomenωn . . investigantur* . . ., London, 1669.

[7] *Lectiones Geometricae*, London, 1670. There is a translation with commentary by J. M. Child, Chicago, 1916.

[8] For a good summary, see Ball, *Hist. Math.*, 6th ed., p. 311.

[9] Born at Woolsthorpe, near Grantham, Lincolnshire, December 25, 1642; died at Kensington, March 20, 1727. One of the best books with which to begin a study of Newton's life is A. De Morgan, *Essays on the Life and Work of Newton*, 2d ed., Chicago, 1914, not so much because of the essays themselves as for the bibliographical and critical notes of the editor, P. E. B. Jourdain. The best biography of Newton is that of Sir David Brewster, *The Memoirs of Newton*, 2d ed., Edinburgh, 1860, hereafter referred to as Brewster, *Memoirs of Newton*. A new biography of Newton and a definitive edition of his complete works are both sadly needed.

NEWTON

From a mezzotint after Seeman's painting, one of the most refined portraits of Newton extant

CROKER'S MEDAL OF NEWTON

The medal of which Brewster says: "In the beginning of 1731, a medal was struck at the Tower in honour of Sir Isaac Newton. It had on one side the head of the philosopher, and on the reverse a figure representing mathematics, with the motto, *Felix cognoscere causas.*" The figure is that of winged Science holding a tablet upon which appears the solar system. The date, MDCCXXVI, is that of Newton's death according to the old calendar. The medal is by John Croker

ability and (1660) was sent to Trinity College, Cambridge. Not even then did he seem to have developed any particular strength, and there is no record of any unusual achievement until a little before he attained the degree of B. A., in 1665. By the time he was awarded the degree of M. A., however, in 1668, his genius had so asserted itself that he was considered the most promising mathematician and physicist in England. In 1664 he investigated infinite series, and about this time he showed that the ordinary rule for expanding $(a + b)^n$ was valid for all values of n, subject, as is now more generally recognized, to considerations of convergency. In 1665 he began his work on the calculus, described by him as the theory of fluxions, and used this theory in finding the tangent and the radius of curvature at any point on a curve. In 1666 he worked on the theory of light, began his great investigations into the theory of gravity, and applied the method of fluxions to the study of equations.

Newton's work was so highly appreciated that about this time (1668) he was invited to revise the lectures of Barrow, his former teacher. In 1669 the latter resigned the Lucasian professorship of mathematics at Cambridge, and Newton was appointed his successor. Thus at the age of twenty-seven he had begun his great work on the calculus and mathematical physics, and held already one of the highest academic honors in the world.

For nearly sixty years after receiving his professorship he was looked upon as one of the greatest leaders in the fields of physics and mathematics, receiving all the honors that could be hoped for by a man in academic life. He was elected fellow of the Royal Society in 1672, was chosen in 1689 to represent the university in parliament, and was appointed warden of the mint in 1696 and master in 1699. He was again elected to parliament in 1701, but he took no interest in politics. He was

ROETTIERS'S MEDAL OF NEWTON

Of the numerous portrait medals of Newton, those by Croker and by Roettiers (first struck in 1739) are the best. This particular medal was struck in 1774. The reverse bears the motto from Vergil's Æneid (Lib. V, 378), "Another is sought for him." Its significance appears from the context: "Another is sought for him, nor does any one from so great a band dare approach the man, and draw the gauntlets on his hands"

knighted by Queen Anne in 1705. The latter part of his life was spent in London and was mathematically unproductive.

Newton always hesitated to publish his discoveries. His greatest work, the *Principia*,[1] was begun in 1685 but was not published until 1687, and then only under pressure from his friend Halley, the astronomer. In this work he sets forth his theory of gravitation, "indisputably and incomparably the greatest scientific discovery ever made,"[2] following the methods of the Greek geometry as being more easily understood by students of his time.[3] It was the third of the three great discoveries in the field of mathematical astronomy,— the heliocentric theory, finally established by Copernicus; the elliptic orbits of the planets and the laws relating thereto, finally established by Kepler; and the law of universal gravitation, with which Newton's name will always be connected.

Referring to this great work, there is humor as well as justice in the remark of Lagrange to the effect that Newton was the greatest genius that ever lived, and the most fortunate, since we can find only once a system of the universe to be established. Whether Einstein's Theory of Relativity invalidates this statement remains to be seen.

Newton's *Arithmetica Universalis*, a work on algebra and the theory of equations, was written in lecture form in the period 1673–1683 but was not published until 1707.[4] He wrote a work on analysis by series[5] in 1669, but it was not published until 1711. His work on the quadrature of curves[6] was written in 1671 but was not published until 1704, and his other

[1] *Philosophiae Naturalis Principia Mathematica*, London, 1687, reissued the same year with a new title-page; 2d ed. by Roger Cotes, Cambridge, 1713; 3d ed. by Pemberton, London, 1726. For a list of editions of all of Newton's works see G. J. Gray, *A bibliography of the Works of Sir Isaac Newton*, Cambridge, 1888; 2d ed., Cambridge, 1907.

[2] W. Whewell, *History of the Inductive Sciences*, Bk. VII, ii, § 5.

[3] For a résumé of the work see W. W. R. Ball, *Essay on the Genesis, Contents, and History of Newton's Principia*, London, 1893.

[4] *Arithmetica universalis sive de compositione et resolutione arithmetica liber*, Cambridge, 1707; English translation, London, 1720.

[5] *Analysis per Æquationes numero terminorum infinitas*, London, 1711.

[6] *Tractatus de quadratura curvarum*, printed in the first edition of his *Opticks*, London, 1704.

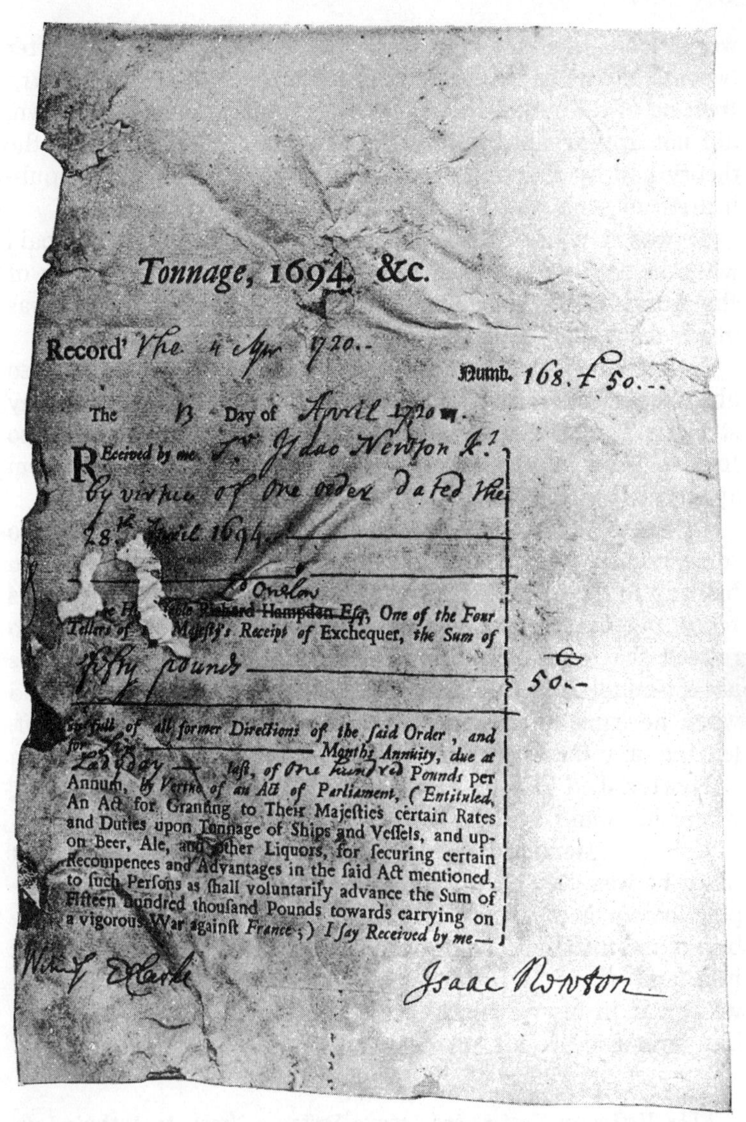

Tonnage, 1694, &c.

Record' the 4 Apr 1720..

Numb. 168. £ 50...

The 13 Day of April 1720

Received by me Sr Isaac Newton Kt by virtue of One order dated the 28th April 1694

Ld Onslow

[illegible] Richard Hampden Esq, One of the Four Tellers of [illegible] Majeſty's Receipt of Exchequer, the Sum of

fifty pounds — £ 50..

in full of all former Directions of the ſaid Order, and for Six Months Annuity, due at Ladyday laſt, of One hundred Pounds per Annum, by Vertue of an Act of Parliament, (Entituled, An Act for Granting to Their Majeſties certain Rates and Duties upon Tonnage of Ships and Veſſels, and upon Beer, Ale, and other Liquors, for ſecuring certain Recompences and Advantages in the ſaid Act mentioned, to ſuch Perſons as ſhall voluntarily advance the Sum of Fifteen hundred thouſand Pounds towards carrying on a vigorous War againſt France;) I ſay Received by me—

Witness (?) E Clarke (?)

Isaac Newton

AUTOGRAPH OF NEWTON

The document was written after he had retired from his work at Cambridge and is interesting politically

works were similarly held from the reading public until after Newton had given up his lectures at Cambridge. His *Fluxions*,[1] treating of the subject by which he is probably the best known, did not appear until nine years after his death, although the theory had by that time become well known through the publication of such works as that of Charles Hayes (1704).

It was a noble and a generous tribute that Leibniz paid when he said that, taking mathematics from the beginning of the world to the time when Newton lived, what he did was much the better half.

Newton had the eccentricities of genius, always being so absorbed in his work as to be oblivious of life about him. Many stories are related of his absent-mindedness, some of them also told of other mathematicians and very likely most of them apocryphal.

Of these stories one relates that, when giving a dinner to some friends, he left the table to fetch a bottle of wine; but on his way to the cellar he forgot all about the errand, went to his room, put on his surplice, and ended up in chapel. It is also related that once in riding he dismounted and started in his absent-minded way to lead his horse up the hill, but found when he came to remount that the horse had wandered off, leaving only the bridle in Newton's hand.

Newton died in 1727 and was buried in Westminster Abbey, where his tomb is still seen.

Voltaire attended the funeral, and we shall see later that when he was very old he did much to make Newton's philosophy known in France. It is said that "his eye would grow bright and his cheek flush" when he said that he had once lived in a land where "a professor of mathematics, only because he was great in his vocation," had been buried "like a king who had done good to his subjects."[2]

[1] *The Method of Fluxions and Infinite Series . . . from the Author's Latin Original not yet made publick . . .*, London, 1736, translated and edited by John Colson. There was another translation published shortly after this, London, 1737.

[2] S. G. Tallentyre, *Life of Voltaire*, p. 57. New York, n. d.

Halley. In his youth Edmund Halley[1] attended St. Paul's School, London, distinguishing himself in both mathematics and the classics. He entered Queen's College, Oxford, in 1673, and before he was twenty he communicated a paper to the Royal Society. So noteworthy had been his progress that in the very month in which he reached his twentieth birthday (November, 1676) he set out for St. Helena for the purpose of making astronomical observations. On the day before he was twenty-one he made the first complete observation of a transit of Mercury. So remarkable was his work at St. Helena that Flamsteed[2] called him the "southern Tycho" and the Royal Society elected him to a fellowship when he was only twenty-two (1678). Although his supposed materialistic views prevented his election as Savilian professor of astronomy at Oxford in 1691, he followed Wallis as Savilian professor of geometry in 1703. He succeeded Flamsteed as astronomer royal in 1721. He was the first to predict the time of the return of a comet, and the comet of which he announced the period has since then been known by his name.

EDMUND HALLEY

Friend of Newton, illustrious as an astronomer

[1] Born in London, November 8 (October 29, o.s.), 1656; died at Greenwich, January 14, 1742.

[2] John Flamsteed (1646–1719), first astronomer royal of England.

While Halley's tastes were chiefly astronomical, he was deeply interested in geometry,[1] algebra,[2] and the construction of logarithmic tables.[3] He solved the problem of the construction of a conic, given a focus and three points.

It was due to Halley's insistence as well as his financial help that Newton published his *Principia* in 1687. Besides editing the conics of Apollonius (1710) he edited the works of Serenus[4] and Menelaus.[5] He also compiled a set of mortality tables, thus giving a practical basis to the subject of life insurance.

Wallis. Of the contemporaries of Newton one of the most prominent was John Wallis.[6] He studied theology at Emmanuel College, Cambridge, and took the degree of B.A. in 1637 and that of M.A. in 1640, the year in which he was ordained. He became a fellow of Queens' College in 1644. His tastes, however, were in the line of physics and mathematics, and in 1649 he was elected to the Savilian professorship of geometry at Oxford, a position which he held until his death. He was awarded the degree of doctor of divinity in 1653,[7] became chaplain to Charles II in 1660, and was one of the founders of the Royal Society (1663).

Wallis was a voluminous writer, and not only are his writings erudite, but they show a genius in mathematics that would have appeared the more conspicuous had his work not been so overshadowed by that of his great Cambridge

[1] *Apollonii Pergaei Conicorum Libri octo*, Oxford, 1710. His version from the Arabic of the treatise of Apollonius *De Sectione Rationis*, with a restoration of the two lost books *De Sectione Spatii*, appeared at Oxford in 1706.

[2] *A new and exact Method of finding the Roots of any Equation*, London, 1720; "De numero radicum in aequationibus solidis ac biquadraticis," *Phil. Trans.*, 1687; "Methodus nova, accurata et facilis inveniendi radices aequationum," *Phil. Trans.*, 1694. See also the abridgment of the *Phil. Trans.*, 4th ed., I, 63, 68, 81, 137 (London, 1731).

[3] "A most compendious and facile method for constructing the logarithms . . .," *Phil. Trans.*, 1695.

[4] *De Sectione Cylindri et Coni*, Oxford, 1710.

[5] *Menelai Sphaericorum Libri III*, published by Dr. Costard, Oxford, 1758.

[6] Born at Ashford, Kent, November 23, 1616; died at Oxford, October 28. 1703.

[7] According to the *Dict. Nat. Biog.* Older authorities give 1654.

JOHN WALLIS, 1616–1703

From D. Loggan's drawing from life. Published in the *Opera Mathematica*, Oxford, 1695

contemporary. He was one of the first to recognize the significance of the generalization of exponents to include negative and fractional as well as positive and integral numbers. He recognized also the importance of Cavalieri's method of indivisibles, and employed it[1] in the quadrature of such curves as $y=x^n$, $y=x^{\frac{1}{n}}$, and $y=x^0+x^1+x^2+\cdots$. He failed in his efforts at the approximate quadrature of the circle by means of series because he was not in possession of the general form of the binomial theorem. He reached the result, however, by another method. He also obtained the equivalent of $ds=dx\sqrt{1+\left(\frac{dy}{dx}\right)^2}$ for the length of an element of a curve, thus connecting the problem of rectification with that of quadrature. In 1673 he wrote his great work *De Algebra Tractatus; Historicus & Practicus*, of which an English edition appeared in 1685.[2] In this there is seen the first serious attempt in England to write on the history of mathematics, and the result shows a wide range of reading of the classical literature of the science. This work is also noteworthy because it contains the first record of an effort to represent the imaginary number graphically by the method now used. The effort stopped short of success but was an ingenious beginning. Wallis was in sympathy with the Greek mathematics and astronomy, editing parts of the works of Archimedes, Eutocius, Ptolemy, and Aristarchus; but at the same time he recognized the fact that the analytic method was to replace the synthetic, as when he defined a conic as a curve of the second degree instead of a section of a cone, and treated it by the aid of coordinates. His writings include works on mechanics, sound, astronomy, the tides, the laws of motion, the Torricellian tube, botany, physiology, music, the calendar (in opposition to the Gregorian reform), geology, and the compass,—a range too wide to allow of the greatest success in any of the

[1] *Arithmetica Infinitorum, sive Nova Methodus Inquirendi in Curvilineorum Quadraturam, aliaque difficiliora Matheseos Problemata,* Oxford, 1655.

[2] The Latin edition is in Volume II of his *Opera Mathematica,* Oxford, 1693–1695.

lines of his activity. He was also an ingenious cryptologist and assisted the government in deciphering diplomatic messages.[1]

Among his interesting discoveries was the relation

$$\frac{4}{\pi}=\frac{3\cdot3\cdot5\cdot5\cdot7\cdot7\cdots}{2\cdot4\cdot4\cdot6\cdot6\cdot8\cdots},$$

one of the early values of π involving infinite products.[2]

Gregory. Scotland seems frequently to attempt to conceal her great men by a kind of protective coloring. Probably it is because she does not wish to seem to boast of her success. She had in the 17th century, and still has, a distinctive school of mathematical teaching, but she seems reluctant to proclaim the fact. Occasionally, however, she fails to keep her scholars and writers under cover, and so in mathematics we have such names as Napier, Gregory, and Kelvin.

James Gregory[3] was one of the first Scotchmen to make for himself a great name both in mathematics and in physics. He lived for some years in Italy, but in 1668 returned to Scotland to assume the professorship of mathematics at St. Andrews. In 1674 he became professor of mathematics at Edinburgh, where he died a year later. Like most of the leading British mathematicians of this period, he was equally interested in mathematics and physics. In 1661 he invented but did not practically construct the reflecting telescope which bears his name.[4] He also originated the photometric method of estimating the distances of stars. In the field of pure mathematics he expanded $\tan^{-1}\theta$, $\tan\theta$, and $\sec^{-1}\theta$ in series (1667), distinguished between convergent and divergent series, calculated the areas of successive polygons as the number of sides

[1] D. E. Smith, in the *Bulletin of the Amer. Math. Soc.*, XXIV (2), p. 82.

[2] See his *Opera Mathematica*, I, 441, where he explains that he uses the symbol □ for the ratio of the square on the diameter to the area of the circle. In our symbolism this is $4r^2:\pi r^2$, or $4:\pi$. On page 469 he gives the value stated above. See also his Volume II, cap. lxxxiv.

[3] Born at Drumoak, near Aberdeen, November, 1638; died at Edinburgh, October, 1675.

[4] Described in his *Optica promota, seu abdita radiorum reflexorum et refractorum mysteria geometrice enucleata*, London, 1663.

was doubled, and gave an ingenious but unsatisfactory demonstration of the incommensurability of π.[1] The series

$$\theta = \tan\theta - \tfrac{1}{3}\tan^3\theta + \tfrac{1}{5}\tan^5\theta - \cdots$$

is commonly known by his name. He died in the thirty-seventh year of his age, shortly after being stricken with blindness as a result of the strain upon his eyes in carrying on his astronomical observations.

AUTOGRAPH OF LORD BROUNCKER

Part of a document dated December 31, 1661, just before Brouncker's appointment as Chancellor of Queen Catherine

Other British Writers. First among the other British writers of the period made memorable by the great work of Barrow, Wallis, and Newton, was William, Viscount Brouncker.[2] He was one of the founders and was the first president of the Royal Society, a linguist with the genius of a mathematician,

[1] *Vera circuli et hyperbolae quadratura*, Padua, 1667, and again in 1668. *Geometriae pars universalis*, Venice, 1667; "De circuli quadratura," in the *Phil. Trans.*, London, 1668. See also G. Heinrich, *Bibl. Math.*, II (3), 77.

[2] Born at Castle Lyons, Ireland, *c.* 1620; died at Westminster, April 5, 1684. The name is also given as Brounker. He was the second viscount.

and a man who was justly esteemed by the mathematical world of his day. He was interested in the rectification of the parabola[1] and the cycloid[2] and in the quadrature of the hyperbola[3] (1668) and the circle. The last of these investigations led him to state that

$$\frac{4}{\pi} = \frac{1}{1+}\ \frac{1^2}{2+}\ \frac{3^2}{2+}\ \frac{5^2}{2+} \cdots,$$

a form which he derived from the continued product which Wallis had discovered. This was the first use of continued fractions by an English writer. In 1662 he was appointed Chancellor of Queen Catherine, and in 1681 he became Master of St. Catherine's Hospital in London. Aubrey, who knew him, relates that "he was of no university, he told me. He addicted himselfe only to the study of the mathematicks, and was a very great artist in that learning."[4]

John Pell,[5] who was about ten years older than Brouncker, was admitted to Trinity College, Cambridge, at the age of thirteen, and also studied at Oxford. At the age of twenty he had mastered Latin, Greek, Hebrew, Arabic, Italian, French, Dutch, and Spanish. He became a professor of mathematics at Amsterdam (1643–1646) and at Breda (1646–1652), was Cromwell's representative in Switzerland (1654–1658), entered the ministry (1661), was given the degree of D.D. at Lambeth (1663), and in the same year was made a Fellow of the Royal Society. He spent the latter part of his life in parish work and in writing on mathematics.

He wrote on the quadrant (1630), on other matters relating to astronomy, on the magnetic needle, and on the quadrature problem,[6] and computed a table of 10,000 square numbers

[1] "On the Proportion of a curved line of a paraboloid to a straight line," *Phil. Trans.*, III, 645.

[2] "Of the Finding of a straight Line equal to that of a cycloid," *Phil. Trans.*, VIII, 649.

[3] "An algebraical paper on the squaring of the hyperbola," *Phil. Trans.* abridgment, 4th ed., I, 10. London, 1731.

[4] Oxford ed., 1898, I, 128.

[5] Born at Southwick, Sussex, March 1, 1611; died in London, December 12, 1685.

[6] *Refutation of Longomontanus's Pretended Quadrature of the Circle*, 1646; in Latin, 1647.

(1672). He edited (1668) the algebra of Rhonius[1] (1622–1676), which had been translated by Thomas Branker[2] and which contains the first mention of the Anglo-American sign (÷) of division. Through an error on the part of Euler, Pell's name is commonly connected with a certain equation of the form

$$x^2 - Ay^2 = 1,$$

although he had but little to do with it.[3]

One of the most promising pupils of Wallis was William Neile,[4] who gave (1660) the rectification of the semicubical parabola $y^3 = ax^2$. He died too soon, however, to fulfill his early promise.

If it had not been for the great fire of London (1666), Sir Christopher Wren[5] would have been known as a mathematician rather than as the architect of St. Paul's Cathedral. He was educated at Westminster School and in 1649 or 1650 was entered at Wadham College, Oxford. He was graduated B.A. in 1650–1651 and M.A. in 1653. In 1653 he was elected fellow of All Souls College, Oxford, where he resided until 1657. From 1657 to 1660 he was professor of astronomy at Gresham College, London, and from 1661 to 1673 was Savilian professor of astronomy at Oxford. He was a Fellow of the Royal Society and was its president from 1680 to 1682. He wrote on "The law of nature in the collision of bodies,"[6] on the grinding of hyperbolic mirrors, on perspective, and on the rectification

[1] J. H. Rahn, *An Introduction to Algebra,* London, 1668. The book was first published in Zürich in 1659, in German.

[2] Or Brancker. He was a clergyman, born in Devonshire, 1636; died at Macclesfield, 1676. He also taught at Macclesfield. He wrote a work of no merit on astronomy.

[3] E. E. Whitford, *The Pell Equation,* New York, 1912, with a good bibliography; H. Konen, *Geschichte der Gleichung* $t^2 - Du^2 = 1$, Leipzig, 1901; G. Eneström, "Ueber den Ursprung der Benennung Pell'sche Gleichung," *Bibl. Math.*, III (3), 204.

[4] Born at Bishopsthorpe, December 7, 1637; died in Berkshire, August 24, 1670.

[5] Born at East Knoyle, Wiltshire, October 20, 1632; died in London, February 25, 1723.

[6] *Phil. Trans.*, 1669, but originally separately printed in Latin as *Lex naturae de collisione corporum.*

problem, and he discovered (1669) the two systems of generating lines on a hyperboloid of one sheet. After the great fire he took a prominent part in the rebuilding of St. Paul's Cathedral and more than fifty other churches and public buildings in London. His noble epitaph in the cathedral is well known.[1]

Among the other pupils of Wallis who acquired some reputation was John Caswell.[2] He matriculated at Wadham College, Oxford, at the age of sixteen. Six years later he began teaching mathematics in Oxford, and so marked was his success that in 1709 he was elected Savilian professor of astronomy. He published a trigonometry in 1685 and at one time thought of publishing the work of Menelaus on spherics, but gave up the plan.[3]

James Gregory's nephew, David Gregory,[4] was professor of mathematics at Edinburgh from 1684 to 1691, after which he became Savilian professor of astronomy at Oxford. He published a work on geometry[5] when he was only twenty-three years of age, and in a work on optics[6] set forth the possibility of achromatic lenses. He also wrote on the Newtonian theory.[7] In 1703 he brought out at Oxford an elaborate edition of Euclid's works.

Textbook Writers. In the 17th century we reach a time in the development of elementary mathematics when textbooks became so standardized and numerous as to require in this connection the mention of only the most important. Their mission henceforth was to improve the method of presenting theories already largely developed and to adapt the applications of these theories to the needs of the world. From that time on they ceased to be a great factor in the presentation of mathematical discoveries.

[1] "*Lector, si monumentum requiris, circumspice.*"

[2] Born at Crewkherne, Somerset, 1655; died April, 1712.

[3] L. C. Karpinski, *Bibl. Math.* XIII (3), 248.

[4] Born at Kinnairdie, Banffshire, June 24, 1661; died at Maidenhead, Berkshire, October 10, 1708.

[5] *Exercitatio Geometrica de dimensione figurarum*, Edinburgh, 1684.

[6] *Catoptricae et Dioptricae Sphaericae Elementa*, Oxford, 1695.

[7] *Astronomiae Physicae et Geometriae Elementa*, Oxford, 1702.

The most prominent British textbook writer on elementary arithmetic in the 17th century was Edmund Wingate.[1] This is somewhat strange, because he entered the profession of the law after leaving Oxford,[2] went to Paris in 1624, where he taught English to the princess Henriette-Marie, future wife of Charles I, and returned to England in 1650 and became member of parliament for Bedford, none of which activities related to mathematics or even to textbooks. His tastes, however, seem to have led him to look upon mathematics as an avocation, for in the same year that he went to Paris he published a work on Gunter's scale,[3] two years later a work on logarithmic arithmetic,[4] and in 1630 his popular work *Of natural and artificial Arithmetick* (London), a work that went through many editions and was popular for more than a century. He also prepared a set of logarithmic tables[5] and a work[6] on a mathematical instrument which he had invented.

Of the textbook makers of this period in the domain of elementary algebra the best known was John Kersey,[7] a self-made teacher. He was highly esteemed in London as an instructor in mathematics and was a friend of Wingate's. His algebra[8] presents the subject in a logical and teachable manner. One of the most interesting features is the preliminary explanation of the analogies between proportion, which then held a high place, and the modern treatment of equations. The work was, however, altogether too elaborate to meet with great success.

William Leybourn (1626–*c.* 1700) was well known in London in his day, not only as a writer of textbooks but as a

[1] Born at Flamborough, Yorkshire, 1596; died in London, December 13, 1656.

[2] He edited the second edition of Britton's famous *Collections* on English law, London, 1640.

[3] *Construction, description et usage de la règle de proportion*, Paris, 1624.

[4] *Arithmétique logarithmique*, Paris, 1626; English edition, London, 1635.

[5] *Tables of the Logarithms of the sines and tangents*, London, 1633.

[6] *Ludus Mathematicus*, London, 1654.

[7] Born probably at Bodicote, near Banbury, Oxfordshire, 1616 (at any rate he was baptized there November 23, 1616); died May, 1677. See the *Bibl. Math.* XII (3), 263, with evidence for this date.

[8] *The Elements of that Mathematical Art commonly called Algebra*, London, 1673.

teacher of mathematics and a surveyor. He wrote on astronomy,[1] surveying,[2] arithmetic,[3] the logarithmic rule,[4] the Napier computing rods,[5] and mathematical recreations.[6] In 1690 he published a *Cursus Mathematicus* containing the substance of his other works. Three years later he published the best known of his works, a ready reckoner,[7] the first elaborate book of the kind to appear in English.

Leybourn was not, however, as popular as Edward Cocker,[8] who is described in 1657 as living "on the south side of St. Paul's Churchyard, over against St. Paul's Chain . . . where he taught the art of writing and arithmetick in an extraordinary manner." He was also a bibliophile, and in his "public school for writing and arithmetic" where he "takes in boarders" he had a large library of manuscripts and printed books in various languages. He wrote three or four books on arithmetic and penmanship, but it was his *Arithmetick, being a Plain and Easy Method*, edited by John Hawkins in 1678, that fixed the expression "according to Cocker" in the common speech of England. This remarkable book went through upwards of a hundred editions and had great influence upon British textbooks for more than a century. He also wrote on algebra,[9] but there has always been a question as to how much of this work

[1] *Urania Practica*, London, 1648, written with Vincent Wing. One of the first books on astronomy written in English, but not the first, Recorde's *Castle of Knowledge* having appeared in 1551.

[2] *Planometria, or the Whole Art of Surveying of Land . . . by Oliver Wallinby*, London, 1650, the pseudonym being formed by transposing the letters of his name, with minor changes.

[3] *Arithmetick, Vulgar, Decimal, and Instrumental*, London, 1657.

[4] *The Line of Proportion or Numbers, commonly called Gunter's Line, made easie*, London, 1667.

[5] *The Art of Numbering by Speaking-Rods: Vulgarly termed Nepeirs Bones*, London, 1667 and 1685. Leybourn, as will be shown in Volume II, mistakes the etymology of *Rabdologia*, and translates the word as "Speaking-Rods."

[6] *Pleasure with Profit; consisting of Recreations of divers kinds, Numerical, Geometrical, Mechanical, Statical, Astronomical*, London, 1694.

[7] *Panarithmologia, being a Mirror Breviate, Treasure Mate for Merchants, Bankers, Tradesmen, Mechaniks*, . . . London, 1693.

[8] 1631–1675, probably a descendant of the Northamptonshire family of Cokers.

[9] *Algebraical Arithmetic, or Equations*, London, 1684.

as well as the arithmetic was due to Cocker and how much to Hawkins. The world is as well off either way, the important thing being that in them we have two books that represent the popular view of the elementary science at the period in which they were written.

5. Germany

Germany in the Seventeenth Century. In the 16th century Germany moved toward the front in the mathematical progress of Europe, but with Italy always in the lead. In the 17th century she fell behind, producing only a single mathematician of the first class. The first part of the century was the period of the Thirty Years' War (1618–1648), and indeed the whole period was one of unrest in the Teutonic countries,—a period quite unsuited to intellectual progress.

Kepler. Although known chiefly for his work in the domain of astronomy, Johann Keppler,[1] or Kepler, as the name more often appears in English, ranked high as a mathematician. He studied in the cloister school at Maulbronn and at the University of Tübingen. At the age of twenty-two he began teaching mathematics and moral philosophy in the Gymnasium at Gratz in Steyermark.[2] For two years (1599–1601) he was an assistant of Tycho Brahe, and in 1601 became court astronomer to Kaiser Rudolph II and later to his successors Matthias (1612) and Ferdinand III (1615). His life was, however, made almost unendurable through domestic infelicity, troubles at court, and financial difficulties. In his work in mathematics he shows himself an excellent geometer and was also much interested in algebra. He set forth (1604) the idea of continuity in elementary geometry,[3] made (1615) some advance in the use of the infinitesimal,[4] and did much to further the cause of logarithms.[5]

[1] Born at Weil der Stadt, Württemberg, December 27, 1571; died at Regensburg, November 15, 1630. From the family of von Kappel.

[2] Or Graz, in the former Austrian crownland of Styria.

[3] In his *Ad Vitellionem Paralipomena*. See the Frisch edition of his *Opera Omnia*, II, 119, 187 (Frankfort, 1858–1871).

[4] In his *Stereometria*; e. g., see the Frisch edition of his *Opera Omnia*, IV, 583.

[5] His *Chilias Logarithmorum* appeared at Marburg in 1624, with a *Supplementum* in 1625.

Tschirnhausen. Ehrenfried Walther, Graf von Tschirnhausen,[1] often known as Tschirnhaus, is one of the few minor German mathematicians of the time to require special mention. He served for a year or two (1672–1673) with the Dutch army, traveled extensively, and then settled down on his estates, devoting his time largely to mathematics and physics. He is known chiefly for his study of curves,[2] including caustics and catacaustics,[3] and for his work in maxima and minima[4] and the theory of equations.[5]

Leibniz. Gottfried Wilhelm, Freiherr von Leibniz,[6] was the only pure mathematician of the first class produced by Germany during the 17th century. He early showed great proficiency in mathematics, having read the most important treatises on the subject before he was twenty. He studied law at Nürnberg, entered upon a diplomatic career, traveled extensively, and made the acquaintance of leading mathematicians in Holland, France, and England. He finally went to Hannover and became librarian to the duke. He lived in good style, and the visitor to Hannover today may see his palatial house, now used as a museum.

The leisure which his office allowed him gave Leibniz the opportunity to develop the differential and integral calculus. He seems to have begun to think about the subject in 1673,[7]

[1] Born at Kiesslingswalde, near Görlitz, April 10, 1651; died at Dresden, October 11, 1708. H. Weissenborn, *Lebensbeschreibung des Ehrenfried Walther von Tschirnhaus*, Eisenach, 1866.

[2] "Nova methodus tangentes curvarum expedite determinandi," *Acta Eruditorum*, I (1682), and various other memoirs in the same publication.

[3] All published in the *Acta Eruditorum*. The terms are due to Jacques and Jean Bernoulli. See Jacques Bernoulli, *Opera*, I, 466, *et passim*.

[4] "Nova methodus determinandi maxima et minima," *Acta Eruditorum*, II (1683).

[5] "Methodus auferendi omnes terminos intermedios ex data aequatione," *Acta Eruditorum*, II, 204.

[6] Born at Leipzig, June 21, 1646 (O.S.); died at Hannover, November 14, 1716 (N.S.). His father wrote his name Leibnütz. It is often written Leibnitz. The Latin form used by our Leibniz was at first Leibnüzius or Leibnuzius, but later the name appears as Leibnitius. The preferred spelling in modern German scientific works is Leibniz.

[7] At least, this is his claim. See J. M. Child, *The Early Mathematical Manuscripts of Leibniz*, p. 37 (Chicago, 1920).

some years after Newton had explained the fluxional calculus to his pupils. Two years later he had his theory well developed, but it was not until 1684 that he published, in the *Acta Eruditorum*, a description of the method and its possibilities. There is no longer any doubt that Leibniz developed his calculus quite independently, and that he and Newton are each entitled to credit for their respective discoveries. The two lines of approach were radically different, although the respective theories accomplished results that were practically identical. Leibniz knew or could easily have known what Newton was doing, and this may have suggested the line of work; and he knew the contribution already made by Barrow[1] in the form of the "differential triangle," but at any rate he was original in much that he accomplished. In a word, it may be said that he made Cavalieri scientific. He also laid the foundation for the theory of envelopes and defined the osculating circle and showed its importance in the study of curves.

Leibniz was a diplomat in the days when Machiavelli was a model, and this is not a flattering way in which to characterize him. Of no other man does the visitor to the portrait gallery of the mathematical world carry away such varied impressions. Some of the engravings show him as a man of great refinement and dignity, while others show the mean, dishonest, disappointed face of a man whose word would never be accepted. In the old controversy as to the invention of the calculus, the word of Leibniz would not have the weight usually given to the statement of a scholar. Nevertheless it must be repeated that he showed great originality in his theory and in the symbolism of the calculus, and is entitled to a high degree of credit for this work.

Minor Writers. Among the minor writers of the first half of the 17th century Johann Faulhaber[2] is known as a successful teacher of mathematics in Ulm. He published various works on algebra and on the curious phases of elementary mathematics.

[1] J. M. Child, *loc. cit.*, p. 19 n. *et passim*, and his *Geometrical Lectures of Isaac Barrow*, Chicago, 1918.

[2] Born at Ulm, May 5, 1580; died at Ulm, 1635.

GODEFROI GUILLAUME LEIBNITZ
Né le 3 Juillet 1646 mort le 14 Novembre 1716
Il fut dans l'Univers connu par ses Ouvrages,
Et dans son Païs même, il se fit respecter ;
Il instruisit les Rois, il éclaira les Sages,
Plus sage qu'eux il sut douter
VOLTAIRE.

LEIBNIZ

After an engraving by Ficquet

His contemporary, Benjamin Ursinus,[1] professor of mathematics at the University of Frankfort a. d. O., was one of

Leibnitius Wolfio 10 febr 1712
summae potentiarum

Vir celeberrime Fautor & Amice Honoratissime

Gratias ago quod de Libro Ill. Comitis de Herberstein ad me destinato significas; occasionem ad Dn. Abbatem Smidium vel alium Helmstadium aut Brunsvigam mittendi habes, non dubito quin inde recte ad me perveniat.

Nuper summas potentiarum considerans sequentia facile meditata observo. Numeri naturales 0, 1, 2, 3, 4, etc. sint n, potentiae exponens sit e; summa potentiarum exponentis superioris ab omnibus summis inferioris pendet ope hujus theorematis

$$\frac{1}{e+1}\overline{n+1}^{e+1} = \frac{1}{1}\int n^e + \frac{e}{1.2}\int n^{e-1} + \frac{e.e-1}{1.2.3}\int n^{e-2} + \frac{e.e-1.e-2}{1.2.3.4}\int n^{e-3} + etc.$$

Quia autem e est numerus rationalis integer, hinc series est finita, et descendendo pervenitur ad terminos evanescentes. Inveni et talem aequationem independentem a summis inferioribus, sed tamen certos numeros constantes A, B, C, D etc. pro quavis summa, semel calculandos supponentem

$$\frac{1}{e+1}\overline{n+1}^{e+1} = \int n^e + A\,\overline{n+1}^{e} + B\,e\,\overline{n+1}^{e-1} + C\,e.e-1.\overline{n+1}^{e-2} + D\,e.e-1.e-2.\overline{n+1}^{e-3}$$

ubi est $A = \frac{1}{1.2}$, $B = \frac{1}{1.2.3} + \frac{1}{1.2}A$, $C = \frac{1}{1.2.3.4} + \frac{1}{1.2.3}A + \frac{1}{1.2}B$ et

$$D = \frac{1}{1.2.3.4.5} + \frac{1}{1.2.3.4}A + \frac{1}{1.2.3}B + \frac{1}{1.2}C$$

Et ita porro pro E, F etc.

AUTOGRAPH LETTER OF LEIBNIZ

First page of a letter to Christian Wolf, dated February 10, 1712

the first to introduce logarithms into Germany.[2] Still another contemporary, Daniel Schwenter,[3] was professor of

[1] Born at Sprottau, July 5, 1587; died at Frankfort a. d. O., September 27, 1633 or 1634. Ursinus is the Latin form for Behr, his family name.

[2] *Trigonometria logarithmica, . . . cum magno logarithmorum canone,* Frankfort a. d. O., 1618; 2d ed. 1635; *Magnus canon triangulorum logarithmicus,* Köln a. d. Spree, 1624.

[3] Born at Nürnberg, January 31, 1585; died at Altdorf, January 19, 1636.

Hebrew (1608), oriental languages (1625), and mathematics (1628) at the University of Altdorf. He wrote a well-known work on mathematical recreations.[1] A fourth of the minor

AUTOGRAPH LETTER OF LEIBNIZ

Second page of the letter shown on page 420

writers of this period, Peter Roth,[2] was a Rechenmeister at Nürnberg. He wrote an unimportant algebra[3] in which he treats of equations of the third and fourth degrees.

[1] *Deliciae physico-mathematicae oder Mathematische und philosophische Erquickstunden*, Nürnberg, 1636 (posthumous).

[2] Born at Ingolstadt, *c.* 1580; died at Nürnberg, 1617.

[3] *Arithmetica philosophica, oder schöne, neue, wohlgegründete, überaus künstliche Rechnung der Coss oder Algebra*, Nürnberg, 1608.

For his general learning and his great gift to the world through the museum at Rome which bears his name, if not for any important contribution to mathematics, Athanasius Kircher[1] should be mentioned, although he might quite as appropriately be recorded in the list of Italian scholars, since he spent the better years of his life in Rome. He was a Jesuit, professor of mathematics and philosophy, and later of Hebrew and Syriac, at the University of Würzburg, after which he went (1635) to Avignon and then to Rome, where he taught mathematics and Hebrew. He was a voluminous writer on a great variety of subjects and a zealous collector of curios. His study of optics may have led to the invention of the stereopticon, or to its improvement, but the claims of Kircher's friends are questionable. His mathematical works relate to instruments[2] and to the occult in number,[3] and are not to be taken seriously.

6. The Netherlands

Geographical Limits. In speaking of the Netherlands we must again mention the fact that reference is made to geographical rather than political boundaries. The 17th century was a strenuous one for those countries which we now designate as Holland and Belgium. Roughly speaking, the latter was then under the sovereignty of Spain until Louis XIV began his conquest in 1667, and the former was nearly lost to the same invader in 1672. In the midst of these strenuous times, but generally before the wars, the Netherlands produced several mathematicians who stood high in their respective lines.

Snell. There lived at Leyden at the opening of the 17th century a professor of mathematics (1581) and Hebrew by the name of Rudolph Snell.[4] He wrote a few unimportant works on mathematics, all of which were published in the 16th

[1] Born at Geisa, near Fulda, May 2, 1602; died at Rome, November 28, 1680.

[2] *Pantometrum Kircherianum, h. e. Instrumentum geometricum novum*, Würzburg, 1660.

[3] *Arithmologia sive De occultis numerorum mysteriis*, Rome, 1665.

[4] The Latin form, Snellius, is often used. Born at Oudewater, October 8, 1546; died at Leyden, April 2, 1613.

century. He had a son, Willebrord Snell van Roijen,[1] who succeeded him as professor of mathematics at Leyden (1613), devoting himself chiefly to astronomy, physics, and trigonometry. He wrote on the mensuration of the circle,[2] and set forth the properties of the polar triangle in spherical trigonometry.[3] To him is due the name "loxodrome" for the rhumb line in navigation, the latter name being due to the Portuguese navigator and mathematician Nunes (Nonius).

Girard. Little is known of the life of the mathematician Albert Girard.[4] He wrote on trigonometry,[5] fortifications, practical geometry,[6] and algebra,[7] and edited the works of Simon Stevin.[8] He was one of the first to appreciate the significance of the negative sign in geometry, and was successful in his use of imaginary quantities in the theory of equations. He inferred by induction, as others had done, that an equation of the nth degree has n roots, expressed the sum of the first four powers of the roots of an equation as functions of the coefficients, and discussed general polygons, both cross and simply convex.[9]

Huygens. Although he was known chiefly as one of the world's greatest physicists, particularly in relation to the study of the pendulum, the invention of pendulum clocks, and the laws of falling bodies, Christiaan Huygens[10] should be ranked

[1] Born at Leyden, 1581; died at Leyden, October 31, 1626.

[2] *Cyclometricus, de Circuli Dimensione secundum Logistarum Abacos*, Leyden, 1621.

[3] *Doctrinae Triangulorum Canonicae . . . Libri IV*, Leyden, 1627 (posthumously published). Vieta had already given them.

[4] Born at St. Mihiel, Lorraine, 1595; died at The Hague, December 8/9, 1632. He seems to have lived chiefly in Holland.

[5] *Tables de sinus, tangentes et secantes*, The Hague, 1626.

[6] *Géomètrie contenant la théorie et la pratique d'icelle, escrite par Sam. Marolois, revue, augmentée et corrigée*, Amsterdam, 1627. Marolois was a writer on fortifications.

[7] *Invention nouvelle en l'algèbre*, Amsterdam, 1629. Reprinted by Bierens de Haan, Leyden, 1884. [8] This edition appeared posthumously, Leyden, 1634.

[9] Thus he classified quadrilaterals as "la simple, la croisée et l'autre ayant l'angle renversée."

[10] Born at The Hague, April 14, 1629; died at The Hague, June 8, 1695. The name often appears as Huyghens or Hugenius, with several other variants. There is a biography in his *Opera Varia*, Leyden, 1724.

high among those who improved the new geometry and made known the power of the calculus. He introduced the notion of evolutes,[1] rectified the cissoid, determined the envelope of a moving line, investigated the form and the properties of the catenary,[2] wrote on the logarithmic curve, gave in modern form the rule for finding maxima and minima of integral functions, wrote on the curve of descent (1687), published a work on probability, proved that the cycloid is a tautochronous curve, and contributed extensively to the application of mathematics to physics.

CHRISTIAAN HUYGENS

From an engraving by Edelinck, after a drawing by Drevet

Minor Writers. Among the minor writers the name of Grégoire de Saint-Vincent[3] is one of the best known. He was a Jesuit, taught mathematics in Rome and Prag (1629–1631), and was afterwards called to Spain by Philip IV as tutor to his son Don Juan of Austria. He wrote two works on

[1] *Horologium oscillatorium*, Paris, 1673. It is in this treatise that his work appears at its best.

[2] "Solutio problematis de linea catenaria," *Acta Eruditorum*, 1691.

[3] Born at Bruges, 1584; died at Ghent, January 27, 1667. Among the common variants of his name are Gregorius a Sancto Vincentio and Gregorius von Sanct Vincentius. For a biographical sketch by Quetelet, see the *Annales Belgiques,* VII (1821), 253, and one by Bosmans in the *Annales* of the Société Scientifique of Brussels, XXXIV, 1, 174.

geometry,[1] giving in one of them the quadrature of the hyperbola referred to its asymptotes, and showing that as the area increased in arithmetic series the abscissas increased in geometric series.

The Van Schooten family produced three generations of professors of mathematics at Leyden, all sympathetic with the science but no one of them of first rank. The first was Frans van Schooten,[2] with a trigonometric table to his credit (1627). The second, his son, was also called Frans van Schooten.[3] He wrote on mathematics[4] and was a professor in the engineering school at Leyden and the teacher of Huygens. He edited Vieta (1646), wrote on perspective (1660), and is well known for his Latin edition of the geometry of Descartes (1649). His half brother, Petrus van Schooten,[5] occupied the chair of mathematics at Leyden, being later (1669) transferred to the chair of Latin, but he contributed nothing of permanent value in either of his fields of interest.

René François Walter, Baron de Sluze[6] was a man of standing in the Church, with a taste for mathematics. He contributed to the geometry of spirals and the finding of geometric means, and also invented a general method for determining points of inflection of a curve. One of his contemporaries, but a few years his junior, Johann Hudde[7] was an officeholder in Amsterdam. He was interested in the theory of maxima and minima

[1] *Opus geometricum quadraturae circuli et sectionum coni,* 2 vols., Antwerp, 1647; *Opus geometricum posthumum ad mesolabium per rationem,* . . . Ghent, 1668. See also K. Bopp, *Abhandlungen*, XX, 87.

[2] Born at Leyden, 1581; died at Leyden, December 11, 1646. The name often appears in the Latin form Franciscus. There are a number of interesting MSS. of works by the various Van Schootens now in the library at Groningen. For a list of these, see H. Brugmans, *Catalogus codicum*, No. 108 seq. (Groningen, 1898).

[3] Born at Leyden, *c.* 1615; died at Leyden, May 29, 1660.

[4] *Principia Matheseos Universalis*, Leyden, 1651; *Exercitationum Mathematicarum Libri quinque*, Leyden, 1657. See *Bibl. Math.*, XII (3), 156.

[5] Born at Leyden, February 22, 1634; died at Leyden, November 30, 1679.

[6] Born at Visé on the Maas, July 7, 1622; died at Liége, March 19, 1685. The name also appears as Sluse or Slusius.

[7] Latin, Huddensis. Born at Amsterdam, probably in 1628 or 1629; died at Amsterdam, April 16, 1704.

(1658) and the theory of equations (1657), and in his work on the latter subject he separated into factors the polynomial which he equated to zero.

About this time Cornelis van Beugham wrote a *Bibliographia Mathematica* which appeared at Amsterdam in 1688. This seems to have been the first printed book devoted solely to mathematical bibliography.

7. Other European Countries

Countries Considered. Of the other European countries that exerted substantial influence on mathematics in the 17th century Switzerland stands easily at the head. This seems due not to any particular intellectual influences but to the efforts of one of the most interesting families known in the history of science, and to the labors of a man who came near being the inventor of logarithms.[1]

The other countries demanding attention are Spain and Denmark.

The Bernoullis. Students of heredity have called attention to the extraordinary number of distinguished scholars who descended from the protestant population expelled from the catholic countries in the 16th and 17th centuries.[2] Presumably the same result would be found among the descendants of Catholics, Jews, political refugees, or others who maintained their faith in any manner that tries the souls of men.[3] Of those who descended from Belgian stock that was rooted up during the reign of terror of the Duke of Alba were the members of the Bernoulli family, a family that furnishes one of the most remarkable evidences of the power of heredity or of early home influence in all the history of mathematics. No

[1] L. Isely, *Histoire des Sciences Mathématiques dans la Suisse Française*, Neuchâtel, 1901.

[2] A. de Candolle, *Histoire des sciences et des savants*, 2d ed., p. 338. Paris, 1885.

[3] Among similar cases of descendants of religious refugees are those of Jean Trembley, Simon Lhuilier, Georges-Louis Le Sage, Louis Bertrand, and Élie Bertrand.

less than nine of its members attained eminence in mathematics and physics,[1] and four of them were honored by election as foreign associates of the Académie des Sciences of Paris.

Jacques Bernoulli. The first of the family to attain any reputation in mathematics was Jacques Bernoulli.[2] He first studied theology, but his taste was in the direction of astronomy, mathematics, and physics, and he traveled in France, Holland, Belgium, and England for the purpose of devoting his time to these studies and to meeting learned men. He returned to Switzerland in 1682, took up the study of the new calculus as set forth by Leibniz, and in 1687 became professor of mathematics in the University

P. Dupin Sculp.
JACQ. BERNOULLI
Professeur de Mathematique a Basle,
de la Société Royale de Londes et des Academies
des Sciences de Paris et de Berlin.

JACQUES BERNOULLI

The elder of the two brothers who founded the famous Bernoulli family of mathematicians

[1] For the relationships, see P. H. von Fuss, *Correspondance mathématique . . . de quelques célèbres géomètres du XVIII. siècle*, Vol. I, p. xviii (Petrograd, 1843) (hereafter referred to as Fuss, *Correspondance*); also a note in Terquem's *Nouvelles Annales de Math., Suppl. Bulletin,* Vol. 17, p. 85 (Paris, 1858); P. Merian, *Die Mathematiker Bernoulli,* Basel, 1860.

[2] Born at Basel, December 27 (o.s.), 1654; died at Basel, August 16, 1705. Jacobus Bernoulli, often called Jacques (I) Bernoulli to distinguish him from another Jacques (1759–1789) in the 18th century. English writers often call him James and the German writers Jakob. Since, however, he wrote in French or Latin, the preferable form of his given name is Jacques or Jacobus.

of Basel. He wrote (1683–1701) a large number of memoirs for the *Acta Eruditorum*. These memoirs include such lines of research as series (1686), the quadrisection of a general triangle by two normals (1687), conics (1689), lines of descent (1690), mensuration (1691), cycloids (1692, 1698, 1699), transcendent curves (1696), and isoperimetry (1700). He wrote the second book devoted to the theory of probability,[1] although the subject had been studied in Italy and France much earlier than this.

He solved by infinitesimal analysis the problem of the isochronous curve which had already attracted the attention of various writers and had been solved by Huygens, Leibniz, L'Hospital, and Newton. He determined the length of the catenary curve. Because of his study of the logarithmic spiral, $r = a^{\theta}$, he directed that this curve should be engraved upon his tombstone, with the words *Eadem mutata resurgo*,[2] and the visitor to the cloisters at Basel may still see the rude attempt of the stonecutter to carry out his wish.[3]

The fact that his father was emphatically opposed to his study of astronomy and mathematics, placing all possible obstacles in his way, led him to choose for his device Phaethon driving the chariot of the sun, with the legend, *Invito patre sidera verso*.[4]

His brother Jean Bernoulli[5] was thirteen years his junior,[6] which may account for an attitude of superiority on the part of Jacques and one of resentment on the part of Jean which caused the ill feeling that long existed between them. His father learned no wisdom by his failure to have his brother

[1] *Ars conjectandi*, Basel, 1713 (posthumous). With this was published his *Tractatus de Seriebus Infinitis, Earumque summa Finita, et Usu in Quadraturis Spatiorum & Rectificationibus Curvarum*, and a letter *De Ludo pilae reticularis*. His *Opera* in two volumes appeared at Geneva in 1744.

[2] "I shall arise the same, though changed."

[3] There is an engraving of the design in J. J. Battierius, *Vita . . . Jacobi Bernoulli*, p. 40 (Basel, 1705).

[4] "I study the stars against my father's will."

[5] Born at Basel, July 27 (o.s.), 1667; died at Basel, January 1, 1748. The first name also appears as Johann or John, and often as Jean (1).

[6] Jacques was the fifth child in his father's family, and Jean was the tenth, a fact which may interest those who have faith in the theory of primogeniture.

Jacques become a theologian, and so he determined to make Jean a merchant. The latter thought that he preferred medicine or literature, but soon found that his real taste was for mathematics, and so the world was saved the loss of a genius.

Jean first studied medicine and wrote his doctor's dissertation *De effervescentia et fermentatione* (Basel, 1690). As Jacques found theology uncongenial, so Jean found medicine equally so, and each sought in the study of mathematics the mental activity that he required. Jean became professor of mathematics in the University of Groningen in 1695, and on his brother's death (1705) was elected to fill the place thus vacated at Basel. Each was made a foreign member of the Paris Académie des Sciences in 1699.

JEAN BERNOULLI

The younger brother of Jacques (I) Bernoulli

Jean was even more prolific than his brother in his contributions to mathematics, writing on a wide range of topics, including caustic curves (1692), differential equations (1694), the rectification and quadrature of curves by series (1694), the cycloid (1695), catoptrics and dioptrics (1701), the multisection of angles and arcs (1701), isochronous curves and curves of quickest descent (1718), and various related subjects. He wrote on the calculus and was one of the most influential scholars on the Continent in making its power appreciated. His collected works appeared six years before his death.[1]

Je viens de recevoir une lettre de mon second fils à Petersbourg,
dans laquelle il confirme le contentement avec lequel tous les
Professeurs y passent leur temps: il me marque qu'il souhaiteroit
d'entrer en commerce de lettres avec vous si votre commodité
le permettoit, il croit que cela se pourroit faire sans aucune
depense de part ni d'autre par le canal de Mr. de L'isle, qui
entretiendra correspondance avec l'Academie de Paris, ensorte
que vos lettres à mon fils et les siennes à vous pourroient être
envoyées sous la couverture des leurs. J'avois oublié, lorsque je
vous fis ecrire la derniere fois, de vous faire mander l'arrivée
de Mr. de L'isle à Petersbourg, mais je ne doute pas que vous
ne l'ayez déja apprise il y a longtemps. Je suis avec toute
la consideration possible

Monsieur

Bâle
ce 20. Juin 1726.

Votre tres humble et tres
obeissant Serviteur
J. Bernoulli

AUTOGRAPH OF JEAN (I) BERNOULLI

The "second fils à Petersbourg" mentioned in the letter was Nicolas (II) Bernoulli, who died at Petrograd only a little over a month after this letter was written

To him is due the use of the term "integral" in its technical sense in the calculus, the first attempt to construct an integral calculus, and the invention of the exponential calculus.[2] He was the first, except in such obvious cases as those known to Cardan, to obtain real results by the use of $\sqrt{-1}$; for example, in finding $\tan n\phi$ in terms of $\tan \phi$.

[1] *Opera omnia*, 4 vols., Lausanne, 1742.
[2] "Principia calculi exponentialium," *Acta Eruditorum*, 1697.

The Later Bernoullis. The work of the later Bernoullis falls in the 18th century, but it is appropriate to mention it briefly at this time. The leading one of the descendants of the two brothers was Daniel Bernoulli.[1] He was the son of Jean (I) Bernoulli, who, curiously enough, made the same mistake as his own father in trying to force his son into trade. Daniel spent some years (1725–1733) in Petrograd as professor of mathematics in the Academy of Petrograd, but in 1733 returned to Basel, where he became a professor in the university. He was a prolific writer, most of his work appearing in the memoirs of the Academy of Petrograd, but he published one volume on mathematics before going to Russia.[2] Most of the memoirs were upon physical questions, but a few related to pure mathematics, including the computation of trigonometric functions (1772, 1773), continued fractions (1775), and the Riccati problem.

DANIEL (I) BERNOULLI
After a portrait from life

Dr. Hutton[3] relates these incidents concerning him:

He used to tell two little adventures, which he said had given him more pleasure than all the other honours he had received. Travelling with a learned stranger, who, being pleased with his conversation, asked his name; "I am Daniel Bernoulli," answered he with great modesty; "And I," said the stranger (who thought he meant to

[1] Usually designated as Daniel (I) Bernoulli. Born at Groningen, February 9 (January 29, o.s.), 1700; died at Basel, March 17, 1782.

[2] *Exercitationes quaedam mathematicae*, Venice, 1724.

[3] *Philosophical and Mathematical Dictionary*, I, 205. London, 1796.

laugh at him), "am Isaac Newton." Another time having to dinner with him the celebrated Koenig the mathematician, who boasted, with some degree of self-complacency, of a difficult problem he had resolved with much trouble, Bernoulli went on doing the honours of his table, and when they went to drink coffee he presented Koenig with a solution of the problem more elegant than his own.

Nicolas (I) Bernoulli[1] was a nephew of Jacques (I) and Jean (I). In his younger days he was professor of mathematics at Padua (1716–1719), but returned to Basel, where he became a professor in the university. He was trained in the law, and his first mathematical treatise was upon the use of the theory of probability in legal matters.[2] He wrote extensively on differential equations and geometry.

Nicolas (II) Bernoulli[3] was the son of Jean (I). He studied law, traveled extensively, became professor of law at Bern (1723–1725), and was finally called to Petrograd as professor of mathematics. He wrote on the geometry of curves, but his death at the age of thirty-one closed a promising career.

Jean (II) Bernoulli[4] was the youngest son of Jean (I). He studied law but spent his later years as professor of mathematics in his native city. His work was chiefly on physics.

Jean (III) Bernoulli[5] was the son of Jean II. Like his father, he studied law but soon turned to mathematics, becoming director of the mathematics class at the Academy of Sciences at Berlin. He was much interested in the history of astronomy but also wrote on the doctrine of chance (1768), recurring decimals (1771), factoring (1771), and indeterminate equations (1772).

The other Bernoullis who were interested to a greater or less degree in mathematics were Daniel (II) (1751–1834), son

[1] Born at Basel, October 10, 1687; died at Basel, November 29, 1759. A note in Terquem's *Nouvelles Annales, Suppl. Bulletin*, 1858, p. 86, says November 25, 1749, but this is an error.

[2] *De Usu Artis Conjectandi in Jure*, Basel, 1709. He also edited the *Ars Conjectandi* of his uncle Jacques (1713).

[3] Born at Basel, January 27, 1695; died at Petrograd, July 26, 1726.

[4] Born at Basel, May 18, 1710; died at Basel, July 17, 1790.

[5] Born at Basel, November 4, 1744; died at Köpnick, near Berlin, July 13, 1807.

of Jean (II); Jacques (II) (1759–1789), son of Jean (II); Christoph (1782–1863), son of Daniel (II); and Jean Gustave (1811–1863), son of Christoph; but none of these attained great fame. The Bernoulli blood had lost its strength.

Other Swiss Writers. Two other Swiss mathematicians of the 17th century deserve mention,—one a genius, the other a plagiarist. The genius was Jobst Bürgi,[1] from 1579 to 1603 court watchmaker to Landgraf Wilhelm IV of Hesse, and later (until 1622) to Kaiser Rudolph II. He wrote on the proportional compasses[2] and on astronomy, but is best known for his invention of logarithms independently of Napier. He was led to the idea by an entirely different route from that taken by the latter, approaching it through the theory of exponents. He did not publish anything upon the subject until after Napier had made known his discovery, and when he finally concluded to print his work it was in the form of a small table of antilogarithms, issued anonymously at Prag in 1620.[3] The book never attracted any attention and remained practically unknown except to historians of mathematics.

The other Swiss writer was of a different character. He was a professor while Bürgi was a watchmaker; his name has been known for three centuries, while Bürgi's has been almost forgotten; but he was a plagiarist, while Bürgi was a genius. Paul Guldin[4] began his work as a goldsmith. He later entered the Jesuit order, lived for a long time in Rome, and became professor of mathematics at the University of Vienna and

[1] Born at Lichtensteig, February 28, 1552; died at Cassel, January 31, 1632. The first name also appears as Joost and Justus; the last as Burgi, Byrgi, Borgen, and Byrgius. See R. Wolf, "Zwei Kleine Notizen zur Geschichte der Mathematik," *Bibl. Math.*, III (2), 33.

[2] Invented by Galileo. They were also described (1607) by Levinus Hulsius in his *Dritter Tractat der mechanischen Instrumente,* Frankfort a. M., 29 pp., being the third of four tractates, 1603–1615. In the fourth of these tractates Hulsius describes a pedometer ("Instrument Viatorii oder Wegzählers"), the earliest mention of this instrument.

[3] *Arithmetische und Geometrische Progress Tabulen*, Prag, 1620.

[4] Born at St. Gall, June 12, 1577; died at Gratz, November 3, 1643. His name was originally Habakuk Guldin, but he changed it when he went from Protestantism to Catholicism.

later at Gratz. He wrote on physics and mathematics,[1] but is chiefly known for the fact that his name attaches to a theorem of Pappus on the volume of a solid generated by the revolution of a plane about an axis,[2]—a theorem which he included in his works without credit, fully aware that it was in the works of Pappus, to which he is known to have had access.

Denmark. In the 17th century Holstein, then a part of Denmark, produced one and only one mathematician of note,—Nicolaus Mercator.[3] He was one of the leading writers of the time on cosmography,[4] and also wrote on trigonometry,[5] the method of computing logarithms,[6] and astronomy, besides editing Euclid's *Elements*.[7] He lived for some time in London and was one of the first members of the Royal Society. In his *Logarithmotechnia* he gives the series that bears his name,

$$\log(1+x)=x-\frac{x^2}{2}+\frac{x^3}{3}-\frac{x^4}{4}+\cdots.$$

Spain. Whatever may be said for mathematics in the 16th century in Spain, less can be said for the century following. Philip II was a bigot, but he was a great bigot. He commanded the respect and won the loyalty of his people, and his reign was one of great works in art and in letters. After his death (1598), however, there occupied the throne a sorry line of kings, and the ruin of the country began. The population of Madrid fell one half in a century, and Seville's sixteen thousand looms were reduced to less than three hundred. From being a prosperous world power the country became a wreck among nations, and in the general destruction mathematics suffered with the other sciences and with letters and art.

[1] *Problema arithmeticum de rerum combinationibus*, Vienna, 1622.

[2] *Centrobaryca seu de centro gravitatis . . . liber I*, Vienna, 1635; *Centrobarycarum pars altera*, Vienna, 1641.

[3] Born near Cismar, in Holstein, *c.* 1620; died in Paris, February, 1687.

[4] *Cosmographia sive Descriptio coeli et terrae in circulos . . .*, Danzig, 1651.

[5] *Trigonometria sphaericorum logarithmica . . .*, Danzig, 1651.

[6] *Logarithmotechnia, sive Methodus construendi logarithmos nova accurata et facilis . . .*, London, 1668.

[7] *Euclidis Elementa Geometrica Libri VI novo ordine ac methodo fere demonstrata*, London, 1678.

Russia. As for Russia, she had not yet awakened. The first Russian arithmetic to appear with Hindu-Arabic numerals was written by a teacher named Magnitzky and was printed in 1703. Under Peter the Great there was some development of vocational mathematics, but it was only after the founding of the Academy of Sciences at Petrograd in 1725 that pure mathematics had any standing.[1]

8. The Orient

Effect of Western Civilization. The introduction of Western civilization into India, China, and Japan is interesting because of its diverse effects. As to India, mathematics was already stagnant, and the European influence gave it no stimulus. India has always been content to take her time. Not since Bhāskara (12th century) has she produced a single native genius in this field. In the 17th century only one name, that of Raganātha (*c.* 1621), attracted any general attention even in India, and he contributed nothing that was original. China, which had once done so much in algebra, was content in the 17th century to adopt the European astronomy while allowing her own undoubted abilities to lie dormant. Japan alone of all the Orient developed her native mathematics, although with more or less suggestion from France, Belgium, Italy, and Germany, through the Jesuit missionaries in China; from Holland, through the Dutch traders at Nagasaki; and possibly through scholars who secretly visited the universities of the Low Countries. As to Persia and Arabia, mathematics was dead and forgotten.

China. The mathematical feature of importance at the opening of the 17th century in China was the work of the Italian Jesuit Matteo Ricci already mentioned in Chapter VIII. After his death (1610) other missionaries carried on this work, not only in a religious line but also in the introduction of Western

[1] V. Bobynin, "De l'étude sur l'Histoire des mathématiques en Russie," *Bibl. Math.*, II (2), 103; A. N. Peepin, *History of Literature* (in Russian), I, 253 (Petrograd, 1911).

science. Among these were the Jesuit Nicolò Longobardi[1] (who went to China in 1596), Giacomo Rho,[2] Johann Adam Schall von Bell,[3] Smogolenski[4] (1611–1656), and Ferdinand Verbiest.[5] Of these Schall and Verbiest were particularly prominent in astronomical work. Smogolenski made known the use of logarithms,[6] and his pupil, Sié Fong-tsu, published the first Chinese work on the subject (*c.* 1650). It is interesting to observe that Vlacq's tables (1628) were reprinted in Peking in 1713.[7]

In this century there were several Chinese scholars who wrote important works, but they were all inspired by the Jesuits, and their works are based on European models.[8] Mention should also be made of Mei Wen-ting (1633–1721), a profound scholar, well versed in European as well as native science, who wrote on a variety of subjects and to whom we are indebted for much information concerning the history of Chinese mathematics.

The reason for the activity of the Jesuits in teaching the Western astronomy to Chinese scholars is apparent. It is only by establishing an intellectual superiority that a foreign religion

[1] Born at Calatagirone, Sicily, 1565; died at Peking, December 11, 1655. The Chinese name was Lung Hua-ming.

[2] Born at Milan, 1593; died at Peking, April, 1638. He reached China in 1618. The name also appears as Jacomo Russ, whence the Chinese, Lo Ya-ku.

[3] Born at Cologne, 1591; died at Peking, August 15, 1666. His Chinese name was T'ang Jo-wang. He arrived in China in 1622, and in 1630 was called upon by the government to reform the calendar. He and his colleagues wrote a large number of works.

[4] Chinese name, Mu Ni-ko.

[5] Born at Pitthem, near Courtrai, Belgium, October 9, 1623; died at Peking, January 28, 1688. Chinese name, Nan Huai-jên. He arrived in China in 1659. See H. Bosmans, "Ferdinand Verbiest," in the *Revue des Quest. scientifiques,* pp. 195, 375 (Brussels, 1912), and "Le problème des relations de Verbiest avec la Cour de Russie," in *Annales de la Société d'Émulation pour l'étude de l'hist. . . . de la Flandre*, p. 193 (Bruges, 1913). It was under his direction that most of the large astronomical instruments were made (1674) for the emperor.

[6] In his *T'ien-pu Chên-yüan.*

[7] In the *Lü-li Yüan-yüan.* This work also contained a treatment of algebra on European lines.

[8] For example, Tu Chih-ching, who wrote a geometry, *Chi-ho Lun-yüeh*, based on Euclid, and the *Su-hsiao Tao*, based on European mathematics; Huang Tsung-i, who wrote on the calendar; and Ch'ên Chin-mo (*c.* 1650), who gave 3.15025 as the value of π.

ever makes permanent progress. The Jesuits were not long in seeing that the two sciences in which Europe far surpassed the East were geometry and astronomy, and on these they concentrated their attention.[1]

The Intellectual Awakening of Japan. When Japan, in the 17th century, finally awoke to her intellectual possibilities, it was in a blaze of glory not unlike that which characterized her awakening to her national possibilities in the 19th century. Her progress in mathematics was strangely comparable to the remarkable progress that was going on at the same time in Europe, for in this century she developed a native calculus at almost the same time that Newton and Leibniz were working out their epoch-making theories.

Of the pupils of Mōri Kambei Shigeyoshi (p. 352), one of those to achieve renown was Yoshida Shichibei Kōyū, or Mitsuyoshi (1598–1672), whose *Jinkō-ki*[2] was the first great work on arithmetic to appear in Japan. In this the value of π is given as 3.16. So familiar was the name of this work that it was often used subsequently as a synonym for arithmetic.[3]

The second of Mōri's pupils to contribute to mathematics in a noteworthy manner was Imamura Chishō, whose *Jugai-roku*, devoted to stereometry as well as to arithmetic, appeared in 1639. In this work the value of π is given as 3.162, the area of a circle as $\frac{1}{4}\,cd$, and the volume of a sphere of radius $\frac{1}{2}$ as 0.51.

The third of Mōri's celebrated pupils was Takahara Kisshu, or Yoshitane, but he published nothing on mathematics.

The middle of the century saw also a number of minor writers whose works show considerable ability in mathematics. Among these was Isomura or Iwamura Kittoku, known also as Yoshinori. In his *Ketsugi-shō* (1660) there are a number

[1] "Ce fut alors que des Jésuites pénétrèrent dans la Chine pour y prêcher l'évangile. Ils ne tardèrent pas à s'appercevoir qu'un des moyens les plus efficaces pour s'y maintenir . . . étoit d'étaler des connoissances astronomiques." Montucla, *Histoire*, I (2), p. 468.

[2] The full title means "Small number, large number, treatise," that is, a treatise on numbers from the smallest to the largest.

[3] Compare the word *algorismus* as synonymous with arithmetic, and the name "Euclid" as synonymous with geometry.

of interesting problems proposed by Yoshida Kōyū, of which the following, referring to measurements, are typical:

> There is a log of precious wood 18 feet[1] long, whose bases are 5 feet and 2½ feet in circumference. . . . Into what lengths should it be cut to trisect the volume?
>
> A circular piece of land 100 measures in diameter is to be divided among three persons so that they shall receive 2900, 2500, and 2500 measures respectively.[2] Required the lengths of the chords and the altitudes of the segments.

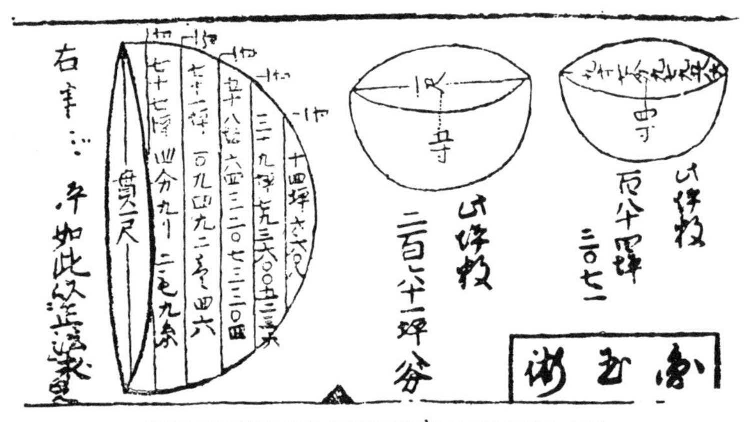

ONE OF ISOMURA KITTOKU'S PROBLEMS, 1660

His *Ketsugi-shō* appeared in 1660. This is from the 1684 edition. It represents the early state of the advanced native mathematics of Japan and shows the rise of a crude integration

There is also in this work a rough approach to an integral calculus.

In his later years Isomura devoted much attention to magic squares, magic circles, and magic wheels.[3] By a very ingenious method he showed that the surface of a sphere, which he at first thought was equal to π^2r^2, has the value πd^2.

[1] In the original, "3 measures." For this and other problems see Smith-Mikami, p. 66.

[2] That is, square measures, by drawing parallel chords.

[3] For particulars, see Smith-Mikami, p. 69.

In 1663 Muramatsu Kudayū Mosei began the publication of a work on arithmetic and mensuration which contributed to the knowledge of the circle and the regular polygons, but only in respect to measurement.

In 1664 Nozawa Teichō published a work called *Dōkai-shō*, in which some ingenious problems in mensuration appear and a step is taken in advance of Isomura in the integral calculus.

In 1666 Satō Seikō wrote the *Kongenki*, a work in which the custom is continued of proposing and solving ingenious problems. This is the first Japanese work in which the ancient Chinese method of solving numerical higher equations appears.

In 1670 Sawaguchi Kazuyuki wrote a work entitled *Old and New Methods in Mathematics*.[1] In this there again appears an approach to an integral calculus, somewhat after the method of Cavalieri, and also a treatment of numerical equations.

Seki Kōwa. The most distinguished Japanese mathematician of the 17th century, and in some respects the most distinguished of all Japanese mathematicians, Seki Shinsuke Kōwa, or Takakazu (1642–1708), was born of a samurai[2] family and showed his great mathematical ability at an early age. He acquired his knowledge to a large extent without the aid of teachers, showing great ingenuity in the affairs of life, in mechanics, in mathematics in general, and in problem-solving in particular. He improved upon the Chinese methods of solving higher equations, systematized the early Chinese use of determinants, possibly invented the circle principle (*yenri* method) which was later developed into a kind of calculus, and proposed numerous problems of an intricate nature. Two problems proposed by Sawaguchi Kazuyuki and solved by Seki Kōwa are substantially as follows:

In a circle three circles are inscribed, each tangent to the other two and to the original circle. They cover all but 120 square units of the circumscribing circle. The diameters of the two smaller circles are equal and each is 5 units less than the diameter of the next larger one. Find the diameters of the three inscribed circles.

[1] *Kokon Sampō-ki.* [2] Feudal lords.

In a certain triangle ABC there is a point P such that $PA = 4$, $PB = 6$, and $PC = 1.447$; such that the sum of the cubes of the longest and shortest sides is 637; and such that the sum of the cubes of the other side and the longest side is 855. Find the lengths of the sides.[1]

The Chinese had some idea of the determinant, as we have seen, but it is to Seki that the honor must be given of expanding a determinant in solving simultaneous equations,—a discovery which anticipated the one made by Leibniz.

Seki's reputation was such as to attract to him a large number of pupils, and his influence upon them was so great as to make itself felt up to the time when the native mathematics became absorbed in the Western science which was so completely adopted in the 19th century.

On the other hand, Seki made no great discovery in mathematics, with the exception of his anticipation of determinants. He was a great teacher, he did pioneer work in the awakening of a scientific spirit in Japan, he showed ingenuity in improving upon the work of his predecessors, but he was not the author of any new method that is now recognized as valuable, and he wrote no great treatise that stands forth today as anything more than a historical document. Because of his efforts to give to his people a knowledge of the mathematical sciences, however, His Majesty the Emperor of Japan very justly paid honor to his memory in 1907 by bestowing upon him the highest posthumous honor ever awarded to such a scholar.

Space does not permit of the further mention of Japanese scholars, with the exception of Nakane Genkei (1661–1733), a contemporary and disciple of Seki, whose works on astronomy (see the illustration on page 441) were influenced by European treatises which had begun to find their way into Japan through the Dutch traders at Nagasaki. He is also known to have been familiar with certain works of the Jesuit missionaries in China and to have recognized their superiority in the astronomical field.

[1]For the problems and suggestions as to solutions see Smith-Mikami, pp. 96, 100.

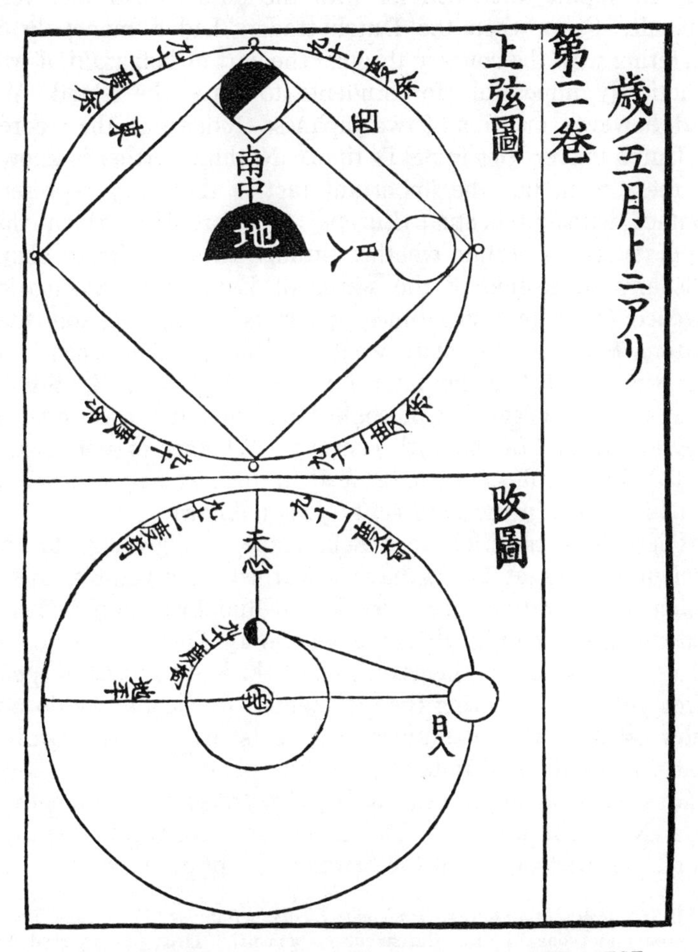

JAPANESE ASTRONOMY INFLUENCED BY THE WEST

From a work by Nakane Genkei, printed in 1696, showing the explanation of the phases of the moon. Influenced by the Dutch astronomy which had begun to be known in Japan

Contact with Europe. There were certain periods in the history of Japan when contact with the outer world was very difficult. Even when the Dutch traders had a monopoly of bartering with the country through the port of Nagasaki, it was practically impossible for students to leave the island. We find, however, mention of two Japanese students on the records of Dutch universities in Seki's time.[1] Nothing further is known of these men, but the important fact is that they represent contact with intellectual Europe. The problem which this suggests is to ascertain whether through these or similar channels any suggestion of the status of European mathematics reached Japan in Seki's time. There is a tradition, too, that Hatono Sōha, a physician, went to "the Spanish lands"[2] at this time, and that he later returned to Japan. If this is true, European learning of some kind entered the country in the second half of the 17th century. Whether or not it suggested the calculus, which reached its highest native development in Japan in the 18th century, is unknown.

There is a tradition that Seki made a pilgrimage to the ancient shrines at Nara, having learned of certain treatises which were carefully preserved in the Buddhist temples there, and which no one was able to understand. These proved to be Chinese works on mathematics, and Seki is said to have spent three years in mastering their contents. But again, we do not know whether they contained fragments of Western learning that the Jesuits had brought from Europe, or the ancient algebraic science of the Chinese, or possibly traces of Hindu astrology. No doubt, however, the Japanese scholars will in due time search out the sources of the mathematics of Seki's school.

[1] In the *Album Studiosorum Academiae Lugduno Batavae* (The Hague, 1875) it appears that one "Petrus Hartsingius Japonensis," aged 31, was studying philosophy at Leyden in 1654. He is also mentioned by van Schooten in his *Tractatus de concinnandis demonstrationibus geometricis ex calculo algebraico*, in Descartes's *La Géométrie*, 1661 and 1683 editions, p. 413. He also appears on the roll in 1660, as a student of medicine, and again in 1669. In the *Album* there is also the entry, under the date September 4, 1654, "Franciscus Carron Japonensis." Of course the names are not Japanese, and Franciscus Carron is that of a Christian missionary of a century earlier.

[2] Which were then interpreted by the Japanese as including Holland.

TOPICS FOR DISCUSSION

1. Conditions particularly favorable to the development of mathematics in the 17th century.

2. Causes of the decline of Italy's position in mathematics in the 17th century.

3. Forerunners of Newton and Leibniz in the development of the calculus.

4. Cases of relatively late development of mathematical ability.

5. Cases of the influence of inheritance or of early environment upon the development of mathematical power.

6. The relative standing of mathematics in France and in England in the 17th century.

7. The rise of analytic geometry.

8. The greatest mathematical discovery in the 17th century, with statements showing why it was the greatest.

9. The four greatest mathematical books of the century.

10. From the standpoint of the individual and his life, the most interesting mathematical personage of the century.

11. The steps taken in the 17th century to improve geometry.

12. The five most interesting mathematicians of France in the 17th century.

13. The five most interesting British mathematicians in the 17th century.

14. The influence of Newton's English predecessors upon him and upon his work.

15. Certain early steps in the use of continued fractions, infinite series, and infinite products in the 17th century.

16. The rise of books on mathematical recreations.

17. The influence of astronomy and mathematics upon each other in the 17th century.

18. Men who were prominent in both mathematics and physics in the 17th century.

19. The six leading mathematical countries in the 17th century, arranged according to their importance, with reasons justifying the arrangement.

20. The introduction of Western mathematics into the East.

21. The work, the general standing, and the influence of the Japanese scholars in the 17th century.

CHAPTER X

THE EIGHTEENTH CENTURY AND AFTER

1. General Conditions

Status of Elementary Mathematics. Since this work is concerned primarily with the history of elementary mathematics, it would be quite justifiable to set its limit at the close of the 17th century. By that time arithmetic as we ordinarily speak of it, referring to the operations with numbers for commercial and industrial purposes, was practically what it is today. We have changed the way of teaching it, and we have added new applications from time to time as the requirements of business dictated; but the mathematical part of the subject has been very nearly static. We even preserve certain traditional topics and methods that might profitably have been discarded long ago, rarely recognizing that logarithms are more easily handled than roots, and that the algebraic equation is superior to the method of proportion, which we still retain for certain purposes where it might better be discarded.

The algebra that is taught in the secondary schools and in the freshman course in college was practically all in use before 1700. The symbolism has changed but little, and although the elementary textbook is more extensive, it contains no mathematics that was not generally known before that date. The changes that have been made relate chiefly to methods of teaching and to the applications of the subject.

Elementary geometry as ordinarily taught to beginners has made no advance, although, scientifically speaking, the foundations have been explored with far-reaching results. The pupil who studies geometry in a secondary school today is not getting as good mathematics as the one who studied it in the 17th century, simply because it was the selected boy who took the work at that time. Euclidean geometry is what it was then; it has been rearranged for educational purposes, but the modern textbook of the popular type is not mathematically as scientific as its predecessor. Geometry has made giant strides, but not in the field that teachers generally cultivate in the secondary schools.

Elementary trigonometry and analytic geometry were well known to the mathematical world at the close of the 17th century, and even our modern geometry had made some progress in the work of Desargues.

The calculus has been greatly improved with respect to its foundation principles and the method of presentation, but the elementary calculus that is taught in our colleges, both differential and integral, with its most important applications, was familiar by the year 1700.[1]

The elementary theory of equations, the solution of numerical equations, the symmetric functions of the roots, such forms as continued fractions, the actual handling of complex numbers, the use of infinite series, and even the elementary use of determinants,—all these and various similar topics were well understood before the 18th century. From the standpoint of elementary mathematics, therefore, a large part of the history of the subject closes with the year 1700.

It would be a mistake, however, to suppose that the further progress of the science has an interest only to the student of higher mathematics. Great achievements in any line of work are always stimulating, and some knowledge of these achievements and of the men who made them is necessary to the well-informed teacher or student of the science.[2]

Limitation of the Study. It is the purpose, in this chapter, to limit the study chiefly to a consideration of those mathematicians whose achievements were so noteworthy that everyone who is interested in mathematics should be informed concerning

[1] On the mathematics of the 18th century, particularly in Great Britain, consult J. Leslie, "Dissertation Fourth . . . The Progress of Mathematical and Physical Science, chiefly during the eighteenth century," in the seventh edition of the *Encyclopædia Britannica*.

[2] For a summary of the history of mathematics in the 19th century see J. Pierpont, *Bulletin of the Amer. Math. Soc.*, XI, 136.

them. In the mathematical world names like De Moivre, d'Alembert, Euler, Laplace, Lagrange, Legendre, Gauss, Monge, Galois, Poncelet, von Staudt, and Steiner are so often seen that all teachers of mathematics should know something of the achievements of the men. The student of the higher branches will know this in connection with his researches, but for the teacher in the elementary field of the science a brief résumé will be helpful. It should be understood, however, that no mention will ordinarily be made of certain names which, had they been met in the formative period of elementary mathematics, would have found place. Various men who might properly be mentioned in this connection will be referred to in Volume II of this work.

Royal Patronage. With the 18th century the king of France no longer stands out as the sole royal patron of science.[1] Queen Anne bestows knighthood upon Newton; George I shows an interest in scientific laboratories; Peter the Great is at pains to meet with learned men and founds (1724) an academy at Petrograd to which there later come such mathematicians as the Bernoullis and Euler; George III, in spite of his parsimony, endows the observatory of Herschel; and Frederick II calls to the Berlin Academy Maupertuis, d'Alembert, a Bernoulli, and Lagrange, not to speak of Voltaire, who, as we shall see, had some claim to the title of mathematician. In spite of all this, France held high place in the fostering of all the sciences,—perhaps the highest; and no other nation could boast, at the close of the 18th century, such a galaxy of stars in the mathematical firmament.

2. Great Britain

Nature of the Work. As would naturally be expected, the influence of Newton determined to a large extent the nature of the work in Cambridge, Oxford, London, and Edinburgh in the 18th century. The improvement of the calculus and the widening of the range of applications of the subject were the

[1] A. Rambaud, *Histoire de la Civilisation Française*, 12th ed., II, 473. Paris, 1911.

characteristic features. A few of the leading names connected with this movement will be mentioned, together with a brief statement of the contributions of each.

It should be observed, also, that the 18th century saw mathematics made popular for the first time in Great Britain. Schools were established for the poor to attend on Sunday, the only day that they could attend them at all, and this was done much against the opposition of many in the Church; circulating libraries were established; printing ceased to be largely a London monopoly and spread throughout the country; elementary handbooks appeared; and such popularizers of mathematics as the *Ladies' Diary* (1704–1840) and the *Gentleman's Diary* (1741–1840) had a wide circulation. Even such classical works as Newton's *Principia* were printed in the vernacular. Mathematics had ceased to be aristocratic; democracy had begun to assert its rights in intellectual as well as political matters.

Cotes. Newton is credited with the statement, "If Cotes had lived, we had known something," a remark that might, it would seem, be made with greater force with respect to certain others who, like Pascal, Galois, and Clifford, died at a relatively early age. Nevertheless it is true that few scholars showed such powers of analysis before the age of thirty-four, the age at which Roger Cotes[1] died. He had shown a taste for mathematics when only about twelve years of age. He was educated at St. Paul's School, London, where he also developed a taste for metaphysics, philosophy, and divinity, and thence proceeded to the mathematical Mecca of England, Trinity College, Cambridge. When only twenty-four years of age he was appointed (1706) to the Plumian professorship of astronomy, the first to fill the chair which had just been established (1704) by Dr. Plume, archdeacon of Rochester. In 1713 he published at Cambridge the second edition of Newton's *Principia*. Only two of his memoirs appeared during his lifetime, but most of his writings were collected and published,

[1] Born at Burbage, Leicestershire, July 10, 1682; died at Cambridge, June 5, 1716.

shortly after his death, by his cousin, Dr. Robert Smith (1689–1768),[1] who succeeded him in the Plumian professorship.

He discovered an important theorem on the *n*th roots of unity, partly anticipated the method of least squares, and discovered a method of integrating rational fractions with binomial denominators. His theorem on the harmonic mean between the segments of a secant to a curve of the *n*th order, reckoned from a fixed point, is well known.[2]

The Calculus in English. The first work on the Newtonian calculus to appear in the English language was published in London in 1704.[3] It was written by Charles Hayes,[4] a member of Gray's Inn, London. The purpose of the work is set forth in the preface:

> The Author has been well assur'd that there are in England as many Lovers of the Mathematicks as in any part of the World; . . . that in other Nations the best pieces of Learning are written in their own mother Tongues, for the good of their Country which we seem purposely to slight, seeking a little empty applause by writing in a Language not easily attain'd.

The work is clearly written, but was overshadowed later by such treatises as those of James Hodgson[5] and John Rowe.[6] Hayes also wrote on the finding of longitude (1710) and began but did not live to complete a *Chronographia Asiatica et Aegyptica.*

[1] *Harmonia Mensurarum, sive Analysis et Synthesis . . .*, Cambridge, 1722. The second part of the volume comprised his *Opuscula Mathematica.*

[2] W. W. R. Ball, *A History of the Study of Mathematics at Cambridge*, p. 88 (Cambridge, 1889), hereafter referred to as Ball, *Hist. Math. Cambridge*; Chasles, *Aperçu*, p. 147.

[3] *A Treatise of Fluxions: or, an Introduction to Mathematical Philosophy*, London, 1704.

[4] Born 1678; died in London, December 18, 1760.

[5] Master of the mathematical school in Christ's Hospital, London. He was born in 1672 and died in London, June 25, 1755. Among his several works was *The Doctrine of Fluxions*, London, 1736, with an edition in 1758.

[6] *Introduction to the Doctrine of Fluxions*, London, MXCCLI (*sic* for MDCCLI, 1751), with editions in 1757, 1767, and 1809. This was the first really popular presentation of the subject in English.

In connection with these works Joseph Raphson's[1] history of the calculus[2] should be mentioned. The purpose of the book is thus stated:

To assert the Principal Inventions of this Method, to their First and Genuine Authors; and especially those of Sir Isaac Newton, who has vastly the Advantage of all others as well in respect of Priority of Time, as the Great and Noble Nature of his Discovery.

Such a book, written in the heat of the controversy as to the priority of the works of Newton and Leibniz, would naturally be open to the charge of partisanship.

James Stirling. A brief list of some of the other important 18th century writers will tend to show the nature of the work being done in Great Britain.

James Stirling[3] was educated at Glasgow and at Balliol College, Oxford. He left Oxford (1715), partly on account of his relations to the Jacobites, and went to Venice to accept a professorship. In Italy he formed the acquaintance of Nicolas Bernoulli, who was then at Padua, and was probably encouraged by him to write his well-known work on lines of the third order.[4] Of these lines he added four to those already discussed by Newton, and he also wrote a paper on the differential method in the treatment of infinite series.[5] He returned to London in 1725, devoting himself to mathematics, meeting with Newton in the latter's closing years, writing several important memoirs, and corresponding with many noted mathematicians of the day. His sojourn in Venice gave him the nickname "the Venetian," and by this he was commonly

[1] Or Ralphson. He wrote an *Analysis aequationum universalis*, London, 1690, and died before 1715.

[2] *The History of Fluxions, Shewing in a compendious manner The first Rise of, and various Improvements made in that Incomparable Method*, London, 1715 (posthumous).

[3] Born at Garden, Stirlingshire, 1692; died at Edinburgh, December 5, 1770.

[4] *Lineae tertii ordinis Newtonianae sive illustratio tractatus Newtoni de enumeratione linearum tertii ordinis*, Oxford, 1717.

[5] *Methodus differentialis sive Tractatus de summatione et interpolatione serierum infinitarum*, London, 1730. An English translation by Francis Holliday appeared in 1749.

known to his friends. He made an important survey of the Clyde, and in later life (1735) became the manager of a mining company in Lanarkshire; and, strange to say of such a scholar, he made a great success of this venture.

De Moivre. Although born in France, Abraham de Moivre[1] spent his life from the age of eighteen in London, and may properly be ranked with the English school of mathematicians. Compelled by narrow circumstances to forego the life of a student, he supported himself by private teaching, by lecturing, and by giving answers to mathematical puzzles. It is said that he passed most of his time in a London coffee house, where his genius in solving problems brought to him sufficient return for his humble needs. Having come by chance upon a copy of Newton's *Principia*, he discovered his weakness in the higher range of mathematics and by assiduous application soon became recognized as a man of genuine ability in research. He was admitted to membership in the Royal Society and into the academies of Paris and Berlin. His work was chiefly on trigonometry,[2] probability,[3] and annuities.[4] In his discussion of trigonometry he gave the theorem which bears his name, $(\cos x + i \sin x)^n = \cos nx + i \sin nx$, a relationship already stated in substance by Cotes,[5] and one which leads to numerous interesting identities in connection with complex numbers. Indeed, it stands as one of the basic propositions in the theory of such numbers. He is also known for having given the various quadratic factors of $x^{2n} - 2kx^n + 1$, for having stated the rule for finding the probability of a compound event, for his work on recurring series, and for his extension of the quadrature of the lunes of Hippocrates.[6]

[1] Born at Vitry, Champagne, May 26, 1667; died in London, November 27, 1754.

[2] *Miscellanea Analytica, de seriebus & quadraturis* . . ., London, 1730.

[3] *Doctrine of Chances*, London, 1718, with later editions in 1738 and 1756. It was dedicated to Newton. There was an Italian edition, Milan, 1776.

[4] *Annuities upon Lives*, London, 1725, with later editions.

[5] See Volume II, Chapter IV.

[6] See the summaries of his papers in the *Phil. Trans.* abridgment, 4th ed., I, 1, 29, 81, 90, *et passim*; IV, 3, 25, 77 (London, 1731).

There is often told a story of his death, to the effect that he had declared it to be necessary to sleep a quarter of an hour longer each day than on the preceding one. If he was sleeping six hours a day when he began this series, it is evident that the first day thereafter he would sleep $6\frac{1}{4}$ hours, and on the 73d day he would reach the limit.

Whiston. Born in the same year as De Moivre, but under seemingly more favorable stars, and dying only two years before him, William Whiston[1] lived a life as ideal as De Moivre's was discouraging. He received many honors, wrote numerous scientific and theological works, and held the Lucasian chair of mathematics in Cambridge (1703–1710); and yet he left a name that is today by no means so well known in the history of mathematics as that of his humbler contemporary. His chief interests were in astronomy.

Brook Taylor. The name of Brook Taylor[2] is familiar to every student who knows the rudiments of the calculus, Taylor's Theorem being one of the first instruments that he uses. The discoverer of this theorem was educated at St. John's College, Cambridge. He early gave great promise of success in mathematics, wrote various papers for the *Philosophical Transactions*, was admitted to the Royal Society, and became its secretary. When only thirty-four, however, he gave up his secretaryship and devoted himself to writing. In 1715 he published a work[3] in which is contained his well-known proposition,

$$f(x+h)=f(x)+hf'(x)+\frac{h^2}{2!}f''(x)+\cdots,$$

and some treatment of the calculus of finite differences, of interpolation, and of the change of the independent variable. He published two works on perspective,[4] giving the first

[1] Born at Norton, Leicestershire, December 9, 1667; died in London, August 22, 1752.

[2] Born at Edmonton, August 18, 1685; died in London, December 29, 1731.

[3] *Methodus Incrementorum Directa et Inversa*, London, 1715. He had announced the discovery in 1712.

[4] *Linear Perspective*, London, 1715; *Principles of Linear Perspective*, London, 1719.

general enunciation of the principle of vanishing points. He was also the author of various memoirs on physics, logarithms, and series.[1]

BROOK TAYLOR

From an engraving by J. Dudley

Maclaurin. Associated with Taylor's name, on account of the theorem above mentioned, is that of Colin Maclaurin,[2] a Scotch mathematician, who entered the University of Glasgow (1709) at the age of eleven. He soon showed a taste for

[1]For a biography see the preface to his posthumous work, *Contemplatio Philosophica*, London, 1793.

[2]Born at Kilmodan, Argyllshire, February, 1698; died at York, June 14, 1746. See C. Tweedie, *Math. Gazette*, IX, 303.

mathematics, and at the age of twelve, having accidentally run across a copy of the work, he mastered the first six books of Euclid in only a few days. At the age of fifteen he took the degree of M.A., publicly defending with much success a thesis on the power of gravity. At the age of nineteen he was elected to the chair of mathematics in the Marischal College, Aberdeen, and at the age of twenty-one (1719) he took to his London printer his first important work.[1] After traveling for some time as tutor to the son of Lord Polwarth he became (1725) an assistant at the University of Edinburgh, finally being elected to a professorship.

COLIN MACLAURIN

Known chiefly for the formula which bears his name

To him is due a method of generating conics which bears his name. His treatise on fluxions[2] contains the well-known identity,

$$f(x) = f(0) + xf'(0) + \frac{x^2}{2!}f''(0) + \cdots,$$

a relationship easily deduced from Taylor's Theorem, and one which had been announced by James Stirling twelve years earlier. Maclaurin greatly generalized the theory of the mystic hexagram and was the first to publish a work on the subject, for Pascal's essay, although written more than a century earlier, did not appear in print until 1779. He wrote an algebra

[1] *Geometria Organica: sive Descripto Linearum Curvarum Universalis*, London, 1720. Part of the propositions were worked out in his 16th year.

[2] *Treatise of Fluxions*, 2 vols., Edinburgh, 1742.

that was published posthumously,[1] and various memoirs on geometry and physics. Ball[2] has very well summed up his influence in these words:

> Maclaurin was one of the most able mathematicians of the 18th century, but his influence on the progress of British mathematics was on the whole unfortunate. By himself abandoning the use both of analysis and of the infinitesimal calculus, he induced Newton's countrymen to confine themselves to Newton's methods, and it was not until about 1820, when the differential calculus was introduced into the Cambridge curriculum, that English mathematicians made any general use of the more powerful methods of modern analysis.

Saunderson. Nicholas Saunderson[3] deserves mention as one of the mathematicians of this period, not so much because of his great achievements in advancing the science as on account of the inspiration that his history offers to those who labor under difficulties such as discourage most men and lead them early to abandon hope.

When only one year of age he became blind through an attack of smallpox. He was a pupil of Whiston's, who was then Lucasian professor of mathematics at Cambridge, and succeeded him in 1711. He was created doctor of laws in 1728 by command of George II, and became a Fellow of the Royal Society in 1736. He was very successful as a teacher, and is especially known for his *Algebra*, published posthumously in 1740–1741 and translated into French by É. de Joncourt (Amsterdam, 1756). His *Method of Fluxions* also appeared after his death (1751). Saunderson counted among his friends such well-known scholars as Newton, Cotes, and De Moivre, and did much to make the philosophy of Newton known to the mathematicians of his time. He could carry on long and complicated mathematical problems mentally, which partly accounts for his success in spite of his misfortune.[4]

[1] *Treatise of Algebra*, London, 1748, with several later editions. On his attempted proof of Taylor's Theorem see *Bibl. Math.*, I (3), 438.

[2] *Hist. Math.*, 6th ed., 388.

[3] Born at Thurlston, Yorkshire, January, 1682; died at Cambridge, April 19, 1739.

[4] For his biography consult the *Algebra*, mentioned above.

NICOLAS SAUNDERSON
Profesſeur en Mathematiques
dans l'Universite de Cambridge
mort le 19 Avril 1739 âgé de 56 ans
Vanderbanck pinx
C.F. Fritzsch Sc

NICHOLAS SAUNDERSON

After Vanderbanck's painting

Other British Writers of the Century. Among those who contributed to the advance of the Newtonian philosophy at this time Humphrey Ditton[1] deserves mention. He was a man without university training but interested in Church work. He left this work, however, and devoted the latter part of his life to the study of mathematics and to teaching. He was much esteemed by Newton, on whose recommendation he was elected mathematical master at Christ's Hospital, London. He published a number of memoirs on mathematics and physics, a work on fluxions,[2] a revision of an algebra,[3] and a work on perspective (1712).

All students of the history of geometry will recognize the name of Robert Simson,[4] who, although educated as a physician, became professor of mathematics (1711) at the University of Glasgow. He was a thorough student of Greek mathematics,[5] and most of the English editions of Euclid are based upon his edition of the *Elements*. He was averse to the use of algebraic analysis in geometry, and his methods are those of the Greeks.

Another well-known Scotch writer of this period, Matthew Stewart,[6] entered the University of Glasgow in 1734, coming under the instruction of Simson. He also attended Maclaurin's lectures at Edinburgh, and succeeded him in his professorship in 1747. He was particularly interested in geometry and in the introduction of the simple form of the Greek synthetic demonstration into modern higher mathematics.[7] He also devoted

[1] Born at Salisbury, May 29, 1675; died in London, October 15, 1715.

[2] *An Institution of Fluxions,* London, 1706. There was an enlarged edition in 1726.

[3] *Synopsis Algebraica* of John Alexander, London, 1709.

[4] Born at Kirktonhall, Ayrshire, October 14, 1687; died at Glasgow, October 1, 1768. See William Trail, *Account of the Life and Writings of Robert Simson*, Bath, 1812.

[5] *Sectionum Conicarum Libri V*, Edinburgh, 1735, with an enlarged edition in 1750; *Apollonii Pergaei Locorum Planorum libri II, restituti . . .*, Glasgow, 1749; *Euclid's Elements*, Glasgow, 1756, with numerous editions. Some of his works were published posthumously at Glasgow in 1776.

[6] Born at Rothesay, Isle of Bute, 1717; died at Edinburgh, January 23, 1785.

[7] *General Theorems*, Edinburgh, 1746; *Tracts, Physical and Mathematical,* Edinburgh, 1761.

much of his energy to astronomy, particularly to the problem of the sun's distance from the earth.[1] Several propositions of modern geometry bear his name.[2]

Contemporary with these Scotch writers, but living in England, there was that strange mathematical genius, Thomas Simpson.[3] He was brought up by his father to be a weaver, and hence his early education was confined to the reading and writing of English. Since Thomas persisted in reading beyond what his father thought necessary, resulting in a vigorous paternal protest, the boy decided to run away from home. A peddler having given him a copy of Cocker's arithmetic containing an appendix on algebra, he began the study of mathematics. His life was a turbulent one, and it suffices for our purposes to say that he struggled against poverty in London with sufficient success to allow him to publish works on the calculus,[4] probability,[5] algebra,[6] and various other subjects, and to prepare numerous monographs of importance. He was a man of undoubted genius, and his abilities were recognized in his election as professor of mathematics at the Woolwich Military Academy in 1743 and as a Fellow of the Royal Society in 1745. As a teacher he was a failure; equally was he a failure in the home; and, as with many other human failures, drink finally asserted the mastery.

Another self-made mathematician of this period appeared in the person of John Landen,[7] known for his work on residual analysis,[8] a theory by which he solved a number of problems more simply than had been done by fluxions, and for his memoir

[1] See *Essays of the Phil. Soc. of Edinburgh*, 1756.

[2] Chasles, *Aperçu*, 173.

[3] Born at Market Bosworth, Leicestershire, August 20, 1710; died at Market Bosworth, May 14, 1761.

[4] *A new Treatise of Fluxions*, London, 1737; *The Doctrine and Application of Fluxions*, London, 1750.

[5] *A Treatise on the Nature and Laws of Chance*, London, 1740; *The Doctrine of Annuities and Reversions*, London, 1742.

[6] *An Elementary Treatise of Algebra*, London, 1745; 2d ed., 1755.

[7] Born at Peakirk, near Peterborough, January 23, 1719; died at Milton, near Peterborough, January 15, 1790.

[8] *Discourse concerning Residual Analysis*, London, 1758; *The Residual Analysis*, London, 1764.

(1755) on the rectification of the arc of a hyperbola. He also wrote on astronomy, series, elliptic transcendents (1771), and physics, and was one of the early contributors to the theory of the top. He was admitted to the Royal Society in 1766.

Charles Hutton,[1] one of the best-known English writers on mathematics at the close of the 18th century, owed his prominence more to his perseverance than to his scientific ability. He was professor of mathematics at the Military Academy at Woolwich (1772–1807) and is chiefly known for his mathematical tables[2] and his dictionary.[3]

Among the contemporaries of Hutton one of the intellectual leaders, but by no means one of the best-known, was William Wallace.[4] As a young man he was a printer, later becoming interested in bookselling. Meantime he developed a taste for mathematics, gave private lessons, and at the age of twenty-six became a teacher in the Perth Academy. He finally became a professor in the University of Edinburgh. He wrote a number of memoirs on logarithms, trigonometry, the pantograph, and geodesy, but his chief work was in relation to the quadrature of the hyperbola and to hyperbolic functions.[5]

Contemporary with Landen and Hutton there was a writer who did very much less but possibly gained somewhat more in the estimation of many people, the accident of a single discovery having given his name a place in the history of the number theory. John Wilson[6] was one of those men who did one thing well in the field of mathematics and then failed to

[1] Born at Newcastle upon Tyne, August 14, 1737; died in London, January 27, 1823.

[2] *Mathematical Tables, containing Common, Hyperbolic, and Logistic Logarithms, with other Tables, and a large and original History of the Discoveries and Writings relating to those Subjects*, London, 1785.

[3] *Mathematical and Philosophical Dictionary*, London, 2 vols., 1795, 1796, with a second edition in 1815.

[4] Born at Dysart, Fifeshire, September 23, 1768; died at Lauriston, near Edinburgh, April 28, 1843. S. Günther, *William Wallace, ein Vorläufer der Lehre von den Hyperbelfunktionen*, Prog., Ansbach, 1880.

[5] Günther, *loc. cit.*, and the *Trans. of the Royal Soc. of Edinburgh*, VI, 269, 271, 302, etc.

[6] Born at The How, Westmoreland, August 6, 1741; died at Kendal, October 18, 1793. See a note on his life by Cantor, *Bibl. Math.*, III (3), 412.

meet the expectations of his contemporaries. He entered Peterhouse, Cambridge, in 1759, and while still an undergraduate he discovered that if p is a prime number, then $1+(p-1)!$ is a multiple of p,—a fact already known to Leibniz but not published. The statement has generally been known as Wilson's Theorem.[1] He was senior wrangler in 1761 and became a Fellow of the Royal Society in 1782.

The Early Nineteenth Century. Of British writers born in the 18th century but whose work was done in the century following, relatively few stand out as brilliant mathematicians. Among those who contributed in some noteworthy way to the progress of their science, one of the earliest was Robert Woodhouse,[2] who was educated at Caius College, Cambridge, became Plumian professor in the university, and did much to replace the calculus of fluxions by the differential calculus. He sought, moreover, to put the latter subject on a firm scientific foundation, and it is due in no small degree to his efforts that this was done.

Probably the name best known of all this group, in books on elementary mathematics, is that of William George Horner.[3] Although not a man of great ability as a mathematician, he succeeded in making for himself a name that is well known to students of algebra. He was a teacher at Bath when he came independently upon an ancient Chinese method of approximating the roots of a numerical equation. This method, which had been practically forgotten in China, was made known in a paper read before the Royal Society in 1819, and since that time Horner's Method has become familiar in all parts of the English-speaking world.[4]

His contemporary, George Peacock,[5] was a very different type of man. A student at Trinity College, Cambridge, he

[1] On the history of the theorem see Dickson, *Hist. Th. Numb.*, I, 59.

[2] Born at Norwich, April 28, 1773; died at Cambridge, December 23, 1827.

[3] Born in 1786; died at Bath, September 22, 1837.

[4] Republished in the *Ladies' Diary* for 1838 and again in revised form in *The Mathematician* for 1843. His proof (1826) of Euler's Theorem (see Volume II, Chapter I) should be mentioned as showing his ability in number theory. As to the validity of the Chinese claim, see Volume II, Chapter VI.

[5] Born at Denton, April 9, 1791; died at Ely, November 8, 1858.

came to represent the solid, substantial mathematics of England and to do much to improve its status. He was appointed Lowndean professor of astronomy and geometry in 1836, and three years later became dean of Ely cathedral, spending the last twenty years of his life there. He was interested in the movement to introduce the differential notation into the work in the calculus, in the founding of the observatory at Cambridge, and in the preparation of scholarly treatises on elementary mathematics. He was one of the prime movers in all mathematical reforms in England during the first half of the 19th century, although contributing no original work of particular value.

One of his contemporaries, and also a Trinity man, Charles Babbage[1] was a worthy representative of the output of Cambridge in this period. He became Lucasian professor (1828–1839), assisted in founding the Astronomical Society (1820), the British Association for the Advancement of Science (1831), and the Statistical Society of London (1834), and did much to introduce the differential notation into British mathematics. He worked on an elaborate calculating machine, the most noteworthy effort in this direction, from the point of view of originality, since that of Pascal.[2]

Perhaps the British mathematician of this period who showed the greatest genius, or at any rate the greatest perseverance, was Peter Barlow.[3] Born of humble parents, he became one of the leading writers of England on the theory of numbers,[4] professor of mathematics at the Woolwich Military Academy, and a Fellow of the Royal Society (1823). His contributions to the magnetic theory, the strength of materials, and optics were also noteworthy, and his mathematical dictionary[5] is still a valuable source of information. His tables of factors,

[1] Born at Teignmouth, Devonshire, December 26, 1792; died in London, October 18, 1871.

[2] His *Calculating Engines*, a work including much historical information, was edited by his son, General H. P. Babbage, and published at London in 1889.

[3] Born at Norwich, October 13, 1776; died March 1, 1862.

[4] *An Elementary Investigation of the Theory of Numbers*, London, 1811.

[5] *A New Mathematical and Philosophical Dictionary*, London, 1814.

reciprocals, powers, roots, hyperbolic logarithms, and primes should also be mentioned as the best of the earlier publications of the kind, being still looked upon as standards.[1]

Sir William Rowan Hamilton. Of those whose work added to the prestige of Great Britain and Ireland in the mid-Victorian period, only a few need be mentioned in a work devoted chiefly to elementary mathematics. Of these, one of the best known is Sir William Rowan Hamilton.[2] He was one of the great mathematical products of Ireland in the 19th century. Although descended from Scotch stock, he was proud to proclaim himself an Irishman. He was one of the infant prodigies that occasionally arise in the history of mathematics, usually, as we have seen, with disappointing results. At the age of three he read English fluently and was somewhat advanced in arithmetic; at five he could read Latin, Greek, and Hebrew; at eight he could also write Latin and read French and Italian; at ten he was studying Arabic and Sanskrit; at twelve he had a working knowledge of all these languages, together with Syriac, Persian, Hindustani, and Malay, and was contesting with the American prodigy, Zerah Colburn, in long mental calculations; and at thirteen he had written an algebra, which, fortunately no doubt, was not offered for publication. At fourteen he was able to write Persian, and at sixteen he had made known an error in one of the demonstrations in the *Mécanique Céleste* of Laplace. At Trinity College, Dublin, he continued his brilliant record, receiving his appointment to the professorship of astronomy while still an undergraduate. Hamilton was knighted in 1835. In 1843 he made his great discovery of quaternions, but his first work on the subject was not published until 1853. His second work appeared posthumously.[3] If the theory has not led to the results anticipated by Hamilton and his friends,

[1] *New Mathematical Tables,* London, 1814. De Morgan published an edition in 1856.

[2] Born at Dublin, August 3, 1805; died at Dublin, September 2, 1865. For biography, see A. Macfarlane, *Ten British Mathematicians,* p. 34 (New York, 1916); hereafter referred to as Macfarlane, *Ten Brit. Math.*

[3] *Lectures on Quaternions,* Dublin, 1853; *Elements of Quaternions,* London, 1866 (posthumous).

both it and the *Ausdehnungslehre* of Grassmann have greatly extended the vision of both mathematicians and physicists.

Salmon, De Morgan, and Boole. Among the prominent mathematicians produced by Ireland in the 19th century there should also be mentioned George Salmon,[1] whose works on conics, higher plane curves, analytic geometry of three dimensions, and higher algebra are recognized as standard authorities.

Less well known than Hamilton for any single achievement, less well known than Salmon for his thoroughness in science, familiar to a much larger circle of readers because of his wide range of interest and his skill in popularizing the science, was that eccentric but brilliant teacher, Augustus De Morgan.[2] Educated at Trinity College, Cambridge, he became professor of mathematics in London in 1828, displaying unusual gifts as a teacher and scattering his energies recklessly. His *Trigonometry and Double Algebra* (1849) contained certain features of quaternions, but he did not follow this or any other theory to the conclusion that seemed within his reach. He wrote various textbooks, each a mine of information for the teacher and entirely hopeless for the pupil. His contributions to the theory of probability still rank as among the best in English, and the same may be said for his contributions to logic. He devoted considerable attention to the history of mathematics, but his articles are not only eccentric but unreliable. His best work in this line is to be found in Smith's *Dictionary of Greek and Roman Biography* (London, 1862–1864), the *Penny Cyclopædia* (London, 1833–1843), the *Companion to the Almanac* for various years, and his *Arithmetical Books* (London, 1847). His *Budget of Paradoxes*, edited by Mrs. De Morgan after his death,[3] is an interesting satire on circle squarers and their kind. Had he been able to confine himself to one line, he might have been a much greater though a less interesting man.

If poverty, delayed education, and general lack of early advantages were a bar to progress in abstract science, George

[1] Born at Cork, September 25, 1819; died at Dublin, January 22, 1904.
[2] Born at Madura, India, June 27, 1806; died in London, March 18, 1871.
[3] London, 1872; 2d ed., by D. E. Smith, Chicago, 1915.

Boole[1] would never have been professor of mathematics at Queen's College, Dublin, the theory of invariants and covariants would not have been what it is today, and the mathematical theory of logic might not have reached its present position. Boole's circumstances did not permit of his beginning any serious study of mathematics until he was twenty, although he picked up by himself some knowledge of Latin and Greek and was able to do a little teaching to help him on with his scholastic work. When he was twenty years old he decided that he was able to open a school of his own, and this he did, using the small income to assist his aged parents and to buy books for himself. Thus, without any university training, he advanced, and in 1849 was appointed to the chair of mathematics in Queen's College. His *Mathematical Analysis of Thought* (1847), *Laws of Thought* (1854), *Differential Equations* (1859), and *Finite Differences* (1860) are still looked upon as standard authorities.

Sylvester. There were two men, companions for many years but, like many companions, of very different character, who stand out with special prominence at this period. The first of these was James Joseph Sylvester.[2] Educated at St. John's College, Cambridge, and one of the most gifted members of his class, he was not allowed to take a degree because of his Jewish faith, and for the same reason he was barred from a fellowship. For the degree he went to Dublin, but after the abolition of the theological tests in 1872 the University of Cambridge awarded him both the bachelor's and master's degrees. Soon after leaving the university he was appointed (1837) professor of natural philosophy in University College, London, and two years later (1839) was elected a Fellow of the Royal Society. Opportunities for advancement not being promising in England, he accepted an appointment (1841) as

[1] Born at Lincoln, November 2, 1815; died at Cork, December 8, 1864. See Macfarlane, *Ten Brit. Math.*, p. 52.

[2] Born in London, September 3, 1814; died in London, March 15, 1897. See F. Franklin, in *Bulletin of the Amer. Math. Soc.*, III, 299; M. Noether, *Math. Annalen*, L, 133; P. A. MacMahon, *Nature*, March 25, 1897.

professor of mathematics in the University of Virginia. His election took place on July 3, 1841, and he began his work in the autumn. He made a failure of his teaching, had a serious personal encounter with a student who is said to have attacked him, and hurriedly left the university in the following March. The official records of March 22, 1842, contain the following resolution:

JAMES JOSEPH SYLVESTER

His influence was strongly exerted in establishing university research in mathematics in the United States

> Resolved, That the resignation of Mr. Sylvester be accepted, to take effect from and after the 29th day of the present month or at any earlier period that he may elect;—that a copy of this resolution be forthwith communicated to him by the Secretary, and that he be informed that in accepting his resignation the Board has not deemed it necessary to investigate the merits of the matter in difference between himself and the student Ballard, and does not mean to impute to Mr. Sylvester any blame in the matter.

Evidently, therefore, the university authorities were convinced that Sylvester's further relations with the faculty were not desirable.

He seems to have returned to London about three years after leaving Virginia. Here he took up actuarial work, became a student in the Inner Temple (1846), and was called to the bar (1850). He became professor of mathematics at the Military Academy in Woolwich in 1855 and remained there until his forced retirement on account of age (1869). In 1877 he was called to Johns Hopkins University and did more than any other man of his time to establish graduate work in mathematics in America. Among his other contributions to the advance of the science in this country was his founding of the *American Journal of Mathematics.* In 1883 he was elected to succeed H. J. S. Smith in the Savilian professorship at Oxford, but his lectures were not popular and in 1892 he gave place to a deputy professor and spent his last years in London.

Sylvester was often looked upon as unsystematic, domineering, impractical, conceited, and unhappy, but those who knew him well have testified to his genial nature and his enthusiasm in his work with students. His contributions show that he was a genius in mathematical investigation, his chief line of interest being in higher algebra, including the study of invariants.

ARTHUR CAYLEY

From a photograph made in 1870

Cayley. The companion of Sylvester already referred to was Arthur Cayley.[1] He was the son of an English merchant who had settled in Petrograd and who looked forward to his son's taking part in the business which he had established. Soon after young Cayley, at the age of fourteen, was sent to King's College School, London, it was found that he showed such ability in mathematics that his father decided that he should proceed to Cambridge. He accordingly entered Trinity College at the age of seventeen, and his progress was such that he graduated with the highest honors, secured a fellowship, and

[1] Born at Richmond, Surrey, August 16, 1821; died at Cambridge, January 26, 1895. For biography, see Macfarlane, *Ten Brit. Math.*, p. 64.

devoted himself to the preparation of a number of important memoirs. Forced to find some remunerative employment, he then took up the law, and for fourteen years made a specialty of conveyancing, devoting his leisure to the preparation of further scientific memoirs. Sylvester was at this time an actuary in London, and the two were close friends and were in frequent consultation. About 1860 the Sadlerian professorship of pure mathematics was established at Cambridge,[1] and Cayley

the circle can be described under the action
of the single force $(SP)^{-\frac{1}{2}(n-1)}(PV)^{-\frac{1}{2}(n+1)}$, or (therefore) under
the action of the forces to F, F'. Would you please
forward the problems to Miller. Believe
me, yours sincerely
A. Cayley

Cambridge 20th Nov.

FROM AN AUTOGRAPH LETTER FROM CAYLEY TO SYLVESTER

The Miller referred to is W. J. C. Miller, for many years the editor of the mathematical columns in the *Educational Times*, London

was the first to occupy the chair (1863). Although he wrote but one extensive work, the *Treatise on Elliptic Functions* (Cambridge, 1876), he contributed a large number of important memoirs to various scientific publications. In 1889 the Cambridge University Press began the publication of his papers, nearly a thousand in number, in collected form. Seven volumes appeared under his own editorship, the remaining six volumes being published under the supervision of Professor Forsyth.

Cayley's papers cover a very wide range, but it may be said that his chief interest was in the fields of elliptic functions, the theory of invariants, and analytic geometry. Of these the theory of invariants was the one which he did most to advance.

[1] Through funds bequeathed originally by Lady Sadler to found certain lectureships (1710).

America is indebted to him for his course of lectures at Johns Hopkins University in 1882 on the Abelian and theta functions, whereby he again coöperated with Sylvester, who was then helping to place mathematics in this country on a university basis. Well might Sylvester say of him that whatever he touched he embellished.[1]

H. J. S. Smith. Less well known because working in a narrower field, but in the same class of genius as Sylvester and Cayley, Henry John Stephen Smith[2] began his education under the care of his mother, a woman of unusual ability. In 1841 he went to Rugby, where he came under the influence of Dr. Arnold. From Rugby he went (1844) to Balliol College, Oxford, later spending some time at the Sorbonne and the Collège de France. In 1860 he became Savilian professor of geometry at Oxford and a year later was made a Fellow of the Royal Society. His time was so taken up in public duties of various kinds that he did not achieve the success in mathematics of which he was capable. His interest in this science was chiefly in the direction of the theory of numbers and the study of binary and ternary quadratic forms. It was probably in relation to a problem in this theory that he is said to have remarked, "It is the peculiar beauty of this method, gentlemen, and one which endears it to the really scientific mind, that under no circumstances can it be of the smallest possible utility." His various writings were collected by Dr. Glaisher and published in 1894.[3]

Clifford. Among the most promising mathematicians produced in England in the 19th century, but one whose early death prevented the maturing of his genius, was William Kingdon Clifford.[4] He was educated at King's College, London

[1] "Cayley, of whom it may so truly be said, whether the matter he takes in hand be great or small, 'nihil tetigit quod non ornavit.'" *Phil. Trans.*, XVII (1864), 605.

[2] Born at Dublin, November 2, 1826; died at Oxford, February 9, 1883. For biography see Macfarlane, *Ten Brit. Math.*, p. 92.

[3] *The Collected Mathematical Papers of H. J. S. Smith*, 2 vols., Oxford, 1894.

[4] Born at Exeter, May 4, 1845; died in Madeira, March 3, 1879.

(1860–1863) and Trinity College, Cambridge. At the university his mathematical genius was at once recognized, and in 1868 he was elected to a fellowship at Trinity. In 1871 he was made professor of applied mathematics at University College, London, and three years later was elected a Fellow of the Royal Society. He was among the first to protest against the analytic bias of the Cambridge mathematicians, and he assisted in introducing into England the graphic methods of Möbius and other German writers. His most important works were in relation to Riemann's surfaces, biquaternions, and the classification of loci. His *Common Sense of the Exact Sciences* is a classic on the foundations of mathematics, and suggests, as other works (including those of Copernicus and Kepler) had already done, the idea of relativity in all physical measurements. His *Mathematical Papers*, edited by R. Tucker, appeared in 1882.[1]

WILLIAM KINGDON CLIFFORD
From a photograph made shortly before his death

Todhunter. Of all the English mathematicians of this period the one most widely known to elementary students and to teachers is Isaac Todhunter.[2] His name is familiar to everyone who has studied the history of textbook making in the 19th century, although he was much more than a textbook writer. As a young man he attended evening classes at the University of London, where he came under De Morgan's influence, receiving the B.A. degree in 1842 and the M.A. degree two years later. He then entered St. John's College, Cambridge, and

[1] These contain (pp. xv, xxxiii) a brief biography.

[2] Born at Rye, Sussex, November 23, 1820; died at Cambridge, March 1, 1884. See Macfarlane, *Ten Brit. Math.*, p. 134.

in 1848 took his second B.A. degree. He remained at Cambridge until 1864, beginning the textbook writing which made him financially independent and resulted in a series of works that exercised great influence on education in all the English-speaking world. He also wrote (1865) a work on the history of probability and one (1873) on the history of the mathematical theories of attraction, each a classic in its line. He was a good mathematician but not a great one, an excellent linguist, and a man who stood for sound scholarship. As is commonly the case with men in his line of work, numerous stories are told of him, one being that he used to remark that he knew two tunes, the first of which was "God save the Queen" and the second wasn't, and that he recognized the former by the fact that people stood up when it was sung.

Other British Mathematicians. Among the other British mathematicians of prominence in this period it is possible at this time to mention only a few of those whose names should be familiar to the general student of mathematics. Others will be found in the second volume of this work. Dr. George Berkeley (1684–1753) is known in the history of the calculus for his work, *The Analyst* (London, 1734), in which he attacked the foundations of the new science; Edward Waring (1734–1798) was interested in the theory of numbers, and a theorem relating to powers is known by his name; Sir James Ivory (1765–1842), with mathematics as an avocation, did much to advance the progress of analytic methods in England and contributed to the theory of attraction; James Booth (1806–1878) wrote on modern geometry; Sir John Frederick William Herschel (1792–1871) contributed to the study of analysis but finally followed in his father's steps and devoted himself to astronomy; James MacCullagh (1809–1846) contributed to the theory of quadric surfaces; Thomas Penyngton Kirkman (1806–1895) made the attempt to extend the theory of quaternions and was interested in the subject of "analysis situs"; George Biddel Airy (1801–1892), Astronomer Royal of England, contributed to the lunar and planetary theory; John Couch Adams (1819–1892), independently of the French astronomer Leverrier, determined mathematically the position of the planet Neptune; Sir George Howard Darwin (1845–1912), son of Charles Darwin the naturalist, contributed to the theory of three bodies; Sir Robert Stawell Ball (1840–1913), Astronomer Royal of Ireland and later Lowndean Professor of Astronomy and Geometry at Cambridge, wrote on the theory of screws (1876); Peter Guthrie Tait (1831–1901), sometime professor at Belfast and later at Edinburgh, is well known for his work in quaternions and physics; Lord Kelvin (William Thompson, 1824–1907) contributed extensively to the application of mathematics to physical problems; James Clerk Maxwell (1831–1879) is especially known for his application of mathematics to the study of electricity; and Lord Rayleigh (John William Strutt, 1842–1919) is similarly known with respect to mathematics and the study of vibrations.[1]

3. France

Nature of the Work. France took the lead again in the 18th century, as she did in the first half of the century preceding. She may have had no more brilliant intellects than England, but she found in the differential and integral calculus a set of tools that she could use more deftly than the British mathematicians could use the heavy machinery of the calculus of fluxions. It was perhaps as well that the two nations should experiment on different lines of approach, and it is evident that each cultivated its own peculiar power of attack by so doing, but the total results secured in France show for more than those produced in Great Britain.

Early Eighteenth Century Writers. There will first be mentioned a few names of those who, although born in the 17th century, completed their work after its close. Among the first of these writers was Pierre Varignon,[2] professor of mathematics at the Collège Mazarin (1688) and later at the Collège Royal, and

[1] For a list of biographies of mathematicians dying between 1881 and 1900, see G. Eneström, *Bibl. Math.*, II (3), 326, covering all European countries.

[2] Born at Caen, 1654; died in Paris, December 22, 1722.

a member of the Académie des Sciences at Paris. Although intended for the Church, he accidentally came across a copy of Euclid, and thus was led, as so many others have been, to the study of mathematics. He then read Descartes's *Géométrie*, and thereafter devoted himself to the mathematical sciences, with special emphasis upon physical problems. He was one of the first of the French scholars to recognize the value of the new calculus. His chief contributions were to the science of mechanics, although he wrote upon pure mathematics as well.[1]

A little younger than Varignon, although dying before him, Pierre-Rémond de Montmort[2] rose to a position of some prominence. He was born to fortune and thus had ample means to enable him to follow his tastes in the study of law and philosophy. He was made a canon of Notre-Dame at a time when piety was not the chief qualification, but finally married and devoted the rest of his life to travel, to Paris, and to mathematics. He was chiefly interested in the doctrine of chance,[3] a subject which brought him into cordial relations with De Moivre and with Jean and Nicolas Bernoulli. He also wrote on infinite series[4] and summed to n terms the series

$$S = na + \frac{n(n-1)}{2!}\Delta a + \frac{n(n-1)(n-2)}{3!}\Delta^2 a + \cdots.$$

Antoine Parent,[5] a private teacher of mathematics in Paris and (1699) a member of the Académie des Sciences, was another of the minor writers who helped to advance his subject. His interest was chiefly in mechanics and physics, although he also wrote on arithmetic, the cycloid, geometry, and perspective. He is known in the history of mathematics for his

[1] *Éclaircissements sur l'analyse des infiniment petits, Paris*, 1725 (posthumous); *Éléments de mathématiques*, Paris, 1731 (posthumous); *Manière de trouver une infinité de portions de cercle toutes quarrables moyennant la seule géométrie d'Euclide*, Paris, 1703; and other works.

[2] Or Monmort. Born in Paris, October 27, 1678; died in Paris, October 7, 1719. His family name was Rémond, the "de Montmort" being assumed from his estates.

[3] *Essai d'anlyse sur les jeux de hasard*, Paris, 1708; 2d ed., 1714.

[4] "De seriebus infinitis tractatus," in the *Philosophical Transactions* for 1717.

[5] Born in Paris, September 16, 1666; died in Paris, September 26, 1716.

work in analytic geometry of three dimensions.[1] His most important contributions were published in his collected works in 1705.[2]

Among Parent's contemporaries, Joseph Saurin,[3] a priest, wrote on the determination of tangents at multiple points of an algebraic curve, on the curve of least descent, and on various other geometric questions.

There was also Thomas-Fantel de Lagny,[4] who gave up the law for the purpose of devoting himself to mathematics. He wrote on new methods of extracting roots (1692), the cubature of the sphere (1702), binary arithmetic (1703), and methods of solving problems.[5] The story is told that Maupertuis, called to his deathbed and finding him in a comatose state, asked him suddenly for the square of 12; whereupon De Lagny started up, gave the answer, and at once passed away.

Less of a mathematician but contributing worthily to the science, Amédée François Frézier,[6] a French infantry officer (1702–1707) and later an engineer in South America and San Domingo, by means of his works on stereometry as applied to stone cutting and architecture[7] laid part of the foundation for the theory of descriptive geometry.

Among those who formed a brilliant group in Paris at this time was one of the youthful prodigies that, as we have seen, arise from time to time,—François Nicole.[8] He was a boy of unusual promise, having shown his genius in geometry by rectifying the cycloid at the age of nineteen. His interest in the

[1] In a paper read before the Académie in 1700. See also Chasles, *Aperçu historique*, p. 138.

[2] *Recherches de mathématiques et de physique*, Paris, 1705; rev. ed., 3 vols., Paris, 1713.

[3] Born at Courthézon, Vaucluse, September 1, 1655; died in Paris, December 29, 1737.

[4] Born at Lyons, November 7, 1660; died in Paris, April 12, 1734.

[5] *Analyse Générale, ou Méthodes Nouvelles pour résoudre les Problèmes de tous les Genres et de tous les degrez à l'Infini*, Paris, 1733.

[6] Born at Chambéry, 1682; died at Brest, October 16, 1773.

[7] *La théorie et la pratique de la coupe des pierres et des bois, ou Traité de Stéréotomie*, Strasburg, 1738; 2d ed., 3 vols., Paris, 1754, 1768, 1769; *Éléments de Stéréotomie, à l'usage de l'architecture*, Paris, 1750–1760.

[8] Born in Paris, December 23, 1683; died in Paris, January 18, 1758.

study naturally led to a consideration of roulettes in general,[1] a subject in which he showed great insight. He also wrote on the calculus of finite differences,[2] lines of the third order (1729), probability (1730), conics (1731), cubics (1738, 1741, 1743), and the trisection problem (1740). As in most cases of unusually early development, however, his work was not of the highest order.

PIERRE LOUIS MOREAU DE MAUPERTUIS

For some years a favorite at the court of Frederick the Great, interested chiefly in geodesy

Maupertuis. The work of Pierre Louis Moreau de Maupertuis[3] was of a more stable kind. In his younger days (1718) he was a captain of dragoons in the French army, but he later retired to private life and devoted himself to the study of mathematics. He was made a member of the Académie des Sciences in 1731, directed the measurement of a meridian degree in Lapland in 1736, became president of the physical class in the Berlin Academy (1745–1753), basked in the sunshine of the favor of Frederick the Great for the usual brief period, learned that he could "climb, but heights are cold," and after falling from favor spent the last six years of his life in his native country. His chief work was in astronomy and geodesy,

[1] *Méthode générale pour déterminer la nature des courbes formées par le roulement de toutes sortes de courbes sur une autre courbe quelconque*, Mém., Paris, 1707; *Manière de déterminer la nature des roulettes formées sur la superficie convexe d'une sphère*, Paris, 1708, 1732.

[2] *Traité du calcul des différences finies*, Paris, 1717, 1723, 1724.

[3] Born at St. Malo, July 17, 1698; died at Basel, July 27, 1759.

but he also wrote on maxima and minima (1724), quadrature problems (1727), curves in general (1727–1730), and various physical questions.[1] He taught mathematics to his friend the Marquise du Châtelet and, considering her friendship for Voltaire, deserved something better than the harsh treatment which the latter gave him in his *Diatribe du Dr. Akakia*, written with

Je m'estois proposé dans cet ouvrage de donner une courte exposition du systeme de l'univers, depouillée des figures de la Geometrie et des calculs de l'Algebre. J'y ay fait preceder divers recherches metaphysiques sur les loix generales de la Nature, qui a deja paru dans les memoires de l'Academie. Je serois infiniment flatté Sire si ce Livre avoit l'approbation de votre Majesté. Ce sera toujours un hommage que mon esprit et mon coeur vous rendent.

Je suis avec le plus profond respect

Sire

De votre Majesté

Potsdam 4 Fevr. 1750

Le tres humble et tres obeissant serviteur

Maupertuis

AUTOGRAPH LETTER OF MAUPERTUIS

This letter was written to Frederick the Great in 1750, while Maupertuis was still president of the physical class in the Berlin Academy

the desire to defend a learned but indiscreet Swiss mathematician, Samuel Koenig (died 1757), who had accused Maupertuis of plagiarism. One of the biographers of Voltaire speaks of Maupertuis as "the pompous and touchy mathematician," and the phrase is probably appropriate.

Minor Writers. Alexis Fontaine des Bertins[2] was more promising in his youth but less successful in his later years than his contemporary, Maupertuis. He was a man of means, was

[1] *Œuvres de Mr. de Maupertuis*, 4 vols., Paris, 1752; Lyons, 1768.

[2] Born at Bourg-Argental, Loire, *c.* 1705; died at Cuiseaux, August 21, 1771.

therefore able to devote his life uninterruptedly to study, and became a member of the Académie des Sciences in 1733, when only about twenty-eight years of age. His early promise was not fulfilled, however, his efforts not being directed in lines of probable success. He wrote on tautochronous curves (1734 and 1768) and differential equations, and proposed various problems in geometry and astronomy. He suggested the common notation of partial derivatives of a function of several variables.

Among his contemporaries, and a man of considerable influence in the scientific circle of Paris, there should be mentioned Jean Paul de Gua de Malves.[1] He belonged to a family that had been impoverished by John Law's Mississippi scheme, and, seeing no career open to him, he entered the Church and secured a benefice which enabled him to live comfortably and to devote his life to study. He wrote a work on the Cartesian analysis[2] which gave him admission to the Académie des Sciences (1740) and a professorship (1743) of philosophy in the Collège de France. He seems to have suggested the idea of the *Encyclopédie* which was finally carried out by Diderot, d'Alembert, and Voltaire. He perfected the proof of Descartes's Law of Signs (1741) and wrote on geometry and trigonometry.

Clairaut Family. One of several noteworthy instances in the history of mathematics, showing the influence of heredity or early environment, is seen in the case of the Clairaut family. Jean Baptiste Clairaut[3] was a teacher of mathematics in Paris about the middle of the 18th century. He was a correspondent of the Berlin Academy and published three memoirs on geometry in the *Miscellanea* (1734, 1737, 1743). One of his sons was Alexis Claude Clairaut,[4] the most prominent member of the family, an infant prodigy who read l'Hospital's *Analyse des*

[1] Born at Carcassonne, *c.* 1712; died in Paris, June 2, 1786.

[2] *Usage de l'analyse de Descartes pour découvrir, sans le secours du calcul différentiel, les propriétés . . . des lignes géométriques de tous les ordres*, Paris, 1740.

[3] Died soon after 1765. The name is also spelled Clairault.

[4] Born in Paris, May 7, 1713; died in Paris, May 17, 1765.

infiniment petits and his *Traité des sections coniques* at the age of ten, presented a paper on geometry before the Académie des Sciences when he was thirteen, and was admitted to membership in the Académie and published a work on curves of double curvature[1] when he was only eighteen. His solutions of the

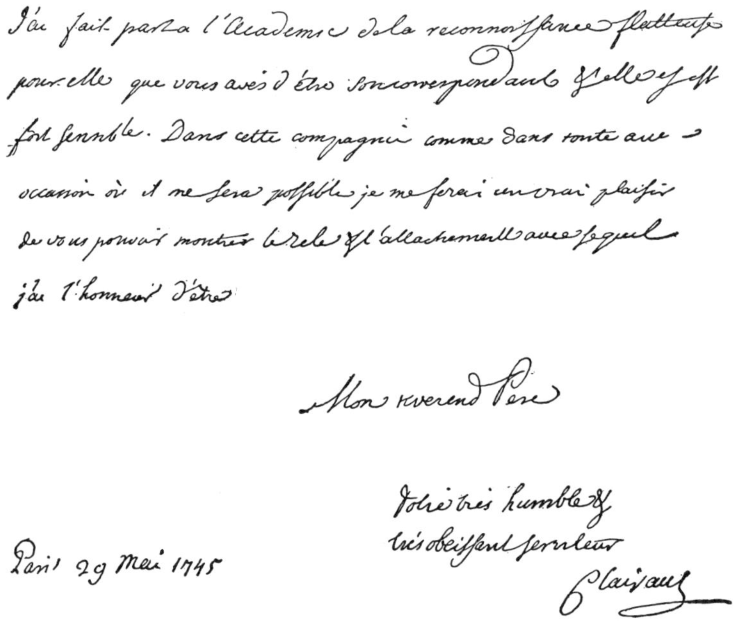

J'ai fait part à l'Academie de la reconnoissance flatteuse pour elle que vous avés d'être son correspondant et elle y est fort sensible. Dans cette compagnie comme dans toute autre occasion où il me sera possible je me ferai un vrai plaisir de vous pouvoir montrer le zele et l'attachement avec lequel j'ai l'honneur d'être

Mon reverend Pere

Votre tres humble et
tres obeissant serviteur
Clairaut

Paris 29 Mai 1745

AUTOGRAPH OF ALEXIS CLAUDE CLAIRAUT

Written about twenty years before his death and while he was working on his algebra

problem of tangents drawn to such curves and of the quadrature of the curves themselves are still found in current treatises. Clairaut was only twenty-three when he was made a member of the commission which went to Lapland to measure the length of a degree. He now began to devote most of his attention to problems of celestial mechanics, but still found time to write on geometry (1741), algebra (1746), algebraic

[1] *Recherches sur les courbes à double courbure*, Paris, 1731.

curves on a cone (1732), maxima and minima (1733), the calculus (1739), and similar topics, most of his contributions being in the form of memoirs presented to the Académie. He demonstrated Newton's theorem that all curves of the third order are projections of one of five parabolas. Possibly it was his interest in Newton that first brought him under the spell of the Marquise du Châtelet in the months just preceding her death, when she was hastening to complete her translation of the *Principia*.

Arago, in his eulogy of Laplace, remarked:

> Five geometers—Clairaut, Euler, d'Alembert, Lagrange, and Laplace—shared among them the universe of which Newton had disclosed the existence. They explored it in all directions, penetrated into regions which had been thought inaccessible, pointed out there a multitude of phenomena which observation had not yet detected, and finally, and herein lies their imperishable glory, they brought within the domain of a single principle, a single law, all that is most refined and mysterious in the movements of the celestial bodies.

Alexis had a brother[1] who died when he was sixteen, but who at the age of fourteen had read a memoir on geometry before the Académie des Sciences, and who published a work on geometry[2] when he was only fifteen.

Voltaire and the Marquise du Châtelet. The world does not often connect the name of Voltaire with mathematics, and when it connects that of the Marquise du Châtelet with the science, it is largely by courtesy. Each, however, did something to make the Newtonian theory known, and each absorbed enough mathematics to make the labor fairly serious.

François Marie Arouet,[3] known to the world as Voltaire and as the foremost leader of the 18th century in the contest for human liberty,[4] was interested in mathematics chiefly because he was interested in all things English, was interested in Newton,

[1] Born in Paris, 1716; died in Paris, 1732.

[2] *Traité des quadratures circulaires et hyperboliques*, Paris, 1731.

[3] Born in Paris, November 21, 1694; died in Paris, May 30, 1778.

[4] Among his monographs is the *Essai sur les Probabilités en fait de Justice*.

was interested in getting out a work on Newton's philosophy,[1] and was interested in Émilie, Marquise du Châtelet.[2] Daughter of the Baron de Breteuil, the marquise married at nineteen, and turned her brilliant mind to Euclid, to Newton, to the literary classics of Greece and Rome, to Locke, and to Voltaire. She had studied mathematics under Maupertuis and Koenig, read Newton and understood him,—at least in part,—and in due time translated the *Principia*,[3] completing it a few days before her death. Frederick the Great, who loved an epigram far more than he loved the courtesies of life, suggested this epitaph: "Here lies one who lost her life in giving birth to an unfortunate child and to a treatise on philosophy." Madame du Châtelet also wrote on physics, but at best she was only an amatrice in science.[4]

ÉMILIE, MARQUISE DU CHÂTELET

After a lithograph from a contemporary drawing by N. H. Jacob

[1] *Élémens de la philosophie de Neuton*, Amsterdam, 1738.

[2] Gabrielle Émilie Le Tonnelier de Breteuil; born in Paris, December 17, 1706; died at Commercy, September 10, 1749.

[3] It was published posthumously at Paris in 1759. There is a bibliography of her works in A. Rebière, *Les femmes dans la science,* 2d ed., p. 65 (Paris, 1897).

[4] Voltaire, in one of his many epigrams about her, wrote:

"Son esprit est très philosophe,
Mais son cœur aime les pompons."

In his work on Newton he addresses a poem to her, beginning:

"Tu m'appelles à toi, vaste & puissant Génie,
Minerve de la France, immortelle Emilie,
Disciple de Neuton, & de la Vérité."

D'Alembert. There are certain names in the history of mathematics to which there attaches a special human interest apart from the mere recital of a list of discoveries. One of these is d'Alembert.[1]

JEAN BAPTISTE LE ROND D'ALEMBERT
From an engraving made by P. Maleuvre in 1775 after a drawing made by A. Pujos in 1774

On the night of November 16, 1717, a gendarme, while making his rounds in Paris, found near the church of Saint-Jean le Rond a newly born infant who had been abandoned to the fate of winter, and had him hurriedly christened with the name of his first resting place, Jean Baptiste le Rond. Foster parents were found and Jean grew up, known but unrecognized by his mother, pitied and somewhat helped by his father, and soon showed remarkable intellectual powers that spoke for intellectual parentage. His mother, Madame de Tencin, sister of a cardinal, has been described by one of d'Alembert's biographers as "small, keen, alert, with a little sharp face like a bird's, brilliantly eloquent, bold, subtle, tireless, a great minister of intrigue, and insatiably ambitious." His father, General Destouches, was a man of large heart, and at his death in 1726 left enough to provide for the boy's education. When Jean was eighteen (1735) he took his bachelor's degree and soon, for reasons unknown, adopted the name of d'Alembert. He prepared for the bar, then took up medicine, and

[1] Born in Paris, November 16, 1717; died in Paris, October 29, 1783.

finally devoted his life to mathematics. Friend of Voltaire, collaborator on the *Encyclopédie*, admirer of Madame du Deffand, and lover of her companion Mademoiselle Julie de Lespinasse, he knew those in France who were best worth knowing and experienced all the joys and sorrows that Paris affords. One of his biographers says:

> In himself d'Alembert was always rather a great intelligence than a great character. To the magnificence of the one he owed all that has made him immortal, and to the weakness of the other the sorrows and the failures of his life. For it is by character and not by intellect the world is won.[1]

D'Alembert wrote upon mathematics in general,[2] the calculus and its applications,[3] the theory of differential equations,[4] and dynamics.[5]

Minor Writers. Among the minor writers of the middle of the 18th century one of the best known is Johann Heinrich Lambert.[6] Born in humble surroundings, leading a roving life, acting as bookkeeper, secretary, private tutor, and architect, and living in Germany, Holland, France, Italy, and Switzerland, he was, in spite of such an unsettled life, a voluminous writer, his fields of interest being as varied as his occupations and his places of abode. He wrote on perspective, light, astronomy, logarithms, pyrometry, transcendent quantities, theory of equations, the slide rule, psychology, ballistics, photometry, and a variety of other subjects, most of his efforts displaying respectable mediocrity,—all save one, hyperbolic trigonometry, and this gave him an enduring place in history.[7]

[1] S. G. Tallentyre, *The Friends of Voltaire*, chap. i, London, 1907.

[2] *Opuscules mathématiques*, 8 vols., Paris, 1761–1768.

[3] In the memoirs of the Berlin Academy, 1746 and other dates.

[4] Cantor, *Geschichte*, II, chap. 118. [5] J. Bertrand, *D'Alembert*, Paris, 1889.

[6] Born at Mulhouse, Alsace, August 26, 1728; died at Berlin, September 25, 1777. Mulhouse was then Swiss territory, so that he may also properly be ranked among Swiss scholars.

[7] For bibliography see Engel and Stäckel, *Die Theorie der Parallellinien*, p. 151 (Leipzig, 1895). Consult also F. Rudio, *Archimedes, Huygens, Lambert, Legendre*, Leipzig, 1892; D. Huber, *Johann Heinrich Lambert*, Basel, 1829; J. Lepsius, *Johann Heinrich Lambert*, Munich, 1881; F. Schur, *Johann Heinrich Lambert als Geometer*, Karlsruhe, 1905.

Alexandre Théophile Vandermonde,[1] member (1771) of the Académie des Sciences at Paris and director (1782) of the Conservatoire des Arts et Métiers, was another of the relatively minor writers of this period. He contributed to the theory of equations through two memoirs (1771, 1772), and to the general theory of determinants.

Étienne Bézout[2] was also one of the writers of this period on the theory of equations. He was an examiner for the navy and is known for several memoirs and textbooks. He was among the first to recognize the value of determinants. His method of elimination by the aid of symmetric functions (1764 and 1779) is well known to students in the theory of equations.

ANTOINE-NICOLAS CARITAT
MARQUIS DE CONDORCET
After an engraving from a drawing from life

During the Reign of Terror the revolutionists spared most of those whose mathematical genius is now recognized, but they did so reluctantly in the case of Marie-Jean-Antoine-Nicolas Caritat, Marquis de Condorcet.[3] Brought up under the Jesuits, he admired their learning and hated their doctrines. He was admitted to the Académie des Sciences when only twenty-six (1769) and at thirty became its secretary. At an age when all his family

[1] Born in Paris, February 28, 1735; died in Paris, January 1, 1796. On his given name see *Zeitschrift* (Hl. Abt.), XLI, 83.

[2] Born at Nemours, March 31, 1730; died in Paris, September 27, 1783.

[3] Born at Ribemont, near Saint-Quentin, September 17, 1743; died near Bourg-la-Reine, in the vicinity of Paris, March 29, 1794. See Arago's *Biographie*, read before the Académie des Sciences, December 28, 1841.

was demanding that he should be a captain of cavalry he was making for himself a name by his essay on the integral calculus and by his work on the problem of the three bodies.[1] He then took up the theory of differential equations, wrote extensively on the calculus, applied himself to the study of probability, wrote various eulogies on deceased academicians which are still read as classics in French literature, and lived the life of a scholar and, in Voltaire's words, of "the man of the old chivalry and the old virtue." D'Alembert spoke of him as a volcano covered with snow; in other words, he was an intellectual aristocrat, and that was enough to condemn him in the days of 1794. "If I have one night before me," he had said, "I fear no man; but I will not be taken to Paris." When the jailer to whom the gendarme had taken him for the night opened the door of his cell, Condorcet was dead, with an empty poison ring by his bed. He had kept the faith with himself.

Lagrange. Joseph Louis, Comte Lagrange,[2] was Tourangean by descent, Italian by birth, German by adoption, and Parisian by choice. He began his teaching as professor of mathematics in the artillery school at Turin (1755) when only nineteen years of age, succeeded Euler (1766) as mathematical director in the Berlin Academy, and was called to Paris (1787), where he became a member of the Académie des Sciences and, somewhat later (1795), was professor of mathematics at the newly founded École Normale and (1797) at the École Polytechnique. Under Napoleon he was made a senator and a count, and was awarded other honors appropriate to his genius.

Lagrange was not one of the infant prodigies in mathematics. Indeed, it is said that he showed no interest in the

[1] *Essai sur le calcul intégral,* Paris, 1765; *Analyse de la solution du problème des trois corps,* Paris, 1768, the memoir which first called attention to his powers.

[2] Born at Turin, January 25, 1736; died in Paris, April 10, 1813. As an Italian by birth, the name might be given as Giuseppe Luigi; but the family was originally French, affiliated to that of Descartes, and Lagrange spent the best years of his life in France. His complete works were published in Paris, 14 vols. (1866–1892), with a biography by Delambre in Volume I. See also G. Loria, "G. L. Lagrange nella vita e nelle opere," in the *Annali di Matematica pura ed applicata,* XX (3), p. ix.

subject until he was seventeen; but from that time on he made such marvelous progress that in a few years he became recognized as the greatest living scholar in his science. When he was twenty-three years old he published two memoirs[1] which at once attracted attention. Euler wrote (October 2, 1759) an enthusiastic letter to him about the problem of isoperimetry which is here solved and on which the great Swiss mathematician had long been working, and d'Alembert was equally appreciative of its importance. It is here that we find the beginning of the calculus of variations, and it is here that Lagrange took the first step toward Berlin and Paris, although it was not until 1766 that Frederick the Great wrote that "the greatest king in Europe" wanted "the greatest mathematician of Europe" at his court. As a result of this letter, Lagrange went to Berlin and remained there more than twenty years. At about the time that Frederick was urging him to go to Berlin he solved Fermat's problem relating to the equation $nx^2 + 1 = y^2$, n being integral and not a square,[2] an intellectual feat that added greatly to his reputation. He now began a series of investigations on partial differential equations, numerical equations, the theory of numbers, the calculus of variations, and the application of mathematics to physical problems, and made some progress in the theory of elliptic functions. To one of his memoirs (1773) may be traced the first important step in the theory of invariants, and in another there is evidence that the notion of a group was in his mind. At this time, too, he composed his monumental work on analytic mechanics,[3] although this was not published until a year after he left Berlin.

The death of Frederick (1787) brought many changes to Prussia. Lagrange, whose frail constitution had never found the climate of Berlin salutary, and whose sensitive nature now

[1] *Recherches sur la méthode de maximis et minimis*, Turin, 1759; *Sur l'intégration d'une équation différentielle à différences finies, qui contient la théorie des suites récurrentes*, Turin, 1759. These appeared in the *Miscellanea Taurinensia*.

[2] "Sur la solution des problèmes indéterminés du second degré," published in the *Miscellanea Taurinensia* in 1767.

[3] *Mécanique analytique*, Paris, 1788.

found the intellectual atmosphere far from agreeable, decided to accept the invitation of Louis XVI to take up his residence in Paris. It was about the time of the agitation for the

6

Les membres du jury pour les écoles centrales de paris, proposent a l'administration du departement comme professeurs et bibliotecaires

pour les mathematiques le cit-yen Costas
pour la phisique, le citoyen Libes
pour les langues anciennes, le citoyen villars membre de l'institut
pour l'histoire le citoyen Maherault
pour les deux bibliotheques des écoles centrales qui n'en ont point encore, Les citoyens Niccleau & Eymar

Laplace Lagrange

Eymer ... St Michel No 785

Le Cn Stange Pr de Belles lettres Rue St apolline No 6 destiné pour l'abbaye St Martin des Champs (Cette maison ... servir à l'École centrale, ... demande être destiné pr la 1re École qui ouvrira)

DOCUMENT SIGNED BY LAGRANGE

This official document was written by Laplace and was signed by Lagrange and himself

metric system, and Lagrange was made president of the commission to carry out the work. The value of such an undertaking could appeal even to a Sans-culotte, and so, although all foreigners were banished from France, the Committee of Public Safety expressly excepted Lagrange from the decree. Nevertheless the fate of Lavoisier and Bailly, both of whom met their death by the guillotine, led Lagrange to decide on leaving France. He spoke bitterly to Delambre of the death of Lavoisier, saying that the mob had removed in an instant a head that it would take a century to reproduce. Prussia knew his genius and seriously wanted him back; Paris knew his name and vaguely wished him to remain. The Prussia of that day wished to be scientific; the Paris of that day merely wished to be thought so. But just as Lagrange was reaching a decision, new forces were created in France, and these forces were more potent than any fear of the guillotine, than any discouragement at the acts of the revolutionary leaders, or than any call of the successor of Frederick. France had decided to establish a school with the humble name of École Normale, and two years later she established a second one, the École Polytechnique, and to each school Lagrange was called and to each he gave a mathematical impetus that it has never lost. In the first he saw a chance to found the training of teachers on the most thorough scholarship, and no similar institution either before or since that time has so thoroughly recognized the value of this principle, and none has ever stood so high in the esteem of the world. Similarly, in the École Polytechnique he saw the opportunity for basing the technical work on a foundation of the highest type of mathematical skill, and this institution, like the other, has ever since been a constant inspiration to the world of science. It was at the École Normale that he gave those lectures on algebra and arithmetic[1] that he had to temper to the revolutionary demand before he could bring the work up to the standard that permitted him to present the calculus of functions.[2] It was at the École Polytechnique that he lectured on analytic functions,[3] setting forth in new fashion the differential and integral calculus and the calculus of fluxions. Here, too, he expounded his noteworthy work on numerical higher equations.[4]

[1] *Leçons d'arithmétique et d'algèbre*, 1794–1795.

[2] *Leçons sur le calcul des fonctions*, Paris, 1801.

[3] *Théorie des fonctions analytiques contenant les principes du calcul différentiel*, Paris, 1797.

[4] *Traité de la résolution des équations numériques de tous degrés*, Paris, 1798.

It is probable that his work more profoundly influenced later mathematical research than did that of any of his contemporaries, although it was an era of giants in this field.

Laplace. Pierre-Simon, Marquis de Laplace,[1] was born in poverty and owed his early education to the interest which his promise excited in men of intellectual power. Of these days of struggle he never spoke. Almost the first reliable records that we have of his life show him studying and afterwards teaching mathematics in the military school at Beaumont and making such a reputation as to lead to his call to succeed (1784) Bézout as examiner of the artillery corps. He later took part in the organization of the École Polytechnique and the École Normale. Napoleon made him a count and appointed him minister of the interior (1799). After standing his eccentricities for six months, the consul dismissed him with the remark that he carried into his work the spirit of the infinitesimal.[2] Laplace then entered the senate but made no worthy record. After the restoration, Louis XVIII raised him to the peerage and (1817) made him a marquis.

PIERRE-SIMON LAPLACE
From Goutière's engraving after a painting by Naigeon

[1] Born at Beaumont-en-Auge, Calvados, March 23, 1749; died in Paris, March 5, 1827. For Arago's eulogy on Laplace, see the English translation in the Smithsonian Institution *Report* for 1874, p. 129 (Washington, 1875).

[2] "L'esprit des infiniment petits."

Laplace was a political opportunist. At heart he was a royalist, but for his personal interests he became a follower of Napoleon. He was a friend of the people but not a believer in the people's judgment.

His name is chiefly connected with astronomy and celestial mechanics,[1] but he also wrote on probability,[2] the calculus, differential equations, and geodesy.[3] As a master of the theory of celestial mechanics he stands unrivaled.[4] As to his style of exposition, Nathaniel Bowditch (1773–1838), the self-made American astronomer, remarked: "I never come across one of Laplace's 'Thus it plainly appears' without feeling sure that I have hours of hard work before me to fill up the chasm and find out and show how it plainly appears."[5]

Legendre. The third of the great trio, of which the first two were Lagrange and Laplace, appeared in the person of Adrien-Marie Legendre.[6] He was educated at the Collège Mazarin in Paris, where he early showed his taste for mathematics, and with the help of his teacher, the Abbé Marie,[7] and of d'Alembert, he became (1775) professor of mathematics in the École Militaire at Paris, resigning in 1780. Two years later (1782) he won the prize of the Berlin Academy for his essay on the path of a projectile.[8] In elementary mathematics

[1] *Exposition du système du monde,* 2 vols., Paris, 1796; *Traité de mécanique céleste,* 5 vols. and suppl., Paris, 1799–1825.

[2] *Théorie analytique des probabilités,* Paris, 1812; *Essai philosophique sur les probabilités,* Paris, 1814.

[3] His collected works were published in seven volumes in Paris in 1843–1847. A later and better edition was published in fourteen volumes by the Académie des Sciences, 1878–1912.

[4] With that felicity of speech which characterizes the French, Fourier enumerated his great discoveries, and added: "Voilà des titres d'une gloire véritable, que rien ne peut anéantir. Le spectacle du ciel sera changé; mais à ces époques reculées, la gloire de l'inventeur subsistera toujours; les traces de son génie portent le sceau de l'immortalité."

[5] For an appreciation of Laplace and for his influence upon the century following, see R. S. Woodward, *Bulletin of the Amer. Math. Soc.,* V, 133.

[6] Born at Toulouse, September 18, 1752; died in Paris, January 10, 1833.

[7] Joseph François Marie (1738–1801), who edited Lacaille's *Tables de Logarithmes* (1768) and his *Leçons élémentaires de Mathématiques* (1798), which had appeared in 1760 and 1741 respectively.

[8] *Recherches sur la trajectoire des projectiles dans les milieux résistants.*

he is known chiefly for his geometry,[1] a work which had a generous reception in various countries and which justly ranks as one of the best textbooks ever written upon the subject. In it he sought to rearrange the propositions of Euclid, separating the theorems from the problems and simplifying the proofs, without lessening the rigor of the ancient methods of treatment. To Legendre is largely due the abandoning of Euclid as a textbook in American schools.

ADRIEN-MARIE LEGENDRE
After a lithograph by Delpech

In higher mathematics Legendre is known for his works on the theory of numbers[2] and on elliptic functions.[3] He is also known for his treatises on the calculus, higher geometry, mechanics, astronomy, and physics. To him is due the first satisfactory treatment of the method of least squares,[4] although Gauss had already discovered the method. In his theory of numbers appears the law of quadratic reciprocity which Gauss called the "gem of arithmetic." The treatise on elliptic functions appeared almost simultaneously with the works by Abel and Jacobi on the same subjects; and although Legendre had spent thirty years on the theory, he

[1] *Éléments de géométrie*, Paris, 1794.

[2] *Essai sur la théorie des nombres*, Paris, 1798; 2d ed., Paris, 1808, with supplements in 1816 and 1825; 3d ed. under the title *Théorie des nombres*, Paris, 2 vols., 1830.

[3] *Mémoire sur les transcendantes elliptiques*, Paris, 1794; *Traité des fonctions elliptiques et des intégrales eulériennes*, Paris, 1827–1832.

[4] On the history of this subject see M. Merriman, *Method of Least Squares*, p. 182 (New York, 1884); *Transactions of Connecticut Academy*, IV, 151 (1877), with complete bibliography to that date; Todhunter, *Hist. Probability*, Cambridge, 1865.

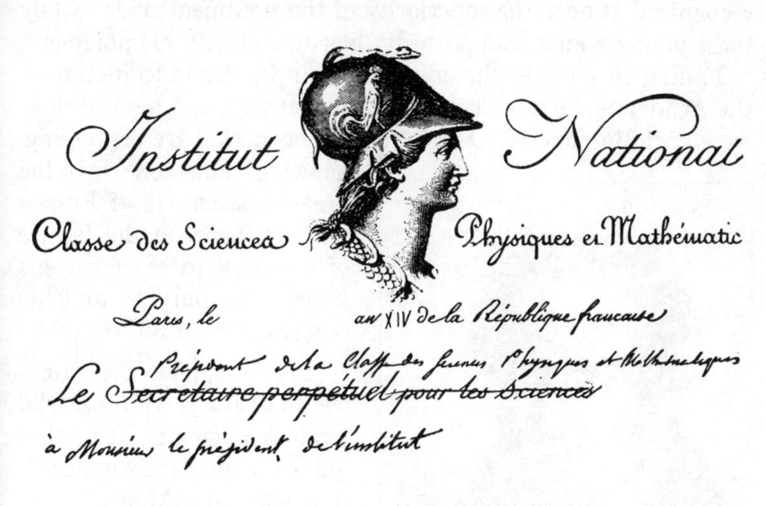

Institut National

Classe des Sciences Physiques et Mathématiques

Paris, le an XIV de la République française

Le ~~Secrétaire perpétuel pour les Sciences~~ Président de la Classe des Sciences Physiques et Mathématiques

à Monsieur le président de l'institut

Monsieur et cher confrère

Je m'empresse de vous informer que la Classe des Sciences physiques et Mathématiques a adopté unanimement le projet d'arrêté délibéré dans l'assemblée des Bureaux. Elle a pensé en même temps qu'il suffisait que cet objet fut rapporté à l'assemblée générale prochaine de l'institut sans convoquer aucune séance extraordinaire ad hoc.

Agréez, Monsieur et cher confrère, l'assurance de la considération distinguée avec laquelle j'ai l'honneur de vous saluer

Le Gendre

AUTOGRAPH LETTER OF LEGENDRE

In some of his letters the form "Le Gendre" appears, as in this case. In general the name is spelled Legendre

recognized at once the superiority of the treatment given to it by these younger men, and posterity has agreed with his judgment.

Failing to yield to the government in its desire to dictate to the Académie, he was deprived of his pension, and his last days were spent in poverty. His letters of this period are depressing, showing how one of the greatest scientists of France had lost heart at the failure of a nation to recognize his honesty of purpose and his powers of intellect.

GASPARD MONGE, COMTE DE PÉLUSE
After a lithograph of a drawing by Hesse

Beginning of the Nineteenth Century. Of those who made France a great mathematical center in Napoleon's day, Gaspard Monge[1] was, after Lagrange, Laplace, and Legendre, one of the leaders. He was the son of an itinerant tradesman, and was one of many in the history of science who early showed promise of success. At the age of fourteen he constructed a fire engine which was put into service, and at sixteen he was teaching in a secondary school (*collège*) in Lyons. At twenty-two he was professor of mathematics in the military school at Mézières, and from this time on he continued to progress, with the excitement of just escaping the guillotine, until he reached a professorship (1794) in the École Polytechnique. He also became a member of Napoleon's staff in Egypt, a senator

[1] Born at Beaune, May 10, 1746; died in Paris, July 28, 1818. See M. Brisson, *Notice historique sur Gaspard Monge*, Paris, 1818; F. Arago, "Gaspard Monge," in Arago's *Œuvres complètes*, II, 427 (Paris, 1854); Ch. Dupin, *Essai historique sur . . . Monge*, Paris, 1819.

(1799), and, as Comte de Péluse, a member of the nobility. With the restoration, however, all his honors were taken from him, and his last years were a period of disappointment.

He is known chiefly for his elaboration of descriptive geometry, a theory which, as we have seen, was suggested by Frézier in 1738, but which Monge worked out independently while at Mézières. It was some years before anything was published on the subject, the idea being held as a military secret of great value in the designing of fortifications. The opening of the École Polytechnique gave him an opportunity to lecture upon the theory, and finally, in 1799, he published his treatise upon it.[1] He also wrote numerous memoirs on differential equations, curves on various surfaces, and physical problems.

Among the most unfortunate scholars on the roll of the world's mathematicians there will always rank the name of Pierre-François-André Méchain,[2] a man who rose from the position of private tutor to become one of the leading astronomers of France, charged with duties of greatest importance in connection with the metric system, a collaborator with the great Delambre in the field work on which the units were based, and one who was recognized as a scientist of genuine ability. It was his duty to measure that part of the meridian lying between Rodez and Barcelona. After his report was sent to Paris he discovered that he had made an error of 3″ in the latitude of Barcelona. In his endeavor to conceal this error, which he knew would ruin his scientific reputation, he sought to extend the meridian, cutting out Barcelona altogether, but died from yellow fever while carrying out the plan. Instead of being known as a scientist of repute he is thought of as the man who made the chief mistake in the determining of the standard meter. It should be said, however, that the fault was not really his, for the obstacles placed in his way were such as to make accurate observations almost impossible.

[1] *Géométrie descriptive. Leçons données aux Écoles normales, l'an 3 de la République* (1794–1795), Paris, l'an VII (1798–1799).

[2] Born at Laon, August 16, 1744; died at Castellon de la Plana, near Valencia, September 20, 1804.

Sylvestre François Lacroix,[1] whose work falls in this period, was one of those men who succeed by persevering rather than by distinguished scientific ability. In his early years he occupied various positions in the naval and military schools, but finally became connected with the École Normale, the École Polytechnique, and the Collège de France, positions of highest prominence. He was a voluminous writer on higher algebra, geometry, probability, and the calculus, but he is not known for the original development of any great theory. The translation (1816) of his calculus into English by Charles Babbage, Sir John Frederick William Herschel, and George Peacock, however, did much to introduce the Continental methods and notation into the work of the Cambridge school of mathematicians.

JEAN-BAPTISTE-JOSEPH DELAMBRE

After a drawing made two years before Delambre's death

Delambre. Jean-Baptiste-Joseph Delambre[2] furnishes one of the interesting cases of a man who turned late to the study of mathematics and yet rose to be a leader in the science. As a young student in Amiens he was steeped in the classics, and it was not until he was thirty-six years of age that he seems to have even begun the serious study of astronomy, a subject which required him at the same time to begin his mathematical work. He was forty before he published anything on the

[1] Born in Paris, 1765; died in Paris, May 25, 1843. See the eulogies pronounced at the Institut in 1843 by Libri and Despretz, and various biographical articles of the time.

[2] Born at Amiens, September 19, 1749; died in Paris, August 19, 1822.

mais l'on peut aux erreurs suffit le faire sin MQm ou
sinus de la deviation de la lunette du quart de cercle dirigée sur M =

$$\frac{\text{distance du quart de cercle à la lunette}}{\text{distance de la lunette à la mire.}}$$

Soit donc LQ = 4 po = $\frac{2}{3}$ de pied, log $\frac{2}{3}$ 9.82391
MQ = 1000T........ log. 1000 .... 3.00000
sin. MQm = sin 2'.17" 6.82391
Compl. sin 1" 5.31443
MQm = 2'. 17".31 ..... 2.13834
log. 15 .... 1.17609
MQm entiers = 8".967 .... 0.962 25

il suffiroit donc de donner au cercle un mouvement azimuthal de
2'.17".5 vers m pour le conduire dans le méridien
en general voici le petit calcul log. comp arc ... 5.31443
log. distance du q.d.c. à la lunette 4 p... 6.60206
Compl. arithm. log. distance mire 6000 pieds ... 6.22185
déviation horizontale 2'.17".31 .... 2.13834

ici le calcul est en pieds, et il peut paraître un peu plus commode
aux distances de pieds et 6000 pieds on substituera les distances
veritables si on connoit à peu près la distance de la mire.
si on ne la connoit pas, il faudra mettre une seconde mire m à
une distance Mm = LQ: mais la distance Mm doit être prise sur
la perpendiculaire à la méridienne. si le bâtiment avoit une
déviation sensible il faudroit y avoir égard. si par exemple le
bâtiment avoit la direction Mm' la distance Mm' seroit plus
grande que Mm. en général Mm' = $\frac{Mm}{\text{Cos. } mMm'}$
voila je crois tout ce qu'on peut dire sur ce petit Probleme.
agréez tous mes remercimens et ceux de ma femme. je suis votre bien serviteur

Delambre

AUTOGRAPH LETTER OF DELAMBRE

With computations relating to the survey made for establishing the standard meter

subject,[1] and it was some years later that he was awarded a prize by the Académie for his tables of Uranus.[2] From this time on he was known as one of the leading astronomers of France, and his various works on the history of astronomy are still looked upon as authorities. In the history of elementary mathematics he is chiefly known for his work in measuring the arc of the meridian between Dunkirk and Barcelona for the purpose of establishing the basis for the metric system. He was a scholar, a persistent worker, and a man of highest character.[3]

Carnot. Another interesting illustration of the development of mathematical talent rather late in life is seen in the case of Lazare-Nicolas-Marguerite Carnot,[4] a member of an old and respected family of France. After the manner of so many sons of the well-to-do landowners, he studied for the army, and was thus led to the military school at Mézières. Here he came under the influence of Monge, and thus his tastes were turned toward geometry. He developed into one of the great military leaders of France, held various important offices, voted for the execution of Louis XVI, suffered in the general upheavals of the Revolution, was exiled by Napoleon, and spent his later years in Magdeburg.

His scientific work showed itself in various lines, but especially in his contributions to geometry.[5] It was in these contributions that he assisted in laying the foundations for modern synthetic geometry.

[1] *Tables de Jupiter et Saturne,* Paris, 1789.

[2] Published with other tables in 1792.

[3] The great scientist Cuvier, in his address at the burial of Delambre in the cemetery of Père-Lachaise, said of him: "Qu'il me soit permis, au moment où je vous dis ce triste et dernier adieu, de rendre témoignage à cet admirable caractère que, pendant vingt ans de liaison intime et de rapports journaliers, je n'ai pas vu se démentir un instant. Jamais, pendant ce long intervalle, un seul mouvement n'a troublé votre inaltérable douceur, . . . il ne vous est échappé une parole qui ne fût dictée par la justice et la raison."

[4] Born at Nolay, Côte d'Or, May 13, 1753; died at Magdeburg, August 2, 1823. F. Arago, *Biographie de . . . Carnot,* eulogy delivered before the Académie des Sciences, August 21, 1837.

[5] *Géométrie de Position,* Paris, 1803; *Sur la relation qui existe entre les distances respectives de cinq points quelconques pris dans l'espace, suivi d'un Essai sur la théorie des transversales,* Paris, 1806.

Gergonne and his Time. With this work is also connected the name of Joseph-Diez Gergonne.[1] In his younger days he was lieutenant in the artillery, then becoming a teacher of mathematics at Nîmes and later a professor at Montpellier. His great work, however, was as editor (1810–1831) of the mathematical journal[2] which commonly bears his name. In his later years he gave himself up to the life of a retired student. He was a prolific writer, chiefly on questions of geometry, the terms 'polar' (1810) and 'class' of a curve (1827) originating with him.

While not, like Gergonne and many other men of Napoleon's time, himself an army man, Siméon-Denis Baron Poisson[3] was the son of a soldier. He showed unusual abilities in mathematics when very young, and on this account was sent (1798) to the École Polytechnique, where his powers came to the attention of men like Lagrange and Laplace. Soon after finishing his prescribed work he was given a place on the faculty and devoted the rest of his life to teaching there and in the university, and to contributing to the literature of mechanics, mathematical physics in general, and pure mathematics. In the field of mathematics his chief contributions were to the theory of probability, algebraic equations, differential equations, definite integrals, surfaces, and the calculus of variations.

There have been several instances in the history of mathematics where a man's name has become known for a single discovery, not in itself remarkable, but striking in its peculiar interest. Such an instance is seen in the case of Charles-Julien Brianchon,[4] a student in the École Polytechnique (1804) and later (1808) an artillery officer. Brianchon had the ingenuity, when only twenty-three (1806), to take the dual of Pascal's proposition concerning a hexagon inscribed in a conic. The

[1] Born at Nancy, June 19, 1771; died at Montpellier, May 4, 1859. M. A. Lafon, *Gergonne, Sa vie et ses travaux*, reprint (n. d.) from the *Extraits des Mém. de l'Acad. de Stanislas.*

[2] *Annales de Mathématiques pures et appliquées.*

[3] Born at Pithiviers, June 21, 1781; died at Sceaux, April 25, 1840.

[4] Born at Sèvres, December 19, 1783; died at Versailles, April 29, 1864. J. Boyer, "Charles-Julien Brianchon d'après des documents inédits," *Revue scientifique*, I (4), 592.

result is Brianchon's Theorem with respect to the concurrence of lines joining opposite points of a circumscribed hexagon.[1] He became a professor in the artillery school and wrote several memoirs on geometry, particularly on curves of the second degree (1806) and lines of the second order (1817).

Poncelet. The life of Jean-Victor Poncelet[2] illustrates the military activity of many mathematicians of the disturbed Napoleonic period. A pupil of Monge's in the École Polytechnique (1807–1810), he entered the army (1812) as lieutenant of engineers. On the French retreat from Moscow he was captured by the Russians and was taken to Saratoff, on the Volga River. Here he devoted his time to the contemplation of certain possibilities in the domain of mathematics, and on his return to Metz (1814) he began to put the results of his thoughts into form for publication. The result was his great contribution to the theory of projective geometry.[3] He devoted the latter part of his life to military duties, his leisure being given to writing on mechanics, hydraulics, series, and geometry. He was one of the founders of modern geometry, probably the most important one. The Germans were more strongly influenced by his works, however, than were his own countrymen. It was Chasles who awakened France to the importance of his contributions and to a recognition of his genius.

Cauchy. The great technical and military schools founded or encouraged by Napoleon began at this time to enroll the most brilliant scientists of France. Among these was Augustin-Louis Cauchy,[4] who entered the École Polytechnique in Paris at the age of sixteen, proceeding thence to the École des Ponts et Chaussées. After a certain amount of engineering experience he was elected to the chair of mechanics in the École Polytechnique and to membership in the Académie des Sciences. On

[1] Chasles, *Aperçu*, p. 370.

[2] Born at Metz, July 1, 1788; died in Paris, December 23, 1867.

[3] *Traité des propriétés projectives des figures*, Paris, 1822; *Applications d'analyse et de géométrie*, 2 vols., Paris, 1862, 1864.

[4] Born in Paris, August 21, 1789; died at Sceaux, May 23, 1857. J. Bertrand, "Éloge" in the *Mémoires* of the Académie d. Sci., Paris, Vol. 47, pp. clxxxiii-ccv.

account of the political situation he went to Turin in 1830, where he became professor of mathematics in the university. Two years later he went to Prag and in 1838 returned to Paris and taught in certain Church schools. In 1848 he was made professor of mathematical astronomy in the university. His life was one of unrest on account of his own marked eccentricities as well as because of the changing political situation in France; but in spite of this fact he published upwards of seven hundred memoirs on mathematics and showed himself a man of uncommon scientific ability. Although usually displaying an affable manner, he was not a man of good breeding, being possessed of an unfortunate conceit, narrow in his views, and disposed to argue endlessly over trifles. He was an indefatigable worker, and his contributions to mathematics include researches into the theory of residues, the question of convergence, differential equations, the theory of functions, the elucidation of the imaginary, operations with determinants, the theory of equations, the theory of probability, the foundations of the calculus, and the applications of mathematics to physics. He was one of the first to use the imaginary as a fundamental instead of a subsidiary quantity, was the first to use Gauss's word "determinant" in its present sense, did much to establish the modern theory of convergence, and perfected the theory of linear differential equations and the calculus of variations.

AUGUSTIN-LOUIS CAUCHY
One of the foremost mathematicians of France in the 19th century

Chasles. Michel Chasles,[1] one of the leading French geometers of the 19th century, was, like his leading contemporaries in the field of mathematics, a student at the École Polytechnique (1812–1814). He went into business for a time but again returned to scientific work. He began publishing important memoirs on geometry as soon as he left school (1814), but it was his semihistorical work on the development of geometry[2] and his treatise on higher geometry[3] that gave him a world-wide reputation. These works were followed by various important memoirs on the different branches of geometry.[4] Chasles became professor of geometry and mathematics at the École Polytechnique in 1841 and professor of geometry in the faculty of sciences in 1846. In 1867 he prepared a noteworthy report on the progress of geometry in France.[5] He also received (1865) the Copley medal of the Royal Society for his work in conics.

Galois. The mathematician has not always been as conservative or as engrossed in his studies as the world seems to think. As an illustration of this fact one of the most interesting is that of Évariste Galois.[6] Educated at the Lycée Louis-le-Grand and the École Normale, at Paris, a rabid republican, twice imprisoned for his political views, a hot-blooded lover who fought a duel at twenty which cost him his life, he was able in the space of three or four years, even in his boyhood, to make for himself a lasting reputation as a genius. His life was mentally brilliant, but physically, politically, and morally it was a failure.

[1] Born at Épernon, November 15, 1793; died in Paris, December 18, 1880. For an obituary notice see Boncompagni's *Bullettino,* XIII, 815.

[2] *Aperçu historique sur l'origine et développement des méthodes en géométrie,* Paris, 1837; 2d ed., Paris, 1875; 3d ed., Paris, 1889; German ed., Halle, 1839. Some of the editions of his works bear the imprint of both Brussels and Paris.

[3] *Traité de géométrie supérieure,* Brussels, 1852; *Traité des sections coniques,* Paris, 1865.

[4] For example, "Construction de la courbe du troisième ordre déterminée par neuf points" (1853), *Journal de math. pures et appliquées,* XIX (1854).

[5] *Rapport sur les progrès de la géométrie,* Paris, 1871.

[6] Born in Paris, October 26, 1811; died in Paris, May 30, 1832. P. Dupuy, "La vie d'Évariste Galois," *École normale, Annales,* XIII (3), 197; G. Sarton, *The Scientific Monthly,* XIII, 363.

To him is due, however, one of the first important modern advances in the theory of groups, and hence to him we owe much of our modern theory of algebraic equations of higher degree. His most important memoir[1] was written the year before his death but was not published until 1846.[2]

Poincaré. Of the French mathematicians of the close of the 19th century no one ranked so high in the estimation of his contemporaries as Henri Poincaré.[3] There was hardly a branch of mathematics, pure or applied, to which he did not contribute in one way or another. His reputation was first made in his treatment of Abelian functions and in the more general type to which he gave the name of Fuchsian functions. His memoirs on these subjects began to appear in the *Comptes rendus* in 1880 and in the first volume of the *Acta Mathematica.* His contributions to elliptic functions, modular functions, double integrals, and the general theory of analysis are well known. He is equally well known for his important contributions to astronomy and physics and for his profound researches in the field of philosophy.

Other Contributors. Among the many others who added to the reputation of France in the field of mathematics during this period there may be mentioned Jean-Baptiste-Marie-Charles Meusnier de la Place (1754–1793), usually known as Meusnier, who wrote on the theory of surfaces; Jean-Baptiste Biot (1774–1862), who successfully applied mathematics to problems in physics and astronomy; Jean-Nicolas-Pierre Hachette (1769–1834), who wrote on algebra and geometry; Sophie Germain (1776–1831), known for her work on the theory of elastic surfaces; Louis Poinsot (1777–1859), who

[1] "Mémoire sur les conditions de résolubilité des équations par radicaux," Liouville's *Journal*, XI (1846). His *Manuscrits,* edited by J. Tannery, appeared at Paris in 1908.

[2] On the history of the group theory see the bibliography by C. Alasia in the *Rivista di fisica, matematica e scienze naturali,* XVIII–XXII. See also Miller, *Introduction*, p. 97.

[3] Born at Nancy, April 29, 1854; died in Paris, July 17, 1912. E. Lebon, *Henri Poincaré*, Paris, 1909; V. Volterra, "Henri Poincaré," *Rice Institute Pamphlets,* I, 133 (Houston, Texas).

contributed to the theory of numbers, to geometry, and to mechanics; Gabriel Lamé (1795–1870), primarily a physicist but writing on probability and surfaces; Théodore Olivier (1793–1853), especially concerned with descriptive geometry; Louis Arbogast (1759–1803), whose *Calcul des Dérivations* appeared in 1800; Jean Robert Argand (1768–1822), who wrote on the graphic representation of $\sqrt{-1}$; Joseph Fourier (1768–1830), known for his work in series, particularly with respect to Fourier's series, which is used in studying the flow of heat; Charles Dupin (1784–1873), prominent because of his works on mechanics and differential geometry; Georges-Henri Halphen (1844–1889), who contributed to the theory of invariants; Jean Gaston Darboux (1842–1917), contributor to differential geometry, one of the editors of the *Bulletin des sciences mathématiques et astronomiques*, and permanent secretary of the Académie des Sciences; Edmond Laguerre (1834–1886), a contributor to the theory of equations; Charles Hermite (1822–1901), who proved the transcendence of *e* and who wrote on the theory of functions; Joseph Liouville (1809–1882), long the editor of Liouville's *Journal*; Charles Méray (1835–1911), original in his ideas of the foundations upon which elementary geometry should be built; Joseph Alfred Serret (1819–1885), best known for his *Cours d'algèbre supérieure* (1849) but also a prolific writer on the function theory, groups, and differential equations; Joseph-Louis-François Bertrand (1822–1900), professor of mathematical physics in the Collège de France and secretary of the Académie des Sciences, a writer on the theory of probability, the calculus of variations, and differential equations as applied to dynamics; Pierre Duhem (1861–1916), contributor to mathematical physics and especially, by his study of original sources, to the history of science; and Louis Couturat (1868–1914), writer upon the interrelation of mathematics and logic. Any such list is necessarily fragmentary, and the student who wishes to carry his investigations farther should consult such works as the French or German editions of the encyclopedia of mathematics.

4. Germany

General Survey. Germany began to show her real strength in mathematics at the close of the 18th century. Theretofore she had depended largely on imported men, such as Euler, the Bernoullis, and Lagrange. Now she produced Gauss, and his influence on German mathematics made Göttingen a focus for scholars; it placed Germany among the leading nations in the cultivation of this science, and gave her a position of supremacy during part of the 19th century.

CHRISTIAN VON WOLF
After a mezzotint by Jacob Haid

Of the work accomplished in the 18th century a fair example is that of Freiherr Christian von Wolf,[1] a philosopher of merit and a mathematician of erudition if not of brilliancy. He took his master's degree at Leipzig in 1703 and at once became Dozent in the university. Soon after this (1706) he went to Halle as professor of mathematics and, somewhat later, of physics. Because of his religious views he was banished from the university in 1723, but was immediately invited to accept the professorship of philosophy at Marburg. He was recalled to Halle by Frederick the Great, to whom a religious question was not a matter of much moment, became chancellor of the

[1] Born at Breslau, January 24, 1679; died at Halle, April 9, 1754.

university in 1743, and was raised to the rank of baron (Freiherr) in 1745. He was a member of various learned societies, did much to popularize the theories of his friend Leibniz, and was a voluminous writer but not an original thinker. His most extensive works are his *Elementa* and *Anfangsgründe*,[1] but he also prepared an unimportant set of logarithmic tables (1711) and wrote a mathematical dictionary (1716).[2]

In the field of elementary education, Germany produced a number of important writers, but few whose names can be rated as international. Among the most industrious of the group was Christian Pescheck,[3] who wrote a large number of textbooks and was one of the first of the German writers to consider seriously the methods of teaching the subject.

Gauss. The real founder of modern German mathematics, however, is Carl Friedrich Gauss,[4] one of the many mathematicians who rose to highest eminence from very humble birth.

C F Gauß, z. Dir. d. k. G. d. W.

AUTOGRAPH OF GAUSS

Signed as Director of the Royal Society of Sciences at Göttingen

The son of a day laborer, his abilities showed themselves so early as to attract attention, and he was sent to the Carolineum at Braunschweig (1792–1795) and thence to the University of Göttingen (1795–1798). During his university career he conceived the idea of the theory of least squares,[5] discovered the

[1] *Elementa matheseos universae*, 4 vols., Halle, 1713, with later editions; *Anfangsgründe aller mathematischen Wissenschaften*, 4 vols., Halle, 1710, with several editions.

[2] J. C. Gottsched, *Historische Lobschrift des . . . Christians . . . Freyherrn von Wolf*, Halle, 1755; W. Arnsperger, *Christian Wolff's Verhältnis zu Leibniz*, Weimar, 1897.

[3] Born at Zittau, July 31, 1676; died at Zittau, October 28, 1747.

[4] Born at Braunschweig (Brunswick), April 30, 1777; died at Göttingen, February 23, 1855. In his autographs the first name begins with C. The name was originally Johann Carl Friedrich Gauss.

[5] Legendre (1805) was the first to write upon the subject, introducing it in his work on the orbit of comets. The deduction of the law was effected by an Irish-American writer, Adrain, in 1808.

CARL FRIEDRICH GAUSS

The greatest mathematician of Germany, professor of mathematics at Göttingen

celebrated proposition that a circle can be divided into 17 equal arcs by Euclidean methods, and began his great work on the theory of numbers.[1] Thereafter he devoted his attention largely to the problems of astronomy, goedesy, and electricity,[2] but found time to write on the theory of surfaces, complex numbers, least squares, congruences, hyperbolic geometry, and substantially every leading field of mathematics. He was among the first to give serious thought to the question of a non-Euclidean geometry. He asserted that "mathematics is the queen of the sciences, and the theory of numbers is the queen of mathematics." His first work on celestial mechanics[3] led Laplace to recognize him as the leading mathematician of Europe, and this recognition was general from that time until his death.[4] Kronecker said of him that "almost everything which the mathematics of our century has brought forth in the way of original scientific ideas, is connected with the name of Gauss," and it was only with such exaggeration as is entirely excusable that the elder Bolyai spoke of him as "the mathematical giant who from his lofty heights embraces in one view the stars and the abysses."

Modern Geometers. First among the distinguished pupils of Gauss may be mentioned August Ferdinand Möbius.[5] He studied at Leipzig, Halle, and Göttingen, first giving his attention to law, but through the influence of Gauss he finally decided to devote himself to mathematics and astronomy. He became a Privatdozent at Leipzig in 1815 and in the following year was made professor[6] of astronomy. The observatory

[1] *Disquisitiones arithmeticae*, Leipzig, 1801; French translation by A. C. M. Poullet-Delisle, Paris, 1807.

[2] With the physicist Weber he laid the foundations for telegraphy.

[3] *Theoria motus corporum coelestium in sectionibus conicis solem ambientium*, Hamburg, 1809.

[4] His collected works, with biography, were published at Göttingen and Gotha, in eight volumes, 1863–1900. For an interesting personal document see F. Klein, "Gauss' Wissenschaftliches Tagebuch, 1796–1814," *Mathematische Annalen*, LVII, 1.

[5] Born at Schulpforta, November 17, 1790; died at Leipzig, September 26, 1868. German writers often prefer the equivalent form "Moebius."

[6] Ausserordentlicher Professor in 1816; ordentlicher Professor in 1844.

having been built after his plans (1818–1821), he was made the first director. While his chief writings were on astronomy, he contributed in a very important manner to the theory of modern geometry.[1]

Karl Georg Christian von Staudt,[2] prominent among the founders of modern geometry, is another man who developed rather late as a mathematician of recognized genius. He began teaching in the Gymnasium at Würzburg when only twenty-four (1822), being at the same time a Privatdozent in the university. He then (1827) taught in the Gymnasium and the polytechnic school at Nürnberg, not becoming a university professor until he went to Erlangen in 1835. Von Staudt's great work (1847) consists in his rejection of all assistance of analysis in his study of geometry, employing only the properties of pure position.[3] The value of the work was not immediately appreciated, but is now fully recognized.

Julius Plücker[4] was another of the leaders in the development of modern geometry. He became Privatdozent (1825) and professor of mathematics (1829) in the University of Bonn, professor (1833) in a Gymnasium at Berlin, professor in the University of Halle (1834), and thereafter (1836) professor at Bonn. His first important work was on analytic geometry;[5] his greatest work was on algebraic curves[6] and included his well-known "six equations" relating to the singularities of higher plane curves; his third work of importance was on the analytic geometry of space;[7] and his fourth was on modern

[1] *Der barycentrische Calcul*, Leipzig, 1827. His *Gesammelte Werke*, 4 vols., appeared in Leipzig in 1885–1887.

[2] Born at Rothenburg ob der Tauber, January 24, 1798; died at Erlangen, June 1, 1867. M. Noether, "Zur Erinnerung an Karl Georg Christian von Staudt," *Festschrift der Univ. Erlangen*, 1901.

[3] *Geometrie der Lage*, Nürnberg, 1847; *Beiträge zur Geometrie der Lage*, Nürnberg, 1856, 1857, 1860.

[4] Born at Elberfeld, July 16, 1801; died at Bonn, May 22, 1868. A. Clebsch, *Abhandl. der k. Gesellsch. der Wissensch.*, XV, Göttingen.

[5] *Analytisch-geometrische Entwickelungen*, 2 vols., Essen, 1828–1831; *System der analytischen Geometrie*, Berlin, 1835, dealing with cubic curves.

[6] *Theorie der algebraischen Curven*, Bonn, 1839.

[7] *System der Geometrie des Raumes in neuer analytischer Behandlungsweise*, Düsseldorf, 1846; 2d ed., 1852.

pure geometry of space,[1] a work containing his theory of complexes and congruences. In addition to these important contributions Plücker published a large number of notable memoirs on mathematical and physical subjects.

CARL GUSTAV JACOB JACOBI
After a lithograph from an original drawing

Later Writers. One of the influential contemporaries of Gauss, although not himself a genius in mathematics, August Leopold Crelle,[2] was primarily an engineer and architect. He planned the first railway route in Germany, from Berlin to Potsdam, and was also connected with the educational ministry. He wrote various mathematical works, none of which would rank him as a genius, but he made for himself a worthy reputation through his tables[3] for multiplying and dividing, and through the *Journal für die reine und angewandte Mathematik*, which was founded by him in 1826.[4]

Of the later contributors to the science, only a few of the most prominent can be mentioned. Carl Gustav Jacob Jacobi[5] was educated (1821–1825)

[1] *Neue Geometrie des Raumes gegründet auf die Betrachtung der geraden Linie als Raumelement*, Leipzig, 1868–1869.

[2] Born at Eichwerder, March 11, 1780; died at Berlin, October 6, 1855.

[3] *Rechentafeln, welche alles Multipliciren und Dividiren mit Zahlen unter 1000 ganz ersparen*, 2 vols., Berlin, 1820.

[4] He also founded the *Journal für die Baukunst* in 1828.

[5] Born at Potsdam, December 10, 1804; died at Berlin, February 18, 1851. L. Königsberger, *Carl Gustav Jacob Jacobi*, Leipzig, 1904.

at the University of Berlin, became Privatdozent in 1825, and two years later was made professor of mathematics at Königsberg. He was a prolific contributor in various lines of mathematics, but his chief work was in the fields of elliptic functions,[1] determinants, the theory of numbers,[2] differential equations, the calculus of variations, and infinite series.

Belonging to about the same period as Jacobi, Peter Gustav Lejeune-Dirichlet[3] was educated at Göttingen, studied under Gauss, and became professor of mathematics at Breslau, Berlin, and Göttingen. He was chiefly interested in algebra, the number theory, and quadratic forms.

That a university professorship is not a *sine qua non* to success in mathematics is a fact again illustrated in the case of Hermann Günther Grassmann,[4] who was the son of a teacher of mathematics in the Gymnasium at Stettin and himself occupied a similar position in the same school. The father wrote some textbooks of no particular moment, and both he and the son gave much attention to physical questions. There was also another son, Robert, with whom Hermann collaborated in writing an arithmetic (1860). The entire output of the family, however, was as nothing compared with Hermann's *Ausdehnungslehre*.[5] In this he set forth a theory that covered much the same ground as the theory of quaternions, then being independently developed by Sir William Rowan Hamilton.

Ernst Eduard Kummer[6] is another instance of a man of genius who spent some years as a Gymnasium teacher before being called to a university chair. Educated for theology as well as mathematics, he began his teaching at Sorau, afterwards going to Liegnitz, where he taught for ten years

[1] *Fundamenta nova theoriae functionum ellipticarum*, Königsberg, 1829.

[2] *Canon arithmeticus*, Berlin, 1839.

[3] Born at Düren, February 13, 1805; died at Göttingen, May 5, 1859.

[4] Born at Stettin, April 15, 1809; died at Stettin, September 26, 1877. See F. Engel, *Jahresbericht* of the *Deutsche Math.-Verein.*, XIX, 1.

[5] *Die Wissenschaft der extensiven Grösse oder die Ausdehnungslehre*, Leipzig, 1844; completed in 1862. A list of his works may be found in the *Mathematische Annalen*, XIV, 43.

[6] Born at Sorau, Nieder-Lausitz, January 29, 1810; died at Berlin, May 14, 1893.

(1832–1842) in the Gymnasium, having Kronecker for one of his pupils. He then became (1842) professor of mathematics in the University of Breslau, later (1855) being transferred to Berlin, where he remained until 1884. In Crelle's *Journal* may be found his valuable contributions to the theory of hypergeometric (Gaussian) series (1836), the Riccati equation (1834), the question of the convergency of series (1835), the theory of complex numbers (1844, 1850), and cubic and biquadratic remainders (1842, 1848). He created the theory of ideal prime factors of complex numbers (1856) and laid down the principles applicable to Kummer surfaces.[1]

His contemporary, Georg Friedrich Bernhard Riemann,[2] also proved himself a genius in the study of surfaces. He studied at Berlin and Göttingen, receiving his doctorate at the latter university in 1851. His dissertation[3] has since been recognized as a genuine contribution to the theory of functions. Three years later (1854) he became a Privatdozent in Göttingen and in 1857 became a professor[4] of mathematics in the university. His introduction of the notion of geometric order into the theory of Abelian functions, and his invention of the surfaces which bear his name, led to a great advance in the function theory. He also set forth (1854) a new system of non-Euclidean geometry,[5] and wrote on partial differential equations,[6] elliptic functions,[7] and physics.[8]

[1] Surfaces of the fourth degree, 16 knot points, 16 singular tangent planes. See "Allgemeine Theorie der gradlinigen Strahlensysteme," Crelle's *Journal*, LVII (1860), 189. For a biographical sketch and a list of his works see the *Jahresbericht* of the *Deutsche Math.-Verein.*, III, 13.

[2] Born at Breselenz, Hannover, September 17, 1826; died at Selasca, Lago Maggiore, July 20, 1866.

[3] *Grundlagen für eine allgemeine Theorie der Functionen einer veränderlichen complexen Grösse*, Göttingen, 1851; 2d ed., Göttingen, 1867.

[4] He succeeded Dirichlet as ordentlicher Professor in 1859.

[5] *Ueber die Hypothesen welche der Geometrie zu Grunde liegen*, Leipzig, 1867. Like several of his works, this appeared posthumously.

[6] *Partielle Differentialgleichungen*, Braunschweig, 1869; 2d ed., Hannover, 1876; 3d ed., Braunschweig, 1882; 4th ed., Braunschweig, 1900–1901.

[7] *Elliptische Functionen, Vorlesungen mit Zusätzen*, Leipzig, 1899.

[8] See his *Gesammelte mathematische Werke und wissenschaftlicher Nachlass*, Leipzig, 1876; 2d ed., Leipzig, 1892; French translation, Paris, 1898; contains a biography.

One of the most brilliant and promising mathematicians of Germany in the middle of the 19th century appeared in the person of Ferdinand Gotthold Max Eisenstein.[1] Gauss, in a moment of enthusiasm, and without sufficiently weighing his words, said of him: "There have been but three epoch-making mathematicians,—Archimedes, Newton, and Eisenstein." He was brought up in poverty, showed no particular taste for mathematics until he was nineteen, died at the age of twenty-nine, and yet in ten years developed powers so remarkable as to place him in the first rank of scholars. His most important contributions were to the theory of ternary and quadratic forms, the theory of numbers, and the theory of functions. He has been spoken of by his countrymen as the real founder of the theory of invariants.

Weierstrass. As a type of those great leaders who, towards the close of the 19th century, made Germany a great gathering place for scholars there may be mentioned Karl Weierstrass.[2] He studied law and finance at Bonn (1834), taught in various Gymnasien, went to Berlin in 1856 as a teacher in the Gewerbe-institut, and became ordentlicher Professor of mathematics in the University of Berlin in 1864. Here he became one of the great leaders in the theory of elliptic and Abelian functions, in the theory of functions in general, and in the development of the theory of irrational numbers. The Berlin Akademie der Wissenschaften began the publication of his collected works in 1894.[3]

Dedekind, Cantor, and Fuchs. Julius Wilhelm Richard Dedekind[4] stands out as one of the most prominent contributors of the 19th century to the theory of algebraic numbers. He studied at Göttingen and in 1854 became a Dozent in the university. In 1858 he went to the polytechnic school at

[1] Born at Berlin, April 16, 1823; died at Berlin, October 11, 1852. F. Rudio, "Eine Autobiographie von Gotthold Eisenstein," *Abhandlungen*, VII, 145.

[2] Born at Ostenfelde, October 31, 1815; died at Berlin, February 19, 1897.

[3] On his life, see *Acta Mathematica*, XXI, 79; XXII, 1.

[4] Born at Braunschweig, October 6, 1831; died at Braunschweig, February 12, 1916.

Zürich, and four years later became a professor in a similar institution at Braunschweig. He wrote various important memoirs on the binomial equation and on the theory of modular and Abelian functions, but is best known for his treatises *Was sind und was sollen die Zahlen?* (1888) and *Stetigkeit und irrationale Zahlen* (1872).[1] In the latter work he set forth his idea of the *Schnitt* (cut) in relation to irrational numbers,—an idea which he had in mind as early as 1858.

Although Georg Cantor[2] was the son of a Danish merchant and was born in Russia, he should properly be ranked among the German mathematicians, having spent the greater part of his life in German universities. He studied at Zürich, Göttingen, and Berlin, and became a Dozent at Halle in 1869 and a professor three years later. He was a man of original ideas, and the theory of assemblages is practically his creation. His researches on this subject were first published in the *Annalen* in 1879.

Emmanuel Lazarus Fuchs[3] became professor of mathematics at Greifswald in 1869 and afterwards occupied similar positions at Göttingen, Heidelberg, and Berlin. His earlier labors were in the fields of higher geometry and the theory of numbers, but he attained his highest reputation in his work on linear differential equations.

Other Writers. Among the other German writers of the 19th century not many were at the same time leaders in advanced research and contributors to elementary mathematics. A few of the best-known names of those who extended the boundaries of mathematics should, however, be mentioned. These are Johann Friedrich Pfaff (1765–1825), professor at Helmstädt (1788) and Halle (1810), known for his work in astronomy, geometry, and analysis; Ludwig Otto Hesse (1811–1874), one of the foremost writers on modern pure geometry, analytic geometry, and determinants; Christoph Gudermann (1798–1852), who wrote on hyperbolic functions;

[1] English translations by Professor W. W. Beman, Chicago, 1901.
[2] Born at Petrograd, March 3, 1845; died at Halle, January 6, 1918.
[3] Born at Moschin, May 5, 1833; died at Berlin, April 26, 1902.

Johann August Grunert (1797–1872), editor of the *Archiv*; Ernst Ferdinand August (1795–1870), known for his work on mathematical physics; Rudolff Friedrich Alfred Clebsch (1833–1872), professor at Carlsruhe, Giessen, and Göttingen, a contributor to modern geometry; Hermann Ludwig Ferdinand von Helmholtz (1821–1894), a contributor to many fields of scientific research including that of non-Euclidean geometry; Leopold Kronecker (1823–1891), a leading writer on the theory of equations and on elliptic functions; Friedrich Wilhelm Bessel (1784–1846), one of the leading astronomers of the century, a physicist, and well known for the functions which bear his name; Paul Du Bois-Reymond (1831–1889), known for his work on Fourier's series, the problem of convergence, the calculus of variations, and integral equations; Siegfried Heinrich Aronhold (1819–1884), professor in the technical high school at Berlin, well known for his work on invariants; and Karl Theodor Reye (1837–1919), whose *Geometrie der Lage* (3d ed. 1886–1892) is one of the best-known textbooks on the subject.

5. ITALY

Nature of the Work. In the 18th and 19th centuries Italy produced a worthy line of mathematicians, but until recently she has not made a serious effort to regain her earlier standing. Her scholars seemed to work in isolation during much of this period,—a result due in some measure, no doubt, to the lack of political homogeneity in Italy herself. A few names in the 18th and the early 19th century, however, deserve our attention.

Ceva Brothers. About the middle of the 17th century there were born two brothers, Giovanni Ceva[1] and Tommaso Ceva,[2] each of whom contributed to geometry and physics. The latter was a teacher of mathematics in the Jesuit college at Milan, while the former was in the service of the Duke of Mantua. Tommaso wrote on the cycloid, the mechanical trisection of an

[1] Born at Mantua, December, 1647; died May 13, 1736. These dates are uncertain.

[2] Born at Milan, December 20, 1648; died at Milan, February 3, 1737.

angle,[1] and mathematics in general,[2] but Giovanni was the more original and prolific. To him is due a well-known theorem which asserts that if three lines from the vertices A, B, C of a triangle ABC are concurrent in P and meet the opposite sides in X, Y, Z respectively, then

$$\frac{AZ}{ZB}\cdot\frac{BX}{XC}\cdot\frac{CY}{YA}=1.$$

This theorem was published by him in 1678.[3] His work extended into the 18th century.[4]

Manfredi Brothers. Another interesting example of a family devoted to mathematics is seen in the case of the Manfredi brothers. Eustachio Manfredi[5] was a jurist who at the age of twenty-five (1699) became professor of mathematics in the University of Bologna. He was one of the founders of the Institute of Bologna, and is particularly known for his elaborate ephemerides[6] and his general works on astronomy. He contributed also to the textbook literature of geometry.[7]

Gabriele Manfredi[8] was professor of mathematics (1720) in the University of Bologna and wrote on differential equations[9] and geometry. Eraclito Manfredi[10] was professor of medicine and later of geometry in the University of Bologna, his major interest being in mechanics.

Riccati Family. Still another family that showed decided genius in mathematics and physics was that of Jacopo Francesco, Conte Riccati,[11] a man of private fortune, who studied

[1] *Acta Eruditorum*, 1695. [2] *Opuscula mathematica*, Milan, 1699.

[3] *De lineis rectis se invicem secantibus statica constructio*, Milan, 1678.

[4] *Tria problemata geometris proposita*, Mantua, 1710; *De re numeraria, quod fieri potuit, geometrice tractata*, Mantua, 1711. Chasles, *Aperçu*, note vii, p. 294; T. Perelli, in A. Fabbroni, *Vitae Italorum doctrina excellentium*, Vol. XVIII (Pisa, 1778–1805).

[5] Born at Bologna, September 20, 1674; died at Bologna, February 15, 1739.

[6] Bologna, 1714, 1715.

[7] *Elementi della geometria piana e solida, e della trigonometria*, Bologna, 1755 (posthumous).

[8] Born at Bologna, March 25, 1681; died at Bologna, October 13, 1761.

[9] *De Constructione Æquationum Differentialium Primi Gradus*, Bologna, 1707, a work highly esteemed by Leibniz.

[10] Born at Bologna, *c.* 1682; died at Bologna, September 15, 1759.

[11] Born at Venice, May 28, 1676; died at Treviso, April 15, 1754.

in Padua and spent the latter part of his life in Venice and Treviso. He wrote on philosophy, physics, differential equations, mensuration, and related subjects, and did much to make the theories of Newton known. His collected works were published after his death.[1] He is known chiefly for his elaborate study of the so-called Riccati Equation $\frac{d\eta}{dx} = A + B\eta + C\eta^2$, where A, B, C are functions of x, of which he gave solutions for certain special cases. The equation had already been studied by Jacques Bernoulli. His second son was Vincenzo Riccati,[2] a Jesuit, professor of mathematics in the college of his order at Bologna.[3] His line of interest was quite like that of his father, his work in mathematics including the study of differential equations (1752), series (1756), quadrature problems (1767), and the hyperbolic functions.[4] Jacopo's third son was Giordano Riccati,[5] who wrote on the Newtonian philosophy (1764, 1777), geometry (1778,

JACOPO FRANCESCO, CONTE RICCATI

After an engraving by Comirato. Jacopo was the first of the well-known mathematical family of Riccati

[1] *Opere del Conte Jacopo Riccati*, 4 vols., Treviso, 1758.

[2] Born at Castelfranco, province of Treviso, January 11, 1707; died at Treviso, January 17, 1775.

[3] Collegio di San Francesco Saverio.

[4] His *Opuscula ad Res Physicas et Mathematicas pertinentia*, 2 vols., appeared at Bologna in 1757–1762.

[5] Born at Castelfranco, February 25, 1709; died at Treviso, July 20, 1790. D. M. Federici, *Commentario sopra la Vita e gli Studj del Conte Giordano Riccati*, Venice, 1790.

1779, 1790), cubic equations (1784), and physical problems (1763, 1764, 1767). The fifth son was Francesco Riccati,[1] who wrote on geometry as applied to architecture.

Other Writers. In order to give some further idea of the general nature of Italian mathematics at this time it will suffice to record the names of a few scholars who did the most to advance their science. Among the earliest of these in the 18th century was Luigi Guido Grandi,[2] a member of the order of the Camaldolites, professor of philosophy (1700) and later (1714) of mathematics at the University of Pisa, and author of a number of works on geometry,[3] in which are considered the analogies of the circle and the equilateral hyperbola, curves of double curvature on a sphere, and the quadrature of certain parts of a spherical surface.

Slightly younger than Guido Grandi, Giulio Carlo Fagnano de Fagnani, Marchese de' Toschi e S. Onorio,[4] commonly known as Fagnano, a man of private means and a savant at sixteen, was an important contributor to mathematics.[5] He was chiefly interested in geometry and in the study of algebraic equations. One of his interesting discoveries (1719) was the relation

$$\frac{c}{4} = 2 \log \left(\frac{1 - \sqrt{-1}}{1 + \sqrt{-1}} \right)^{\frac{1}{2}\sqrt{-1}}$$

where $c = 2\pi$. He further showed that the integral which expresses the arc of a lemniscate has properties analogous to those of the integral which represents the arc of a circle. Euler, in extending the theory of elliptic functions, gives Fagnano credit for having first directed attention to it.

[1] Born at Castelfranco, November 28, 1718; died July 18, 1791.

[2] Born at Cremona, October 7, 1671; died at Pisa, July 4, 1742.

[3] *Geometrica demonstratio Vivianeorum problematum*, Florence, 1699; *Geometrica demonstratio theorematum Hugenianorum circa logisticam seu logarithmicam*, Florence, 1701; *Quadratura circuli et hyperbolae per infinitas hyperbolas et parabolas geometrice exhibita*, Pisa, 1703; *Istituzioni geometriche*, Florence, 1741.

[4] Born at Senigallia, December 6, 1682; died September 26, 1766. *Opere Matematiche*, ed. Volterra, Loria, and Gambioli, 3 vols., Milan, 1911, 1912.

[5] His *Produzioni Matematiche* appeared at Pesaro, 2 vols., 1750.

Giulio had a son, Giovanni Francesco (1715–1797) or Gianfrancesco, who deduced from his father's formula the relations[1]

$$\frac{c}{4} = \sqrt{-1}\log(-\sqrt{-1})$$

and
$$\frac{c^2}{4} = \log(+1)\cdot\log(-1).$$

To students of geometry, one of the most familiar names of this period is that of Giovanni Francesco Giuseppe Malfatti.[2] He was educated in his early youth at a Jesuit college in Verona, but at the age of seventeen (1748) he went to Bologna and studied under Vincenzo Riccati. In 1771 he became professor of mathematics at Ferrara, where he spent the rest of his life. He wrote a variety of works, but his name is chiefly remembered for a problem which he published[3] in 1803: In a triangular prism to inscribe three cylinders of altitude equal to that of the prism and of maximum volume and so that the remaining volume of the prism shall be a minimum. This at once reduces itself to the inscription, in a given triangle, of three circles, each tangent to the others and to two sides of the triangle. Malfatti solved the problem analytically, Steiner solved it geometrically, and it has been the subject of extended study by various other scholars.[4]

Students of higher algebra will possibly be more familiar with the name of Paolo Ruffini,[5] who taught mathematics and medicine at Modena and contributed extensively to the theory of equations. He wrote several works and important

[1] Montucla, *Histoire*, III, 285.

[2] Gianfrancesco Malfatti. Born at Ala di Trento, 1731; died at Ferrara, October 9, 1807.

[3] *Memoria sopra un problema stereotomico*, in the *Memorie di . . . Società Italiana delle Scienze*, Modena, X, parte Iª., p. 235.

[4] M. Baker, "The History of Malfatti's Problem," in the *Bulletin of the Philosophical Society of Washington*, II, 113; Boncompagni's *Bullettino*, IX, pp. 361, 383; A. Wittstein, *Geschichte des Malfatti'schen Problems*, Diss., Munich, 1871.

[5] Born at Valentano, September 23, 1765; died at Modena, May 10, 1822. G. Bianchi, *Elogio*, pronounced at Modena, November 25, 1822.

monographs on the subject.[1] In connection with his work on equations he made a beginning in the theory of groups.[2]

In the second half of the century the study of the geometry of the compasses, which had already attracted attention at various times, was successfully undertaken by Lorenzo Mascheroni.[3] He was one of those who succeeded in making a name for himself in mathematics although beginning the study relatively late. His first interests were in the humanities, and he taught Greek and poetry at the liceo in his native town and at Pavia. He also took holy orders and became an abbot, which did not hinder him, however, from being a good fighter for the Cisalpine republic. After his experience in teaching the humanities he took up the study of geometry and became professor of mathematics at Pavia. He wrote on physics, the calculus, and the proposed metric system, but is known chiefly for his geometry of the compasses.[4] In the last-named work he showed that the ordinary constructions of elementary

LORENZO MASCHERONI

After an engraving by F. Redenti, from a drawing made from life

[1] Among them, *Teoria generale delle equazioni*, 2 vols., Bologna, 1799.

[2] H. Burkhardt, *Abhandlungen*, VI, 119.

[3] Born at Castagneta (Castagnetto), near Bergamo, May 14, 1750; died in Paris, July 30, 1800. See G. Mangili, *Elogio di Lorenzo Mascheroni*, Milan, 1812; G. Savioli, *Memorie . . . Mascheroni*, Milan, 1801; F. Landi, *Elogio di Lorenzo Mascheroni*, Modena, 1804. These writers disagree slightly as to the date of his birth.

[4] *Geometria del Compasso*, Pavia, 1797; new ed., Palermo, 1901. There was a French translation by A. M. Carette, 1798; 2d ed., Paris, 1828. For bibliography see *L'Intermédiaire*, XIX, 92.

geometry can be performed by the use of the compasses alone; that is, that all critical points of an elementary figure can be found with no assistance from the straightedge. The idea seems to have been suggested to him by Benedetti's work of 1553, already mentioned.

There were in this period a number of Italian mathematicians who, following the French school, made notable advance in mathematical astronomy. Among these was Ruggero Giuseppe Boscovich[1] who, as the family name indicates, descended through his father from a Herzegovina family, although his mother was a native of Ragusa, Italy. He entered the Jesuit order and in 1740 became professor of mathematics and philosophy in the Collegio Romano at Rome. After traveling extensively in various countries of Europe he became a professor (1764–1770) at Pavia. He afterwards lived in Paris for some years, returning to Italy in 1783 and taking up his residence in Milan. He wrote extensively on astronomy, giving the first geometric solution of the problem of determining the equator of a planet by three observations of a spot and determining its orbit from three observations of the body.[2] He also worked on the general problem of the determination of the orbits of comets and wrote a general treatise on mathematics.[3]

In the field of pure geometry perhaps the most original of the Italian writers of this period was Giusto Bellavitis,[4] professor of mathematics at Padua and Vicenza, who contributed to the theory of projective geometry and was the first to set forth the method of equipollence.[5]

Luigi Cremona (1830–1903) is even better known in this field, having written an important textbook on projective

[1] Born at Ragusa, May 18, 1711; died at Milan, February 13, 1787. G. Dionisi, *Ruggero G. Boscovich*, Zara, 1887; F. Ricca, *Elogio . . . Ruggiero Giuseppe Boscovich*, Milan, 1789.

[2] *De determinanda Orbita Planetae*, Rome, 1749.

[3] *Elementa universae matheseos*, 3 vols., Rome, 1754.

[4] Born at Bassano, near Padua, November 22, 1803; died at Padua, November 6, 1880.

[5] *Saggio d' applicazioni del calcolo delle equipollenze*, Padua, 1837; *Metodo delle equipollenza*, Padua, 1837. See O. Brentari, *Biografia di Giusto Bellavitis*, Bassano, 1881.

geometry. He was, however, more than a textbook writer, having contributed to the theory of cubic surfaces (1866) and to the theory of transformation of curves.

Eugenio Beltrami (1835–1900) was a prolific writer on the theory of higher geometry and on physical problems, higher algebra, and invariants, most of his memoirs appearing in the *Annali di Matematica.* He was professor of mathematics at Pisa, Bologna, Pavia, and Rome, and was president of the Accademia dei Lincei in Rome. He was also minister of public instruction (1862) and a senator. Giuseppe Battaglini (1826–1894) was well known for his interest in the theory of groups, his work in line geometry, and his editorial work on the *Giornale di Matematiche* (Naples, beginning in 1863). His contemporary, Barnaba Tortolini (1808–1874), founded (1850) the *Annali di Scienze Matematiche e Fisiche* and was a voluminous writer on analysis. Of the later writers, Francesco Brioschi (1824–1897), professor at Pavia, contributed to the study of mathematical physics; Enrico Betti (1823–1892), professor at Pisa, added to the theory of binary forms; Felice Casorati (1835–1890), professor at Pavia, did much to advance the study of analysis in Italy; and Ulisse Dini (1845–1918) wrote upon the theory of the functions of a real variable (1877) and contributed to the theory of series.

MARIA GAETANA AGNESI

Professor of mathematics at Bologna and a writer on analytic geometry

Among the women of Italy who have added to the store of the world's knowledge of mathematics the most erudite one of this period was Maria Gaetana Agnesi,[1] who occupied for a time the chair of mathematics in the University of Bologna. Her work on analytic geometry is well known.[2]

6. SWITZERLAND

Nature of the Work. The influence of the Bernoullis showed itself in Switzerland in the 18th century, but with the exception of two great scholars no one arose to maintain the standard which they had set. There was certainly no Swiss school of mathematicians, hardly even a Basel school. To have produced in one century both Euler and Steiner is, however, quite enough of glory for any country.

Minor Writers. Among the minor writers the earliest of any considerable note was Nicolas Fatio de Duillier.[3] At various times a resident of Geneva, Paris, The Hague, and London, pilloried for his relations with the mystic sect of Camisards, a Fellow of the Royal Society, an extensive traveler, and a friend of Giovanni Domenico Cassini, Huygens, and Leibniz, his life was filled with adventure rather than with study. Nevertheless he found time to write on the curve of least descent,[4] on navigation, and on astronomy.

Jean Pierre Crousaz[5] also belongs to the period between the 17th and 18th centuries. Although born in Switzerland and spending the last fifteen years of his life there, he was also professor of mathematics at Groningen for four years (1724–1728) and spent some time in Germany. He wrote on the

[1] Pronounced än yä′zē. Born at Milan, March 16, 1718; died at Milan, January 9, 1799.

[2] *Istituzioni analitiche ad uso della gioventù Italiana*, Milan, 1748. There is an English translation, 2 vols., London, 1801, with a brief biography.

[3] Facio, Faccio. Born at Basel, February 16, 1664; died at Maddersfield, Worcestershire, May 10, 1753.

[4] *Lineae brevissimi descensus investigatio geometrica duplex* . . ., London, 1699, which helped to precipitate the contest between Newton and Leibniz.

[5] Born at Lausanne, April 13, 1663; died at Lausanne, March 22, 1750. The name also appears as Crouzas.

geometry of lines and surfaces (Amsterdam, 1718), on algebra (Paris, 1726), and on the new calculus (Paris, 1721). He also wrote an essay on arithmetic, one of the latest to give a complete discussion of the ancient finger reckoning.[1]

Of the pupils of the Bernoullis only a few rose even to moderately high rank. One of the first to acquire considerable reputation, although now generally forgotten, was Jacob Hermann.[2] He was a pupil of Jacques (I) Bernoulli and was professor of mathematics in the University of Padua (1707–1713), at Frankfort a. d. O. (1713–1724), and at Petrograd (1724–1731), after which he became professor of moral philosophy at Basel. He was one of the early writers on the differential calculus (1700) and contributed to the *Acta Eruditorum* (from 1702) and other periodicals numerous memoirs on the application of the subject to geometric problems.

Among the Swiss mathematicians of the 18th century Gabriel Cramer[3] was much better known than Hermann. He belonged to a Holstein family which settled for a time at Strasburg but finally went to Geneva. At the age of twenty (1724) he was made professor of mathematics, becoming a colleague of Giovanni Ludovico Calendrini[4] at the University of Geneva. Cramer's work related chiefly to physics, but he also wrote on geometry (1732), the history of mathematics (1739, 1741, 1748, 1750), and algebraic curves,[5] most of these contributions appearing in the form of memoirs. He had a wide personal acquaintance with the mathematicians of England, Holland, and France. The revival of the subject of determinants, already suggested by Leibniz in Europe and by Seki in Japan, is due largely to him (1750).

Euler. Basel had achieved enough glory in the history of mathematics through being the home of the Bernoullis, but she

[1] *Réflexions sur l'utilité des mathématiques . . . avec un nouvel essai d'arithmétique démontrée,* Amsterdam, 1715.

[2] Born at Basel, July 16 (o.s.), 1678; died at Basel, July 11, 1733.

[3] Born at Geneva, July 31, 1704; died at Bagnols, near Nîmes, January 4, 1752.

[4] Born at Geneva, 1703; died at Geneva, 1758. He wrote on Newton's theories and on conics.

[5] *Introduction à l'Analyse des Lignes Courbes Algébriques,* Geneva, 1750.

doubled her glory when she produced Léonard Euler.[1] While the Bernoullis had generally been destined by their parents for commerce or the law, but had finally entered the field of mathematics, Euler had been taught mathematics by his father, who had himself studied under Jacques (I) Bernoulli. Léonard studied theology, but also put himself under the instruction of Jean (I) Bernoulli. To his study of theology and mathematics he added that of medicine, the oriental languages, astronomy, and physics, and became not only the greatest mathematician and astronomer of his generation but one of its most all-round savants. He went to Petrograd in 1727 and taught mathematics and physics there. In 1735, through excessive work, he lost the use of his right eye. "J'aurai moins de distractions" was his comment,—the

LÉONARD EULER
From a contemporary engraving

[1] Born at Basel, April 15, 1707; died at Petrograd, September 18, 1783. Since he wrote chiefly in French when not employing, as he usually did, the international Latin, the French spelling of his Christian name has been retained. In German it appears as Leonhard. In his Latin books he uses Leonhardus Eulerus. For a bibliography of recent articles, see *Bibl. Math.*, X (3), 284. See also P. Stäckel, in the *Vierteljahrschrift der Naturf. Gesellsch. in Zürich*, Jahrg. 54, and the recent edition of his works. Condorcet's *Éloge de M. Euler* was pronounced before the Académie des Sciences in 1785 and is well known, as is also N. Fuss, *Lobrede auf Herrn Leonhard Euler*, Basel, 1786.

comment of a practical philosopher with a sense of humor. In 1766 he lost the use of the other eye, but continued his labors without complaint to the day of his death. Few writers ever contributed so extensively or so fruitfully. Indeed, he was in a sense the creator of modern mathematical expression. In his lifetime there appeared about six hundred important memoirs from his hand, besides various treatises. He published three monumental works on analysis,[1] in which he "freed the analytic calculus from all geometric bonds, thus establishing analysis as an independent science."[2] He also wrote on algebra,[3] arithmetic, mechanics, music,[4] and astronomy, and had an extensive knowledge of botany, chemistry, medicine, and *belles-lettres*. It is said that he could repeat the *Æneid* from beginning to end.[5] Arago said of him that he calculated without effort, just as men breathe and as eagles sustain themselves in the air.

Several algebraic expressions bear his name. Euler's constant is the limit of $1 + \frac{1}{2} + \frac{1}{3} + \cdots + \frac{1}{n} - \log n$ as $n \to \infty$, the value being 0.577215664901532 8 as found by him, but since carried much farther. Euler's Equation is $\frac{dx}{\sqrt{X}} = \frac{dy}{\sqrt{Y}}$, where X and Y are two quartic functions of x and y, differing only in the variable. From it is derived a formula that is of great importance in the theory of elliptic functions. The equation $e^{ix} = \cos x + i \sin x$ is also called by his name.

De Morgan relates an interesting anecdote concerning Euler's meeting with Diderot at the Russian court. Diderot

[1] *Introductio in analysin infinitorum*, 2 vols., Lausanne, 1748; German edition, Berlin, 1788–1790; *Institutiones calculi differentialis*, Berlin, 1755; German edition, 3 vols., Berlin, 1790–1793; *Institutiones calculi integralis*, Petrograd, 1768–1770, with Vol. IV in 1794 (posthumous), and later editions. Eneström has published a complete list of his works (Leipzig, 1910–1913).

[2] H. Hankel, *Die Entwickelung der Mathematik in den letzten Jahrh.*, p. 13. Tübingen, 2d ed., 1884.

[3] *Vollständige Anleitung zur Algebra*, 2 vols., in Russian, Petrograd, 1768; German edition, Petrograd, 1770; French edition by Lagrange in 1774.

[4] Fuss remarked that his *Tentamen novae theoriae musicae* (Petrograd, 1729) "contained too much geometry for musicians and too much music for geometers."

[5] D. Brewster, *Letters of Euler*, I, 24. New York, 1872.

had somewhat displeased the Czarina by his antireligious views, and so she persuaded Euler to assist her in suppressing him.

Diderot was informed that a learned mathematician was in possession of an algebraical demonstration of the existence of God, and would like to give it him before all the Court, if he desired to hear it. Diderot gladly consented: though the name of the mathematician

Monsieur

Je Vous suis bien obligé de la communication des intentions de Mr. D'Alembert à l'egard de son memoire. Il m'est fort indifferent, qu'il soit imprimé dans nos Memoires, ou supprimé. J'aurois cru que Mr. D'Alembert après y avoir mieux reflechi n'en demanderoit plus l'impression. Mais puisqu'il promet à l'Academie un Memoire de sa façon, en cas qu'on imprime son Memoire je ne voudrois pas en priver l'Academie. Voilà donc ici ma reponse, de la solidité de laquelle je suis autant assuré que Mr. D'Alembert le peut être de la sienne, et je souhaiterois qu'elle fut imprimée après la sienne. Je consens aussi très volontiers, que Vous Lui en envoyiez d'avance une copie. Si le Copiste de l'Academie ne peut pas la faire, je m'offre de la faire copier par mon fils: peut être qu'il change de sentiment après l'arrivée quoique j'en aye fort peu d'esperance. L'autre Memoire que j'ai l'honneur d'ajouter ici, est apparemment celui de Mr. Castillon

L. Euler

le 6 Xbr. 1756.

AUTOGRAPH LETTER OF EULER

Written during the period of his residence in Petrograd

was not given, it was Euler. He advanced toward Diderot, and said gravely, and in a tone of perfect conviction:

"Monsieur, $\frac{a + b^n}{n} = x$, donc Dieu existe; répondez!"

Diderot, to whom algebra was Hebrew, was embarrassed and disconcerted: while peals of laughter rose on all sides. He asked permission to return to France at once, which was granted.[1]

[1] *Budget of Paradoxes*, 2d ed., II, 4. Chicago, 1915.

His death was as he would have wished,—a discussion on mathematics, a dinner, a cup of tea, a pipe, and in an instant, in the words of Condorcet, "il cessa de calculer et de vivre."

Euler's eldest son, Johann Albrecht Euler,[1] was also a scientist of high attainments, particularly in the line of physics.[2]

Steiner. Johann Heinrich Pestalozzi[3] is not generally thought of as a mathematician, although no man of his time did so much as he to improve the teaching of elementary arithmetic. He deserves a place in the history of mathematics, however, for having taken a poor Swiss boy who could not write a single word before reaching the age of fourteen, admitting him at the age of seventeen to his school at Yverdon, and giving him a love for mathematics. The boy was Jacob Steiner,[4] the last one who would have been selected, by those who knew him in childhood, as destined to become one of the greatest geometers of modern times. Not long after leaving Yverdon he went to the University of Heidelberg (1818), where he soon showed his ability in mathematics, and in 1821 he began giving private lessons in Berlin, becoming soon after

JACOB STEINER
After a photograph

[1] Born at Petrograd, November 16 (o.s.), 1734; died September 6 (o.s.), 1800.

[2] For a list of his works see P. Stäckel, in the *Vierteljahrschrift der Naturforschenden Gesellsch. in Zürich*, 1910.

[3] Born at Zürich, January 12, 1746; died at Brugg, Aargau, February 17, 1827.

[4] Born at Utzensdorf, March 18, 1796; died at Bern, April 1, 1863.

(1825) a teacher in the Gewerbeakademie. In 1834 he became a professor in the University of Berlin, and from that time on he was a prolific contributor to geometry. The later years of his life were spent in Switzerland. He wrote several treatises of highest rank[1] and stands out as one of the prominent men in his field of work.[2] He extended Carnot's treatment of the complete quadrilateral to the n-gon in space, discussed the properties of ranges and pencils, and perfected the theory of curves and surfaces of the second degree.[3]

Lhuilier. Of the later Swiss contributors to elementary mathematics one of the best known was Simon-Antoine-Jean Lhuilier.[4] In his younger days he was a private tutor in Warsaw, after which he spent some time in Tübingen, then (1795) becoming professor of mathematics in Geneva. He wrote upon geometry while in Warsaw (1780) but first attracted attention by winning (1786) the prize offered by the Academy of Berlin on the nature of infinity. He wrote several elementary works but is known chiefly for his numerous memoirs on the measure and construction of polygons and polyhedrons and on the analogy between rectilinear and spherical right-angled triangles.[5] In the latter study he showed that the theorem analogous to the Pythagorean is

$$\sin^2 \tfrac{1}{2} a = \sin^2 \tfrac{1}{2} b \cdot \cos^2 \tfrac{1}{2} c + \sin^2 \tfrac{1}{2} c \cdot \cos^2 \tfrac{1}{2} b.$$

His contemporary, Jean Trembley (1749–1811), also a native of Geneva, a member of a family that contributed not a little to the appreciation of mathematics in Switzerland, wrote extensively on the calculus and its applications.

[1] *Systematische Entwickelung der Abhängigkeit geometrischer Gestalten von einander*, Berlin, 1832; *Die geometrischen Construktionen, ausgeführt mittelst der geraden Linie und eines festen Kreises*, Berlin, 1833; *Vorlesungen über synthetische Geometrie*, Leipzig, 2 vols., 1867.

[2] See his *Gesammelte Werke*, 2 vols., Berlin, 1881–1882.

[3] J. H. Graf, *Der Mathematiker Jakob Steiner von Utzensdorf*, Bern, 1897; J. Lange, *Jacob Steiners Lebensjahre in Berlin, 1821–1863*, Berlin, 1899; Cajori, *History of Math.*, 2d ed., p. 290.

[4] Born at Geneva, April 24, 1750; died March 28, 1840.

[5] "Analogie entre les triangles rectangles, rectilignes et sphériques," Gergonne's *Annales Math.*, I (1810–1811).

7. Other European Countries

Countries Considered. The Low Countries, Russia, Hungary, and Norway all added to the roll of eminent mathematicians in the 18th and 19th centuries, although space allows for the mention of only a few names.

Holland. Of the mathematicians of Holland in the 18th century Wilhelm Jacob Storm van s'Gravesande[1] may be taken as a representative. Beginning with the law, he later took up the teaching of mathematics, first (1717) at The Hague and then (1734) at Leyden. His was another case of the early display of mathematical ability, his essay on perspective having attracted attention when he was only nineteen years old. His first important publication[2] was devoted to an exposition of Newton's philosophy. His mathematical work consisted rather in making known the new theories of the calculus together with its applications than in original contributions,[3] but in physics his inventions and discoveries were considerable.

Belgium. As Belgium's representative may be selected Lambert-Adolphe-Jacques Quételet.[4] He studied in Ghent and was only eighteen when (1814) he began teaching mathematics in the lyceum in his native city, an institution that was converted into a university in 1815. In the same year he received the degree of doctor of science, the first to be conferred by the new faculty. In 1818 he became director of the observatory at Brussels, an empty honor at the time because there was no building. In 1826 the erection of the observatory was decided upon and Quételet's plans were adopted. Owing to political troubles it was not completed until 1832. In 1836 he became professor of astronomy and geodesy in the military school in

[1] Born at Herzogenbusch (Bois-le-Duc), September 27, 1688; died at Leyden, February 28, 1742.

[2] *Physices elementa mathematica, experimentis confirmata, sive Introductio ad philosophiam Newtonianam*, 2 vols., Leyden, 1720–1721.

[3] *Œuvres philosophiques et mathématiques,* Amsterdam, 1774 (posthumous).

[4] Born at Ghent, February 22, 1796; died at Brussels, February 17, 1874. Ed. Mailly, *Essai sur la vie et les ouvrages de Quételet*, Brussels, 1875; English abstract, Smithsonian Institution *Report* for 1874, p. 169 (Washington, 1875).

Ghent. He was much interested in various phases of geometry, including spherical polygons, conics, three-dimensional figures, caustics, and stereographic projections; in astronomy and terrestrial magnetism; and, most of all, in the theory of probability,[1] a subject in which he was a recognized leader. He also wrote the leading works on the history of mathematics in Belgium.[2]

Scandinavia. Scandinavia produced a number of prominent mathematicians in this period, but the one of greatest genius was Niels Henrik Abel.[3] Like Eisenstein, Cotes, Clifford, Pascal, and Galois, he died young and yet attained very high rank in the domain of mathematical research. His memoir on elliptic functions did more for the theory than Legendre was able to accomplish in a long life. His name also attaches to certain functions which have since his time been the object of extended study. To him is due the first proof that an algebraic solution of the general equation of the fifth degree is impossible.[4]

Marius Sophus Lie[5] was the most prominent Scandinavian mathematician of his generation. He spent some time in France, but in 1872 became a professor at Christiania and from 1886 to 1898 taught at Leipzig. He is known chiefly for his work in differential equations, the theory of transformation groups (1873), differential geometry, and the theory of infinite continuous groups.

The Bolyais. The rise of non-Euclidean geometry is so closely connected with the name of Bolyai that a few words are appropriate concerning the family. The name is found in

[1] *Instructions populaires sur le calcul des probabilités*, Brussels, 1828; *Théorie des probabilités*, Brussels, 1845.

[2] *Histoire des sciences mathématiques et physiques chez les Belges*, Brussels, 1864; *Sciences mathématiques et physiques chez les Belges au commencement du XIX siècle*, Brussels, 1866.

[3] Born at Findöe, Norway, August 5, 1802; died at Arendal, April 6, 1829. L. Sylow, *Discours* at the centenary of his birth, Christiania, 1902.

[4] C. A. Bjerknes, *Niels-Henrik Abel*, French translation, Paris, 1885. This work is taken chiefly from articles in the *Nordisk Tidsskrift*, Stockholm, 1880.

[5] Born at Christiania, December 17, 1842; died at Christiania, February 18, 1899. G. Darboux, *Bulletin of the Amer. Math. Soc.*, V, 367; F. Engel, *Bibl. Math.*, I (3), 166; M. Noether, *Mathem. Annalen*, LIII, 1.

the Magyar records as early as the 13th century, the family belonging to the landed gentry, its estate lying in Bolya, a small town in Hungary. Farkas[1] Bolyai, a professor in a college[2] at Maros-Vásárhely,[3] had a son Farkas Bolyai,[4] who, after finishing his preparatory work, went to Göttingen (1796), where Kästner was closing his somewhat mediocre career and Gauss was beginning his brilliant one. Here he and Gauss became very intimate, exchanging ideas, taking long walks together, and together indulging in the few social recreations that they allowed themselves. Circumstances compelled Bolyai to return to his home in 1799, much against his personal desires. In 1804 he became professor of mathematics, physics, and chemistry in the college at Maros-Vásárhely, and here he remained until 1851. Here he wrote to Gauss two letters[5] on geometry, and here he published (1830) a little work in the Magyar language on arithmetic[6] and also (1832) his work on elementary mathematics.[7] The letters outline a book on geometry and show that he was interested in the subject of parallels. The book itself includes both algebra and geometry, and raises the question of the validity of Euclid's postulate of parallels. The general ideas of this book were in his mind when he went to Göttingen, 1796, and for a generation he had pondered upon the foundations of geometry. Among them appears the principle of the permanence of equivalent forms,[8] which English writers assign to Peacock (1830) and the Germans to Hermann Hankel (1867).

[1] Latin Wolfgangus; German Wolfgang. The standard authority on the subject is P. Stäckel, *Wolfgang und Johann Bolyai*, 2 vols., Leipzig, 1913, with excellent bibliography. There is also a "Vita di Giovanni Bolyai," by M. Darvai, in the *Atti* of the *Congresso internaz. di sci. storiche*, XII, 45.

[2] Corresponding to the German Gymnasium.

[3] *I.e.*, "market-place on the Maros" River.

[4] Born at Bolya, February 9, 1775; died at Maros-Vásárhely, November, 1856. Referred to in German works as Wolfgang Bolyai, and in other works occasionally in the Magyar form, Bolyai Farkas.

[5] September 16, 1804, and December 27, 1808.

[6] *Az arithmetica eleje* (*Elements of Arithmetic*), Maros-Vásárhely, 1830.

[7] *Tentamen Juventutem Studiosam in Elementa Matheseos Purae . . . introducendi*, 2 parts, Maros-Vásárhely, 1832, 1833. The book bears the Imprimatur of 1829.

[8] Stäckel, *loc. cit.*, I, 35.

To the work of Farkas there was an appendix[1] written by his son, János Bolyai,[2] of whom the father had written to Gauss (1816) that this boy of fourteen already had a good knowledge of the differential and integral calculus and could apply it to mechanics, to the tautochronism of the cycloid, and to other lines of work, and that he knew Latin and astronomy. János went to the engineering school in Vienna when he was sixteen, and at twenty-one entered the army. In 1825 or 1826 he worked out the theory of parallels which he set forth in the appendix (1832) above mentioned, and in this is a clear discussion of the validity of Euclid's postulate of parallels and a presentation of a non-Euclidean geometry.[3]

The last years of Farkas Bolyai were unhappy ones, owing to the loss of his wife and the estrangement of his son. He wrote several other works,[4] however, including some on poetry.

Lobachevsky. In connection with the Bolyais it is natural to mention the work of Nicolai Ivanovitch Lobachevsky.[5] Although the son of a Russian peasant, he early showed a remarkable genius. He studied at the University of Kazan and when only twenty-one became professor of mathematics in that institution. In 1826 he made known through his lectures his conception of a geometry which should not depend upon the Euclidean postulate of parallel lines. These ideas were published in 1829, and in various later works.[6]

[1] *Appendix. Scientiam spatii absolute veram exhibens: a veritate aut falsitate Axiomatis XI Euclidei (a priori haud decidenda) independentem. . . .*

[2] Born at Klausenburg (in Magyar, Kolozsvár), December 15, 1802; died at Maros-Vásárhely, January 27, 1860. He also uses the Latin form, Johannes Bolyai.

[3] For a further discussion see Volume II, Chapter V.

[4] For a list of his works see Stäckel, *loc. cit.*, I, 205.

[5] Born at Nijni-Novgorod, November 2 (October 22, o.s.), 1793; died at Kazan, February 24 (February 12), 1856. The name is transliterated from the Russian in various ways, such as Lobatschevskij, Lobatschewsky, and Lobatcheffsky.

[6] *Ueber die Principien der Geometrie*, Kazan, 1829–1830; *Geometrische Untersuchungen zur Theorie der Parallellinien*, Berlin, 1840; French translation by Hoüel, 1866; English by Halsted, 1891; *Pangéométrie ou Précis de géométrie fondée sur une théorie générale et rigoureuse des parallèles*, Kazan, 1855. He also published various memoirs on other subjects. Vassilief's eulogy on Lobachevsky has been translated into French (1896) and English (1894).

Of the independent discovery of the non-Euclidean geometry by Lobachevsky and Bolyai there can be no doubt. The subject was in the general intellectual atmosphere of the time. Gauss, who was considering the question as early as 1792, had doubtless stimulated the elder Bolyai to study the problem, and no doubt had been stimulated in return. Both Lobachevsky and the younger Bolyai had been influenced by the Göttingen school. Each in his own way had attacked the question, and each had worked out his theory at about the same time (1825–1826); Lobachevsky published his theory first (1829), but János Bolyai published his independently (1832).[1]

NICOLAI IVANOVITCH LOBACHEVSKY

After a contemporary drawing

Kovalévsky. Among the Russian mathematicians of the latter part of the 19th century none was better known in western Europe than Sónya Krukovsky, who married Vladímir Kovalévsky and is commonly called by the name of Sophia Kovalévsky.[2] She was a pupil of Weierstrass, took her doctor's degree at Göttingen, and became professor of mathematics at Stockholm, where she was associated with Mittag-Leffler in the study of the function theory. Her own work was largely connected with the theory of differential equations,—a theory to which she made contributions of recognized value.

[1] On Bolyai's appreciation of Lobachevsky's work, see Stäckel, *loc. cit.*, I, 140.

[2] The date of her birth is unknown, but it was probably 1850. She was married in 1868 and died at Stockholm, February 10, 1891. See her autobiography, translated by Isabel F. Hapgood, New York, 1895.

Wronski. In the 19th century Poland produced only one mathematician who succeeded in attracting much attention abroad, and this was Hoëné Wronski.[1] He spent most of his life in France and wrote on the philosophy of mathematics. His *Introduction to a Course in Mathematics* appeared in London in 1821.

In Bohemia the subject of infinite series was studied by Bernhard Bolzano (1781–1848), but there were few other writers of prominence in the field of mathematics in this part of the world during the period now under consideration.

HOËNÉ WRONSKI

After an etching by Mme Frédérique O'Connell and autographed by Wronski

8. United States

Brief History. The United States inherited its first mathematics almost wholly from Great Britain. Early in the 19th century there was an influx of French mathematics.[2] Until Johns Hopkins University brought Professor

[1] Born August 24, 1778; died August 9, 1853. Since he wrote chiefly in French, the French spelling of his name is used. S. Dickstein has various references to and articles upon him in the second series of the *Bibliotheca Mathematica*, particularly VI (2), 48. See also his *Catalogue des œuvres imprimées et manuscrites de Hoëné Wronski*, Cracow, 1896.

[2] F. Cajori, *The Teaching and History of Mathematics in the United States*. Washington, 1890; J. Pierpont, "The History of Mathematics in the Nineteenth Century," *Bulletin of the Amer. Math. Soc.*, XI, 156; R. S. Woodward, "The Century's Progress in Applied Mathematics," *ibid.*, VI, 133; T. S. Fiske, *ibid.*, XI, 238; C. J. Keyser, *Educ. Rev.*, XXIV, 346.

Sylvester to this country (1876) for his second sojourn, and gave him the facilities for graduate work, little effort was made to encourage the study of modern mathematics. Several mathematicians of genuine ability had, however, developed before that time. Among these may be mentioned Benjamin Peirce,[1] who became a professor at Harvard at the age of twenty-four. He was for some time in charge of the U. S. Coast Survey, but is best known for his work on linear associative algebra (1870). Nathaniel Bowditch,[2] a self-made mathematician, was a generation older than Peirce. At the age of seventeen he began the study of Latin for the purpose of reading Newton's *Principia*, and later became proficient in French, Spanish, Italian, and German in order to study mathematics in these languages. His *New American Practical Navigator* (1802) and his translation of Laplace's *Mécanique céleste* (published in 1829–1839) gave him an international reputation.

It is proper at this time to mention Robert Adrain,[3] who contributed in a noteworthy manner to the progress of mathematics in his adopted country. His work on the form of the earth and on the theory of least squares showed that he was possessed of genuine mathematical ability. He founded the *Analyst* (1808) and the *Mathematical Diary* (1825).

In the domain of mathematical astronomy the most prominent of our native scholars in the 19th century was George William Hill.[4] He was for many years connected with the Nautical Almanac, but is known chiefly for his work in the lunar theory. When Poincaré visited this country he met Hill, and his first words were, "You are the one man I came to America to see," and he meant them.

Simon Newcomb (1835–1909) was a native of Nova Scotia but spent most of his life in the United States. He was

[1] Born at Salem, Massachusetts, April 4, 1809; died at Cambridge, October 6, 1880.

[2] Born at Salem, Massachusetts, March 26, 1773; died at Boston, March 16, 1838.

[3] Born at Carrickfergus, Ireland, September 30, 1775; died at New Brunswick, New Jersey, August 10, 1843.

[4] Born at Nyack, New York, 1838; died at Nyack, August 17, 1916.

connected with the Nautical Almanac and was also professor at Johns Hopkins. He was largely a self-made mathematician and was possessed of undoubted ability in the science.

Of the new school of mathematicians the best-known representative among those who have passed away was Maxime Bôcher (1867–1918). His special fields of research were those of linear differential equations, higher algebra, and the function theory. He was educated at Harvard and Göttingen and was called to the mathematical faculty at Harvard in 1891. In 1913–1914 he lectured at Paris on "les méthodes de Sturm dans la théorie des équations différentielles linéaires et leurs développements modernes."[1]

Josiah Willard Gibbs (1839–1903) was one of the most original of the American physicists of his time. He received his doctor's degree at Yale in 1863 and afterwards studied in Paris, Berlin, and Heidelberg. His interest in mathematics was great, particularly in the application of the science to mechanics.[2]

In the theory of probability, and in actuarial science in particular, Emory McClintock (1840–1916) is the best-known American contributor. He was for a time an assistant professor in Columbia College (later Columbia University) and then became a leader in life insurance work. He was the first president of the American Mathematical Society, was one of the founders of its *Transactions*, and was a frequent contributor to various mathematical journals.

9. The Orient

Nature of the Work. Except for the case of Japan, the Orient lost its initiative in mathematics when Western science was made known. India produced nothing that was distinctive in the 18th or 19th century. China, while occasionally protesting against Western mathematics, in reality sacrificed on the altar of the Jesuit missionaries her own originality in the science.

[1] W. F. Osgood, *Bulletin of the Amer. Math. Soc.*, XXV, 337.
[2] P. F. Smith, *Bulletin of the Amer. Math. Soc.*, X, 34.

Japan, owing to her policy of isolation, had little besides herself on which to depend, at least until the close of the 18th century, and so she continued to show her remarkable ingenuity in the application of her native calculus. This policy was unchanged until her doors were opened to Western civilization, when she, like China, lost that power of initiative along the lines of native research which her scholars had shown in the 17th and 18th centuries.

平方表

四	三	二	一	
一六〇〇 八〇	九〇〇 六〇	四〇〇 四〇	一〇〇 二〇	〇
一六八一 八二	九六一 六二	四四一 四二	一二一 二二	一
一七六四 八四	一〇二四 六四	四八四 四四	一四四 二四	二
一八四九 八六	一〇八九 六六	五二九 四六	一六九 二六	三
一九三六 八八	一一五六 六八	五七六 四八	一九六 二八	四
二〇二五 九〇	一二二五 七〇	六二五 五〇	二二五 三〇	五
二一一六 九二	一二九六 七二	六七六 五二	二五六 三二	六
二二〇九 九四	一三六九 七四	七二九 五四	二八九 三四	七
二三〇四 九六	一四四四 七六	七八四 五六	三二四 三六	八
二四〇一 九八	一五二一 七八	八四一 五八	三六一 三八	九

CHINESE TABLES OF SQUARES

This work of the 18th century, by Chang Tsu-nan, Fan Ching-fu, and Chiung Lin-tai, also contains a table of logarithms. It illustrates the early stages of modern tables in China

China. Chinese mathematics in the 18th and 19th centuries was at first characterized by the continued appreciation and study of the science introduced by the Jesuits. With this, however, there went a parallel study and appreciation of the native mathematics, and later in the 18th century there was a revulsion of feeling against the Jesuits, with the result that native mathematics came even more into favor, although with no appreciable progress. Among those who were most influential in the European movement was the Jesuit missionary Pierre Jartoux (1670–1720), who went to China in 1700. He was in correspondence with Leibniz and is known to have interested the Chinese scholars in certain algebraic series. Among these scholars was Ch'ên Shih-jen (fl. 1715), who was skillful in the use of series of this kind. The century is noteworthy also for the general introduction into China of tables of various kinds, including logarithmic tables. In 1799 Yüan Yüan (1764–1849) published his great work[1] on the biographies of astronomers and mathematicians,—the most valuable book on the history of mathematics that has appeared in China. Supplements were afterwards added which made it more nearly complete.

Japan. In the 18th century the mathematics of Japan was enriched by a native calculus known as the *yenri*, or "circle principle," a name which may have been suggested by the title of the Chinese work of Li Yeh (1248), the *Tsĕ-yüan Hai-ching*.[2] Tradition asserts that Seki, the greatest mathematician of Japan in the 17th century, discovered the method, and there is strong circumstantial evidence to confirm this belief, but no works of his that are now known make any mention of the principle.

One of the pupils of Seki, Takebe Hikojirō Kenkō (1664–1739), may, however, have been the discoverer of the *yenri*, for he was an excellent mathematician and was probably familiar with certain European books. At any rate, he wrote upon the subject, the chief problem being to express an arc of a circle in terms of the versed sine. In Takebe's work $(\sin^{-1}x)^2$ is expressed in terms of versin x in three ways by means of three different infinite series. The question of the further development of the theory in Japan is not as interesting as that of the possible source of these series. Takebe himself does not seem to have understood their development, for his attempted explanations are very obscure. The series resemble those given by Wallis,[3] but they have not as yet been identified with any of the latter. In the form given by Ōyama (or Awayama)

[1] *Ch'ou-jen Ch'uan.*

[2] *Tsĕ-yüan* means "to measure the circle," and *Hai-ching* means "mirror ot sea." See Smith-Mikami, 49, 143, and consult this work on the entire topic.

[3] *De Algebra Tractatus*, cap. xcvi (Oxford, 1693).

Shōkei in 1728, a modification of the most general form given by Takebe, the square of the arc may be expressed as

$$a^2 = 4\,dh\left[1 + \sum_1^{\infty} \frac{2^{2n+1}(n!)^2}{(2n+2)!}\cdot\left(\frac{h}{d}\right)^n\right],$$

where a is the length of the arc, d the diameter, and h the height of the arc (versed sine). This series is said to have been known to the Jesuit missionary Pierre Jartoux, already mentioned. There seems, therefore, to be good reason for believing that the *yenri* principle was suggested by a study of Jartoux, who in turn received it from European sources.

ARIMA'S PROBLEM OF INSCRIBED SPHERES

From Arima Raidō's *Shūki Sampō* of 1769. This shows the curious Japanese method of indicating a sphere by a circle with a lune on one side. In this work of Arima the value of π is given as

$$\frac{42822\ 45933\ 49304}{13630\ 81215\ 70117},$$

which is correct to twenty-nine decimal places

Among those who were intrusted with the secrets of the Seki school, since mathematical knowledge was still held as a kind of Pythagorean mystery, was Matsunaga Ryōhitsu (died 1744), who received them from his teacher Araki Hikoshirō Sonyei (1640–1718), a pupil of Seki's. Matsunaga did much toward the improvement of the native algebra, and by means of the *yenri* method computed the value of π to fifty figures.

Among the other Japanese writers of importance there was Miyake Kenryū, who wrote a well-known book of problems (1716) and another work in which he treated of the prismoid. About the same time Baba Nobutake (1706) wrote a work on astronomy that had much influence on the study of spherics.

Nakane Genjun (1701–1761), together with his father, Nakane Genkei (p. 440), contributed to the study of mathematics and astronomy. Each was influenced by the Chinese writings of the time, which in turn reflected the European sciences which the Jesuit missionaries had made known in the East. One of the pupils of the younger Nakane was Murai Chūzen, who wrote (1765) a work[1] on numerical higher equations. In a later work (1781) he used the Pascal Triangle in expressing the coefficients of the terms in the expansion of a binomial. Arima Raidō (1714–1783), Lord of Kurume, was the first to publish the secret theory of algebra developed in the Seki school—a theory known as the *tenzan* algebra. As was the usual custom, Arima set forth his contributions in the form of problems. These problems related to indeterminate equations, the various roots of an equation, the application of algebra to geometry, the inscription of spheres within spheres, maxima and minima, binomial series, and stereometry. The most prominent of his protégés was Honda Teiken, better known by his later name Fujita Sadasuke (1734–1807). He wrote a notable work on algebra (1779),[2] together with various other treatises which show that he was an excellent teacher but a mathematician of no originality. Among his contemporaries was Aida Ammei (1747–1817), who gave a number of interesting series which he used in connection with his work in mensuration. Another contemporary and a far

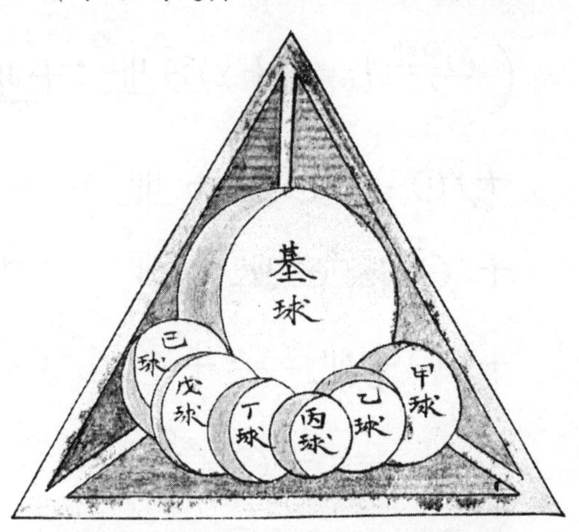

JAPANESE PROBLEM OF SPHERES TANGENT TO A TETRAHEDRON

This is typical of the native Japanese problems and shows the great patience of the Japanese scholar. The illustration is from an undated MS. by one Iwasaki Toshihisa (*c.* 1775)

[1] *Kaishō Tempei Sampō.* [2] *Seiyō Sampō.*

better mathematician, Ajima Chokuyen (1739–1798), was a contributor to the theory of indeterminate equations and to geometry. Among his solutions is an algebraic one of the Malfatti Problem. He also advanced the *yenri* theory by taking equal divisions of the chord instead, as his predecessors had done, of the arc.

$$\left(\frac{6\ominus\text{业}}{2}\right) = +9\ominus^2\text{业}^2 : +\text{业} = +9\ominus^2$$

$$+9\ominus^2 + 6\ominus\text{业} + \text{业}^2 = +\ominus^2 + 9\ominus^2$$

$$+9\ominus^2 + 6\ominus\text{业} + \text{业}^2 = +10\ominus^2$$

$$+3\ominus + \text{业} = +\ominus\sqrt{10}$$

$$+\text{业} = +\ominus\sqrt{10} - 3\ominus$$

$$+\text{业} = -\ominus\sqrt{10} - 3\ominus$$

WESTERN MATHEMATICS REPLACES THE *WASAN*

From a Japanese MS. of the middle of the 19th century, when European and American mathematics was beginning to replace the *wasan*, the native mathematics of Japan

The most interesting feature of Japanese mathematics in the 18th century is the slow penetration of Western theories into the domain of the *wasan*, or native mathematics. Japan showed great ingenuity in problem solving, but she never developed a great theory. Her nearest approach to originality in this respect was in her work on the *yenri*, and even here she seems to have had her start through contact with China and the Jesuit missionaries.

The 19th century opened with a general geodetic survey of the whole empire. By this time the European sources were beginning to be available, but the problems were still of the ingenious native variety. In 1856 an arithmetic written on the European plan was published, and in 1859 an American work on the calculus, by Elias Loomis (1811–1899), was translated. Japanese mathematics assumed a new character from that time on. For better or worse the *wasan* was dead, and European science reigned in its stead. Whether Japan will ever be able to adapt her remarkable ingenuity to Western methods, or vice versa, remains to be seen.

10. The Historians of Mathematics

Seventeenth Century. Before referring to a few of the writers on the history of mathematics in the 18th century a word should be said concerning two of their predecessors. One of these was Bernardino Baldi,[1] an abbot, a native of Urbino, who wrote on mechanics but whose chief contribution was on mathematical biography,[2] his *Cronica* serving as a source to which writers on the history of mathematics have been much indebted. He was a linguist of unusual attainments and thus was able to secure his materials from several languages.

The second of these historians was John Wallis (1616–1703), whose contributions to mathematics in general have already been mentioned. His erudite *Treatise of Algebra*[3] contains a wealth of historical material and constitutes the beginning of the serious study of the history of mathematics in England.

Eighteenth Century. The first work bearing the name of history of mathematics was written by Johann Christoph Heilbronner,[4] a man who had given considerable attention to

[1] Born 1553; died 1617. The family name was Cantagallina.

[2] *Cronica de Matematici*, Urbino, 1707. His manuscripts on the subject were printed in Boncompagni's *Bullettino,* V, 427; VII, 337; XII, 420; XIX, 335, 437, 521; XX, 197. G. Zaccagnini, *Della vita . . . Baldi*, Reggio Emilia, 1918.

[3] English ed., London, 1685; Latin ed., Oxford, 1693. Some of the "additional treatises" in the English edition have dates 1684.

[4] Born at Ulm, 1706; died at Leipzig, *c.* 1747. His first work was the *Versuch einer Geschichte der Mathematik und Arithmetik*, Frankfort, 1739. His leading work is the *Historia Matheseos Universae a mundo condito ad seculum post Chr. Nat. XVI*, Leipzig, 1742.

theology and the mathematical sciences. He did some teaching in Leipzig, but his only work of importance was in the line of the history of mathematics. His style is sometimes prolix, and our present knowledge of details is naturally better than that of any writer of the middle of the 18th century, but the work has much value even today on account of its erudition, its extracts from early writers, and particularly its list of manuscripts then to be found in various libraries and its list of early printed books.

JEAN ÉTIENNE MONTUCLA

After an engraving by P. Viel, from a miniature

To Jean Étienne Montucla[1] are due two noteworthy treatises on the history of mathematics.[2] Of these the one on the quadrature of the circle was the first publication of the kind; and while the subject has greatly changed since Montucla wrote, the work is still a classic on the early history. His larger work was the first modern history of mathematics that may be called a classic. Montucla was a man of erudition, he wrote an excellent style, and there are no early histories more highly esteemed than his.

Of less importance in this field is the work of Charles Bossut,[3] a Jesuit, mathematical examiner at the École du Génie, Mézières, and later an examiner at the École Polytechnique.

[1] Born at Lyons, September 5, 1725; died at Versailles, December 18, 1799.

[2] *Histoire des recherches sur la quadrature du cercle*, Paris, 1754; 2d ed., by Lacroix, Paris, 1831; *Histoire des mathématiques*, 2 vols., Paris, 1758; 2d ed. by Lalande, 4 vols., Paris, 1799–1802.

[3] Born at Tarare, near Lyons, August 11, 1730; died in Paris, January 14, 1814.

He was primarily a writer of textbooks, but he also wrote various monographs relating to geometry. His history[1] is of unequal value, the part relating to ancient mathematics being much better than that on modern times. He shows a prejudice in favor of English mathematics,—a fact which may account for the translation and publication (1803) of the work in London.

Of the various Italian writers of the same period the best known is Pietro Cossali,[2] who taught both physics and astronomy at the University of Parma and was later (1806) professor of mathematics at Padua. He wrote various memoirs on mathematical questions, chiefly algebraic, but is best known for his history of algebra,[3] a work of considerable scholarship and even yet of service to the student of this phase of the subject.

AUTOGRAPH LETTER OF MONTUCLA

Written May 24, 1789, ten years before his death

The German historian of mathematics in the same period was Abraham Gotthelf Kästner,[4] Dozent (1739) and professor (1746) of mathematics in the University of Leipzig, and later

[1] *Essai sur l'histoire générale des mathématiques*, Paris, 2 vols., 1802; 2d ed., Paris, 1810; English translation, London, 1803.

[2] Born at Verona, June 29, 1748; died at Padua, December 20, 1815. G. Avanzini, *Elogio*, Modena, 1822.

[3] *Origine, trasporto in Italia, primi progressi in essa dell' Algebra,* 2 vols., Parma, 1797. On the contributions of the Italians to the more general history of mathematics see P. Riccardi, in the *Memorie della R. Accad. delle Scienze dell' Istituto di Bologna*, VI (5).

[4] Born at Leipzig, September 27, 1719; died at Göttingen, June 30, 1800.

(1756) at Göttingen. He was the first mathematician of prominence to write a work devoted entirely to the history of his subject.[1] Naturally, however, the best histories of mathematics have never been written by the best mathematicians, and Kästner's contribution in this line is not as important as his numerous though by no means brilliant works and memoirs on equations, geometry, hydrodynamics, and various other branches. The work is poorly arranged, contains no index, and shows the effect of hurried compilation from notes taken in the course of a wide reading.

Nineteenth Century. The 19th century has seen so many important works on the history of mathematics that it is possible to mention only a few of the earlier writers, together with some of the most prominent of the later ones who have already passed away.

Among the first in point of time to write upon the subject in this century was Pietro Franchini,[2] a priest whose life was devoted chiefly to the teaching of mathematics in various secondary schools of Italy. He was a mathematician of considerable power, writing several works on the various branches of the science and a number of essays of some originality on analysis. He wrote three works on the history of mathematics,[3] no one of which is of special importance to the student,[4] although each has some value in connection with the development of the science in Italy. Nothing of Franchini's, however, can be said to compare with Libri's work, which is mentioned later.

Arthur Arneth[5] was one of the minor German writers of the middle of the century. He was a teacher of mathematics and

[1] *Geschichte der Mathematik*, 4 vols., Göttingen, 1796–1800.

[2] Born at Partigliano, near Lucca, April 24, 1768; died at Lucca, January 26, 1837.

[3] *Saggio sulla storia delle matematiche*, Lucca, 1821, with *Supplementi al Saggio*, Lucca, 1824; *La storia dell' algebra e de' suoi principali scrittori*, Lucca, 1827, with a *Supplemento*, the same year; *Dissertazione sulla storia matematica dell' antica Nazione Indiana*, Lucca, 1830.

[4] G. Barsotti, *Nelle esequie fatte al Prof. Pietro Franchini*, with a list of his works (Lucca, 1837).

[5] Born at Heidelberg, September 19, 1802; died at Heidelberg, December 16, 1858.

physics in a Lyceum at Heidelberg and a Privatdozent in the University. He wrote a few memoirs of no special value, a work on geometry (1840), and a history[1] that may, at the best, be described as mediocre.

Hermann Hankel[2] might have been one of the greatest historians of mathematics had he enjoyed the ordinary span of life, for he not only knew mathematics remarkably well,[3] being a professor of the subject (1867) at the University of Leipzig, but he was versed in various Oriental languages and in those of ancient and modern Europe. He left a number of fragments on the history of mathematics which showed such power that they were collected after his death and published[4] by his father. Although made up of scattered notes, this work will well repay the student's attention.

Denmark produced in the 19th century one prominent writer on the history of mathematics, Hieronymus Georg Zeuthen.[5] He studied at Copenhagen, received the degree of doctor of philosophy in 1865, and became (1883) professor in the university. He was a member of various scientific societies and was one of the leading writers on the general history of Greek mathematics.

Of the British writers in the 19th century, George Johnston Allman[6] was one of the most scholarly. He became professor of mathematics in Queen's College, Galway, in 1853 and is well known for his *History of Greek Geometry from Thales to Euclid* (Dublin, 1889), a work unsurpassed in its line until the appearance of Sir Thomas Heath's history in 1921.

[1] *Die Geschichte der reinen Mathematik*, Stuttgart, 1852.

[2] Born at Halle, February 14, 1839; died at Schramberg, Schwarzwald, August 29, 1873.

[3] See his *Zur allgemeinen Theorie der Bewegung der Flüssigkeiten*, Göttingen, 1861; *Vorlesungen über die komplexen Zahlen und ihre Funktionen*, Leipzig, 1867 (Vol. I); *Die Elemente der projektivischen Geometrie*, Leipzig, 1875.

[4] *Zur Geschichte der Mathematik in Alterthum und Mittelalter*, Leipzig, 1874. He also wrote *Die Entwickelung der Mathematik in den letzten Jahrhunderten*, Tübingen, 1869; 2d ed., 1884.

[5] Born at Grimstrup, February 15, 1839; died at Copenhagen, January 6, 1920.

[6] Born at Dublin, September 28, 1824; died in 1904.

France, which led the world in the history of mathematics in the 18th century, produced few prominent writers on the subject in the century following. Libri, who wrote in France, was an Italian, and the only native writer of profound scholarship in the general field may be said to have been Paul Tannery.[1] Although he was "professor remplaçant" of Greek and Latin philosophy at the Collège de France (1892), he was never worthily recognized in academic circles. He was connected with government service, and devoted much of his leisure to writing upon the history of Greek mathematics, his essays easily ranking as the best produced in France at the close of the 19th century.

Of the earlier French writers Michel Chasles, whose name has already been mentioned (p. 498), was among the most important. Of the later biographers of mathematicians, Maximilien Marie (died May 8, 1891) was among the most pretentious writers. His work cannot, however, be recommended for its scholarship, nor can it be considered an accurate source of information.

Unquestionably the best-known work on the history of mathematics in the 19th century was that of Moritz Benedict Cantor.[2] He studied at Heidelberg and received his doctor's degree in 1851. His works include not merely the *Vorlesungen über Geschichte der Mathematik* (4 vols., 1880–1908), but also a work on Roman surveyors (1875), one on the mathematics of primitive peoples (1863), and a large number of monographs. He was also connected with various journals and was the founder and for a long time editor of the *Abhandlungen zur Geschichte der Mathematik* (Leipzig, from 1877) and editor of the "Historisch-literarische Abtheilung" of the *Zeitschrift für Mathematik und Physik* (from 1875). He began his work as Dozent in the University of Heidelberg in 1853 and became professor in 1877. The fourth volume of his *Geschichte* was chiefly the work of various contributors who carried out his plans.

[1] Born at Mantes, December 20, 1843; died November 27, 1904.
[2] Born at Mannheim, August 23, 1829; died at Heidelberg, April 10, 1920.

Of the Austro-German writers, Moritz Steinschneider[1] deserves special mention for his numerous essays on Hebrew and Arabic mathematics.

The best known of the Italian historians is probably Libri.[2] He was educated at Pisa and for some time was professor in the university of that city. He went to Paris as a political refugee in 1830, was naturalized in 1833, and became professor of analysis at the Sorbonne. His *Histoire des sciences mathématiques en Italie* was published at Paris (4 vols., 1838–1841). On the day when the printing of the first volume was completed he called at the printer's and carried away a few copies. An hour later all the rest of the edition was burned, and the work had to be reset.[3] Libri was accused (1847) of stealing valuable books and manuscripts, and indeed did, in some way not satisfactorily explained to his critics, acquire certain rare works for his large library. He fled to England but was (1850) found guilty in Paris and was sentenced to ten years' imprisonment. The latter part of his life was spent in exile.

Much more important as a contributor to the history of the science, however, Principe Baldassare Boncompagni[4] became well known through his *Scritti di Leonardo Pisano* (1857–1862), although he had already written on the same subject. His greatest contribution was his *Bullettino di Bibliografia e Storia delle Scienze matematiche e fisiche* (Rome, 1868–1887). In spite of the fact that this monumental series has a vast amount of infinitesimal detail and that pages were frequently changed in type as the printing progressed, it will always remain as a standard work of reference.

[1] Born at Prossnitz, Moravia, March 30, 1813; died at Berlin, January 24, 1907.

[2] Conte Guglielmo Bruto Icileo Timoleon Libri-Carucci dalla Sommaja. Born at Florence, January 2, 1803; died near Fiesole, September 28, 1869. G. Loria, "Guglielmo Libri come storico della scienza," *Atti della Soc. Ligustica di Sci. Nat. e Geog.*, XXVIII, No. 3, with excellent bibliography; A. Stiattesi, *Commentario storico-scientifico sulla vita e le opere del Conte Guglielmo Libri*, Florence, 2d ed., 1879.

[3] One of the original copies with Libri's corrections was given by him to Professor Jacoli of Venice and by him to the present writer.

[4] Born at Rome, 1821; died at Rome, 1894. See G. Codazza, "Il principe Boncompagni," *Politecnico*, Vol. XX (Milan).

The best known of the historians of special branches is Anton von Braunmühl,[1] who is naturally thought of as a German by birth because of his long professorship at Munich, although he was born on Russian soil. His history of trigonometry (Leipzig, 2 vols., 1900, 1903) is the standard work on the subject.

In addition to these writers there are many others who might properly be mentioned but whose names are not, in the main, so familiar to the general reader of mathematical literature. A few names will be added, however, for the purpose of showing the general interest taken in the subject during the 19th century.

Richard Baltzer (1818–1887), professor at Giessen, wrote various notes upon special points in mathematical history; David Bierens de Haan (1822–1895), professor at Leyden, contributed to the history of mathematics in the Netherlands; Gumersindo Vicuña (1840–1890), a native of Havana, wrote upon the history of Spanish mathematics; William Whewell (1794–1866) wrote upon the history of the inductive sciences (1837); Heinrich Suter (born in 1848) wrote upon the general history of mathematics, but his most valuable contribution was to the biography of Arab mathematicians; E. L. W. M. Curtze (born in 1837) published a large number of memoirs upon the history of the science; G. Milhaud (1858–1918) published several essays upon the achievements of the Greek scientists and left a work on Descartes which appeared posthumously (1921); Alexander Macfarlane (1851–1913) published a series of lectures upon ten of the leading British mathematicians of the 19th century; A. M. Rebière (1841–1900) wrote the lives of woman mathematicians; Paul Mansion (1844–1919), professor at Ghent and editor of *Mathesis*, published a history of the calculus; Max Simon (1844–1918) published various essays and works upon the history of mathematics; J. H. Graf (1852–1918) published a history of mathematics in Switzerland.

Some of these names have already been mentioned, and the brief list here given has been supplemented by many references in the footnotes. It serves to show, however, that the 19th century developed a new interest in the history of mathematics.

[1] Born at Tiflis, December 22, 1853; died at Munich, March 9, 1908.

TOPICS FOR DISCUSSION

1. Meaning of elementary mathematics and the approximate date at which, for most practical purposes, it was brought to its present state of development.

2. A comparison of the work of those mathematicians who ranked as infant prodigies with that of those who developed later in life.

3. Effect of Newton upon the mathematics of Great Britain.

4. Evidence of opposition to Newton's mathematical theories.

5. The contest in Great Britain between the fluxional and the differential notation.

6. A comparison of the general nature of the British mathematics in the 18th century with that of the French.

7. The general nature of the problems which were suggested by the calculus and which attracted the attention of mathematicians in the 18th century.

8. The three most noteworthy mathematicians of France whose work fell chiefly in the 18th century, with a sketch of the life and contributions of each.

9. The rise of the metric system.

10. A consideration of the statement that the close of the 18th century was an era of giants in the field of mathematics.

11. The new branches of mathematics developed in the 18th century, with the names of those chiefly involved in their development.

12. The rise of modern pure geometry.

13. The general nature of the work done by Italian mathematicians in the 18th and 19th centuries.

14. The three most noteworthy Italian mathematicians whose work fell chiefly in the 18th century, with a sketch of the life and contributions of each.

15. The contributions of Switzerland in the 18th century.

16. The general nature of the Oriental mathematics in the 18th century, and particularly that of Japan.

17. The contributions of Germany in the 19th century.

18. The general nature of the mathematics of the 19th century compared with that of the centuries immediately preceding.

19. The general nature of the standard works on the history of mathematics, with a comparison of their purposes, their scholarship, and their methods.

CHRONOLOGICAL TABLE

This table includes the more important mathematical names mentioned in this volume up to the year 1850.

Through the year 1800 the relative importance of the names, judged chiefly by their influence upon elementary mathematics, is indicated by three sizes of type, although it is evident that such a distinction is often a matter on which opinions might reasonably differ. In order to assist the student in placing mathematical events in their proper relation to world history, a number of well-known historical events (with their dates), through the year 1804, have also been inserted, italic type being used for this purpose.

In general the dates are only approximations, as is seen in the large number of items under centennial and semicentennial years. Students who wish the precise dates can easily find them, where they are known, by consulting the pages mentioned in the Index.

After each mathematical name there is given a brief statement of the major interest of the individual, particularly with respect to elementary mathematics, although it is evident that in the cases of men like Newton, Euler, Gauss, and Lagrange the statement cannot give any satisfactory idea of the range of work accomplished. Where the major interest is given as astronomy or physics, the individual applied mathematics in his work.

In this table, dates before the Christian Era are, in the left-hand column, indicated by a minus sign.

The significance of the styles of type will be understood from the following scheme:

1. *Historical names and events.*
2. Mathematical names and events.
3. **Important mathematical names and events.**
4. ***Mathematical names and events of greatest importance.***

-50000. *Early Stone Age begins. (Rough approximation of date.) Means of making fire discovered.*
-15000. *Middle Stone Age begins. Works of art appear.*
-5000. *Late Stone Age begins.*
-4700. Possible beginning of Babylonian calendar.
-4241. Egyptian calendar introduced.

-4000. *Metal discovered.*
-3500. *Writing in use.*
-3000. *Early Babylonia. Sargon I, 2750 B.C.; Hammurabi, 2100 B.C. Earliest stone masonry.* Egyptian reliefs refer to taxes.
-2900. *Great Pyramid built.*
-2852. *Fuh-hi, reputed first emperor of China.* Astronomical observations.
-2700. *Huang-ti reigns in China.* Astronomy and arithmetic.
-2400. Babylonian tablets of Ur record measures.
-2350. *Yau reigns in China.* Astronomy.
-2200. Date of many mathematical tablets found at Nippur.
-2100. *Hammurabi king of Babylon.* The calendar.
-2000. *Postal system in Asia.*
-1850. *Amenemhat (Amenemmes) III reigns in Egypt.* Surveying, leveling. Oldest astronomical instrument.
-1650. **Ahmes (Rhind) papyrus.**
-1500. Oldest Egyptian sundial.
Reliefs show tax lists.
Babylonians knew simple rules of mensuration.
Queen Hatshepsut reigns in Egypt.
-1350. Date of later mathematical tablets found at Nippur.
Rollin papyrus with elaborate problems about bread.
Seti I reigns in Egypt.
-1347. Rameses II (Sesostris) said to have redivided land in Egypt.
-1180. Harris papyrus with list of temple wealth in Egypt under Rameses III.
-1150. Wön-wang may have written the I-king.
-1122. First historical period of Chinese mathematics.
Chow Dynasty begins with the reign of Wu Wang.
-1105. **The Chóu-peï, Chinese classic in mathematics, may have been written. Also (but possibly as early as the 27th century B.C.) the K'iu-ch'ang Suan-shu.**
-1055. *David becomes king of Israel.*
-1032. First historical record of rules for Chinese currency by weight.
-1000. *Vedic literature begins (c. 1000–c. 800 B.C.).*
-753. *Rome founded.*
-750. *Assyrian Empire (c. 750–606 B.C.).*
-670. Knife money appears as a Chinese coinage.
-660. Tradition of Japanese numeration to high powers of 10.
Coins appear in China in circular form.
-650. Coins struck in Lydia, Asia Minor.
-606. *Chaldean Empire (606–539 B.C.).*
-600. ***Thales. Demonstrative geometry.***
Solon. Calendar.
-575. Anaximander. Gnomon.

-561. *Reign of Nebuchadnezzar closes.*
-550. Ameristus. Geometry.
-542. Bamboo rods used in China for calculating.
-540. ***Pythagoras. Geometry, theory of numbers. Earth spherical.***
-539. *Cyrus captures Babylon.*
-530. Anaximenes. Astronomy.
-517. Hecatæus. Maps.
-500. Śulvasūtras (date very uncertain). Pythagorean numbers.
-485. *Xerxes begins to reign.*
-480. *Battle of Thermopylæ.*
-475. Dispersion of Pythagorean brotherhood.
-470. Agatharcus, Athenian. Perspective.
-465. **Œnopides of Chios. Geometry.**
-460. **Hippocrates of Chios. Quadrature.**
Parmenides. Astronomy.
-450. Zeno. Paradoxes of motion.
Herodotus the historian.
-444. *Pericles becomes supreme in Athens.*
-443. *Phidias begins the Parthenon*
-440. Leucippus. Atomic theory.
Anaxagoras. Geometry.
-432. Meton, Phæinus, Euctemon. Astronomy.
-430. **Antiphon. Method of exhaustion.**
-425. **Hippias of Elis. Quadratrix.**
Theodorus of Cyrene. Irrationality.
Philolaus. Gnomon.
Socrates. Induction and definition.
-410. **Democritus. Atomic theory, irrationals.**
-400. **Archytas. Proportion.**
-380. Leodamas. Analytic proof.
Plato. Foundations of mathematics.
-375. **Theætetus. Geometry.**
Callippus. Greek astronomer.
Chinese coins with weight or value stamped.
-370. **Eudoxus. Proportion.**
-350. **Menæchmus. Conics.**
Deinostratus. Quadratrix.
Philippus Medmæus. Geometry.
Theophrastus. History of mathematics.
Chinese compass.
Xenocrates. Theory of numbers. History of geometry.
-340. ***Aristotle. Applications of mathematics, logic.***
Speusippus. Proportion.

-336. *Alexander the Great begins his reign.*
-335. Eudemus. History of mathematics.
-332. *Alexandria founded.*
-330. Autolycus. Geometry.
-323. *Alexander the Great dies. Ptolemy Soter begins to reign in Egypt.*
-320. Aristæus. Solid loci. Conics.
Dicæarchus. Mensuration.
-300. ***Euclid. Geometry.***
-260. Aristarchus. Astronomy.
Conon. Astronomy. Spiral of Archimedes.
-250. Nicoteles. Conics.
Berosus introduces Chaldean astronomy into Greece.
-247. *Ptolemy Euergetes begins his reign in Egypt. Patron of learning.*
-230. **Eratosthenes. Prime numbers, geodesy.**
-225. ***Apollonius. Conics.***
Archimedes. Geometry, infinite series, mechanics.
Ch'êng Kiang Chen. Knotted cords.
-213. *Shï Huang-ti, emperor of China, burns all the books.*
-200. Ch'ang Ts'ang revises the "Nine Sections."
-180. **Hypsicles. Astronomy, number theory.**
Nicomedes. Conchoid.
Diocles. Cissoid.
Zenodorus. Isoperimetry.
-150. Perseus. Sections of an anchor ring.
-140. **Hipparchus. Astronomy, trigonometry.**
-106. *Pompey and Cicero born.*
-100. *Caesar (100–44 B.C.) born.*
Wu-ti (140–87 B.C.) opens communications with the West.
-77. Poseidonius. Geometry, cosmography.
-75. *Cicero discovers the tomb of Archimedes.*
-63. *Augustus (63 B.C.–14 A.D.) born.*
-60. Geminus. History of mathematics.
P. Nigidius Figulus. Astronomy.
Marcus Terentius Varro. Mensuration.
-51. *Caesar completes the subjugation of Gaul.*
-50. Dionysodorus. Geometry.
-46. Cæsar, assisted by Sosigenes, reforms the calendar.
-40. Cleomedes. Astronomy, arithmetic.
-20. Marcus Vitruvius Pollio. Applied mathematics.
-8. Diodorus of Sicily. History of mathematics.
-4. *Probable date of the birth of Jesus.*
1. *Rag paper known in China.*
10. Strabo. In his geography considerable history of mathematics.

25. Columella. Surveying.
50. Heron of Alexandria. Geodesy, mathematics. (Possibly *c.* 200.)
Serenus of Antinoopolis. Cylindric sections.
Sun-tzï wrote the Wu-ts'ao Suan-king.
64. *Buddhism introduced into China.*
66. Liu Hsing devises a new Chinese calendar.
75. Pliny the Elder. His Natural History is valuable for the study of Roman numerals.
Pan Ku. Bamboo rods in use in Chinese computation.
100. Nicomachus. Theory of numbers.
Menelaus. Spherics, anharmonic ratio.
Ch'ang ch'un-ch'ing. Commentary on the Chóu-peï.
Theodosius. Geometry, astronomy.
Balbus. Surveying.
Frontinus. Surveying.
120. Hyginus. Surveying.
125. Theon of Smyrna. Theory of numbers. History of Pythagoras.
Ch'ang Höng. Astronomy, geometry, $\pi = \sqrt{10}$.
150. Claudius Ptolemaeus (Ptolemy). Astronomy, trigonometry, geodesy.
Marinus of Tyre. Geodesy, geography.
166. *Marcus Aurelius sends an embassy to China.*
180. Nipsus. Surveying.
190. Ts'ai Yung. Chinese calendar.
200. Epaphroditus. Surveying, theory of numbers.
Domitius Ulpianus. Mortality table.
Quintus Sammonicus Serenus. General mathematics.
220. Sextus Julius Africanus. Encyclopedia with some history of mathematics.
235. Censorinus. Astronomy.
245. Wang Pi. On the I-king.
250. Siu Yo. Arithmetic.
Hsü Yüeh. Commentary on Siu Yo.
263. Liu Hui. Wrote the Hai-tau Suan-king.
265. Wang Fan. Astronomy; $\pi = \frac{142}{45}$.
275. ***Diophantus. Algebra, theory of numbers.***
Sporus of Nicæa. History of mathematics.
280. Anatolius. Astronomy.
Porphyrius. Life of Pythagoras.
289. Liu Chih. Possibly the one who gave $\pi = 3.125$.
300. Pappus. Geometry.
323. *Constantine becomes sole emperor.*
324. *Constantinople founded.*
325. Iamblichus. Theory of numbers.

325. Metrodorus. (Possibly, but see 500.)
340. Julius Firmicus Maternus. Astrology.
372. *Buddhism introduced into Korea.*
379. *Theodosius the Great, emperor of the East.*
390. Theon of Alexandria. Geometry.
400. Fa-hién, Chinese Buddhist in India. Hindu mathematics becomes known in China.
Sūrya Siddhānta written in the 4th or 5th century.
410. Hypatia of Alexandria. Geometry and astronomy.
Synesius. Astrolabe.
Sack of Rome by Alaric.
425. Wang Jong. Arithmetic.
Tun Ch'üan wrote the San-töng-shu.
433. *Attila, king of the Huns.*
440. P'i Yen-tsung. Circle measure.
450. Ho Ch'êng-t'ien. Astronomy.
Wu, a Chinese geometer. $\pi = 3.1432+$.
Domninus. Theory of numbers.
Victorius. Computus.
455. *Sack of Rome by Genseric.*
460. **Proclus. Geometry.**
Capella. Encyclopedia.
470. Tsu Ch'ung-chih. $\pi = \frac{355}{113}$.
485. Marinus of Flavia Neapolis. On Proclus.
500. Metrodorus. Arithmetical epigrams in the Greek Anthology.
505. Varāhamihira. Hindu astronomy.
510. **Boethius. Geometry, theory of numbers.**
Āryabhaṭa the Elder. General mathematics. $\pi = 3.1416$.
Damascius. Geometry.
518. *Hui-sing, Chinese Buddhist, visits India.*
520. Cassiodorus. Computus, encyclopedia.
522. *Buddhism introduced into Japan (522–552).*
525. Dionysius Exiguus. Christian calendar.
Anthemius. Architecture, conics.
527. *Justinian's reign begins.*
529. *St. Benedict founds the monastery at Monte Cassino.*
535. Ch'ön Luan wrote the Wu-king Suan-shu.
550. Hsia-hou Yang wrote his Suan-king.
Codex Arcerianus written probably about the 6th century. Surveying.
554. Korean scholars introduce Chinese mathematics into Japan.
560. Eutocius. History of geometry.
575. Ch'ang K'iu-kien. Arithmetic.
Men. $\pi = 3.14$.

600. Prince Shōtoku Taishi. Arithmetic.
602. Korean priests bring works on the calendar to Japan.
610. Stephen of Alexandria. Astronomy and general mathematics.
Isidorus. Encyclopedia.
615. *Arab ambassadors visit China.*
622. *Flight of Mohammed from Mecca. Beginning of the Hegira era.*
625. Wang Hs'iao-t'ung. Numerical cubic equations.
628. Brahmagupta. Geometry, algebra.
629. *Hüan-tsang goes to India. Translates Hindu works.*
635. Asclepias of Tralles. On Nicomachus.
636. *Priest from Rome visits China.*
640. Joannes Philoponus. Astrolabe, on Nicomachus.
642. *Library of Alexandria burned.*
650. Sebokht. Hindu numerals.
670. *Emperor Tenchi (Tenji) reigns, 668–672.* Observatory established. Arithmetic.
710. Bede. Calendar, finger reckoning.
711. *Saracens invade Spain.*
713. *Arab ambassadors visit China and foreign ships sail to Canton.*
727. I-hsing. Chinese calendar, indeterminate equations.
732. *Battle of Tours. Charles Martel defeats the Saracens.*
750. Akhmim Papyrus written *c.* 7th or 8th century.
756. *Cordova made the seat of the western caliphate.*
762. *Bagdad founded by al-Mansûr, c. 762–763.*
766. The Sindhind translated into Arabic. Hindu numerals.
770. Geber. Alchemy, astrolabe.
771. *Charlemagne's sole reign begins (771–814).*
775. Alcuin called to the court of Charlemagne. Mathematical problems.
Ya'qûb ibn Ṭâriq. The sphere.
Abû Yaḥyâ. Translated Ptolemy.
Kia Tan. Geographer.
790. *Harun al-Rashid (reigned 786–808/9).* Patron of mathematics.
Al-Fazârî. Mathematical instruments.
800. Jacob ben Nissim. Theory of numbers.
Messahala. The astrolabe.
Al-Tabarî. Astronomy.
Harun al-Rashid sends an embassy to China.
820. Mohammed ibn Mûsâ al-Khowârizmî. Algebra.
Hrabanus Maurus. Computus.
Al-Nehâvendî. Astronomy.
Al-Hajjâj. Greek mathematics.
Al-Mâmûn (reigned 809–833). Patron of mathematics.

830. Al-'Abbâs. Greek mathematics.
Al-Astorlâbî. The astrolabe.

840. **Honein ibn Isḥâq. Greek mathematics.**
Walafried Strabus (Strabo). Teacher.

850. **Mahāvīra. Arithmetic, algebra, mensuration.**
Sahl ibn Bishr. Astronomy, arithmetic, algebra.
Al-Arjânî. Greek mathematics.
Abû'l-Ṯaiyib. Trigonometry.

860. **Alchindi. Astronomy, optics, proportion.**
Almâhânî. Trigonometry, cubic equation.
Al-Mervazî. Astronomy.

870. **Ṯâbit ibn Qorra. Conics. Greek mathematics.**
The Three Brothers. Geometry, astronomy.

871. *Alfred the Great begins his reign.*

880. Al-Ḥimṣî. Greek mathematics.
Albumasar. Astronomy.

890. Aḥmed ibn al-Taiyib. Algebra.
Aḥmed ibn Dâ'ûd. Algebra.
Tenjin. Japanese patron of mathematics.

900. **Abû Kâmil. Geometry, algebra.**
Isḥâq ibn Ḥonein ibn Isḥâq. Greek mathematics.
Remigius of Auxerre. On Capella.
Muslim ibn Aḥmed al-Leiṯî. Arithmetic.
Al-Qass. On Euclid.
Qosṭâ ibn Lûqâ. On Diophantus.
Al-Miṣrî. Geometry.

910. **Al-Nairîzî. Geometry.**
Al-Faraḍî. Arithmetic.

915. Sa'id ibn Ya'qûb. Greek mathematics.

920. **Rhases. Geometry.**
Albategnius. Astronomy.
Odo of Cluny (879–*c.* 942). Abacus.

925. Al-Ḥasan ibn 'Obeidallâh. On Euclid.
Aethelstan's reign begins in England. Learning fostered.

940. Al-Fârrâbî. On Euclid and Ptolemy.

950. Hasan. (Date very doubtful.) The calendar.
Bakhshālī manuscript. Algebra. (Date very doubtful).

960. Abû Ja'far al-Khâzin. Geometry.

970. Hrotsvitha, a nun. Number theory.

975. Al-Ḥarrânî. On Euclid.

980. **Abû'l-Wefâ. Trigonometry.**
Abbo of Fleury. Computus.

987. **Abû'l-Faradsh. The Fihrist.**

993. Bernward. Theory of numbers.
Al-Masîḥî. On Ptolemy.
1000. Mohammed ibn al-Leiṯ. Geometry.
Al-Majrîṭî. Theory of numbers.
Ḥâmid ibn al-Khiḍr. Astrolabe, algebra.
Al-Ḥasan (al-Ḥaiṯam) of Baṣra. Algebra, geometry.
Manṣûr ibn 'Alî. Trigonometry.
Gerbert (Sylvester II). Arithmetic.
Byrhtferth. Calendar.
Ibn Yûnis. Astronomy.
Avicenna. Geometry, arithmetic.
Alberûnî. On Hindu mathematics.
1020. **Al-Karkhî. Algebra.**
Bernelinus. Arithmetic.
Śrīdhara. Arithmetic.
Firdusi, Persian poet, dies.
1025. **Al-Nasavî. Greek mathematics.**
Ibn al-Ṣaffâr. Astronomical tables.
1028. Guido of Arezzo (Aretinus). Arithmetic.
1042. *Edward the Confessor becomes king.*
1050. **Hermannus Contractus. Arithmetic, astrolabe.**
Ch'ön Huo. Astronomy.
Ibn al-Zarqâla. Astronomy.
Wilhelm of Hirschau. Teacher.
1066. *Norman Conquest.*
1075. Psellus. Quadrivium.
Franco of Liège. Arithmetic, geometry.
1076. *Turks capture Jerusalem.*
1077. Benedictus Accolytus. Rithmomachia.
1083. China prints Liu Hui's classic. Block book.
1084. China prints Ch'ang K'iu-kien's arithmetic.
1090. Fortolfus. Rithmomachia.
1095. *First Crusade proclaimed.*
1100. **Savasorda. Geometry.**
Omar Khayyam, the poet. Algebra, astronomy.
Abû'l-Ṣalt. Geometry.
Walcherus. Geometry, arithmetic, astronomy.
1115. China prints the Huang-ti K'iu-ch'ang.
1120. **Plato of Tivoli. Translates from the Arabic.**
Adelard of Bath. Translates from the Arabic.
1125. Radulph of Laon. Arithmetic.
1130. Jabir ibn Aflah. Trigonometry.
1137. Gerland of Besançon. Computus.

1140. **Abraham ben Ezra (Rabbi ben Ezra). Theory of numbers, magic squares, calendar.**
Avenpace. Geometry.
Johannes Hispalensis. Translates from the Arabic.
Robert of Chester. Translates from the Arabic.
1144. Rudolph of Bruges. Translates Ptolemy.
1146. *Second Crusade proclaimed.*
1150. **Gherardo of Cremona. Translates from the Arabic.**
Bhāskara. Algebra.
Fujiwara Michinori. Mensuration.
Gherardo da Sabbionetta. Translates from the Arabic.
N. O'Creat. Arithmetic.
1175. **Averroës. Astronomy, trigonometry.**
Maimonides. Astronomy.
Samuel ben Abbas. Arithmetic.
Al-Ḥaṣṣâr. Arithmetic.
1180. Ts'ai Yüan-ting. On the I-king.
1200. Ta'âsîf. On Euclid.
Ibn Yûnis. Conics.
Ibn al-Yâsimîn. Algebra.
Al-Râzî. Geometry.
Daniel Morley. Translates from the Arabic.
Alpetragius. Astronomy.
Ibn al-Kâtib. Geometry.
Al-Ṭûsî. Geometry, algebra.
1202. **Leonardo Fibonacci. Algebra, arithmetic, geometry.**
1220. *Chinghiz Khan's great expedition to Europe.*
1225. **Jordanus Nemorarius. Algebra.**
Michael Scott. Translates from Greek and Arabic.
Genshō. Arithmetic.
1230. Ye-lü Ch'u-ts'ai. Astronomy.
Barlaam. Algebra, Euclid.
1240. Jehuda ben Salomon Kohen. On Ptolemy.
Alexandre de Villedieu. Arithmetic.
Robert Greathead. Geometry, computus.
John of Basingstoke. Translates from the Greek.
1250. **Sacrobosco. Numerals and the sphere.**
Naṣîr ed-dîn. Trigonometry.
Roger Bacon. Astronomy, general mathematics.
Ch'in Kiu-shao. Higher numerical equations.
Liu Ju-hsieh. Algebra.
William of Moerbecke. Translates from the Greek.
Li Yeh. General mathematics.

1250. Isaac ben Sid. Astronomical tables.
Albertus Magnus. Astronomy, physics.
Vincent de Beauvais. Quadrivium.
Guglielmo de Lunis. (Period very uncertain.) Translates algebra from the Arabic.
Prophatius. Translates Euclid and Menelaus.
Rise of European universities.
Europeans visit the Mongol court.

1260. **Campanus. Translates Euclid.**
Ibn al-Lubûdî. Algebra, Euclid.
Kublai Khan's reign begins.

1261. Yang Hui. On the "Nine Sections."

1265. Petrus de Maharncuria. Magnet, general mathematics.

1270. Bar Hebræus. On Euclid.
Witelo. Perspective.

1271. *Marco Polo begins his travels.*

1275. Oldest algorism in French.
Liu I. General mathematics.
Arnaldo de Villa Nova. Computus.
Alfonso X. Astronomical tables.
Cimabue, the Italian painter.
Guelphs and Ghibellines in Florence (1183–1295).

1280. John Peckham (*c.* 1230–1295). Perspective.

1286. Friar Odoric goes to Canton.
Western learning.

1290. Kóu Shóu-king. Mensuration.

1297. Bartolomeo da Parma. Geometry, astronomy.

1299. **Chu Shï-kié. Algebra.**

1300. **Albanna. Algebra, proportion.**
Pachymeres. General mathematics.
Cecco d'Ascoli. On Sacrobosco.
Hauk Erlendssön. Algorism.
Pietro d'Abano. Astrolabe.
Andalò di Negro. Arithmetic, astronomy.

1315. *Dante (1265–1321).*

1320. John Manduith. Trigonometry.
Kalonymos ben Kalonymos. On Nicomachus.

1325. Petrus de Dacia. Geometry.
Thomas Bradwardine. Geometry, arithmetic.
Walter Burley. Greek mathematician.

1330. Joannes Pediasimus. Geometry.
Levi ben Gerson. Arithmetic.
Isaac ben Joseph Israeli. Geometry.

1330. Richard of Wallingford. Trigonometry.
1340. Maximus Planudes. On Diophantus, arithmetic.
Johannes de Lineriis. Alfonsine tables, arithmetic.
Paolo Dagomari. Arithmetic.
Master Sven. The sphere.
1341. Nicholas Rhabdas. Arithmetic, finger symbols.
1345. Richard Suiceth. Coordinates.
1346. *Battle of Crécy. Attack on feudalism.*
1349. *Black Death destroys large per cent of European population.*
1350. Joannes de Muris. Arithmetic, calendar.
Chunrad von Megenberg. The sphere.
Ibn al-Shâṭir. Trigonometry.
Boccaccio (1313–1375).
Petrarch (1304–1374).
1360. Nicole Oresme. Exponents, proportion, coordinates.
Walter Bryte. Arithmetic.
Jacob Poël. Astronomy.
Imanuel ben Jacob. Astrolabe.
1365. Heinrich von Hessen. Geometry.
Albert of Saxony. Geometry.
1369. *Tamerlane's reign begins.*
1375. Simon Bredon. Geometry.
Jacob Carsono. Astronomy.
1380. Rafaele Canacci. Algebra.
Joseph ben Wakkar. Astronomy.
1383. Antonio Biliotti. Arithmetic.
1384. *Wycliffe's English Bible completed.*
1390. Uniform weights and measures in most of England.
1392. Moschopoulus. Magic squares.
1400. Ibn al-Mejdî. Trigonometry.
Matteo, Luca, and Giovanni da Firenze. Arithmetic.
Petrus de Alliaco. Computus.
Conrad von Jungingen. Geometry.
Biagio da Parma. Perspective.
1410. Prosdocimo de' Beldamandi. Algorism, geometry.
1420. *Era of the Medici in Florence.*
1424. Rollandus. Theory of numbers, algebra.
1425. Leonardo of Cremona. Trigonometry.
1430. Johann von Gmünden. Trigonometry.
Jacob Caphanton. Arithmetic.
1431. *Joan of Arc burned.*

1435. Ulugh Beg. Astronomy.
John Killingworth. Algorism, astronomy.
1440. *Donatello the artist, Florence (1386–1466).*
Al-Kashî. Geometry, arithmetic, astronomy.
1449. Jacob of Cremona. Translates Archimedes.
1450. **Nicholas Cusa. Geometry, theory of numbers.**
Jehuda Verga. Arithmetic.
George of Trebizond. Translates Ptolemy.
Printing from movable type.
1453. *Fall of Constantinople.*
1460. **Georg von Peurbach. Trigonometry.**
Benedetto da Firenze. Arithmetic.
1462. *Sack of Nassau affects printing.*
1469. *Lorenzo the Magnificent, Florence.*
1470. **Regiomontanus. Trigonometry.**
1475. **Al-Qalasâdî. Theory of numbers.**
Pietro Franceschi. Regular solids.
Georgius Valla. Geometry, arithmetic.
1478. First printed arithmetic, Treviso, Italy.
1481. Giorgio Chiarino. Commercial arithmetic.
1482. First printed edition of Euclid, Venice.
First printed German arithmetic, Bamberg.
1484. **Nicolas Chuquet. Algebra.**
Piero Borghi. Arithmetic.
1490. **Johann Widman. Algebra, arithmetic.**
1491. Calandri. Arithmetic.
1492. Pellos. Arithmetic.
Lanfreducci. Arithmetic.
Columbus discovers America.
1493. *Maximilian I, emperor of Germany.*
1494. **Pacioli. General mathematics.**
1500. **Leonardo da Vinci. Optics, geometry.**
Jacques le Fèvre d'Estaples. Geometry, arithmetic.
Georgius de Hungaria. Arithmetic.
Charles de Bouelles. Geometry, theory of numbers.
Johann Stöffler. Astronomical tables.
Elia Misrachi. Arithmetic.
Clichtoveus. On Boethius.
1503. Gregorius Reisch. Encyclopedia.
1505. Ciruelo. Arithmetic.
1506. Scipione del Ferro. Cubic equation.
Antonio Maria Fior. Cubic equation.

1510. **Albrecht Dürer. Geometry of curves.**
Raphael (1483–1520).
1512. Juan de Ortega. Geometry, arithmetic.
1513. Blasius (Sileceus). Arithmetic.
1514. Böschensteyn. Arithmetic.
1515. The Taglientes. Commercial arithmetic.
Gaspar Lax. Proportion, arithmetic.
Giel Vander Hoecke. Arithmetic.
1518. **Adam Riese. Arithmetic.**
1520. **Jakob Köbel. Arithmetic.**
Copernicus. Astronomy, trigonometry.
Feliciano da Lazesio. Arithmetic.
Estienne de la Roche. Arithmetic.
Ghaligai. Arithmetic.
1521. *Luther excommunicated.*
1522. **Tonstall. First arithmetic printed in England.**
1525. **Stifel. Algebra, arithmetic.**
Rudolff. Algebra, decimals.
Buteo. Algebra, geometry, arithmetic.
Erasmus. Thinker.
Oronce Fine. Geometry.
1527. Apianus. Pascal Triangle in print, astronomy, arithmetic.
1530. Zuanne de Tonini da Coi. Cubic equation.
Ringelbergius. Geometry, arithmetic.
Francesco dal Sole. Arithmetic.
Schoner. Arithmetic.
1534. Sfortunati. Arithmetic.
Claude de Boissière. Rithmomachia, arithmetic.
Jesuit order founded by Loyola.
1535. Jean Fernel. Proportion, astronomy.
Grammateus. Algebra, arithmetic.
Suryadaśa. Hindu algebra.
Ganeśa. Hindu algebra.
Giovanni Mariani. Arithmetic.
Glareanus. Geometry, arithmetic.
1536. *Calvin goes to Geneva.*
Tyndale is burned.
1540. **Gemma Frisius. Arithmetic.**
Melanchthon. General mathematics.
Camerarius. On Nicomachus.
1541. *De Soto discovers the Mississippi.*
1542. **Robert Recorde. Algebra, geometry, arithmetic.**
1543. Copernican system published.

1545. **Ferrari. Biquadratic equation.**
Tartaglia. Cubic equation, general mathematics.
Cardan. Cubic equation, general mathematics.
1550. **Rhæticus. Trigonometry.**
Maurolico. Geometry.
Johann Scheubel. Algebra.
Commandino. Greek mathematics.
Cosimo Bartoli. Geometry.
T'ang Shun-ki. On the circle.
K'u Ying-hsiang. Algebra, geometry.
Simon Jacob. Arithmetic.
Ramus. Geometry, optics, arithmetic.
François de Foix-Candale. On Euclid.
Jacobus Micyllus. Arithmetic.
Nunes (Nonius, Núñez). Algebra, geometry, navigation.
Mohammed ibn Ma'rûf. Algebra, spherics, arithmetic.
Titian (1477–1576) and Michelangelo (1475–1564).
1558. *Elizabeth becomes queen of England (died 1603).*
1560. Peletier. Algebra, arithmetic.
1562. Juan Perez de Moya. Algebra, arithmetic.
1565. Trenchant. Arithmetic.
Richard de Benese. Surveying.
1566. Jerónimo Muñoz. Euclid, arithmetic.
1568. Humphrey Baker. Arithmetic.
1570. Billingsley and Dee. First English translation of Euclid.
Menher de Kempten. Arithmetic.
Neander. Metrology, spherics.
Xylander. On Diophantus.
Forcadel. Greek mathematics.
Benedetti. Theory of numbers.
Belli. Geometry.
Dasypodius. Euclid, lexicon.
1572. **Bombelli. Algebra.**
Digges, father (died 1571) and son (died 1595). Arithmetic, geometry.
1573. Otto. $\pi = \frac{355}{113}$ (old Chinese value).
1577. Herbestus. Polish arithmetic.
Gir̊jka Go̊rla z Go̊rlssteyna. Arithmetic.
1580. **François Viète. Algebra.**
Ludolf van Ceulen. On π.
Francesco Barozzi. On Proclus.
1583. **Clavius. Geometry, algebra, arithmetic, the calendar.**
Petrus Bongus. Mystery of numbers.
1587. Fyzi. Persian translation of the Lilāvati.

1590. **Cataldi. Continued fractions.**
Stevin. Decimal fractions.
Hsin Yun-lu. Calendar.
Van der Schuere. Arithmetic.
Thomas Masterson. Algebra, arithmetic.
1592. Mōri Kambei Shigeyoshi. Abacus.
1593. Adriaen van Roomen. Value of π.
Ch'êng Tai-wei. Arithmetic.
1594. Thomas Blundeville. Trigonometry, cosmography.
1595. **Pitiscus. Trigonometry.**
Magini. Geometry, astronomy, trigonometry.
1600. **Thomas Harriot. Algebra, analytic geometry.**
Jobst Bürgi. Logarithms.
Galileo. Geometry, astronomy, mechanics.
Behâ Eddîn. The sphere, arithmetic.
Ghetaldi. Geometry, algebra.
Bernardino Baldi. History of mathematics.
Shakespeare (1564–1616).
1603. **Matteo Ricci, Hsü Kuang-ching, and Li Chi Ts'ao translate Euclid into Chinese.**
James I proclaimed king of Great Britain.
1608. *Telescope invented.*
1610. **Kepler. Astronomy, geometry.**
1612. **Bachet de Méziriac. On Diophantus, recreations.**
1614. ***Napier. Logarithms.***
1615. **Henry Briggs. Logarithms.**
1618. Nicolò Longobardi and Giacomo Rho. European astronomy in China.
1619. Savilian professorships (Oxford) founded.
1620. Gunter. Logarithms.
Paul Guldin. Geometry.
Faulhaber. Series.
Snell. Geometry, trigonometry.
Ursinus. Trigonometry, logarithms.
Daniel Schwenter. Recreations.
Francis Bacon, Novum Organum published.
1621. Raganātha. Hindu mathematics.
1630. **Mersenne. Greek mathematics, theory of numbers, geometry.**
Oughtred. Algebra, slide rule, logarithms.
Mydorge. Geometry, recreations.
Gellibrand. Logarithms.
Albert Girard. Algebra, trigonometry.
Denis Henrion. Logarithms.
Claude Richard. Greek mathematics.

1634. Hérigone. Algebra.

1635. ***Fermat. Analytic geometry, theory of numbers.***

Cavalieri. Indivisibles.

Yoshida Shichibei. General mathematics.

Richelieu founds the French Academy.

1637. ***Descartes. Analytic geometry.***

1639. Imamura Chishō. Geometry.

1640. **Desargues. Projective geometry.**

Florimond de Beaune. Cartesian geometry.

Torricelli. Geometry, physics.

Borelli. Greek mathematics.

Bernard Frénicle de Bessy. Geometry.

Antoine de la Loubère. Curves.

Roberval. Geometry.

Velasquez (1599–1660).

1649. *Charles I executed. England declared a commonwealth.*

1650. ***Pascal. Geometry, probability, theory of numbers.***

John Wallis. Algebra, series, history of mathematics.

Frans van Schooten. Edited Descartes and Viète.

Grégoire de Saint-Vincent. Geometry.

John Kersey. Algebra.

Wingate. Arithmetic.

Nicolaus Mercator. Trigonometry, logarithms.

John Pell. Algebra.

Athanasius Kircher. Instruments.

Smogolenski. Logarithms in China.

Sié Fong-tsu. Logarithms in China.

Milton and Hobbes.

1654. *Louis XIV crowned.*

1659. Ferdinand Verbiest. Astronomy in China.

1660. René François Walter de Sluze. The calculus, geometry.

Isomura (Iwamura) Kittoku. Problems.

Viviani. Geometry.

Dechales. On Euclid.

Brouncker. Series.

1663. Lucasian professorship (Cambridge) founded.

Muramatsu Kudayū Mosei. Geometry.

1665. Nozawa Teichō, Satō Seikō, and Sawaguchi Kazuyuki. Geometry and the native Japanese integration.

Neile. Geometry.

1670. **Barrow. Geometry.**

James Gregory. Series.

William Leybourn. Surveying.

1670. **Huygens. Geometry, physics, astronomy.**
Edward Cocker. Arithmetic.
Sir Christopher Wren. Geometry, astronomy, architecture.
John Locke (1632–1704) and Spinoza (1632–1677).
1671. Giovanni Domenico Cassini. Astronomy.
1675. Greenwich observatory founded.
Mei Wen-ting. Algebra, history of Chinese mathematics.
John Bunyan.
1680. **Seki Kōwa. The calculus.**
Sir Isaac Newton. Fluxional calculus, physics, astronomy, entire field of mathematics.
Johann Hudde. Algebra.
Barrême. Arithmetic.
1681. *John Dryden.*
Pennsylvania granted to William Penn.
1682. ***Leibniz. The calculus.***
1690. **Marquis de l'Hospital. Applied calculus.**
Halley. Astronomy, life insurance, physics.
Jacques Bernoulli. Applied calculus, geometry, probability.
De Lahire. Geometry.
John Caswell. Trigonometry.
Tschirnhausen. Optics.
1696. Nakane Genkei. Japanese calculus.
1700. **Jean Bernoulli. Applied calculus.**
Michel Rolle. Equations.
Pierre Nicolas. Geometry.
Giovanni and Tommaso Ceva. Geometry.
Fatio de Duillier. Geometry.
Varignon. The calculus.
David Gregory. Optics, geometry.
Peter the Great (died 1725).
1702. *Anne becomes queen of England.*
1704. Charles Hayes. The calculus in English.
1710. **Roger Cotes. Geometry, analysis, the calculus.**
De Montmort. Probability, series.
Pierre Jartoux. Astronomy and analysis in China.
Humphrey Ditton. The calculus.
Saurin. Geometry.
De Lagny. Analysis.
Parent. Solid analytic geometry.
1715. Miyake Kenryū and Nakane Genjun. Problems.
Raphson. History of fluxions.

1720. Brook Taylor. Series.
De Moivre. Complex numbers, probability.
Nicolas (II) Bernoulli. Geometry.
Manfredi brothers. Geometry.
Christian von Wolf. General mathematics.
Pescheck. Textbooks.
Crousaz. Geometry.
Jacob Hermann. The calculus.
Fagnano. Curves, elliptic functions.
Guido Grandi. Geometry.
1722. Takebe. Geometry, π to 41 figures.
1730. Nicolas (I) Bernoulli. Differential equations, probability.
Saunderson. Algebra.
Van s'Gravesande. The calculus.
Nicole. Finite differences.
Maupertuis. Geodesy.
Matsunaga. Geometry, π to 50 figures.
1736. James Hodgson. The calculus in English.
1740. Colin Maclaurin. Algebra, series, conics.
Gabriel Cramer. Determinants, equations, curves.
George Berkeley. Attacks fluxional calculus.
Gua de Malves. Analytic geometry.
Frézier. Descriptive geometry.
Frederick the Great (Frederick II) becomes king of Prussia.
1745. Voltaire. On Newton.
Marquise du Châtelet. On Newton.
1750. *Léonard Euler. Analysis, physics, astronomy.*
Montucla. History of Mathematics.
James Stirling. Geometry, series.
Robert Simson. Geometry.
Matthew Stewart. Geometry.
Jean (II) Bernoulli. Physics.
Riccati family. Differential equations.
Boscovich. Geometry, astronomy.
Daniel (I) Bernoulli. Physics.
Thomas Simpson. Algebra, geometry, the calculus.
1751. John Rowe. The calculus in English.
1760. D'Alembert. Differential equations, astronomy, physics.
John Landen. Elliptic integrals.
Alexis Claude Clairaut. Geometry, geodesy.
Seven Years' War (1756–1763). Lessing (died 1781), Burke (died 1797), Rousseau (died 1778), Voltaire (died 1778).

1765. Murai Chūzen. Equations.
1770. **Lambert. Hyperbolic trigonometry.**
Malfatti. Geometry.
Maria Gaetana Agnesi. Geometry.
Kästner. History of mathematics.
1775. Vandermonde. Algebra.
Bézout. Algebra.
1776. Pestalozzi. Arithmetic.
United States independence. Washington (died 1799), Jefferson (died 1826), Lafayette (died 1834).
1780. ***Lagrange. Theory of numbers, analysis, elliptic functions, astronomy.***
Condorcet. Analysis, probability.
Ajima Chokuyen. Indeterminate equations.
Aida Ammei. Series.
Fujita Sadasuke. Algebra.
Charles Hutton. Tables, dictionary, recreations.
John Wilson and Edward Waring. Theory of numbers.
Méchain. Metric system.
1789. *French Revolution.*
1790. Meusnier. Surfaces.
1795. École normale supérieure and École polytechnique founded about this time.
1800. ***Gauss. Theory of numbers, geometry, analysis, physics, astronomy, general field of mathematics.***
Laplace. Astronomy, physics, least squares.
Legendre. Elliptic functions, theory of numbers, geometry.
Carnot. Modern geometry.
Monge. Descriptive geometry.
Delambre. Astronomy, geodesy.
Lacroix. Analysis.
Mascheroni. Geometry of the compasses.
Pfaff. Astronomy, analysis.
Jean (III) Bernoulli. Probability.
Lhuilier. Geometry.
Ruffini. Algebra.
Bossut, Cossali, and Franchini. History of mathematics.
Trembley. The calculus.
James Ivory. Analytic methods.
Arbogast. Differential equations, variations, series.
1804. *Napoleon made emperor. Battle of Waterloo fought in 1815.*
1810. Hachette. Algebra, geometry.
Jean Robert Argand. Complex numbers.

1810. Fourier. Series, physics.
William Wallace. Hyperbolic functions.
Gergonne. Editor of Annales.
Woodhouse. Differential calculus.
Robert Adrain. Least squares.
1819. Horner. Numerical equations.
1820. Peter Barlow. Tables.
Poinsot. Geometry.
Sophie Germain. Elastic surfaces.
Bolzano. Series.
Poisson. Definite integrals, series, physics.
Crelle. Tables, editor of the Journal.
Brianchon. Geometry.
1825. Abel. Elliptic functions.
The Bolyais and Lobachevsky. Non-Euclidean geometry.
Nathaniel Bowditch. Celestial mechanics.
1830. Babbage. Calculating machine.
George Peacock. Differential calculus, algebra.
Möbius. Geometry.
Carl Gustav Jacob Jacobi. Elliptic functions.
Poncelet. Projective geometry.
Galois. Groups.
Cauchy. Functions, determinants, series.
Dupin. Geometry.
1840. Lamé. Elasticity, surfaces.
Jacob Steiner. Geometry.
Olivier. Descriptive geometry.
Arneth. History of mathematics.
J. F. W. Herschel. Astronomy, analysis.
MacCullagh. Surfaces.
1850. William Rowan Hamilton. Quaternions.
Chasles. Modern geometry.
Salmon. Geometry, algebra.
Biot. Physics, astronomy.
Grunert. Editor of the Archiv.
August. Mathematical physics.
De Morgan. History of mathematics, logic.
George Boole. Logic, differential equations.
Sylvester. Algebra.
Cayley. Invariants.
H. J. S. Smith. Theory of numbers.
Todhunter. History of mathematics, textbooks.
Kirkman. Analysis situs.

1850. Airy. Lunar theory.
Adams and Leverrier. Discover Neptune.
Liouville. Editor of the Journal.
Grassmann. Ausdehnungslehre.
Kummer. Series, surfaces.
Riemann. Surfaces, elliptic functions.
Eisenstein. Invariants.
Bellavitis. Geometry.
Gudermann. Hyperbolic functions.
Von Staudt. Geometry.
Plücker. Geometry.
Lejeune-Dirichlet. Theory of numbers.
Quételet. Statistics, geometry, history of mathematics.
Wronski. Philosophy of mathematics.
Benjamin Peirce. Algebra.
Steinschneider. History of mathematics.
Libri. History of mathematics.

INDEX

Since important names are often mentioned many times in the text, only such page references have been given as are likely to be of considerable value to the reader, the first reference after a proper name being to the biographical note in case one is given. Bibliographical references, in general, give only the page on which a book or essay is first mentioned, and on this page the full title appears, together with the abridged form subsequently used. In arranging the words and names alphabetically, ä, ö, and ü are taken as if written ae, oe, and ue respectively. The prefix Mc is indexed under Mac.

CATALOG OF DOVER BOOKS

BOOKS EXPLAINING SCIENCE AND MATHEMATICS

THE COMMON SENSE OF THE EXACT SCIENCES, W. K. Clifford. Introduction by James Newman, edited by Karl Pearson. For 70 years this has been a guide to classical scientific and mathematical thought. Explains with unusual clarity basic concepts, such as extension of meaning of symbols, characteristics of surface boundaries, properties of plane figures, vectors, Cartesian method of determining position, etc. Long preface by Bertrand Russell. Bibliography of Clifford. Corrected, 130 diagrams redrawn. 249pp. 5⅜ x 8.
T61 Paperbound **$1.60**

SCIENCE THEORY AND MAN, Erwin Schrödinger. This is a complete and unabridged reissue of SCIENCE AND THE HUMAN TEMPERAMENT plus an additional essay: "What is an Elementary Particle?" Nobel Laureate Schrödinger discusses such topics as nature of scientific method, the nature of science, chance and determinism, science and society, conceptual models for physical entities, elementary particles and wave mechanics. Presentation is popular and may be followed by most people with little or no scientific training. "Fine practical preparation for a time when laws of nature, human institutions . . . are undergoing a critical examination without parallel," Waldemar Kaempffert, N. Y. TIMES. 192pp. 5⅜ x 8.
T428 Paperbound **$1.35**

PIONEERS OF SCIENCE, O. Lodge. Eminent scientist-expositor's authoritative, yet elementary survey of great scientific theories. Concentrating on individuals—Copernicus, Brahe, Kepler, Galileo, Descartes, Newton, Laplace, Herschel, Lord Kelvin, and other scientists—the author presents their discoveries in historical order adding biographical material on each man and full, specific explanations of their achievements. The clear and complete treatment of the post-Newtonian astronomers is a feature seldom found in other books on the subject. Index. 120 illustrations. xv + 404pp. 5⅜ x 8. T716 Paperbound **$1.50**

THE EVOLUTION OF SCIENTIFIC THOUGHT FROM NEWTON TO EINSTEIN, A. d'Abro. Einstein's special and general theories of relativity, with their historical implications, are analyzed in non-technical terms. Excellent accounts of the contributions of Newton, Riemann, Weyl, Planck, Eddington, Maxwell, Lorentz and others are treated in terms of space and time, equations of electromagnetics, finiteness of the universe, methodology of science. 21 diagrams. 482pp. 5⅜ x 8. T2 Paperound **$2.00**

THE RISE OF THE NEW PHYSICS, A. d'Abro. A half-million word exposition, formerly titled THE DECLINE OF MECHANISM, for readers not versed in higher mathematics. The only thorough explanation, in everyday language, of the central core of modern mathematical physical theory, treating both classical and modern theoretical physics, and presenting in terms almost anyone can understand the equivalent of 5 years of study of mathematical physics. Scientifically impeccable coverage of mathematical-physical thought from the Newtonian system up through the electronic theories of Dirac and Heisenberg and Fermi's statistics. Combines both history and exposition; provides a broad yet unified and detailed view, with constant comparison of classical and modern views on phenomena and theories. "A must for anyone doing serious study in the physical sciences," JOURNAL OF THE FRANKLIN INSTITUTE. "Extraordinary faculty . . . to explain ideas and theories of theoretical physics in the language of daily life," ISIS. First part of set covers philosophy of science, drawing upon the practice of Newton, Maxwell, Poincaré, Einstein, others, discussing modes of thought, experiment, interpretations of causality, etc. In the second part, 100 pages explain grammar and vocabulary of mathematics, with discussions of functions, groups, series, Fourier series, etc. The remainder is devoted to concrete, detailed coverage of both classical and quantum physics, explaining such topics as analytic mechanics, Hamilton's principle, wave theory of light, electromagnetic waves, groups of transformations, thermodynamics, phase rule, Brownian movement, kinetics, special relativity, Planck's original quantum theory, Bohr's atom, Zeeman effect, Broglie's wave mechanics, Heisenberg's uncertainty, Eigen-values, matrices, scores of other important topics. Discoveries and theories are covered for such men as Alembert, Born, Cantor, Debye, Euler, Foucault, Galois, Gauss, Hadamard, Kelvin, Kepler, Laplace, Maxwell, Pauli, Rayleigh, Volterra, Weyl, Young, more than 180 others. Indexed. 97 illustrations. ix + 982pp. 5⅜ x 8. T3 Volume 1, Paperbound **$2.00**
T4 Volume 2, Paperbound **$2.00**

CONCERNING THE NATURE OF THINGS, Sir William Bragg. Christmas lectures delivered at the Royal Society by Nobel laureate. Why a spinning ball travels in a curved track; how uranium is transmuted to lead, etc. Partial contents: atoms, gases, liquids, crystals, metals, etc. No scientific background needed; wonderful for intelligent child. 32pp. of photos, 57 figures. xii + 232pp. 5⅜ x 8. T31 Paperbound **$1.35**

THE UNIVERSE OF LIGHT, Sir William Bragg. No scientific training needed to read Nobel Prize winner's expansion of his Royal Institute Christmas Lectures. Insight into nature of light, methods and philosophy of science. Explains lenses, reflection, color, resonance, polarization, x-rays, the spectrum, Newton's work with prisms, Huygens' with polarization, Crookes' with cathode ray, etc. Leads into clear statement of 2 major historical theories of light, corpuscle and wave. Dozens of experiments you can do. 199 illus., including 2 full-page color plates. 293pp. 5⅜ x 8. S538 Paperbound **$1.85**

PHYSICS, THE PIONEER SCIENCE, L. W. Taylor. First thorough text to place all important physical phenomena in cultural-historical framework; remains best work of its kind. Exposition of physical laws, theories developed chronologically, with great historical, illustrative experiments diagrammed, described, worked out mathematically. Excellent physics text for self-study as well as class work. Vol. 1: Heat, Sound: motion, acceleration, gravitation, conservation of energy, heat engines, rotation, heat, mechanical energy, etc. 211 illus. 407pp. 5⅜ x 8. Vol. 2: Light, Electricity: images, lenses, prisms, magnetism, Ohm's law, dynamos, telegraph, quantum theory, decline of mechanical view of nature, etc. Bibliography. 13 table appendix. Index. 551 illus. 2 color plates. 508pp. 5⅜ x 8.
Vol. 1 S565 Paperbound **$2.00**
Vol. 2 S566 Paperbound **$2.00**
The set **$4.00**

FROM EUCLID TO EDDINGTON: A STUDY OF THE CONCEPTIONS OF THE EXTERNAL WORLD, Sir Edmund Whittaker. A foremost British scientist traces the development of theories of natural philosophy from the western rediscovery of Euclid to Eddington, Einstein, Dirac, etc. The inadequacy of classical physics is contrasted with present day attempts to understand the physical world through relativity, non-Euclidean geometry, space curvature, wave mechanics, etc. 5 major divisions of examination: Space; Time and Movement; the Concepts of Classical Physics; the Concepts of Quantum Mechanics; the Eddington Universe. 212pp. 5⅜ x 8. T491 Paperbound **$1.35**

THE STORY OF ATOMIC THEORY AND ATOMIC ENERGY, J. G. Feinberg. Wider range of facts on physical theory, cultural implications, than any other similar source. Completely non-technical. Begins with first atomic theory, 600 B.C., goes through A-bomb, developments to 1959. Avogadro, Rutherford, Bohr, Einstein, radioactive decay, binding energy, radiation danger, future benefits of nuclear power, dozens of other topics, told in lively, related, informal manner. Particular stress on European atomic research. "Deserves special mention . . . authoritative," Saturday Review. Formerly "The Atom Story." New chapter to 1959. Index. 34 illustrations. 251pp. 5⅜ x 8. T625 Paperbound **$1.45**

THE STRANGE STORY OF THE QUANTUM, AN ACCOUNT FOR THE GENERAL READER OF THE GROWTH OF IDEAS UNDERLYING OUR PRESENT ATOMIC KNOWLEDGE, B. Hoffmann. Presents lucidly and expertly, with barest amount of mathematics, the problems and theories which led to modern quantum physics. Dr. Hoffmann begins with the closing years of the 19th century, when certain trifling discrepancies were noticed, and with illuminating analogies and examples takes you through the brilliant concepts of Planck, Einstein, Pauli, de Broglie, Bohr, Schroedinger, Heisenberg, Dirac, Sommerfeld, Feynman, etc. This edition includes a new, long postscript carrying the story through 1958. "Of the books attempting an account of the history and contents of our modern atomic physics which have come to my attention, this is the best," H. Margenau, Yale University, in "American Journal of Physics." 32 tables and line illustrations. Index. 275pp. 5⅜ x 8. T518 Paperbound **$1.45**

SPACE AND TIME, Emile Borel. An entirely non-technical introduction to relativity, by world-renowned mathematician, Sorbonne Professor. (Notes on basic mathematics are included separately.) This book has never been surpassed for insight, and extraordinary clarity of thought, as it presents scores of examples, analogies, arguments, illustrations, which explain such topics as: difficulties due to motion; gravitation a force of inertia; geodesic lines; wave-length and difference of phase; x-rays and crystal structure; the special theory of relativity; and much more. Indexes. 4 appendixes. 15 figures. xvi + 243pp. 5⅜ x 8.
T592 Paperbound **$1.45**

THE RESTLESS UNIVERSE, Max Born. New enlarged version of this remarkably readable account by a Nobel laureate. Moving from sub-atomic particles to universe, the author explains in very simple terms the latest theories of wave mechanics. Partial contents: air and its relatives, electrons & ions, waves & particles, electronic structure of the atom, nuclear physics. Nearly 1000 illustrations, including 7 animated sequences. 325pp. 6 x 9.
T412 Paperbound **$2.00**

SOAP SUBBLES, THEIR COLOURS AND THE FORCES WHICH MOULD THEM, C. V. Boys. Only complete edition, half again as much material as any other. Includes Boys' hints on performing his experiments, sources of supply. Dozens of lucid experiments show complexities of liquid films, surface tension, etc. Best treatment ever written. Introduction. 83 illustrations. Color plate. 202pp. 5⅜ x 8. T542 Paperbound **95¢**

SPINNING TOPS AND GYROSCOPIC MOTION, John Perry. Well-known classic of science still unsurpassed for lucid, accurate, delightful exposition. How quasi-rigidity is induced in flexible and fluid bodies by rapid motions; why gyrostat falls, top rises; nature and effect on climatic conditions of earth's precessional movement; effect of internal fluidity on rotating bodies, etc. Appendixes describe practical uses to which gyroscopes have been put in ships, compasses, monorail transportation. 62 figures. 128pp. 5⅜ x 8. T416 Paperbound **$1.00**

MATTER & LIGHT, THE NEW PHYSICS, L. de Broglie. Non-technical papers by a Nobel laureate explain electromagnetic theory, relativity, matter, light and radiation, wave mechanics, quantum physics, philosophy of science. Einstein, Planck, Bohr, others explained so easily that no mathematical training is needed for all but 2 of the 21 chapters. Unabridged. Index. 300pp. 5⅜ x 8. T35 Paperbound **$1.60**

A SURVEY OF PHYSICAL THEORY, Max Planck. One of the greatest scientists of all time, creator of the quantum revolution in physics, writes in non-technical terms of his own discoveries and those of other outstanding creators of modern physics. Planck wrote this book when science had just crossed the threshold of the new physics, and he communicates the excitement felt then as he discusses electromagnetic theories, statistical methods, evolution of the concept of light, a step-by-step description of how he developed his own momentous theory, and many more of the basic ideas behind modern physics. Formerly "A" Survey of Physics." Bibliography. Index. 128pp. 5⅜ x 8. S650 Paperbound **$1.15**

THE NATURE OF LIGHT AND COLOUR IN THE OPEN AIR, M. Minnaert. Why is falling snow sometimes black? What causes mirages, the fata morgana, multiple suns and moons in the sky? How are shadows formed? Prof. Minnaert of the University of Utrecht answers these and similar questions in optics, light, colour, for non-specialists. Particularly valuable to nature, science students, painters, photographers. Translated by H. M. Kremer-Priest, K. Jay. 202 illustrations, including 42 photos. xvi + 362pp. 5⅜ x 8. T196 Paperbound **$1.95**

THE STORY OF X-RAYS FROM RONTGEN TO ISOTOPES, A. R. Bleich. Non-technical history of x-rays, their scientific explanation, their applications in medicine, industry, research, and art, and their effect on the individual and his descendants. Includes amusing early reactions to Röntgen's discovery, cancer therapy, detections of art and stamp forgeries, potential risks to patient and operator, etc. Illustrations show x-rays of flower structure, the gall bladder, gears with hidden defects, etc. Original Dover publication. Glossary. Bibliography. Index. 55 photos and figures. xiv + 186pp. 5⅜ x 8. T662 Paperbound **$1.35**

TEACH YOURSELF ELECTRICITY, C. W. Wilman. Electrical resistance, inductance, capacitance, magnets, chemical effects of current, alternating currents, generators and motors, transformers, rectifiers, much more. 230 questions, answers, worked examples. List of units. 115 illus. 194pp. 6⅞ x 4¼. Clothbound **$2.00**

TEACH YOURSELF HEAT ENGINES, E. De Ville. Measurement of heat, development of steam and internal combustion engines, efficiency of an engine, compression-ignition engines, production of steam, the ideal engine, much more. 318 exercises, answers, worked examples. Tables. 76 illus. 220pp. 6⅞ x 4¼. Clothbound **$2.00**

TEACH YOURSELF MECHANICS, P. Abbott. The lever, centre of gravity, parallelogram of force, friction, acceleration, Newton's laws of motion, machines, specific gravity, gas, liquid pressure, much more. 280 problems, solutions. Tables. 163 illus. 271pp. 6⅞ x 4¼. Clothbound **$2.00**

GREAT IDEAS OF MODERN MATHEMATICS: THEIR NATURE AND USE, Jagjit Singh. Reader with only high school math will understand main mathematical ideas of modern physics, astronomy, genetics, psychology, evolution, etc., better than many who use them as tools, but comprehend little of their basic structure. Author uses his wide knowledge of non-mathematical fields in brilliant exposition of differential equations, matrices, group theory, logic, statistics, problems of mathematical foundations, imaginary numbers, vectors, etc. Original publication. 2 appendixes. 2 indexes. 65 illustr. 322pp. 5⅜ x 8. S587 Paperbound **$1.55**

MATHEMATICS IN ACTION, O. G. Sutton. Everyone with a command of high school algebra will find this book one of the finest possible introductions to the application of mathematics to physical theory. Ballistics, numerical analysis, waves and wavelike phenomena, Fourier series, group concepts, fluid flow and aerodynamics, statistical measures, and meteorology are discussed with unusual clarity. Some calculus and differential equations theory is developed by the author for the reader's help in the more difficult sections. 88 figures. Index. viii + 236pp. 5⅜ x 8. T440 Clothbound **$3.50**

FREE! All you do is ask for it!

A DOVER SCIENCE SAMPLER, edited by George Barkin. 64-page book, sturdily bound, containing excerpts from over 20 Dover books explaining science. Edwin Hubble, George Sarton, Ernst Mach, A. d'Abro, Galileo, Newton, others, discussing island universes, scientific truth, biological phenomena, stability in bridges, etc. Copies limited, no more than 1 to a customer. FREE

THE FOURTH DIMENSION SIMPLY EXPLAINED, edited by H. P. Manning. 22 essays, originally Scientific American contest entries, that use a minimum of mathematics to explain aspects of 4-dimensional geometry: analogues to 3-dimensional space, 4-dimensional absurdities and curiosities (such as removing the contents of an egg without puncturing its shell), possible measurements and forms, etc. Introduction by the editor. Only book of its sort on a truly elementary level, excellent introduction to advanced works. 82 figures. 251pp. 5⅜ x 8. T711 Paperbound **$1.35**

FAMOUS BRIDGES OF THE WORLD, D. B. Steinman. An up-to-the-minute revised edition of a book that explains the fascinating drama of how the world's great bridges came to be built. The author, designer of the famed Mackinac bridge, discusses bridges from all periods and all parts of the world, explaining their various types of construction, and describing the problems their builders faced. Although primarily for youngsters, this cannot fail to interest readers of all ages. 48 illustrations in the text. 23 photographs. 99pp. 6⅛ x 9¼. T161 Paperbound **$1.00**

BRIDGES AND THEIR BUILDERS, David Steinman and Sara Ruth Watson. Engineers, historians, everyone who has ever been fascinated by great spans will find this book an endless source of information and interest. Dr. Steinman, recipient of the Louis Levy medal, was one of the great bridge architects and engineers of all time, and his analysis of the great bridges of history is both authoritative and easily followed. Greek and Roman bridges, medieval bridges, Oriental bridges, modern works such as the Brooklyn Bridge and the Golden Gate Bridge, and many others are described in terms of history, constructional principles, artistry, and function. All in all this book is the most comprehensive and accurate semipopular history of bridges in print in English. New, greatly revised, enlarged edition. 23 photographs, 26 line drawings. Index. xvii + 401pp. 5⅜ x 8. T431 Paperbound **$2.00**

FADS AND FALLACIES IN THE NAME OF SCIENCE, Martin Gardner. Examines various cults, quack systems, frauds, delusions which at various times have masqueraded as science. Accounts of hollow-earth fanatics like Symmes; Velikovsky and wandering planets; Hoerbiger; Bellamy and the theory of multiple moons; Charles Fort; dowsing, pseudoscientific methods for finding water, ores, oil. Sections on naturopathy, iridiagnosis, zone therapy, food fads, etc. Analytical accounts of Wilhelm Reich and orgone sex energy; L. Ron Hubbard and Dianetics; A. Korzybski and General Semantics; many others. Brought up to date to include Bridey Murphy, others. Not just a collection of anecdotes, but a fair, reasoned appraisal of eccentric theory. Formerly titled IN THE NAME OF SCIENCE. Preface. Index. x + 384pp. 5⅜ x 8. T394 Paperbound **$1.50**

See also: **A PHILOSOPHICAL ESSAY ON PROBABILITIES, P. de Laplace; ON MATHEMATICS AND MATHEMATICIANS, R. E. Moritz; AN ELEMENTARY SURVEY OF CELESTIAL MECHANICS, Y. Ryabov; THE SKY AND ITS MYSTERIES, E. A. Beet; THE REALM OF THE NEBULAE, E. Hubble; OUT OF THE SKY, H. H. Nininger; SATELLITES AND SCIENTIFIC RESEARCH, D. King-Hele; HEREDITY AND YOUR LIFE, A. M. Winchester; INSECTS AND INSECT LIFE, S. W. Frost; PRINCIPLES OF STRATIGRAPHY, A. W. Grabau; TEACH YOURSELF SERIES.**

HISTORY OF SCIENCE AND MATHEMATICS

DIALOGUES CONCERNING TWO NEW SCIENCES, Galileo Galilei. This classic of experimental science, mechanics, engineering, is as enjoyable as it is important. A great historical document giving insights into one of the world's most original thinkers, it is based on 30 years' experimentation. It offers a lively exposition of dynamics, elasticity, sound, ballistics, strength of materials, the scientific method. "Superior to everything else of mine," Galileo. Trans. by H. Crew, A. Salvio. 126 diagrams. Index. xxi + 288pp. 5⅜ x 8. S99 Paperbound **$1.65**

A DIDEROT PICTORIAL ENCYCLOPEDIA OF TRADES AND INDUSTRY, Manufacturing and the Technical Arts in Plates Selected from "L'Encyclopédie ou Dictionnaire Raisonné des Sciences, des Arts, et des Métiers" of Denis Diderot. Edited with text by C. Gillispie. This first modern selection of plates from the high point of 18th century French engraving is a storehouse of valuable technological information to the historian of arts and science. Over 2000 illustrations on 485 full page plates, most of them original size, show the trades and industries of a fascinating era in such great detail that the processes and shops might very well be reconstructed from them. The plates teem with life, with men, women, and children performing all of the thousands of operations necessary to the trades before and during the early stages of the industrial revolution. Plates are in sequence, and show general operations, closeups of difficult operations, and details of complex machinery. Such important and interesting trades and industries are illustrated as sowing, harvesting, beekeeping, cheesemaking, operating windmills, milling flour, charcoal burning, tobacco processing, indigo, fishing, arts of war, salt extraction, mining, smelting, casting iron, steel, extracting mercury, zinc, sulphur, copper, etc., slating, tinning, silverplating, gilding, making gunpowder, cannons, bells, shoeing horses, tanning, papermaking, printing, dyeing, and more than 40 other categories. Professor Gillispie, of Princeton, supplies a full commentary on all the plates, identifying operations, tools, processes, etc. This material, presented in a lively and lucid fashion, is of great interest to the reader interested in history of science and technology. Heavy library cloth. 920pp. 9 x 12. T421 Two volume set **$18.50**

DE MAGNETE, William Gilbert. This classic work on magnetism founded a new science. Gilbert was the first to use the word "electricity", to recognize mass as distinct from weight, to discover the effect of heat on magnetic bodies; invent an electroscope, differentiate between static electricity and magnetism, conceive of the earth as a magnet. Written by the first great experimental scientist, this lively work is valuable not only as an historical landmark, but as the delightfully easy to follow record of a perpetually searching, ingenious mind. Translated by P. F. Mottelay. 25 page biographical memoir. 90 figures. lix + 368pp. 5⅜ x 8. S470 Paperbound **$2.00**

CHARLES BABBAGE AND HIS CALCULATING ENGINES, edited by P. Morrison and E. Morrison. Babbage, leading 19th century pioneer in mathematical machines and herald of modern operational research, was the true father of Harvard's relay computer Mark I. His Difference Engine and Analytical Engine were the first machines in the field. This volume contains a valuable introduction on his life and work; major excerpts from his autobiography, revealing his eccentric and unusual personality; and extensive selections from "Babbage's Calculating Engines," a compilation of hard-to-find journal articles by Babbage, the Countess of Lovelace, L. F. Menabrea, and Dionysius Lardner. 8 illustrations, Appendix of miscellaneous papers. Index. Bibliography. xxxviii + 400pp. 5⅜ x 8. T12 Paperbound **$2.00**

A HISTORY OF ASTRONOMY FROM THALES TO KEPLER, J. L. E. Dreyer. (Formerly A HISTORY OF PLANETARY SYSTEMS FROM THALES TO KEPLER.) This is the only work in English to give the complete history of man's cosmological views from prehistoric times to Kepler and Newton. Partial contents: Near Eastern astronomical systems, Early Greeks, Homocentric Spheres of Eudoxus, Epicycles, Ptolemaic system, medieval cosmology, Copernicus, Kepler, etc. Revised, foreword by W. H. Stahl. New bibliography. xvii + 430pp. 5⅜ x 8. S79 Paperbound **$1.98**

A SHORT HISTORY OF ANATOMY AND PHYSIOLOGY FROM THE GREEKS TO HARVEY, Charles Singer. Corrected edition of THE EVOLUTION OF ANATOMY, classic work tracing evolution of anatomy and physiology from prescientific times through Greek & Roman periods, Dark Ages, Renaissance, to age of Harvey and beginning of modern concepts. Centered on individuals, movements, periods that definitely advanced anatomical knowledge: Plato, Diocles, Aristotle, Theophrastus, Herophilus, Erasistratus, the Alexandrians, Galen, Mondino, da Vinci, Linacre, Sylvius, others. Special section on Vesalius; Vesalian atlas of nudes, skeletons, muscle tabulae. Index of names, 20 plates. 270 extremely interesting illustrations of ancient, medieval, Renaissance, Oriental origin. xii + 209pp. 5⅜ x 8. T389 Paperbound **$1.75**

FROM MAGIC TO SCIENCE, Charles Singer. A great historian examines aspects of medical science from the Roman Empire through the Renaissance. Includes perhaps the best discussion of early herbals, and a penetrating physiological interpretation of "The Visions of Hildegarde of Bingen." Also examined are Arabian and Galenic influences; the Sphere of Pythagoras; Paracelsus; the reawakening of science under Leonardo da Vinci, Vesalius; the Lorica of Gildas the Briton; etc. Frequent quotations with translations. New Introduction by the author. New unabridged, corrected edition. 158 unusual illustrations from classical and medieval sources. Index. xxvii + 365pp. 5⅜ x 8. T390 Paperbound **$2.00**

HISTORY OF MATHEMATICS, D. E. Smith. Most comprehensive non-technical history of math in English. Discusses lives and works of over a thousand major and minor figures, with footnotes supplying technical information outside the book's scheme, and indicating disputed matters. Vol I: A chronological examination, from primitive concepts through Egypt, Babylonia, Greece, the Orient, Rome, the Middle Ages, the Renaissance, and up to 1900. Vol 2: The development of ideas in specific fields and problems, up through elementary calculus. Two volumes, total of 510 illustrations, 1355pp. 5⅜ x 8. Set boxed in attractive container. T429, 430 Paperbound, the set **$5.00**

A SHORT ACCOUNT OF THE HISTORY OF MATHEMATICS, W. W. R. Ball. Most readable non-technical history of mathematics treats lives, discoveries of every important figure from Egyptian, Phoenician mathematicians to late 19th century. Discusses schools of Ionia, Pythagoras, Athens, Cyzicus, Alexandria, Byzantium, systems of numeration; primitive arithmetic; Middle Ages, Renaissance, including Arabs, Bacon, Regiomontanus, Tartaglia, Cardan, Stevinus, Galileo, Kepler; modern mathematics of Descartes, Pascal, Wallis, Huygens, Newton, Leibnitz, d'Alembert, Euler, Lambert, Laplace, Legendre, Gauss, Hermite, Weierstrass, scores more. Index. 25 figures. 546pp. 5⅜ x 8. S630 Paperbound **$2.00**

A SOURCE BOOK IN MATHEMATICS, D. E. Smith. Great discoveries in math, from Renaissance to end of 19th century, in English translation. Read announcements by Dedekind, Gauss, Delamain, Pascal, Fermat, Newton, Abel, Lobachevsky, Bolyai, Riemann, De Moivre, Legendre, Laplace, others of discoveries about imaginary numbers, number congruence, slide rule, equations, symbolism, cubic algebraic equations, non-Euclidean forms of geometry, calculus, function theory, quaternions, etc. Succinct selections from 125 different treatises, articles, most unavailable elsewhere in English. Each article preceded by biographical, historical introduction. Vol. I: Fields of Number, Algebra. Index. 32 illus. 338pp. 5⅜ x 8. Vol. II: Fields of Geometry, Probability, Calculus, Functions, Quaternions. 83 illus. 432pp. 5⅜ x 8.
Vol. 1: S552 Paperbound **$1.85**
Vol. 2: S553 Paperbound **$1.85**
2 vol. set, boxed **$3.50**

A HISTORY OF THE CALCULUS, AND ITS CONCEPTUAL DEVELOPMENT, Carl B. Boyer. Provides laymen and mathematicians a detailed history of the development of the calculus, from early beginning in antiquity to final elaboration as mathematical abstractions. Gives a sense of mathematics not as a technique, but as a habit of mind, in the progression of ideas of Zeno, Plato, Pythagoras, Eudoxus, Arabic and Scholastic mathematicians, Newton, Leibnitz, Taylor, Descartes, Euler, Lagrange, Cantor, Weierstrass, and others. This first comprehensive critical history of the calculus was originally titled "The Concepts of the Calculus." Foreword by R. Courant. Preface. 22 figures. 25-page bibliography. Index. v + 364pp. 5⅜ x 8. S509 Paperbound **$2.00**

A CONCISE HISTORY OF MATHEMATICS, D. Struik. Lucid study of development of mathematical ideas, techniques from Ancient Near East, Greece, Islamic science, Middle Ages, Renaissance, modern times. Important mathematicians are described in detail. Treatment is not anecdotal, but analytical development of ideas. "Rich in content, thoughtful in interpretation," U.S. QUARTERLY BOOKLIST. Non-technical; no mathematical training needed. Index. 60 illustrations, including Egyptian papyri, Greek mss., portraits of 31 eminent mathematicians. Bibliography. 2nd edition. xix + 299pp. 5⅜ x 8. T255 Paperbound **$1.75**

See also: **NON-EUCLIDEAN GEOMETRY, R. Bonola; THEORY OF DETERMINANTS IN HISTORICAL ORDER OF DEVELOPMENT, T. Muir; HISTORY OF THE THEORY OF ELASTICITY AND STRENGTH OF MATERIALS, I. Todhunter and K. Pearson; A SHORT HISTORY OF ASTRONOMY, A. Berry; CLASSICS OF SCIENCE.**

PHILOSOPHY OF SCIENCE AND MATHEMATICS

FOUNDATIONS OF SCIENCE: THE PHILOSOPHY OF THEORY AND EXPERIMENT, N. R. Campbell. A critique of the most fundamental concepts of science in general and physics in particular. Examines why certain propositions are accepted without question, demarcates science from philosophy, clarifies the understanding of the tools of science. Part One analyzes the presuppositions of scientific thought: existence of the material world, nature of scientific laws, multiplication of probabilities, etc.: Part Two covers the nature of experiment and the application of mathematics: conditions for measurement, relations between numerical laws and theories, laws of error, etc. An appendix covers problems arising from relativity, force, motion, space, and time. A classic in its field. Index. xiii + 565pp. 5⅝ x 8⅜. S372 Paperbound **$2.95**

WHAT IS SCIENCE?, Norman Campbell. This excellent introduction explains scientific method, role of mathematics, types of scientific laws. Contents: 2 aspects of science, science & nature, laws of science, discovery of laws, explanation of laws, measurement & numerical laws, applications of science. 192pp. 5⅜ x 8. S43 Paperbound **$1.25**

THE VALUE OF SCIENCE, Henri Poincaré. Many of the most mature ideas of the "last scientific universalist" covered with charm and vigor for both the beginning student and the advanced worker. Discusses the nature of scientific truth, whether order is innate in the universe or imposed upon it by man, logical thought versus intuition (relating to math, through the works of Weierstrass, Lie, Klein, Riemann), time and space (relativity, psychological time, simultaneity), Hertz's concept of force, interrelationship of mathematical physics to pure math, values within disciplines of Maxwell, Carnot, Mayer, Newton, Lorentz, etc. Index. iii + 147pp. 5⅜ x 8. S469 Paperbound **$1.35**

SCIENCE AND METHOD, Henri Poincaré. Procedure of scientific discovery, methodology, experiment, idea-germination—the intellectual processes by which discoveries come into being. Most significant and most interesting aspects of development, application of ideas. Chapters cover selection of facts, chance, mathematical reasoning, mathematics, and logic; Whitehead, Russell, Cantor; the new mechanics, etc. 288pp. 5⅜ x 8. S222 Paperbound **$1.35**

SCIENCE AND HYPOTHESIS, Henri Poincaré. Creative psychology in science. How such concepts as number, magnitude, space, force, classical mechanics were developed, and how the modern scientist uses them in his thought. Hypothesis in physics, theories of modern physics. Introduction by Sir James Larmor. "Few mathematicians have had the breadth of vision of Poincaré, and none is his superior in the gift of clear exposition," E. T. Bell. Index. 272pp. 5⅜ x 8. S221 Paperbound **$1.35**

PHILOSOPHY AND THE PHYSICISTS, L. S. Stebbing. The philosophical aspects of modern science examined in terms of a lively critical attack on the ideas of Jeans and Eddington. Discusses the task of science, causality, determinism, probability, consciousness, the relation of the world of physics to that of everyday experience. Probes the philosophical significance of the Planck-Bohr concept of discontinuous energy levels, the inferences to be drawn from Heisenberg's Uncertainty Principle, the implications of "becoming" involved in the 2nd law of thermodynamics, and other problems posed by the discarding of Laplacean determinism. 285pp. 5⅜ x 8. T480 Paperbound **$1.65**

EXPERIMENT AND THEORY IN PHYSICS, Max Born. A Nobel laureate examines the nature and value of the counterclaims of experiment and theory in physics. Synthetic versus analytical scientific advances are analyzed in the work of Einstein, Bohr, Heisenberg, Planck, Eddington, Milne, and others by a fellow participant. 44pp. 5⅜ x 8. S308 Paperbound **60¢**

THE NATURE OF PHYSICAL THEORY, P. W. Bridgman. Here is how modern physics looks to a highly unorthodox physicist—a Nobel laureate. Pointing out many absurdities of science, and demonstrating the inadequacies of various physical theories, Dr. Bridgman weighs and analyzes the contributions of Einstein, Bohr, Newton, Heisenberg, and many others. This is a non-technical consideration of the correlation of science and reality. Index. xi + 138pp. 5⅜ x 8. S33 Paperbound **$1.25**

THE PHILOSOPHY OF SPACE AND TIME, H. Reichenbach. An important landmark in the development of the empiricist conception of geometry, covering the problem of the foundations of geometry, the theory of time, the consequences of Einstein's relativity, including: relations between theory and observations; coordinate and metrical properties of space; the psychological problem of visual intuition of non-Euclidean structures; and many other important topics in modern science and philosophy. The majority of ideas require only a knowledge of intermediate math. Introduction by R. Carnap. 49 figures. Index. xviii + 296pp. 5⅜ x 8. S443 Paperbound **$2.00**

MATTER & MOTION, James Clerk Maxwell, This excellent exposition begins with simple particles and proceeds gradually to physical systems beyond complete analysis: motion, force, properties of centre of mass of material system, work, energy, gravitation, etc. Written with all Maxwell's original insights and clarity. Notes by E. Larmor. 17 diagrams. 178pp. 5⅜ x 8. S188 Paperbound **$1.35**

THE ANALYSIS OF MATTER, Bertrand Russell. How do our senses concord with the new physics? This volume covers such topics as logical analysis of physics, prerelativity physics, causality, scientific inference, physics and perception, special and general relativity, Weyl's theory, tensors, invariants and their physical interpretation, periodicity and qualitative series. "The most thorough treatment of the subject that has yet been published," THE NATION. Introduction by L. E. Denonn. 422pp. 5⅜ x 8. T231 Paperbound **$1.95**

SUBSTANCE AND FUNCTION, & EINSTEIN'S THEORY OF RELATIVITY, Ernst Cassirer. Two books bound as one. Cassirer establishes a philosophy of the exact sciences that takes into consideration newer developments in mathematics, and also shows historical connections. Partial contents: Aristotelian logic, Mill's analysis, Helmholtz & Kronecker, Russell & cardinal numbers, Euclidean vs. non-Euclidean geometry, Einstein's relativity. Bibliography. Index. xxi + 465pp. 5⅜ x 8. T50 Paperbound **$2.00**

PRINCIPLES OF MECHANICS, Heinrich Hertz. This last work by the great 19th century physicist is not only a classic, but of great interest in the logic of science. Creating a new system of mechanics based upon space, time, and mass, it returns to axiomatic analysis, to understanding of the formal or structural aspects of science, taking into account logic, observation, and a priori elements. Of great historical importance to Poincaré, Carnap, Einstein, Milne. A 20-page introduction by R. S. Cohen, Wesleyan University, analyzes the implications of Hertz's thought and the logic of science. Bibliography. 13-page introduction by Helmholtz. xlii + 274pp. 5⅜ x 8. S316 Clothbound **$3.50**
S317 Paperbound **$1.85**

THE PHILOSOPHICAL WRITINGS OF PEIRCE, edited by Justus Buchler. (Formerly published as THE PHILOSOPHY OF PEIRCE.) This is a carefully balanced exposition of Peirce's complete system, written by Peirce himself. It covers such matters as scientific method, pure chance vs. law, symbolic logic, theory of signs, pragmatism, experiment, and other topics. Introduction by Justus Buchler, Columbia University. xvi + 368pp. 5⅜ x 8. T217 Paperbound **$1.95**

ESSAYS IN EXPERIMENTAL LOGIC, John Dewey. This stimulating series of essays touches upon the relationship between inquiry and experience, dependence of knowledge upon thought, character of logic; judgments of practice, data and meanings, stimuli of thought, etc. Index. viii + 444pp. 5⅜ x 8. T73 Paperbound **$1.95**

LANGUAGE, TRUTH AND LOGIC, A. Ayer. A clear introduction to the Vienna and Cambridge schools of Logical Positivism. It sets up specific tests by which you can evaluate validity of ideas, etc. Contents: Function of philosophy, elimination of metaphysics, nature of analysis, a priori, truth and probability, etc. 10th printing. "I should like to have written it myself," Bertrand Russell. Index. 160pp. 5⅜ x 8. T10 Paperbound **$1.25**

THE PSYCHOLOGY OF INVENTION IN THE MATHEMATICAL FIELD, J. Hadamard. Where do ideas come from? What role does the unconscious play? Are ideas best developed by mathematical reasoning, word reasoning, visualization? What are the methods used by Einstein, Poincaré, Galton, Riemann? How can these techniques be applied by others? Hadamard, one of the world's leading mathematicians, discusses these and other questions. xiii + 145pp. 5⅜ x 8. T107 Paperbound **$1.25**

FOUNDATIONS OF GEOMETRY, Bertrand Russell. Analyzing basic problems in the overlap area between mathematics and philosophy, Nobel laureate Russell examines the nature of geometrical knowledge, the nature of geometry, and the application of geometry to space. It covers the history of non-Euclidean geometry, philosophic interpretations of geometry—especially Kant—projective and metrical geometry. This is most interesting as the solution offered in 1897 by a great mind to a problem still current. New introduction by Prof. Morris Kline of N. Y. University. xii + 201pp. 5⅜ x 8. S232 Clothbound **$3.25**
S233 Paperbound **$1.60**

BIBLIOGRAPHIES

GUIDE TO THE LITERATURE OF MATHEMATICS AND PHYSICS, N. G. Parke III. Over 5000 entries included under approximately 120 major subject headings, of selected most important books, monographs, periodicals, articles in English, plus important works in German, French, Italian, Spanish, Russian (many recently available works). Covers every branch of physics, math, related engineering. Includes author, title, edition, publisher, place, date, number of volumes, number of pages. A 40-page introduction on the basic problems of research and study provides useful information on the organization and use of libraries, the psychology of learning, etc. This reference work will save you hours of time. 2nd revised edition. Indices of authors, subjects. 464pp. 5⅜ x 8. S447 Paperbound **$2.49**

THE STUDY OF THE HISTORY OF MATHEMATICS & THE STUDY OF THE HISTORY OF SCIENCE, George Sarton. Scientific method & philosophy in 2 scholarly fields. Defines duty of historian of math., provides especially useful bibliography with best available biographies of modern mathematicians, editions of their collected works, correspondence. Observes combination of history & science, will aid scholar in understanding science today. Bibliography includes best known treatises on historical methods. 200-item critically evaluated bibliography. Index. 10 illustrations. 2 volumes bound as one. 113pp. + 75pp. 5⅜ x 8. T240 Paperbound **$1.25**

MATHEMATICAL PUZZLES

AMUSEMENTS IN MATHEMATICS, Henry Ernest Dudeney. The foremost British originator of mathematical puzzles is always intriguing, witty; and paradoxical in this classic, one of the largest collections of mathematical amusements. More than 430 puzzles, problems, and paradoxes. Mazes and games, problems on number manipulation, unicursal and other route problems, puzzles on measuring, weighing, packing, age, kinship, chessboards, joiners', crossing river, plane figure dissection, and many others. Solutions. More than 450 illustrations. vii + 258pp. 5⅜ x 8. T473 Paperbound **$1.25**

THE CANTERBURY PUZZLES, Henry Ernest Dudeney. Chaucer's pilgrims set one another problems in story form. Also Adventures of the Puzzle Club, the Strange Escape of the King's Jester, the Monks of Riddlewell, the Squire's Christmas Puzzle Party, and others. All puzzles are original, based on dissecting plane figures, arithmetic, algebra, elementary calculus, and other branches of mathematics, and purely logical ingenuity. "The limit of ingenuity and intricacy . . ." The Observer. Over 110 puzzles. Full solutions. 150 illustrations. viii + 225pp. 5⅜ x 8. T474 Paperbound **$1.25**

SYMBOLIC LOGIC and THE GAME OF LOGIC, Lewis Carroll. "Symbolic Logic" is not concerned with modern symbolic logic, but is instead a collection of over 380 problems posed with charm and imagination, using the syllogism, and a fascinating diagrammatic method of drawing conclusions. In "The Game of Logic," Carroll's whimsical imagination devises a logical game played with 2 diagrams and counters (included) to manipulate hundreds of tricky syllogisms. The final section, "Hit or Miss" is a lagniappe of 101 additional puzzles in the delightful Carroll manner. Until this reprint edition, both of these books were rarities costing up to $15 each. Symbolic Logic: Index, xxxi + 199pp. The Game of Logic: 96pp. Two vols. bound as one. 5⅜ x 8. T492 Paperbound **$1.50**

PILLOW PROBLEMS and A TANGLED TALE, Lewis Carroll. One of the rarest of all Carroll's works, "Pillow Problems" contains 72 original math puzzles, all typically ingenious. Particularly fascinating are Carroll's answers which remain exactly as he thought them out, reflecting his actual mental processes. The problems in "A Tangled Tale" are in story form, originally appearing as a monthly magazine serial. Carroll not only gives the solutions, but uses answers sent in by readers to discuss wrong approaches and misleading paths, and grades them for insight. Both of these books were rarities until this edition, "Pillow Problems" costing up to $25, and "A Tangled Tale" $15. Pillow Problems: Preface and introduction by Lewis Carroll. xx + 109pp. A Tangled Tale: 6 illustrations. 152pp. Two vols. bound as one. 5⅜ x 8. T493 Paperbound **$1.50**

DIVERSIONS AND DIGRESSIONS OF LEWIS CARROLL. A major new treasure for Carroll fans! Rare privately published puzzles, mathematical amusements and recreations, games. Includes the fragmentary Part III of "Curiosa Mathematica." Also contains humorous and satirical pieces: "The New Belfry," "The Vision of the Three T's," and much more. New 32-page supplement of rare photographs taken by Carroll. Formerly titled "The Lewis Carroll Picture Book." Edited by S. Collingwood. x + 375pp. 5⅜ x 8. T732 Paperbound **$1.50**

THE BOOK OF MODERN PUZZLES, G. L. Kaufman. More than 150 word puzzles, logic puzzles. No warmed-over fare but all new material based on same appeals that make crosswords and deduction puzzles popular, but with different principles, techniques. Two-minute teasers, involved word-labyrinths, design and pattern puzzles, puzzles calling for logic and observation, puzzles testing ability to apply general knowledge to peculiar situations, many others. Answers to all problems. 116 illustrations. 192pp. 5⅜ x 8. T143 Paperbound **$1.00**

NEW WORD PUZZLES, Gerald L. Kaufman. Contains 100 brand new challenging puzzles based on words and their combinations, never published before in any form. Most are new types invented by the author—for beginners or experts. Chess word puzzles, addle letter anagrams, double word squares, double horizontals, alphagram puzzles, dual acrostigrams, linkogram lapwords—plus 8 other brand new types, all with solutions included. 196 figures. 100 brand new puzzles. vi + 122pp. 5⅜ x 8. T344 Paperbound **$1.00**

MATHEMATICAL RECREATIONS

MATHEMATICS, MAGIC AND MYSTERY, Martin Gardner. Card tricks, feats of mental mathematics, stage mind-reading, other "magic" explained as applications of probability, sets, theory of numbers, topology, various branches of mathematics. Creative examination of laws and their applications with scores of new tricks and insights. 115 sections discuss tricks wtih cards, dice, coins; geometrical vanishing tricks, dozens of others. No sleight of hand needed; mathematics guarantees success. 115 illustrations. xii + 174pp. 5⅜ x 8. T335 Paperbound **$1.00**

MATHEMATICAL EXCURSIONS, Helen A. Merrill. Fun, recreation, insights into elementary problem-solving. A mathematical expert guides you along by-paths not generally travelled in elementary math courses—how to divide by inspection, Russian peasant system of multiplication; memory systems for pi; building odd and even magic squares; dyadic systems; facts about 37; square roots by geometry; Tchebichev's machine; drawing five-sided figures; dozens more. Solutions to more difficult ones. 50 illustrations. 145pp. 5⅜ x 8. T350 Paperbound **$1.00**

CRYPTOGRAPHY, L. D. Smith. Excellent elementary introduction to enciphering, deciphering secret writing. Explains transposition, substitution ciphers; codes; solutions. Geometrical patterns, route transcription, columnar transposition, other methods. Mixed cipher systems; single-alphabet, polyalphabetical substitution; mechanical devices; Vigenere system, etc. Enciphering Japanese; explanation of Baconian Biliteral cipher; frequency tables. More than 150 problems provide practical application. Bibliography. Index. 164pp. 5⅜ x 8. T247 Paperbound **$1.00**

CRYPTANALYSIS, Helen F. Gaines. (Formerly ELEMENTARY CRYPTANALYSIS.) A standard elementary and intermediate text for serious students. It does not confine itself to old material, but contains much that is not generally known, except to experts. Concealment, Transposition, Substitution ciphers; Vigenere, Kasiski, Playfair, multafid, dozens of other techniques. Appendix with sequence charts, letter frequencies in English, 5 other languages, English word frequencies. Bibliography. 167 codes. New to this edition: solution to codes. vi + 230pp. 5⅜ x 8. T97 Paperbound **$1.95**

MAGIC SQUARES AND CUBES, W. S. Andrews. Only book-length treatment in English, a thorough non-technical description and analysis. Here are nasik, overlapping, pandiagonal, serrated squares; magic circles, cubes, spheres, rhombuses. Try your hand at 4-dimensional magical figures! Much unusual folklore and tradition included. High school algebra is sufficient. 754 diagrams and illustrations. viii + 419pp. 5⅜ x 8. T658 Paperbound **$1.85**

PAPER FOLDING FOR BEGINNERS, W. D. Murray and F. J. Rigney. A delightful introduction to the varied and entertaining Japanese art of origami (paper folding), with a full crystal-clear text that anticipates every difficulty; over 275 clearly labeled diagrams of all important stages in creation. You get results at each stage, since complex figures are logically developed from simpler ones. 43 different pieces are explained: place mats, drinking cups, bonbon boxes, sailboats, frogs, roosters, etc. 6 photographic plates. 279 diagrams. 95pp. 5⅝ x 8⅜. T713 Paperbound **$1.00**

CHESS, CHECKERS, GAMES, GO

A TREASURY OF CHESS LORE, edited by Fred Reinfeld. A delightful collection of anecdotes, short stories, aphorisms by and about the masters, poems, accounts of games and tournaments, photographs. Hundreds of humorous, pithy, satirical, wise, and historical episodes, comments, and word portraits. A fascinating "must" for chess players; revealing and perhaps seductive to those who wonder what their friends see in the game. 49 photographs (14 full page plates). 12 diagrams. xi + 306pp. 5⅜ x 8. T458 Paperbound **$1.75**

HYPERMODERN CHESS as developed in the games of its greatest exponent, ARON NIMZOVICH, edited by Fred Reinfeld. An intensely original player and analyst, Nimzovich's extraordinary approaches startled and often angered the chess world. This volume, designed for the average player, shows in his victories over Alekhine, Lasker, Marshall, Rubinstein, Spielmann, and others, how his iconoclastic methods infused new life into the game. Use Nimzovich to invigorate your play and startle opponents. Introduction. Indices of players and openings. 180 diagrams. viii + 220pp. 5⅜ x 8. T448 Paperbound **$1.35**

ONE HUNDRED SELECTED GAMES, Mikhail Botvinnik. Author's own choice of his best games before becoming World Champion in 1948, beginning with first big tournament, the USSR Championship, 1927. Shows his great powers of analysis as he annotates these games, giving strategy, technique against Alekhine, Capablanca, Euwe, Keres, Reshevsky, Smyslov, Vidmar, many others. Discusses his career, methods of play, system of training, 6 studies of endgame positions. 221 diagrams. 272pp. 5⅜ x 8. T620 Paperbound **$1.50**

RUBINSTEIN'S CHESS MASTERPIECES, selected and annotated by H. Kmoch. Thoroughgoing mastery of opening, middle game; faultless technique in endgame, particularly rook and pawn endings; ability to switch from careful positional play to daring combinations; all distinguish the play of Rubinstein. 100 best games, against Janowski, Nimzowitch, Tarrasch, Vidmar, Capablanca, other greats, carefully annotated, will improve your game rapidly. Biographical introduction, B. F. Winkelman, 103 diagrams. 192pp. 5⅜ x 8. T617 Paperbound **$1.25**

TARRASCH'S BEST GAMES OF CHESS, selected & annotated by Fred Reinfeld. First definitive collection of games by Siegbert Tarrasch, winner of 7 international tournaments, and the leading theorist of classical chess. 183 games cover 50 years of play against Mason, Mieses, Paulsen, Teichmann, Pillsbury, Janowski, others. Reinfeld includes Tarrasch's own analyses of many of these games. A careful study and replaying of the games will give you a sound understanding of classical methods, and many hours of enjoyment. Introduction. Indexes. 183 diagrams. xxiv + 386pp. 5⅜ x 8. T644 Paperbound **$2.00**

MARSHALL'S BEST GAMES OF CHESS, F. J. Marshall. Grandmaster, U. S. champion for 27 years, tells story of career; presents magnificent collection of 140 of best games, annotated by himself. Games against Capablanca, Alekhine, Emanuel Lasker, Janowski, Rubinstein, Pillsbury, etc. Special section analyzes openings such as King's Gambit, Ruy Lopez, Alekhine's Defence, Giuoco Piano, others. A study of Marshall's brilliant "swindles," slashing attacks, extraordinary sacrifices, will rapidly improve your game. Formerly "My Fifty Years of Chess." Introduction. 19 diagrams. 13 photos. 250pp. 5⅜ x 8. T604 Paperbound **$1.35**

THE ENJOYMENT OF CHESS PROBLEMS, K. S. Howard. A classic treatise on this minor art by an internationally recognized authority that gives a basic knowledge of terms and themes for the everyday chess player as well as the problem fan: 7 chapters on the two-mover; 7 more on 3- and 4-move problems; a chapter on selfmates; and much more. "The most important one-volume contribution originating solely in the U.S.A.", Alain White. 200 diagrams. Index. Solutions. viii + 212pp. 5⅜ x 8. T742 Paperbound **$1.25**

HOW TO SOLVE CHESS PROBLEMS, K. S. Howard. Full of practical suggestions for the fan or the beginner—who need only know the moves of the chessmen. Contains preliminary section and 58 two-move, 46 three-move, and 8 four-move problems composed by 27 outstanding American problem creators in the last 30 years. Explanation of all terms and exhaustive index. "Just what is wanted for the student," Brian Warley. 112 problems, solutions. vi + 171pp. 5⅜ x 8. T748 Paperbound **$1.00**

CHESS AND CHECKERS: THE WAY TO MASTERSHIP, Edward Lasker. Complete, lucid instructions for the beginner—and valuable suggestions for the advanced player! For both games the great master and teacher presents fundamentals, elementary tactics, and steps toward becoming a superior player. He concentrates on general principles rather than a mass of rules, comprehension rather than brute memory. Historical introduction. 118 diagrams. xiv + 167pp. 5⅜ x 8. T657 Paperbound **$1.15**

WIN AT CHECKERS, M. Hopper. (Formerly CHECKERS). The former World's Unrestricted Checker Champion discusses the principles of the game, expert's shots and traps, problems for the beginner, standard openings, locating your best move, the end game, opening "blitzkrieg" moves, ways to draw when you are behind your opponent, etc. More than 100 detailed questions and answers anticipate your problems. Appendix. 75 problems with solutions and diagrams. Index. 79 figures. xi + 107pp. 5⅜ x 8. T363 Paperbound **$1.00**

GAMES ANCIENT AND ORIENTAL, AND HOW TO PLAY THEM, E. Falkener. A connoisseur's selection of exciting and different games: Oriental varieties of chess, with unusual pieces and moves (including Japanese shogi); the original pachisi; go; reconstructions of lost Roman and Egyptian games; and many more. Full rules and sample games. Now play at home the games that have entertained millions, not on a fad basis, but for millennia. 345 illustrations and figures. iv + 366pp. 5⅜ x 8. T739 Paperbound **$1.85**

GO AND GO-MOKU: THE ORIENTAL BOARD GAMES, Edward Lasker. Best introduction to Go and its easier sister-game, Go-Moku—games new to Western world, but ancient in China, Japan. Extensively revised work by famed chess master Lasker, Go-player for over 50 years, stresses theory rather than brute memory, presents step-by-step explanation of strategy, gives examples of world championship matches, in game which has replaced chess as favorite of many physicists, mathematicians. 72 diagrams. xix + 215 pp. 5⅜ x 8.
T613 Paperbound **$1.45**

FICTION

FLATLAND, E. A. Abbott. A science-fiction classic of life in a 2-dimensional world that is also a first-rate introduction to such aspects of modern science as relativity and hyperspace. Political, moral, satirical, and humorous overtones have made FLATLAND fascinating reading for thousands. 7th edition. New introduction by Banesh Hoffmann. 16 illustrations. 128pp. 5⅜ x 8.
T1 Paperbound **$1.00**

THE WONDERFUL WIZARD OF OZ, L. F. Baum. Only edition in print with all the original W. W. Denslow illustrations in full color—as much a part of "The Wizard" as Tenniel's drawings are of "Alice in Wonderland." "The Wizard" is still America's best-loved fairy tale, in which, as the author expresses it, "The wonderment and joy are retained and the heartaches and nightmares left out." Now today's young readers can enjoy every word and wonderful picture of the original book. New introduction by Martin Gardner. A Baum bibliography. 23 full-page color plates. viii + 268pp. 5⅜ x 8. T691 Paperbound **$1.45**

THE MARVELOUS LAND OF OZ, L. F. Baum. This is the equally enchanting sequel to the "Wizard," continuing the adventures of the Scarecrow and the Tin Woodman. The hero this time is a little boy named Tip, and all the delightful Oz magic is still present. This is the book with the Animated Saw-horse, the Woggle-Bug, and Jack Pumpkinhead. All the original John R. Neill illustrations, 16 in full color. 287pp. 5⅜ x 8. T692 Paperbound **$1.45**

FIVE GREAT DOG NOVELS, edited by Blanche Cirker. The complete original texts of five classic dog novels that have delighted and thrilled millions of children and adults throughout the world with stories of loyalty, adventure, and courage. Full texts of Jack London's "The Call of the Wild"; John Brown's "Rab and His Friends"; Alfred Ollivant's "Bob, Son of Battle"; Marshall Saunders' "Beautiful Joe"; and Ouida's "A Dog of Flanders." 21 illustrations from the original editions. 495pp. 5⅜ x 8. T777 Paperbound **$1.50**

3 ADVENTURE NOVELS by H. Rider Haggard. Complete texts of "She," "King Solomon's Mines," "Allan Quatermain." Qualities of discovery; desire for immortality; search for primitive, for what is unadorned by civilization, have kept these novels of African adventure exciting, alive to readers from R. L. Stevenson to George Orwell. 636pp. 5⅜ x 8.
T584 Paperbound **$2.00**

The Space Novels of Jules Verne

TO THE SUN? and OFF ON A COMET!, Jules Verne. Complete texts of two of the most imaginative flights into fancy in world literature display the high adventure that have kept Verne's novels read for nearly a century. Only unabridged edition of the best translation, by Edward Roth. Large, easily readable type. 50 illustrations selected from first editions. 462pp. 5⅜ x 8.
T634 Paperbound **$1.75**

FROM THE EARTH TO THE MOON and ALL AROUND THE MOON, Jules Verne. Complete editions of two of Verne's most successful novels, in finest Edward Roth translations, now available after many years out of print. Verne's visions of submarines, airplanes, television, rockets, interplanetary travel; of scientific and not-so-scientific beliefs; of peculiarities of Americans; all delight and engross us today as much as when they first appeared. Large, easily readable type. 42 illus. from first French edition. 476pp. 5⅜ x 8.
T633 Paperbound **$1.75**

THE CASTING AWAY OF MRS. LECKS AND MRS. ALESHINE, F. R. Stockton. A charming light novel by Frank Stockton, one of America's finest humorists (and author of "The Lady, or the Tiger?"). This book has made millions of Americans laugh at the reflection of themselves in two middle-aged American women involved in some of the strangest adventures on record. You will laugh, too, as they endure shipwreck, desert island, and blizzard with maddening tranquility. Also contains complete text of "The Dusantes," sequel to "The Casting Away." 49 original illustrations by F. D. Steele. vii + 142pp. 5⅜ x 8. T743 Paperbound **$1.00**

GESTA ROMANORUM, trans. by Charles Swan, ed. by Wynnard Hooper. 181 tales of Greeks, Romans, Britons, Biblical characters, comprise one of greatest medieval story collections, source plots for writers including Shakespeare, Chaucer, Gower, etc. Imaginative tales of wars, incest, thwarted love, magic, fantasy, allegory, humor, tell about kings, prostitutes, philosophers, fair damsels, knights, Noah, pirates, all walks and stations of life. Introduction. Notes. 500pp. 5⅜ x 8. T535 Paperbound **$1.85**

THREE PROPHETIC NOVELS BY H. G. WELLS, edited by E. F. Bleiler. Complete texts of "When the Sleeper Wakes" (1st book printing in 50 years), "A Story of the Days to Come," "The Time Machine" (1st complete printing in book form). Exciting adventures in the future are as enjoyable today as 50 years ago when first printed. Predict TV, movies, intercontinental airplanes, prefabricated houses, air-conditioned cities, etc. First important author to foresee problems of mind control, technological dictatorships. "Absolute best of imaginative fiction," N. Y. Times. Introduction. 335pp. 5⅜ x 8. T605 Paperbound **$1.45**

SEVEN SCIENCE FICTION NOVELS, H. G. Wells. Full unabridged texts of 7 science-fiction novels of the master. Ranging from biology, physics, chemistry, astronomy to sociology and other studies, Mr. Wells extrapolates whole worlds of strange and intriguing character. "One will have to go far to match this for entertainment, excitement, and sheer pleasure . . . ," NEW YORK TIMES. Contents: The Time Machine, The Island of Dr. Moreau, First Men in the Moon, The Invisible Man, The War of the Worlds, The Food of the Gods, In the Days of the Comet. 1015pp. 5⅜ x 8. T264 Clothbound **$3.95**

28 SCIENCE FICTION STORIES OF H. G. WELLS. Two full unabridged novels, MEN LIKE GODS and STAR BEGOTTEN, plus 26 short stories by the master science-fiction writer of all time. Stories of space, time, invention, exploration, future adventure—an indispensable part of the library of everyone interested in science and adventure. PARTIAL CONTENTS: Men Like Gods, The Country of the Blind, In the Abyss, The Crystal Egg, The Man Who Could Work Miracles, A Story of the Days to Come, The Valley of Spiders, and 21 more! 928pp. 5⅜ x 8. T265 Clothbound **$3.95**

DAVID HARUM, E. N. Westcott. This novel of one of the most lovable, humorous characters in American literature is a prime example of regional humor. It continues to delight people who like their humor dry, their characters quaint, and their plots ingenuous. First book edition to contain complete novel plus chapter found after author's death. Illustrations from first illustrated edition. 192pp. 5⅜ x 8. T580 Paperbound **$1.15**

HUMOR

THE WIT AND HUMOR OF OSCAR WILDE, ed. by Alvin Redman. Wilde at his most brilliant, in 1000 epigrams exposing weaknesses and hypocrisies of "civilized" society. Divided into 49 categories—sin, wealth, women, America, etc.—to aid writers, speakers. Includes excerpts from his trials, books, plays, criticism. Formerly "The Epigrams of Oscar Wilde." Introduction by Vyvyan Holland, Wilde's only living son. Introductory essay by editor. 260pp. 5⅜ x 8. T602 Paperbound **$1.00**

A NONSENSE ANTHOLOGY, collected by **Carolyn Wells.** 245 of the best nonsense verses ever written, including nonsense puns, absurd arguments, mock epics and sagas, nonsense ballads, odes, "sick" verses, dog-Latin verses, French nonsense verses, songs. By Edward Lear, Lewis Carroll, Gelett Burgess, W. S. Gilbert, Hilaire Belloc, Peter Newell, Oliver Herford, etc., 83 writers in all plus over four score anonymous nonsense verses. A special section of limericks, plus famous nonsense such as Carroll's "Jabberwocky" and Lear's "The Jumblies" and much excellent verse virtually impossible to locate elsewhere. For 50 years considered the best anthology available. Index of first lines specially prepared for this edition. Introduction by Carolyn Wells. 3 indexes: Title, Author, First lines. xxxiii + 279pp. 5⅜ x 8. T499 Paperbound **$1.25**

THE BAD CHILD'S BOOK OF BEASTS, MORE BEASTS FOR WORSE CHILDREN, and A MORAL ALPHABET, H. Belloc. Hardly an anthology of humorous verse has appeared in the last 50 years without at least a couple of these famous nonsense verses. But one must see the entire volumes—with all the delightful original illustrations by Sir Basil Blackwood—to appreciate fully Belloc's charming and witty verses that play so subacidly on the platitudes of life and morals that beset his day—and ours. A great humor classic. Three books in one. Total of 157pp. 5⅜ x 8. T749 Paperbound **$1.00**

THE DEVIL'S DICTIONARY, Ambrose Bierce. Sardonic and irreverent barbs puncturing the pomposities and absurdities of American politics, business, religion, literature, and arts, by the country's greatest satirist in the classic tradition. Epigrammatic as Shaw, piercing as Swift, American as Mark Twain, Will Rogers, and Fred Allen. Bierce will always remain the favorite of a small coterie of enthusiasts, and of writers and speakers whom he supplies with "some of the most gorgeous witticisms of the English language." (H. L. Mencken) Over 1000 entries in alphabetical order. 144pp. 5⅜ x 8. T487 Paperbound **$1.00**

THE PURPLE COW AND OTHER NONSENSE, Gelett Burgess. The best of Burgess's early nonsense, selected from the first edition of the "Burgess Nonsense Book." Contains many of his most unusual and highly original pieces: 37 nonsense quatrains, the Poems of Patagonia, Alphabet of Famous Goops, and the other hilarious (and rare) adult nonsense that places him in the forefront of American humorists. All pieces are accompanied by the original Burgess illustrations. 123 illustrations. xiii + 113pp. 5⅜ x 8.
T772 Paperbound **$1.00**

THE HUMOROUS VERSE OF LEWIS CARROLL. Almost every poem Carroll ever wrote, the largest collection ever published, including much never published elsewhere: 150 parodies, burlesques, riddles, ballads, acrostics, etc., with 130 original illustrations by Tenniel, Carroll, and others. "Addicts will be grateful . . . there is nothing for the faithful to do but sit down and fall to the banquet," N. Y. Times. Index to first lines. xiv + 446pp. 5 x 8.
T654 Paperbound **$1.85**

DIVERSIONS AND DIGRESSIONS OF LEWIS CARROLL. A major new treasure for Carroll fans! Rare privately published humor, fantasy, puzzles, and games by Carroll at his whimsical best, with a new vein of frank satire. Includes many new mathematical amusements and recreations, among them the fragmentary Part III of "Curiosa Mathematica." Contains "The Rectory Umbrella," "The New Belfry," "The Vision of the Three T's," and much more. New 32-page supplement of rare photographs taken by Carroll. x + 375pp. 5⅜ x 8.
T732 Paperbound **$1.50**

THE COMPLETE NONSENSE OF EDWARD LEAR. This is the only complete edition of this master of gentle madness available at a popular price. A BOOK OF NONSENSE, NONSENSE SONGS, MORE NONSENSE SONGS AND STORIES in their entirety with all the old favorites that have delighted children and adults for years. The Dong With A Luminous Nose, The Jumblies, The Owl and the Pussycat, and hundreds of other bits of wonderful nonsense. 214 limericks, 3 sets of Nonsense Botany, 5 Nonsense Alphabets. 546 drawings by Lear himself, and much more. 320pp. 5⅜ x 8.
T167 Paperbound **$1.00**

PECK'S BAD BOY AND HIS PA, George W. Peck. The complete edition, containing both volumes, one of the most widely read of all American humor books. The endless ingenious pranks played by bad boy "Hennery" on his pa and the grocery man, the outraged pomposity of Pa, the perpetual ridiculing of middle class institutions, are as entertaining today as they were in 1883. No pale sophistications or subtleties, but rather humor vigorous, raw, earthy, imaginative, and, as folk humor often is, sadistic. This peculiarly fascinating book is also valuable to historians and students of American culture as a portrait of an age. 100 original illustrations by True Williams. Introduction by E. F. Bleiler. 347pp. 5⅜ x 8.
T497 Paperbound **$1.35**

FABLES IN SLANG & MORE FABLES IN SLANG, George Ade. 2 complete books of major American humorist in pungent colloquial tradition of Twain, Billings. 1st reprinting in over 30 years includes "The Two Mandolin Players and the Willing Performer," "The Base Ball Fan Who Took the Only Known Cure," "The Slim Girl Who Tried to Keep a Date that was Never Made," 42 other tales of eccentric, perverse, but always funny characters. "Touch of genius," H. L. Mencken. New introduction by E. F. Bleiler. 86 illus. 203pp. 5⅜ x 8.
T533 Paperbound **$1.00**

SINGULAR TRAVELS, CAMPAIGNS, AND ADVENTURES OF BARON MUNCHAUSEN, R. E. Raspe, with 90 illustrations by Gustave Doré. The first edition in over 150 years to reestablish the deeds of the Prince of Liars exactly as Raspe first recorded them in 1785—the genuine Baron Munchausen, one of the most popular personalities in English literature. Included also are the best of the many sequels, written by other hands. Introduction on Raspe by J. Carswell. Bibliography of early editions. xliv + 192pp. 5⅜ x 8. T698 Paperbound **$1.00**

HOW TO TELL THE BIRDS FROM THE FLOWERS, R. W. Wood. How not to confuse a carrot with a parrot, a grape with an ape, a puffin with nuffin. Delightful drawings, clever puns, absurd little poems point out farfetched resemblances in nature. The author was a leading physicist. Introduction by Margaret Wood White. 106 illus. 60pp. 5⅜ x 8.
T523 Paperbound **75¢**

MATHEMATICS, ELEMENTARY TO INTERMEDIATE

HOW TO CALCULATE QUICKLY, Henry Sticker. This handy volume offers a tried and true method for helping you in the basic mathematics of daily life—addition, subtraction, multiplication, division, fractions, etc. It is designed to awaken your "number sense" or the ability to see relationships between numbers as whole quantities. It is not a collection of tricks working only on special numbers, but a serious course of over 9,000 problems and their solutions, teaching special techniques not taught in schools: left-to-right multiplication, new fast ways of division, etc. 5 or 10 minutes daily use will double or triple your calculation speed. Excellent for the scientific worker who is at home in higher math, but is not satisfied with his speed and accuracy in lower mathematics. 256pp. 5 x 7¼. T295 Paperbound **$1.00**

TEACH YOURSELF books. For adult self-study, for refresher and supplementary study.

The most effective series of home study mathematics books on the market! With absolutely no outside help, they will teach you as much as any similar college or high-school course, or will helpfully supplement any such course. Each step leads directly to the next, each question is anticipated. Numerous lucid examples and carefully-wrought practice problems illustrate meanings. Not skimpy outlines, not surveys, not usual classroom texts, these 204- to 380-page books are packed with the finest instruction you'll find anywhere for adult self-study.

TEACH YOURSELF ALGEBRA, P. Abbott. Formulas, coordinates, factors, graphs of quadratic functions, quadratic equations, logarithms, ratio, irrational numbers, arithmetical, geometrical series, much more. 1241 problems, solutions. Tables. 52 illus. 307pp. 6⅞ x 4¼.
Clothbound **$2.00**

TEACH YOURSELF GEOMETRY, P. Abbott. Solids, lines, points, surfaces, angle measurement, triangles, theorem of Pythagoras, polygons, loci, the circle, tangents, symmetry, solid geometry, prisms, pyramids, solids of revolution, etc. 343 problems, solutions. 268 illus. 334pp. 6⅞ x 4¼.
Clothbound **$2.00**

TEACH YOURSELF TRIGONOMETRY, P. Abbott. Geometrical foundations, indices, logarithms, trigonometrical ratios, relations between sides, angles of triangle, circular measure, trig. ratios of angles of any magnitude, much more. Requires elementary algebra, geometry. 465 problems, solutions. Tables. 102 illus. 204pp. 6⅞ x 4¼. Clothbound **$2.00**

TEACH YOURSELF THE CALCULUS, P. Abbott. Variations in functions, differentiation, solids of revolution, series, elementary differential equations, areas by integral calculus, much more. Requires algebra, trigonometry. 970 problems, solutions. Tables. 89 illus. 380pp. 6⅞ x 4¼.
Clothbound **$2.00**

TEACH YOURSELF THE SLIDE RULE, B. Snodgrass. Fractions, decimals, A-D scales, log-log scales, trigonometrical scales, indices, logarithms. Commercial, precision, electrical, dualistic, Brighton rules. 80 problems, solutions. 10 illus. 207pp. 6⅞ x 4¼. Clothbound **$2.00**

See also: **TEACH YOURSELF ELECTRICITY, C. W. Wilman; TEACH YOURSELF HEAT ENGINES, E. De Ville; TEACH YOURSELF MECHANICS, P. Abbott.**

HOW DO YOU USE A SLIDE RULE? by A. A. Merrill. Not a manual for mathematicians and engineers, but a lucid step-by-step explanation that presents the fundamental rules clearly enough to be understood by anyone who could benefit by the use of a slide rule in his work or business. This work concentrates on the 2 most important operations: multiplication and division. 10 easy lessons, each with a clear drawing, will save you countless hours in your banking, business, statistical, and other work. First publication. Index. 2 Appendixes. 10 illustrations. 78 problems, all with answers. vi + 36pp. 6⅛ x 9¼. T62 Paperbound **60¢**

THEORY OF OPERATION OF THE SLIDE RULE, J. P. Ellis. Not a skimpy "instruction manual", but an exhaustive treatment that will save you uncounted hours throughout your career. Supplies full understanding of every scale on the Log Log Duplex Decitrig type of slide rule. Shows the most time-saving methods, and provides practice useful in the widest variety of actual engineering situations. Each operation introduced in terms of underlying logarithmic theory. Summary of prerequisite math. First publication. Index. 198 figures. Over 450 problems with answers. Bibliography. 12 Appendices. ix + 289pp. 5⅜ x 8.
S727 Paperbound **$1.50**

ARITHMETICAL EXCURSIONS: AN ENRICHMENT OF ELEMENTARY MATHEMATICS, H. Bowers and J. Bowers. For students who want unusual methods of arithmetic never taught in school; for adults who want to increase their number sense. Little known facts about the most simple numbers, arithmetical entertainments and puzzles, figurate numbers, number chains, mysteries and folklore of numbers, the "Hin-dog-abic" number system, etc. First publication. Index. 529 numbered problems and diversions, all with answers. Bibliography. 50 figures. xiv + 320pp. 5⅜ x 8. T770 Paperbound **$1.65**

APPLIED MATHEMATICS FOR RADIO AND COMMUNICATIONS ENGINEERS, C. E. Smith. No extraneous material here!—only the theories, equations, and operations essential and immediately useful for radio work. Can be used as refresher, as handbook of applications and tables, or as full home-study course. Ranges from simplest arithmetic through calculus, series, and wave forms, hyperbolic trigonometry, simultaneous equations in mesh circuits, etc. Supplies applications right along with each math topic discussed. 22 useful tables of functions, formulas, logs, etc. Index. 166 exercises, 140 examples, all with answers. 95 diagrams. Bibliography. x + 336pp. 5⅜ x 8. S141 Paperbound **$1.75**

HIGHER MATHEMATICS FOR STUDENTS OF CHEMISTRY AND PHYSICS, J. W. Mellor. Not abstract, but practical, building its problems out of familiar laboratory material, this covers differential calculus, coordinate, analytical geometry, functions, integral calculus, infinite series, numerical equations, differential equations, Fourier's theorem, probability, theory of errors, calculus of variations, determinants. "If the reader is not familiar with this book, it will repay him to examine it," CHEM. & ENGINEERING NEWS. 800 problems. 189 figures. Bibliography. xxi + 641pp. 5⅜ x 8. S193 Paperbound **$2.25**

TRIGONOMETRY REFRESHER FOR TECHNICAL MEN, A. Albert Klaf. 913 detailed questions and answers cover the most important aspects of plane and spherical trigonometry. They will help you to brush up or to clear up difficulties in special areas. The first portion of this book covers plane trigonometry, including angles, quadrants, trigonometrical functions, graphical representation, interpolation, equations, logarithms, solution of triangle, use of the slide rule and similar topics. 188 pages then discuss application of plane trigonometry to special problems in navigation, surveying, elasticity, architecture, and various fields of engineering. Small angles, periodic functions, vectors, polar coordinates, de Moivre's theorem are fully examined. The third section of the book then discusses spherical trigonometry and the solution of spherical triangles, with their applications to terrestrial and astronomical problems. Methods of saving time with numerical calculations, simplification of principal functions of angle, much practical information make this a most useful book. 913 questions answered. 1738 problems, answers to odd numbers. 494 figures. 24 pages of useful formulae, functions. Index. x + 629pp. 5⅜ x 8. T371 Paperbound **$2.00**

CALCULUS REFRESHER FOR TECHNICAL MEN, A. Albert Klaf. This book is unique in English as a refresher for engineers, technicians, students who either wish to brush up their calculus or to clear up uncertainties. It is not an ordinary text, but an examination of most important aspects of integral and differential calculus in terms of the 756 questions most likely to occur to the technical reader. The first part of this book covers simple differential calculus, with constants, variables, functions, increments, derivatives, differentiation, logarithms, curvature of curves, and similar topics. The second part covers fundamental ideas of integration, inspection, substitution, transformation, reduction, areas and volumes, mean value, successive and partial integration, double and triple integration. Practical aspects are stressed rather than theoretical. A 50-page section illustrates the application of calculus to specific problems of civil and nautical engineering, electricity, stress and strain, elasticity, industrial engineering, and similar fields.—756 questions answered. 566 problems, mostly answered. 36 pages of useful constants, formulae for ready reference. Index. v + 431pp. 5⅜ x 8. T370 Paperbound **$2.00**

TEXTBOOK OF ALGEBRA, G. Chrystal. One of the great mathematical textbooks, still about the best source for complete treatments of the topics of elementary algebra; a chief reference work for teachers and students of algebra in advanced high school and university courses, or for the mathematician working on problems of elementary algebra or looking for a background to more advanced topics. Ranges from basic laws and processes to extensive examination of such topics as limits, infinite series, general properties of integral numbers, and probability theory. Emphasis is on algebraic form, the foundation of analytical geometry and the key to modern developments in algebra. Prior course in algebra is desirable, but not absolutely necessary. Includes theory of quotients, distribution of products, arithmetical theory of surds, theory of interest, permutations and combinations, general expansion theorems, recurring fractions, and much, much more. Two volume set. Index in each volume. Over 1500 exercises, approximately half with answers. Total of xlviii + 1187pp. 5⅜ x 8.
S750 Vol I Paperbound **$2.35**
S751 Vol II Paperbound **$2.35**
The set **$4.70**

COLLEGE ALGEBRA, H. B. Fine. Standard college text that gives a systematic and deductive structure to algebra; comprehensive, connected, with emphasis on theory. Discusses the commutative, associative, and distributive laws of number in unusual detail, and goes on with undetermined coefficients, quadratic equations, progressions, logarithms, permutations, probability, power series, and much more. Still most valuable elementary-intermediate text on the science and structure of algebra. Index. 1560 problems, all with answers. x + 631pp. 5⅜ x 8. T211 Paperbound **$2.00**

THE CONTINUUM AND OTHER TYPES OF SERIAL ORDER, E. V. Huntington. This famous book gives a systematic elementary account of the modern theory of the continuum as a type of serial order. Based on the Cantor-Dedekind ordinal theory, which requires no technical knowledge of higher mathematics, it offers an easily followed analysis of ordered classes, discrete and dense series, continuous series, Cantor's transfinite numbers. 2nd edition. Index. viii + 82pp. 5⅜ x 8. S129 Clothbound **$2.75**
S130 Paperbound **$1.00**

A TREATISE ON PLANE AND ADVANCED TRIGONOMETRY, E. W. Hobson. Extraordinarily wide coverage, going beyond usual college level trig, one of the few works covering advanced trig in full detail. By a great expositor with unerring anticipation and lucid clarification of potentially difficult points. Includes circular functions; expansion of functions of multiple angle; trig tables; relations between sides and angles of triangle; complex numbers; etc. Many problems solved completely. "The best work on the subject." Nature. Formerly entitled "A Treatise on Plane Trigonometry." 689 examples. 6 figures. xvi + 383pp. 5⅜ x 8.
S353 Paperbound **$1.95**